LIMNOLOGY OF A SMALL MALAYAN RIVER
SUNGAI GOMBAK

MONOGRAPHIAE BIOLOGICAE

Editor

J. ILLIES

Schlitz

VOLUME 22

SPRINGER-SCIENCE+BUSINESS MEDIA, B.V 1973.

LIMNOLOGY
OF A SMALL MALAYAN RIVER
SUNGAI GOMBAK

by

JOHN E. BISHOP

SPRINGER-SCIENCE+BUSINESS MEDIA, B.V 1973.

ISBN 978-94-010-2694-9   ISBN 978-94-010-2692-5 (eBook)
DOI 10.1007/978-94-010-2692-5

CONTENTS

CHAPTERS' CONTENTS

I. INTRODUCTION

Rivers are recognized as major natural resources, not only as sources of domestic, industrial and agricultural water and hydro-electric power, but also for food production and increasingly for recreation and tourist promotion. This awareness has led to increased study of lotic limnology, particularly with respect to the development of the potentials of river systems. The concern over deterioration of watercourses as a result of poor land utilization and their use as repositories for man's rejectamenta has stimulated research into the basic dynamics of the river environment and its biotic communities. Before any assessment of pollution effects can be made, or rational remedial action suggested, there must be a sound, detailed knowledge of the 'natural' characteristics of regional watercourses as a reference standard. This knowledge can result only from investigating the whole drainage area as a unit ecosystem (EVANS 1956) rather than just the specific conditions in the river itself (cf. SLACK 1955, HYNES 1969). Biotic data for any stream are of little value if divorced from consideration of the geology, hydrochemistry, land use and nutrient cycles of the watershed that generated them.

The study of large rivers, potamobiology in the sense of ILLIES (1955, 1961a), is a specialized area of research merging into lenitic ecology in many aspects and cannot be considered here. Rhithrolimnological studies on stream and small river systems within the above terms of reference are not common as most investigations have been confined to particular sections or problems of a river system or to specific taxonomic groups. The notable works (BERG 1943, BERG et al. 1948, FJERDINGSTAD 1950, ALLEN 1951, ILLIES 1952, ALBRECHT 1953, HARRISON and ELSWORTH 1958, OLIFF 1960, ALLANSON 1961, CHUTTER 1963 et sqq., MINCKLEY 1963, ULFSTRAND 1968a) and others have been reviewed and summarized by MACAN (1961a), ILLIES and BOTOSANEANU (1963), CUMMINS (1966) and HYNES (1970). These and most studies in lotic limnology have generally been restricted to rivers of the high latitudes with little work yet done in equatorial regions.

A few comprehensive studies have been carried out in the tropics, most notably in Amazonia where the relationship between landscape and river ecology has been eminently developed (SIOLI 1963, 1964, 1965, 1967a, b, 1968a, b, FITTKAU 1964, 1967, 1970, ILLIES 1964, KLINGE and OHLE 1964) and summarized in FITTKAU et al. (1968/69). TEMPLETON et al. (1969) have recently begun work in Central America. In Africa, apart from studies on the Nile (see TALLING 1958, RZÓSKA 1961), the notable work in the equatorial region has been done by MARLIER (1951, 1954), VAN SOMEREN (1952), HYNES and WILLIAMS (1962), MALAISSE (1969), RAMANANKASINA (1969) and IMEVBORE (1970). A considerable number of investigations have been made of sub-tropical Indian rivers (see VENKATESWARLU and JAYANTI 1968 and VENKATESWARLU 1969 for references) and in Ceylon (COSTA and FERNANDO 1967), but with the exception of fisheries work on the Mekong (see LE-VAN-DANG 1970) there is a dearth of studies on Southeast Asia. The tropical investigations listed are notable in that, with the exception of VAN SOMEREN and TEMPLETON et al., they were carried out on river systems that are severely affected by seasonal variations and disturbances. Periodic inundation of large parts of the watershed, or extended periods of heavy monsoonal rain followed by dry seasons are the general rule in most of the tropics and sub-tropics. Lotic systems not subject to such extremes of climatological stress in the equatorial regions have not been investigated systematically.

The primary object of this study was to define, as far as possible, the hydrobiological conditions existing in a small unaltered river not subject to severe seasonal changes. It was hoped that as the principles governing the biotic and abiotic features of the river became clear, they would provide a reference against which to separate successional changes and to compare conditions in areas under the influence of exogenous polluting agencies. By making qualitative and quantitative physical, chemical, floral and faunal analyses, it was also hoped to be able to compare the Malayan small river biome with corresponding systems in other geographical areas. Where feasible, the guidelines for limnological investigations as outlined by CUMMINS (1966) and particularly VOLLENWEIDER (1969) for IBP *de novo* investigations on a watershed, were followed for this study.

The field work was begun in September 1968 and continued over two years. Quantitative sampling was carried out for 15 four-week periods from October 1968 to December 1969, covering the annual climatic cycles.

II. THE ENVIRONMENT

1. General information

1.1. THE STUDY AREA

The location chosen for the investigation, the Sungai Gombak (Gombak River), Selangor, West Malaysia (Fig. 1), was considered typical of hill-streams of the central range of Malaya although suitable streams with higher source elevations might have been found farther to the north. Criteria for selection of a suitable survey area were:

a. a permanently-flowing small river that could be sampled by standard stream-benthos techniques, short enough so that studies could be carried on throughout its course.

b. accessibility along most of its length for day/night/all-weather sampling. The main east-west highway, Route II, passes through the Gombak valley and old logging tracks make access to many tributaries possible. The area was sufficiently close to the University of Malaya that water samples for chemical analysis could be processed soon after collection, thus eliminating the need for special preservation procedures.

c. undisturbed conditions over part of the watershed, and that the river should pass through the most commonly encountered land-use zones. Logging operations in the upper catchment, part of the Ulu Gombak Forest Reserve, have been largely suspended for 20 years. There are no permanent residents in the reserve so that only transients, jungle-fruit and insect collectors disturb the upper half of the area. After descending from the forest zone, the river passes successively through areas of rubber plantations and orchards, mixed crop and irrigated rice agriculture, urbanized areas with rubber processing and light industry to a zone of alluvial tin-dredging. With the exception of the urban concentration, this is a common sequence found on many West Malaysian rivers. The built-up area was an added attraction as it allowed investigations of the gross effects of various effluents on the river ecosystem.

d. some prior knowledge of the geology and climate of the area. The proximity of the area to the Federal Capital, Kuala Lumpur, ensured that climatological records were available. Some discharge data were obtainable from flood and water-supply studies. The area has been adequately mapped and geologically surveyed.

1.2. PREVIOUS STUDIES ON THE AREA

ROE (1953), GOBBETT (1964) and ALEXANDER (1968) have included

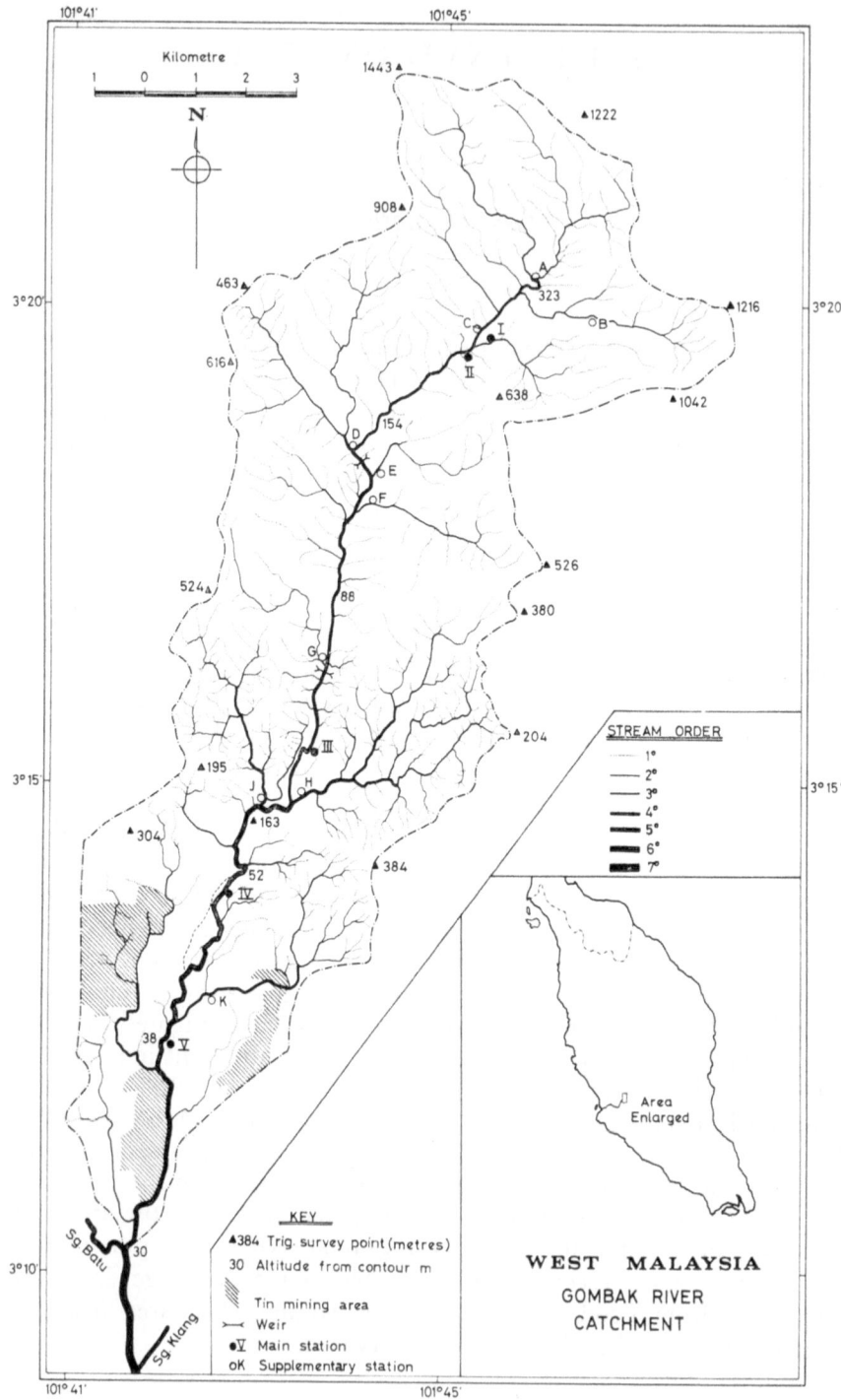

Fig. 1. Gombak River watershed showing drainage system, chief topographical features and distribution of sampling stations.

parts of the area in geological studies. DOUGLAS (1968a, 1970 and in preparation) has worked extensively on denudation dynamics and erosion in the basin. Various technical reports concerned with water-supply have been compiled by the Drainage and Irrigation Department, Malaysia and Binnie and Partners (Malaysia). NORRIS and CHARLTON (1962) conducted a brief survey of pollution in the lower river for the Department of Chemistry, Federation of Malaya. Except for several specialized studies on the cavernicolous fauna of the Batu Caves limestone and the piscivorous avifauna of the area, the only published work with aquatic application has been by BULLOCK and FURTADO (1968) and BULLOCK (1969) who made population assessments of a gerrid and a torrent frog in the upper river and FURTADO (1969) who studied the ecology and ethology of the Odonata.

1.3. CLIMATE

The Malayan climate, although comparatively stable, is not uniform. A seasonal rhythm of rainfall occurs in conjunction with the movements of air masses across the peninsula (DALE 1956); the timing of these varies from year to year. The north-east monsoon, with winds rarely in excess of 40 km/h, prevails from November-December until March and the south-west monsoon of even lighter winds from May until September-October. The inter-monsoonal, transition periods of light or calm winds, are characteristically wetter than the monsoonal months. Peak rainfall in the Western Rainfall Region (DALE 1959, 1960), in which the Gombak watershed is situated, occurs in the October-November transition period, with rather lower intensity during the March to May season. Two maxima therefore occur with intervening minimal precipitation periods in July-August and January-February. However, the monsoons, particularly the NE, do bring some rain, generally as a result of orographic uplift of unstable moist air masses. In addition, convectional rain caused by insolation and differential cooling may occur at any time so that there are rarely extended periods without any precipitation even in the 'dry' season. Temperatures are uniform throughout the year with maxima in the low 30's and minima around 20 °C (mean 26.5 ± 1.5, DALE 1963) at the altitude of the study area. Insolation is not as high as might be expected, averaging 6.5 h/day of bright sunshine over a five-year period (WYCHERLEY 1969), but this, coupled with the high and relatively constant rainfall and high mean temperatures, maintains relative humidity levels near saturation during most of the year.

1.4. GEOLOGY

Events in the geological evolution of the Indo-Malayan region from all available evidence have been summarized by HO (1960) and ALEXAN-

DER (1959, 1962, 1968); therefore, only a brief synopsis is given here.

During the Lower Paleozoic, the whole of the Sundaland area was submerged and sedimentation of considerable magnitude occurred, probably throughout the period from Carboniferous to Triassic. Calcareous and argillaceous materials were deposited contemporaneously in deep areas and arenaceous material in the shallower regions, probably nearly filling the geosyncline by the end of this period. Intense diastrophism followed in the Mesozoic with uplifting and anticlinal folding of the sedimentary layers. VAN BEMMELEN (1949) considers this to have occurred in a progressive series of orogenic waves moving south by southwest from an area in the South China Sea to reach present Malaya by the late Triassic- early Jurassic. These movements culminated in a phase of igneous activity during which intrusion of granitic magma into the anticlinal folds formed the main range batholith. Widespread metamorphism accompanied by faulting converted the arenaceous and argillaceous sedimentary deposits into clastic quartzitic and schistic masses and the calcareous into crystalline limestone beds. Metamorphism was most intense adjacent to the granite intrusions, but also occurred in localized areas where, subsequently, pneumatolytic and hydrothermal reactions took place, with the formation of mineral-bearing bodies and quartz veins and reefs which run through both granitic and sedimentary rocks (ROE 1953). In the late Jurassic, orogenic activity decreased and much of the formation subsided to emerge finally from marine influence in the Cretaceous-Tertiary period as the diastrophic wave moved outward to the west of Sumatra. By the end of the Tertiary, erosional processes had almost completely stripped the Paleozoic deposits from the granite, leaving only isolated, more resistant blocks. The alluviation and redeposition of these deposits and materials sub-aerially denuded from the granites evolved the present topography (RICHARDSON 1947).

2. Physical characteristics

2.1. PHYSIOGRAPHY

Morphometric data were extracted from maps prepared by the Director of National Mapping, Malaysia, printed in 1966, with scale 1:63,360 – Sheet 86 (Kuala Kubu Baharu); Sheet 94 (Kuala Lumpur). Stream length measurements were checked, where feasible, against more recent (1967) aerial photographs to account for course changes since the original surveys on which the maps were based.

2.1.1. Topography

The Gombak River drains a narrow elongated watershed that runs slightly west of south from the steep-sloped main range mountains down

6

through more gently sloping foothills to the alluvial plain in the vicinity of north Kuala Lumpur (Plate 1). Axial length of the drainage basin is 22.2 km, average width 5.5 km and area 123.3 square km. General relief is shown in Fig. 1 on which trigonometric bench marks are indicated as well as elevations at stream level taken from contour readings. The source is at 1051 m and the confluence with the Batu River at 28.3 m altitude. Some tributaries have sources higher than the main river (e.g. 1222 m) and are longer than the headwater stream designated on the map as the Gombak. The river joins the Klang River in the centre of Kuala Lumpur and then flows a further 68.3 km, often canalized between tin-mining bunds, through alluvial plains, lowland swamp-forest and mangrove swamp to the Straits of Malacca. The highest elevation on the watershed is at 1443 m in the northwest corner. Of interest is the depth of the valley in the upper watershed where the altitude at point A on the main river is only 295 m above mouth level while the relief has increased by more than 1400 m.

Location A, where the Gombak forms at the confluence of a number of equal tributaries, is in a broad, almost level area of deposited outwash material, formed as gradient and competence of the torrential tributaries suddenly decrease. This area is small and is surrounded by almost precipitous slopes. Downstream, gorge sections with occasional vertical cascades of 1–2 m and severe valley relief are a feature until the foothill region is reached below Tributary F where the valley opens up, gradient decreases and the relief of the watershed changes from severe to undulating. The profile (Fig. 2) summarizes these features. The division of the watershed into sections, as indicated, is arbitrary, but follows topographic and vegetation patterns and will be discussed in more detail with respect to the selection, location and description of sampling stations. The Upper Zone, including the Upper Tributary Subzone, takes in the undisturbed Forest Reserve areas of the watershed and terminates at the point where the river leaves the steep-sloped hills and enters the more gentle foothill section. The very steep gradient, 173.1‰, for the upper tributaries and the overall gradient for this zone, 78.0‰, give an indication of the severity of the terrain. The Middle and Lower Zones, with gradients of 4.7‰ and 2.2‰ respectively, delimit the foothill and alluvial plain areas and are definable in terms of land use as well as physiography. The gradients are sufficient to generate considerable velocities in the river (up to 250 cm/sec in places), but 'shooting flow' with characteristic Froude numbers $Fr > 1$ (MORISAWA 1968) is rarely maintained for more than a few metres anywhere in the main river. Even in the steep tributaries, alternating pools and cascades are characteristic with little direct running for extended distances. 'Streaming flow' ($Fr < 1$) with eddies and un-ordered turbulent currents (cf. RUTTNER 1963) is the normal character in all zones. Even in the Lower Zone where gradient is slight and pooling common, flow is continuous. Pooled areas are best described as flowing-

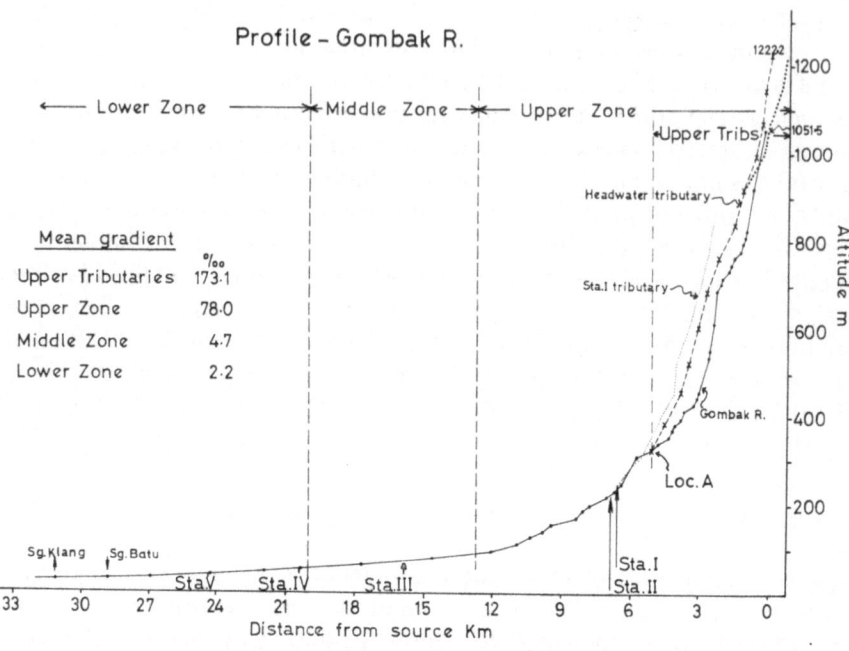

Fig. 2. Profile of the river showing physiographic zones and gradients.

pools with sections of increased gradient and velocity between them. Alternating pool and riffle areas are frequent in the Upper Zone but are often only 5–10 m in length and cover less than half the width of the river. This highly variable pool-riffle mosaic is succeeded in the Middle Zone by long, often 50–100 m, sections of gentle riffle, followed by shorter pooled reaches. The Lower Zone has fewer riffle areas and more extended slow-flowing sections as expected in a lowland river, but dead-water areas rarely occur.

Throughout its length, the river is characterised by alternating areas of erosion and deposition; the length and extent of each area is dependent on gradient. In several extensive reaches of deposition (e.g. Location A mentioned above) and in the Lower Zone, some braiding of the channel is seen when carrying capacity is suddenly lost or where excessive enlargement of the channel during flood peaks has left large areas of constantly shifting sand and gravel bars through which the river finds its course. Considerable undercutting of the banks is seen in the Lower Zone particularly, and bamboo-lattice groynes have been built on many bends in the inhabited areas. In the Upper Zone, floodplain development is limited, 1–2 m of sand and gravel beach in shallow reaches only, and the river rarely exceeds bank capacity. Because of silting and channel variation, some flooding occurs in the Lower Zone with peak flows, particularly in areas where the river is confined by man-made banks. The main river in its upper reaches appeared over the study period to be relatively stable with very few changes in bed configuration. Perhaps these parts of the watercourse are approaching grade (LEOPOLD et al. 1964), as the nick-points in the otherwise relatively smooth profile occur only where the bedrock granite is exposed. Heavy spates altered depositional area dimensions, but these generally reverted to the original in a short time. The small tributaries (< 1 m wide) appeared equally stable, but variations in flow were small and erosive processes would probably not be evident in two years. The larger hill tributaries and the lowland section of the river were relatively unstable, as previously indicated. Changes in some large feeder streams in the hills occurred after every storm and the tributary at Station I over a 100 m section dropped its bed by approximately 20 cm during the field investigation period.

Man-made structures on the river are few. In the Upper Zone, there is a 'V'-notch gauging weir with a drop of about 2 m and in the Middle Zone, below G, a small adjustable barrage that regulates irrigation take-off levels at that point (see Fig. 1). Retention time was estimated to be less than three hours as the impounded area is only 100 m long. At two other locations in this zone irrigation water is abstracted into side channels, but without control structures. The remains of two pre-World War II water supply dams are found just below Station II, but these are completely silted up and have no retention. No stretch is navigable.

2.1.2. Stream order classification

Definitions of river and stream types based on subjective comparison of the physical characteristics between different localities and geographical areas have proved rather unsatisfactory. Classifications by ecologists have usually been based on such criteria as water source, width, velocity, substrate, temperature, gaseous concentrations and pH, or biotic indices or zones (see HARREL et al. 1967 for references), or where possible, combinations of these (HYNES 1960, ILLIES and BOTOSANEANU 1963, CUMMINS 1966). The use of the HORTON (1945) classification system as modified by STRAHLER (1954, 1957) objectively quantifies easily applied parameters that in general correlate closely with the criteria mentioned above. This system is now being used more frequently by ecologists (ABELL 1961, KUEHNE 1962, 1966, HARREL et al. 1967, HARREL and DORRIS 1968, GREEN 1970). These parameters directly define the main structures of a watercourse (LEOPOLD et al. 1964, MORISAWA 1968) and indirectly indicate the status of secondary features that are often difficult to compare between different geographical regions using subjective description. The great drawback to the system is its reliance either on available maps or extensive ground surveys for the status of the minor watercourses. As fully discussed by MORISAWA (op. cit.), the level of

Table 1. Stream order analysis of the Gombak drainage basin

Order	Number of streams	Total length (km)	Average length (km)	Average drainage area (km²)	Bifurcation ratios		Stream length ratios	
1	475	178.0	0.38	—	1 : 2	4.02	2 : 1	1.50
2	118	66.9	0.57	—	2 : 3	3.80	3 : 2	2.61
3	31	46.2	1.49	2.6	3 : 4	5.17	4 : 3	1.78
4	6	15.9	2.65	12.8	4 : 5	3.00	5 : 4	1.68
5	2	8.9	4.45	41.7	5 : 6	2.00	6 : 5	2.92
6	1	13.0	13.0	123.3	Mean	3.60	Mean	2.10

Drainage density 2.67; Stream frequency 5.14.

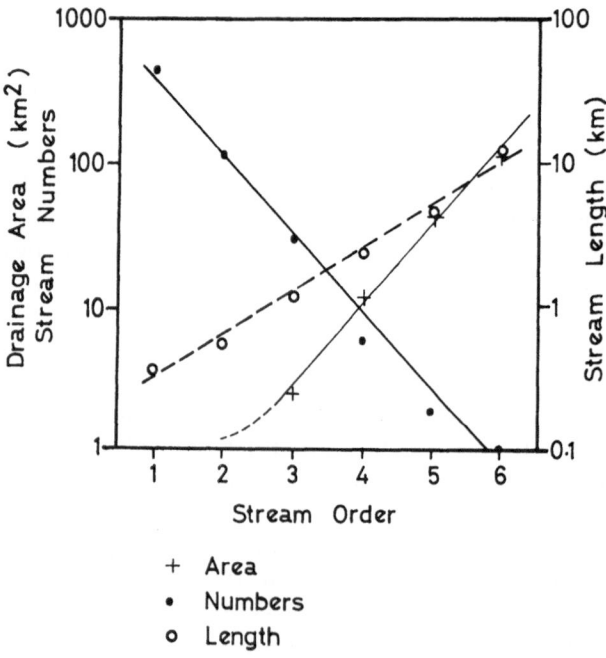

+ Area
• Numbers
o Length

Fig. 3. Numbers, average stream length and drainage basin size in relation to stream order. (Basin size for 3° and larger).

refinement in the method is dependent on the definition of what constitutes the smallest channel size. However, this affects only a watercourse's order number, not its relative ranking (i.e. all channels may lose or gain one order or more depending on the size ultimately designated as first order (1°)). By HORTON's classification, the smallest unbranched tributaries are designated 1°; where two 1° join, a 2° is formed; where two 2° join, a 3° is formed, etc.. Low order adventitious streams directly entering a higher order system have no effect on the ordinal designation. This order ranking is dimensionless so that drainage basins of different measurements can be compared with respect to the defined criteria. According to HORTON (op. cit.), the numbers and average length of streams, watershed area and gradient are geometrically related to stream order. These data for the Gombak are summarized in Table 1. The numbers of streams of each order exhibit an inverse relationship and the average length and drainage area increase exponentially with stream order (Fig. 3). According to LEOPOLD (1962), drainage area increases four to five times with each stream order increase. For the Gombak watershed, these increases are 2.9 to 4.9. The stream frequency (number of streams/unit area) and drainage density (stream length/unit area) with values of 5.14 and 2.67 respectively for the Gombak are within ranges for high relief watersheds. The bifurcation ratio (ratio of number of streams of an order to the number of next higher order) should theoretically be two. However, a large number of adventitious streams in the watershed increases the ratio. STRAHLER (1957) found ratios between three and four for mountainous basins, so the mean value of 3.60 for the Gombak is not exceptional. Stream-length ratios (ratio of average length of streams of an order to average length of next lower order) have low values (1–2) for well-drained watersheds.

Summary: The drainage pattern is a highly developed dentritic system with high bifurcation ratios and the low mean stream-length ratio characteristic of well-drained watersheds. Of interest is the apparent change in drainage density in the Middle Zone where two large tributaries enter the system (see Locations H and J, Fig. 1). The drainage density in these gentle-sloped foothill tributaries is higher than that found in the Upper Zone. DOUGLAS (personal communication) felt that this was an artefact caused by mapping error, but ground survey indicates that on these slopes increased stream bifurcation is genuine. However, as these areas have been under rubber plantation cultivation for a number of years, the natural drainage patterns may have been modified.

2.1.3. Lithology and structure of the catchment

The Gombak catchment contains elements of most of the principal features of the geologic history of the peninsula, giving a complex of lithologies. However, these occur in discrete enough blocks to enable their effects to be distinguished. This will be elaborated further in the section on chemical characteristics.

The upper areas of the catchment (Fig. 4) lie largely on granite of the main range batholith with coarse-grained porphyritic and medium-to coarse-grained muscovite-biotite granites occurring most commonly. In the middle catchment, fine-grained granite-porphyry and microgranite

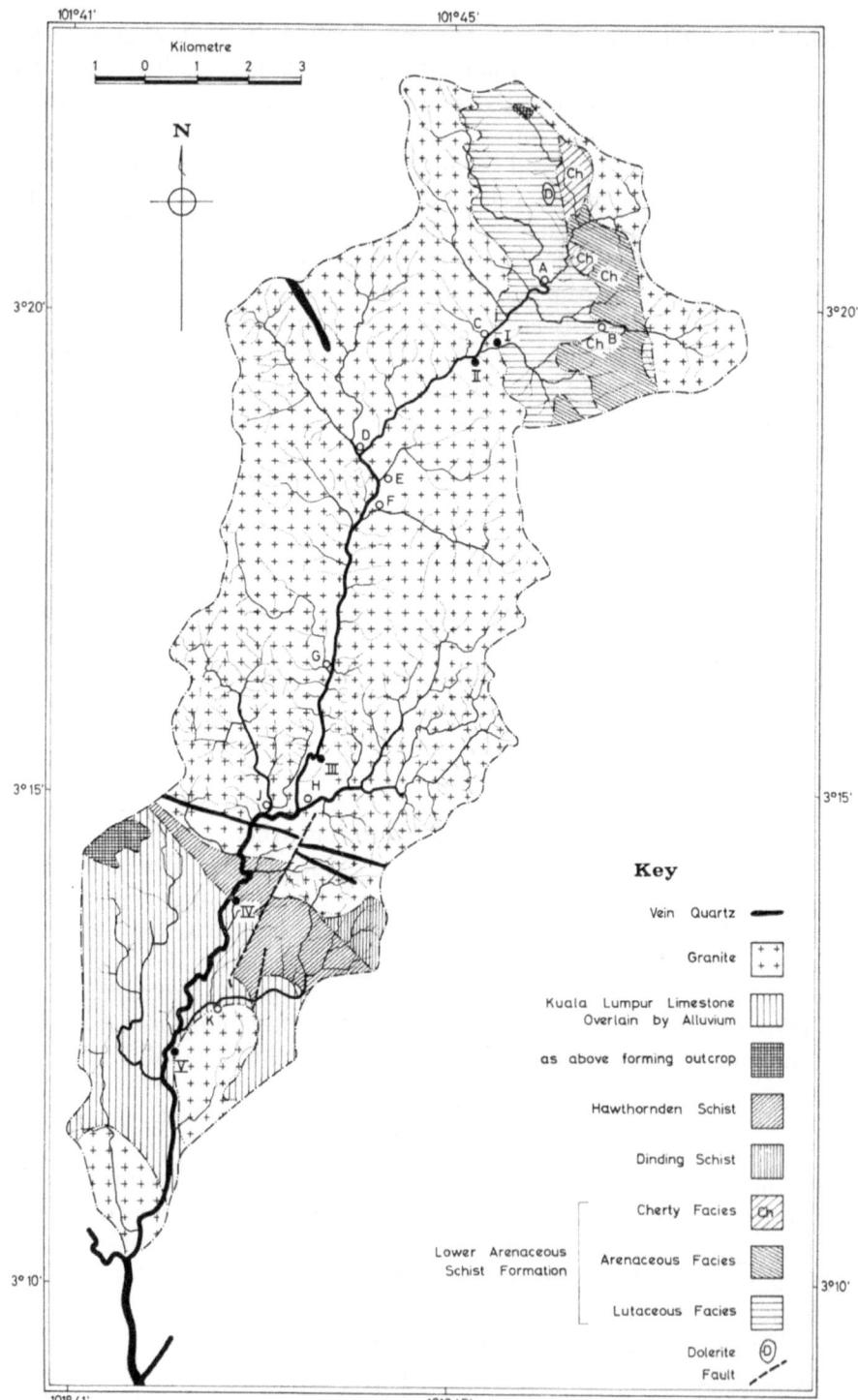

Fig. 4. Geologic formations in the Gombak catchment.

are intruded into the coarser granites (ROE 1953). Crossing the upper part of the watershed is a narrow band of Paleozoic sedimentary rocks of the Lower Arenaceous Series (ALEXANDER 1968) composed of three components: the arenaceous clastic facies of orthoquartzite and sub-greywacke, lutaceous clastic deposits with shale, schists and phyllite layers and the non-clastic chert facies with varying amounts of graphite. A single, very small area of basic intrusive dolerite (post-granitic, probably Tertiary (ALEXANDER 1968)) occurs, cutting through quartzite in the NW corner of this formation, adjacent to the small calcitic outcrop of the Paleozoic Calcareous Series which is prominent in areas west of the catchment. The lower catchment is dominated by a large area of Quaternary alluvial deposits that covers, often to 30 m, the calcitic and dolomitic limestones and dolomite of the Silurian(?) Kuala Lumpur Limestone Formation (GOBBETT 1964), except where the Batu Caves outcrop and isolated granite bosses stand above them. Separating the limestone and granite formations is a strip of sedimentary rocks of the Dinding and Hawthornden Schists, the former mostly quartz-mica schist and quartzite, the latter fine-grained black schist containing pyrites (GOBBETT *op. cit.*).

Through the whole area, in both intrusive and sedimentary rocks, quartz intrusions are common, mainly as small veins (1–20 mm wide), but on the NW watershed and to the north of the Dinding Schist, quartz dykes about 50 m wide have formed prominent ridges. All these intrusions consist of brecciated vein quartz seamed with secondary veins that may contain large well-formed quartz crystals. Severe faulting has affected the alignment of these dykes which control the drainage pattern where they occur (GOBBETT 1964) (see Fig. 4). Percentages of the catchment on various lithologies are shown in Table 2.

Table 2. Percentages of catchment area on various lithologies

Granite	68.1
Chert Facies	0.9
Arenaceous Facies	3.1
Lutaceous Facies	8.6
Hawthornden Schist	3.7
Dinding Schist	0.8
Limestone	0.6
Limestone overlaid by Quaternary Alluvium	13.8
Quartz	0.5

2.1.4. Soils

The formation of tropical soils under high humidity and temperature as found in West Malaysia has been described and references given by

RICHARDS (1957). Soil type depends largely on the parent rock type. The various lithologies previously described give rise to several sedimentary soil types in the catchment. However, local drainage-exposure conditions produce variations over small distances.

a. Granite weathers, often to great depths on moderate slopes, to a reddish-yellow sandy-clay-loam assigned by OWEN (1951) to the Rengam Series and defined as part of the large Red and Yellow Latosol group (PANTON 1964). This has subsequently been redefined as a Red-Yellow Ultisol (LEAMY 1966, Ng 1969) to bring it in line with world soil classification (U.S.D.A. 1960). Such soils are characterized by a very shallow humic horizon, often less than 5 cm but increasing in thickness with altitude; by acid pH, about 5 at the rock face, decreasing towards the surface and with altitude; and by low cation exchange capacity (WHITMORE and BURNHAM 1969) indicating a high degree of leaching, but not necessarily laterization as is often assumed to occur under tropical conditions. A notable feature of this soil group is the bimodal grain size with high proportions of clay and sand, but little silt in the intermediate sizes (DOUGLAS 1968a, WHITMORE and BURNHAM op. cit.) The importance of this will be discussed further with respect to substrate composition of the stream bed. Weathering of the parent granite is by chemical decomposition of the feldspars and micas to clays, leaving *in situ* the quartz fragments. Thus, soil profiles are often uniformly composed of soft, highly permeable saprolite right down to bedrock, with the exception of isolated unweathered core stones. These soils have greater proportions of iron and aluminium sesquioxides than the parent material but lower concentrations of electrolytes, alkaline earths and silica as these are lost through continual leaching. The general paucity of humic material ensures that the iron and aluminium oxides are not transported as complexes but remain in the soil. The colour, red-yellow, depends on the degree of hydration of these oxides (see RICHARDS 1957).

b. The sedimentary and metamorphic rocks yield soils of the same general classification as above but with higher proportions of sands from the quartzitic facies and almost pure clay regolith from some shale formations (DOUGLAS 1968a). Some red lateritic soil may occasionally develop over shale and phyllite facies.

c. The alluvial areas of the lower catchment derived from fluviate deposits of weathered granite have soils in the general Yellow Latosol group (Ultisol), but with some podzolization evident (PATON 1964). These soils often have low clay content and marked nutrient deficiency. However, where the limestone underlying most of these alluvial soils has weathered, a clay cap has often formed which may affect the soils above it. Most of the alluvial region has been worked for tin and the soils completely leached, leaving only disturbed white quartz sand of various sizes with no distinct formation.

d. On the quartz dykes only very thin sandy regolith is formed.

Table 3. Land use as percentage of area

UPPER CATCHMENT	
Rain forest	98.8
Irrigated land	0.1
Paved road concession	1.1
LOWER CATCHMENT	
Rain forest*	34.1
Rubber, orchards and gardens	36.4
Rice and irrigated land	3.8
Kampong-urban and paved road concession	17.8
Tin mining	7.5
Quarry	0.4
TOTAL CATCHMENT	
Rain forest	56.7
Rubber, orchards and gardens	23.7
Rice and irrigated land	2.5
Kampong-urban and paved road concession	11.9
Tin mining	4.9
Quarry	0.3

* Much of this is disturbed as it is not Forest Reserve

2.1.5. Vegetation and general land use

The extent and type of ground cover in specific areas of the watershed exerts considerable influence on the hydrology and chemistry of the river, and has indirect effects on the biota through allochthonous organic production, temperature moderation and, in denuded areas, by elevated removal rates of both inert and nutrient materials. Table 3 and Fig. 5 summarize land use in the catchment. In the Disturbed Lower Catchment, the area of cleared concession for paved roads has been included with the 'urban-kampong' classification. In the Undisturbed Upper Catchment, the single paved road area has been kept separate as there are no permanent settlements of any significant area. No attempt has been made to estimate the areas of unsealed roads or logging tracks. In the lower section these come largely in the area covered by 'urban' and tracks in the forest, although important with respect to erosion, are soon overgrown if not maintained. The small 'rice and irrigated' patch at the top of the disturbed zone is at present used only as a private park, with some perennial irrigation. This area is the bed of an old reservoir destroyed during the Japanese Occupation, and the river is now cutting its way down through the accumulated sediments; in other words, though riceland type is indicated, usage is marginal and conditions are not ideal for wet grains. In the upper NE corner of the watershed is a small area of watercress cultivation on elevated irrigated terraces. This constitutes

15

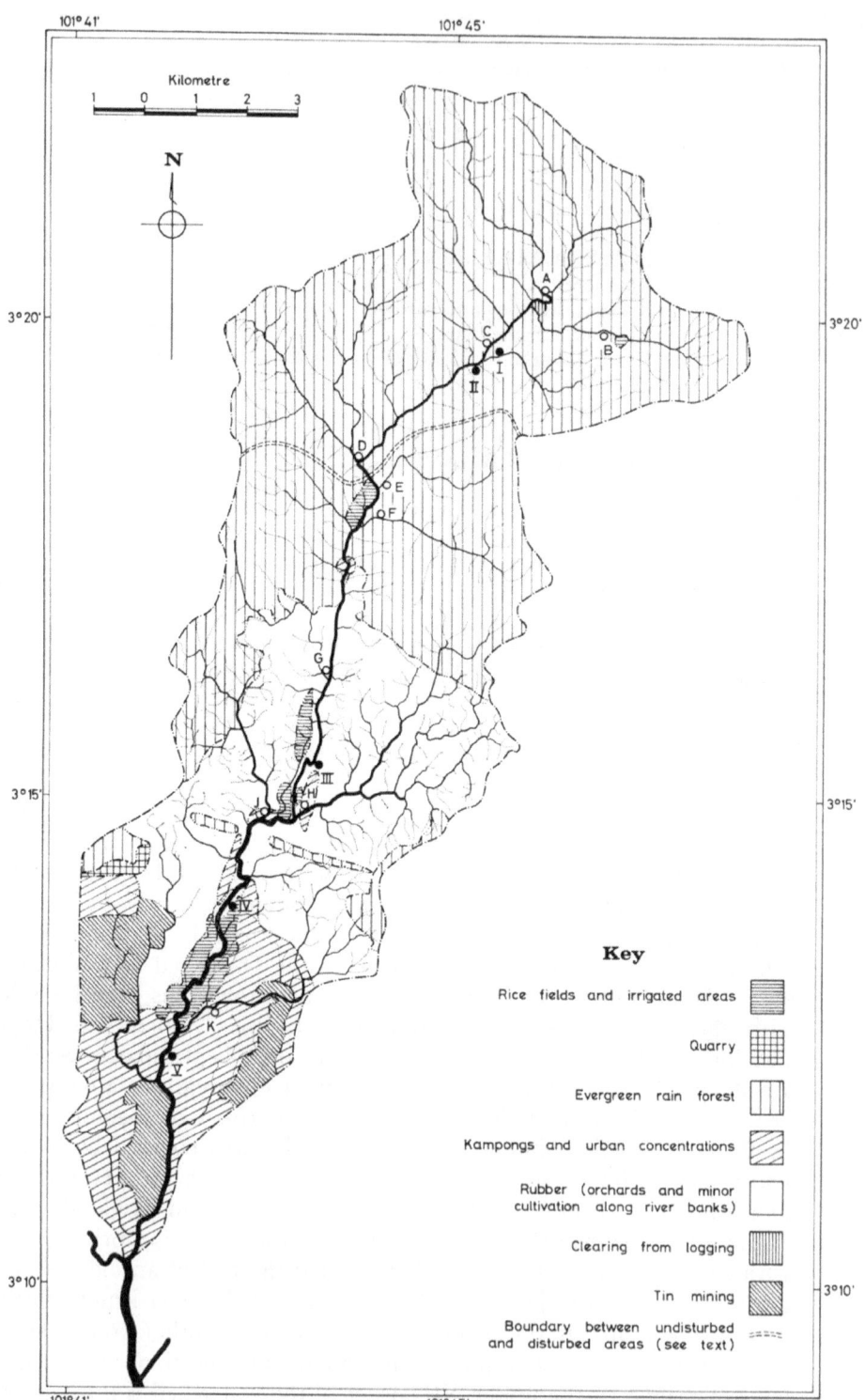

Fig. 5. Vegetation and land use in the watershed.

16

the only continuous agriculture in the forest zone. The small area designated as 'forest clearing' in the figure, in the Upper Catchment near Location A, is an area cleared for a timber camp, long abandoned and now returning to secondary jungle.

The lower watershed is a mixture of alluvial pan, most restrained behind 2–7 m high bunds left by tin mining, and partially reclaimed land. This produces poor, secondary growth and little cover, and high concentrations of eroding inorganic materials at every rainstorm. A large part of this derelict land is occupied by squatters and other subsistence groups who have small vegetable gardens and a few market stock. A large proportion of the southern part of this Disturbed Lower Catchment is taken up with organized urban development with housing estates, river-side kampongs with vegetable gardens and orchards and ribbon growth of light industrial, merchandising and service concerns along the main roads. Much of the area is unserviced for sewage, but piped water is provided up to the vicinity of Station III. Above the urbanized areas, rice cultivation is carried out extensively in the low-lying areas adjacent to the river (Plate 2). Along the banks for up to 0.5 km on both sides are areas of mixed orchard and low-intensity vegetable gardening. Behind these regions and on all slopes > 20° and reaching to the tops of most hills (except the quartz dykes) is plantation and smallholding rubber. Much of the latter is old and only casually attended so that the under-storey is thick, but the plantation areas are maintained. Most of the watershed of the large Tributary H (Sungai Pusu-Bernam) is taken up with a recent land clearance scheme run by the Federal Land Develop-ment Authority and is under rubber. There are some areas of rain forest on the peripheries of this zone and disturbed forest extends to the edges of the Forest Reserve at the zone boundary. The characteristic of the greater part of this Disturbed Lower Catchment is its considerable and continual disruption by man's activities.

The Undisturbed Upper Catchment is almost entirely under evergreen rain forest; the exceptional areas have already been noted. The phyto-sociology of the tropical rain forest is too complex to discuss in this context, but is characterized by very large numbers of species at low densities in a given area (see RICHARDS 1957). Rain forests have been divided into formation type (BURTT DAVY 1938) based on structure and physiognomy. Two types, the Lowland Rain Forest and Lower Montane Forest occur in the Gombak Catchment. These forest types were sub-divided floristically by SYMINGTON (1943) (see also WYATT-SMITH 1964) for Malaya into Lowland Dipterocarp (Lowland Forest) and Upper Dipterocarp and Oak-Laurel (Lower Montane Forest) with demarcations at approximately 300–350 m, 800 m and 1250 m altitude. BURGESS (1969) feels that the lower two zones, despite their heterogeneity, have little consistant floristic difference between them and should be united as Lowland Dipterocarp Forest. BURGESS (op. cit.) and WHITMORE and

BURNHAM (1969) have recently described the characteristic floristic features of these forest zones for the Gombak area. Structurally, the Lowland Dipterocarp Forest, present in the valley bottoms and on lower slopes as far up as Location A (see Plate 3) has a characteristic three-layered canopy 25–50 m high, often little ground cover except fallen leaves, mega- and mesophyllous leaves, numbers of woody climbers, some epiphytes and large buttresses (see GRUBB et al. 1963). The Upper Dipterocarp Forest is similar, but with less distinct canopy layering as only emergent crowns and the lower shrub layer are common. There is more ground cover, less frequent buttressing, fewer lianes and poor development of epiphytes (in contrast to GRUBB et al. who found more epiphytes at this altitude in Ecuador). This forest type, which may contain elements of both Lowland and Oak-Laurel Forests, is seen only on the high ridges ringing the headwaters at altitudes above 800 m. Small areas of Oak-Laurel Forest without emergent crowns of Diptero-carpaceae and with dense ground cover and increased numbers of epiphytes occur on the highest peaks at more than 1250 m.

The abiotic importance of these forest types is the effect they have on the geomorphic cycle. Their influence on evapotranspiration rates and the amount of precipitation reaching the forest floor, through effects of canopy and leaf form, is obvious. The erosional effects of large drop impact on soil surface and trunks and buttresses as foci of downward-rushing water change with the physiognomy of the different forest types. The amount of ground cover and the density of the lower canopy are particularly important in this respect. Rubber plantations approach natural forests in many attributes, but the single canopy layer and frequently denuded ground cover expose the soil surface to erosion.

2.2. PRELIMINARY ZONATION

A provisional zonation of the catchment was made that retained as far as possible the natural divisions of the watershed. This zonation was based on the vegetation and physiographic features discussed in section 2.1. above.

2.2.1. The catchment

The complete watershed falls into two parts as discussed in the vegetation and land use section: the Undisturbed Upper Catchment covering 36.4% of the total area, almost all in the Ulu Gombak Forest Reserve, and considered to be a Bench Mark Catchment as defined for the International Hydrological Decade (BINNIE and PARTNERS 1970); the Disturbed Lower Catchment (63.6%) in which various human activities dominate the landscape. These divisions are convenient for explanations of the hydrologic cycle which involves both the river itself and the surrounding terrestrial watershed. The boundary between the two sections was chosen to coincide with the limits of the drainage area above the gauging weir near Locations D and E (see Fig. 5). This permitted rainfall and discharge comparisons between the two sections, as discharge was also gauged at the lowest point on the river, just above the confluence with the Batu River.

2.2.2. *The river*

The river zones follow slope, substrate and riparian changes.

a. The Upper Zone of mountain torrent is entirely within the rain forest, has considerable gradient and takes in the first 12.6 km from the source. The Upper Tributary Subzone, within the above, includes all the feeder tributaries that descend from sources on the jungle-clad hills and ridges of the upper catchment with even greater slopes (see Fig. 2). At the lower end of this zone are the area of irrigated land mentioned previously and a hospital and large aboriginal settlement. These, however, exert little influence on the river as the jungle-clad banks continue for a further 2 km below the settlement.

b. The transition to the Middle Zone is abrupt. As the river enters the gentle relief of the foothill region, the Forest Reserve ends and partial clearing of the jungle occurs immediately, particularly along the river banks, for subsistence agriculture and rubber growing, with some permanent habitation. This boundary correlates with physical changes in the river. Slope decreases abruptly as the valleys open up, size of substrate decreases and the depositional areas increase in size and frequency. The Middle Zone of foothill stony riffles runs from the 12.6 to the 20.3 km point from the source.

c. The Lower Zone, from the edge of the foothills over the alluvial plain to the confluence with the Batu and Klang Rivers is a further 10.9 km long. The slope decreases by another factor of two and depositional areas become dominant. At the top end of this zone rice cultivation and orchards are the principal land use, while the lower sections are urbanized or mined for tin. The river exhibits the effects of various effluents from these activities.

2.2.3. *Classification of water types*

To facilitate the selection of macrohabitat areas for sampling of biota and such physical agents as substrate, the sampling stations were divided into water types. The definitions of these follow ALLEN (1951) and HARRISON and ELSWORTH (1958) and are those commonly used by stream investigators.

Cascades: water flowing with considerable velocity, usually > 1 m/sec, over large boulders or semi-precipitous bedrock: water depth very variable from a thin film to 0.5 m: occur only in the Upper Zone where gradient is steep. In discussion of the fauna, this type will include the adjacent 'splash zone' of many authors.

Riffles: water of moderate to rapid current, always turbulent, often broken 'white water' flow. During high water, the flow characteristics change to a smoother, less turbulent state. Slow and fast riffle types were distinguished at normal flow, but their characteristics depend largely on stage. Riffles were found in all zones where gradient increased and channel width narrowed. Allen's 'runs' type corresponds to a submerged riffle.

Pools: water deeper than average for the stream with slow to moderate smooth flow or exceptionally, little or no flow. Slow and fast pools, depending largely on depth, were distinguishable. Backwater eddy areas behind obstructions were included in this type. Pools occurred in all zones with increase in area and depth coinciding with increasing stream order.

The Lower Zone also included canalized sections between tin-mining bunds or man-made banks where slow, even current rarely broken by rapids occurred. This type is best classified under the above criteria as a slow-flowing pool type, but should probably be a type of its own. These sections were not sampled extensively and further definition is not necessary.

Such categorization is often difficult to apply as local conditions of gradient and substrate make intergradations common. Changes of type with stage, e.g. pool to slow riffle at low discharge, make absolute characterization meaningless.

Within each water type different gross substrate categories were recognizable: large and small gravel in riffles; coarse and fine sand and silt and organic deposits in depositional areas. The unique habitat afforded by the roots of *Saraca* forming the banks of all water types in the forest zone, the trailing and marginal vegetation bordering riffles and pools in the lower reaches and the deadwater, interstitial habitat of decomposing leaves accumulated behind obstructions, particularly in the Upper and Middle Zones, were additional organic substrate types. These categories, within the framework of the water types, formed the basis for the sampling program in each river zone.

2.3. DESCRIPTION OF THE SAMPLING STATIONS

Stations for chemical and biological study were selected to represent the most obvious areas of the river. Unless a very large number of sites is selected, all habitat types cannot be investigated, but it was hoped that the areas chosen covered the main range of conditions found in the various zones. Five main sampling sites were selected, indicated as I–V on Fig. 1: Station I on an upper tributary; Station II on the main river in the centre of the Upper Zone; Station III on a typical reach of the Middle Zone; Stations IV and V in the Lower Zone, IV above the rice-growing area to provide data on normal successional changes and V below the paddy fields, urban and industrial development and various pollution sources to cover the effects of these on the river. These were sampled at regular weekly intervals throughout the study period for the aspects under investigation (see later sections). In addition, supplementary stations were sampled on a casual basis. These were located at points of variation along the course, particularly in the Upper Catchment where a variety of minor biotopes were evident. These, Locations A–K indicated in Fig. 1, sited on the lower reaches of the principal tributaries, were used to complete hydrochemical characterization of the watershed and as survey sites for the fish investigations. They will not be described in detail, but the characteristics relevant to these studies will be given when needed.

Aid in identification of Bryophyta was given by Dr. A. JOHNSON, Nanyang University, Singapore, and of higher plants by Dr. E. SOE-PADMO, University of Malaya whose assistance is gratefully acknowledged.

Station I was situated at 3°19′36″ N, 101°45′26″ E, altitude 233.2 m, on a 3° tributary characteristic of many that rise high in the ridges and descend with steep gradient to the main river valley (Plates 4, 5). The designated sampling area was 4.0 km below the apparent source and 0.5 km above confluence with the main river. This tributary drained an area of 4.7 sq. km. The 100 m sampling section (average width 2–3 m) included several fairly uniform series of riffle-pool combinations which were used for the faunal studies. Substrate was predominantly small boulders, with stones, gravel and sand between. Because of their importance as factors affecting biotic conditions, temperature, discharge and substrate conditions will be listed and discussed for all stations in subsequent sections. No emergent vegetation was found although various

Table 4. The Bryophyta of the banks and boulders of the forest stream

CLASS MUSCI
Callicostella prabaktiana (C.Muell.) Bosch et Lac.
Calymperes tahitense (Sull.) Mitt.
Fissidens sp.
Himantocladium sp.
Hyophila javanica (Ness et Blum.) Brid.
Micromitrium sp.

CLASS HEPATICAE
Lejeunea sp.
Pycnolejeuna sp.
Radula sp.
Riccardia heteroclada Schffn.

Table 5. Bankside flora of the Upper Zone

		I	II
ARACEAE	*Aglaonema griffithii* Schott	+	++
	Pothos sp.	+	
	Scindapsus hederaceus Schott	+	
COMMELINACEAE	*Commelina nudiflora* Linn.	+	
GRAMINAE	*Axonopus compressus* Beauv.	++	
	Centotheca lappacea Desv.	+	
	Lophatherum gracile Brongn.	R	
	Setaria palmifolia Stapf	R	
HYMENOPHYLLACEAE	*Trichomanes bimarginatum* v.d. Bosch	+	
LEGUMINOSAE	*Mimosa pudica* Mill.	+	
	Saraca thaipingensis Cantley	++	+++
MELASTOMATACEAE	*Clidemia hirta* D.Don	+	
SAPINDACEAE	*Pometia pinnata* Forst.	+++	+++
Ferns and Lianes of various families		+	+

(R = rare; + = present; ++ = common; +++ = abundant)

mosses and liverworts (Table 4) grew on the larger rocks and boulders and on emergent tree-root masses above the modal water line. Lichens covered the aerial parts of many rocks. Algal growth was not conspicuous. There were no clumps of filamentous forms present. However, small patches of diatoms and Cyanophyta were sometimes evident on the upper surfaces of the more stable boulders; these never formed extensive slimes or mats. Bankside vegetation was dominated by the *gapis* tree, *Saraca thaipingensis*, which grew at the edge of the stream with its roots forming a compact mass that constituted an important habitat in itself and that often created marginal eddy areas where silt and detritus built up. Root-mats of various lianes and other epiphytes formed a similar floating habitat. Other common bank plants trailing in the water or overhanging it are listed in Table 5. However, these represent only the more commonly found species from an association of many hundreds of species in the adjacent forest that may occur at the water's edge. A significant habitat was created by the build-up of leaf-stick-flower-fruit accumulations behind rocks and fallen logs. These 'leaf-packets' sometimes considerably impeded flow, particularly at low stage, and caused pooling in what were normally riffle sections. Canopy cover over the tributary was complete with only about 4.5 % of available visible light reaching the water surface. Modal water depth over riffle areas was 10–15 cm, but deeper fast-water sections between boulders were common and depth in average pools was up to 50 cm. Current velocities, depending on stage, were very variable in different areas as alternating pools and riffles prevent the build-up of high or constant velocities. An average passage time through the 100 m section was of the order of 5 min with velocities in short sections up to 150 cm/sec. Over the period of the investigation, considerable

22

deepening of the unstable stream bed occurred at this station, and a logging track ford forming the lower boundary was washed out on several occasions.

Station II was located on the main river as the representative station in the Upper Zone at 3°19′32″ N, 101°45′16″ E. The reach selected for study was immediately upstream from the University of Malaya Field Studies Centre at an altitude of 220 m, 6.8 km from the source, draining an area of 27.7 sq km. Width at this station varied from 7–12 m with narrow (1–2 m) steep flood beaches of silt-sand and pebbles in some places (Plates 6, 7, 8). Large boulders grossly characterized the study location, but the substrate between these was sand, gravel and stones of variable dimensions. There were no aquatic cryptogams or emergents and lichen, moss and liverwort associations were as for Station I. Algae were inconspicuous except for a thin discontinuous film of diatoms and Cyanophyta on stable, submerged boulders. A few short tufts of *Cladophora* were evident in rock- and flood-pools and *Batrachospermum* was sometimes found in these locations, but never in the main channel. Clumps of filamentous *Lyngbya* and *Plectonema* and colonies of *Caloglossa* were a regular, but minor, association with the *Saraca* roots. Bank vegetation was limited to *Saraca thaipingensis*, *Pometia pinnata* and various ferns. At other sites in the zone, the same plant associations as at Station I were to be found (Table 5) and are those common throughout the forested area. At various locations where timber extraction in the past removed the large emergent dipterocarps and created partially open areas, bamboos, notably *Gigantochoa scortechinii* Gamble, had become dominant and at several points formed the cover on the banks. Canopy cover was complete in some places but not so in others, leaving mid-stream sections with short mid-day sunlight periods. On the average, however, only about 7.4 % of the available light reached the water in the study section. Depth was variable depending on bottom configuration, with main stream riffles 15–30 cm deep and some pools up to 1.5 m in depth. Current velocities likewise covered a spectrum from almost zero in the larger pools to 250 cm/sec in short rushing sections between boulders. Passage time through 100 m was of the order of 2–3 min. The bottom was stable over the study period with no visible channel alteration. Build-up and erosion of gravel and sand beds during spates was evident, but in a short period the river returned to the original configuration. The large car-sized boulders that determined the channel form would be moved only by exceptional storm floods and the banks, bound by *Saraca* roots, were not subject to normal erosion.

The Middle Zone foothill *Station III* was situated at 3°15′31″ N, 101°43′33″ E, 15.8 km from the source at an altitude of 70.1 m. Stream width was 8–12 m at the sampling site with low (about 50 cm) banks of alluvial materials; these were rarely flooded. Head-sized and occasionally larger boulders with all sizes of stones, gravel and sand made up the

Table 6. Bankside flora of the Middle Zone

		III
ARACEAE	*Colocasia esculenta* Schott	+
COMPOSITAE	*Mikania cordata* B.L.Robinson	+++
EUPHORBIACEAE	*Manihot utilissima* Pohl	+
	Phyllanthus niruri Linn.	R
GRAMINAE	*Axonopus compressus* Beauv.	+
	Digitario longiflora Pers.	+++
	Eleusine indica Steud.	+
	Imperata cylindrica Beauv.	+
	Panicum auritum Hassk.	++
	Panicum nodosum Kunth	+
	Panicum pilipes Nees et Arn.	+
	Pennisetum purpureum Schum	+++
LEGUMINOSAE	*Mimosa pudica* Mill.	+
POLYGONACEAE	*Polygonum barbatum* Roxb.	++
THELYPTERIDACEAE	*Cyclosorus gongylodes* Link	+
ZINGIBERACEAE	*Catimbium latilabrae* Holltum	+

substrate. No aquatic vegetation, mosses or liverworts occurred here. Diatoms and crustose Cyanophyta often formed thick layers on the bottom stones so that the substrate became too slippery to walk on. Occasionally small clumps of filamentous Chlorophyta developed in the riffle areas. Emergent and bankside vegetation (Table 6) was made up largely of the weed species that invaded the vegetable gardens and orchards which occupied the banks in this section (Plates 9, 10, 11). A few *Saraca* were found in the upper parts of this zone, but were soon replaced, largely by native fruit trees. Canopy cover was generally scanty. The investigated section was completely exposed except at the margins where the taller grasses and *Colocasia* gave some cover. Water depth on the long riffle section was 5–50 cm and in the sandy flowing pools, 20–70 cm at normal flow. Velocities were relatively stable in the range 50–80 cm/sec on the riffles and 5–25 cm/sec in the pools, with local areas of much higher velocity, especially at peak flow. The sandy bottom configuration of the pools was constantly changing, but stony riffles maintained stability as long as channel configuration did not change. Erosion of the banks occurred constantly with considerable loss of cultivated land in some places over the study period. As a result, highly unstable silt and mud banks built up in depositional areas and shifted with each flood. Habitations occurred at regular intervals along the banks in this section with the people using the river for domestic water, bathing and as a latrine. However, there were no concentrations of people and direct effects on the river were insignificant.

Station IV was situated at the top of the Lower Zone, just above a substantial irrigated area and below a large kampong, at 3°13'54" N, 101°42'33" E, 20.4 km from the source at an altitude of 50.3 m, width

Table 7. Bankside flora of the Lower Zone

		IV	V
ARACEAE	*Colocasia esculenta* Schott	+	+++
COMMELINACEAE	*Commelina nudiflora* Linn.		++
	Pollia sp.	+++	
COMPOSITAE	*Mikania cordata* B.L.Robinson		+++
CONVOLVULACEAE	*Merremia hederacea* Hallier	++	++
CYPERACEAE	*Cyperus* sp.	+	+
DENNSTAEDTIACEAE	*Athyrium esculentum* Copel		++
EUPHORBICEAE	*Manihot utilissima* Pohl		+
GRAMINAE	*Axonopus compressus* Beauv.		+
	Brachiaris sp.	+++	
	Eragrostis amabilis Wright et Arn.		+
	Imperata cylindrica Beauv.	++	+
	Panicum spp.		++
	Phragmites communis Trin.		+
	Polytrias amaura O.Ktze	+++	+
	Saccharum spontaneum Linn.		+
HYDROCHARITACEAE	*Hydrilla verticillata* Presl.		R
LEGUMINOSAE	*Mimosa pudica* Mill.	+	R
VERBENACEAE	*Hyptis brevipes* Poit.	+	
ZINGIBERACEAE	*Catimbium latilabrae* Holltum	+	

very variable, but at sampling site 10–17 m. The banks in this section, about 1 m high, were composed of river alluvium subject to erosion during floods. Long deep pool sections on river bends with short riffle areas between were the common bed form. Depositional areas had uniformly sandy bottoms with occasional silty sections in backwater areas; riffles were as at Station III. There were no aquatic or emergent plants or Bryophyta. Some *Cladophora* was present on rocks as very short tufts and filamentous Chlorophyta developed between spates on muddy banksides. Blue-green algae and some diatoms occurred on riffle substrates forming thick slimy masses during low water periods. Marginal vegetation (Table 7) was largely as at Station III but with more grasses (Plate 12). In lower sections of this zone, considerable growths of bamboos 10–15 m high, and *lalang (Imperata cylindrica)* up to 3 m tall, on the banks provided some cover, but in the study section exposure was 100%. A few fruit trees shaded immediate bank areas in some places. Pools were up to 150 cm deep and riffles 10–30 cm with current velocities 0–150 cm/sec. The higher velocities occurred only at low stage when riffle areas were exposed. The riffle areas at the sampling site were stable throughout the two years but shifting sandbanks were constantly changing the profile of the depositional zones. Bank erosion was a serious problem in this zone as rice-field bunds were often undercut and irrigation water was lost from the fields that were generally higher than the river level. Above this station, two diversion canals took off irrigation water. This reduced the flow in the main river by about half during periods of low stage and heavy water demand, thus leaving large areas of exposed sand and gravel. The human inhabitants had some effect on the river here as considerable amounts of sewage from direct defecation, detergents and refuse were added by the large kampong population. The river was still extensively used for bathing and washing, but water from stand-pipes was used for drinking. Irrigation canals in this zone, not considered as part of the river but connected directly to it, had extensive emergent growths of *Nitella*, *Colocasia*, *Blyxa*, *Cyperus* and *Limnocharis* which were periodically cut and dug by the farmers and floated into the river. They did not, however, grow in or along the main stream.

Station V was located in the densely populated, light-industrial area of north Kuala Lumpur at 3°12′24″ N, 101°42′02″ E, below the confluence of a pollution-carrying tributary and the return canals from a large irrigated area, 24.2 km from the source at an altitude of 39.6 m and with width of 16–22 m at the sampling area. Deep (1–2 m) pools were found on most bends with shallower flowing pools and short 10–30 cm deep riffle sections in between. At and below the station there were outcroppings of limestone bedrock, causing short sections of rapids, but normal substrate was sand in all areas with small egg-sized stones added in the faster-flowing areas. Small clumps of *Hydrilla* grew as an emergent at Station V where sufficiently stable bottom conditions were found in association with the bedrock. No Bryophyta were found, but considerable growth, in short (< 1 cm) tufts, of *Cladophora* occurred on hard substrates in the river and some free-floating filamentous forms (*Spirogyra* species) formed mats in slack-water areas between spates. Sparse diatom and cyanophyte growth occurred on the riffle substrate. Bank vegetation was much like Station IV (Table 7), but there were occasional large *Acacia* trees overhanging the water in addition to a few orchard species and some bamboo clumps that shaded narrow regions of river (Plate 13). At the selected site, more than 95 % of the area was totally exposed. Current velocities ranged from 0–125 cm/sec depending on bottom configuration with a flow-through period for the 60 m section of approximately 2 min at normal stage. The bottom was relatively stable in spite of its sand-gravel composition, as flow was smooth with few obstructions to create turbulence. Erosion of the soft clay-mud banks was extensively controlled by groynes and bamboo pilings along river bends. Flooding over the 1 m high banks into the gardens and kampong areas occurred periodically with the more intense storms. Immediately below the station, a small sawmill and large rubber processing factory added considerable organic and chemical waste to the river. Washing was still done in the river at the station but not much further downstream as maladorous rubber production effluents made the water unpleasant. Use of the river as a latrine continued both directly and via privies built over short ditches leading down the banks. Domestic and some commercial refuse was disposed of in the river at the time of the investigation.

2.4. PHYSICAL METHODS

2.4.1. Light

Comparative light readings above the water surface were taken at each station using a barrier layer lux meter equipped with opal filters registering light in the visible range. Readings were taken over 15 sec, 5 cm above the water surface at the sampling stations, and where the station was shaded, in an open area adjacent to it immediately afterwards. This was repeated on each sampling occasion so that all cloud conditions, seasons and degrees of foliage cover were investigated.

Total radiation records were obtained from the Rubber Research Institute Experimental Station at Sungai Buloh, 14.6 km W of the bottom of the catchment, where daily records were made by a Gunn-Bellani radiometer calibrated against a Kipp Solarimeter. No records closer to the study area were available and unfortunately no measurements of total radiation could be made at the higher altitudes of the stations in the upper catchment.

The underwater light conditions at the sites used for algal studies at each station were determined for several turbidity and cloud states to estimate vertical extinction effects. Intensities above the water surface, 1 cm below, and at 10 cm intervals to the available depth were read with the lux meter held vertically. The percentage decrease at each depth compared to the surface reading as control was calculated. No estimate of qualitative change in spectral range was possible as suitable filters were not available.

2.4.2. Temperature

Air temperature records were obtained for the forest area with a maximum-minimum dry bulb mercury thermometer set up at the Field Studies Centre near Station II. For the lower watershed, records from the Forest Research Institute, Kepong, located on the alluvial plain 6 km to the west of the watershed were used. Where the minimum records here were incomplete, values were estimated from records published by the Meteorological Services Malaysia, taken at Kuala Lumpur International Airport, Subang, 15 km SW of the bottom of the watershed.

Water temperatures were obtained by placing maximum-minimum mercury thermometers (Taylor Instrument Company) in wire cages and submerging them in gently turbulent areas at Stations I, II, IV and V. Vandals prohibited the maintenance of any instrument at Station III. All thermometers were calibrated at 20 °C and 30 °C against a reference 0–50 °C mercury thermometer and appropriate correction factors applied to the data obtained. To supplement these readings, spot measurements were made on all sampling days at each station between 10:00 and 12:00 h and 24 h profiles were run on several occasions.

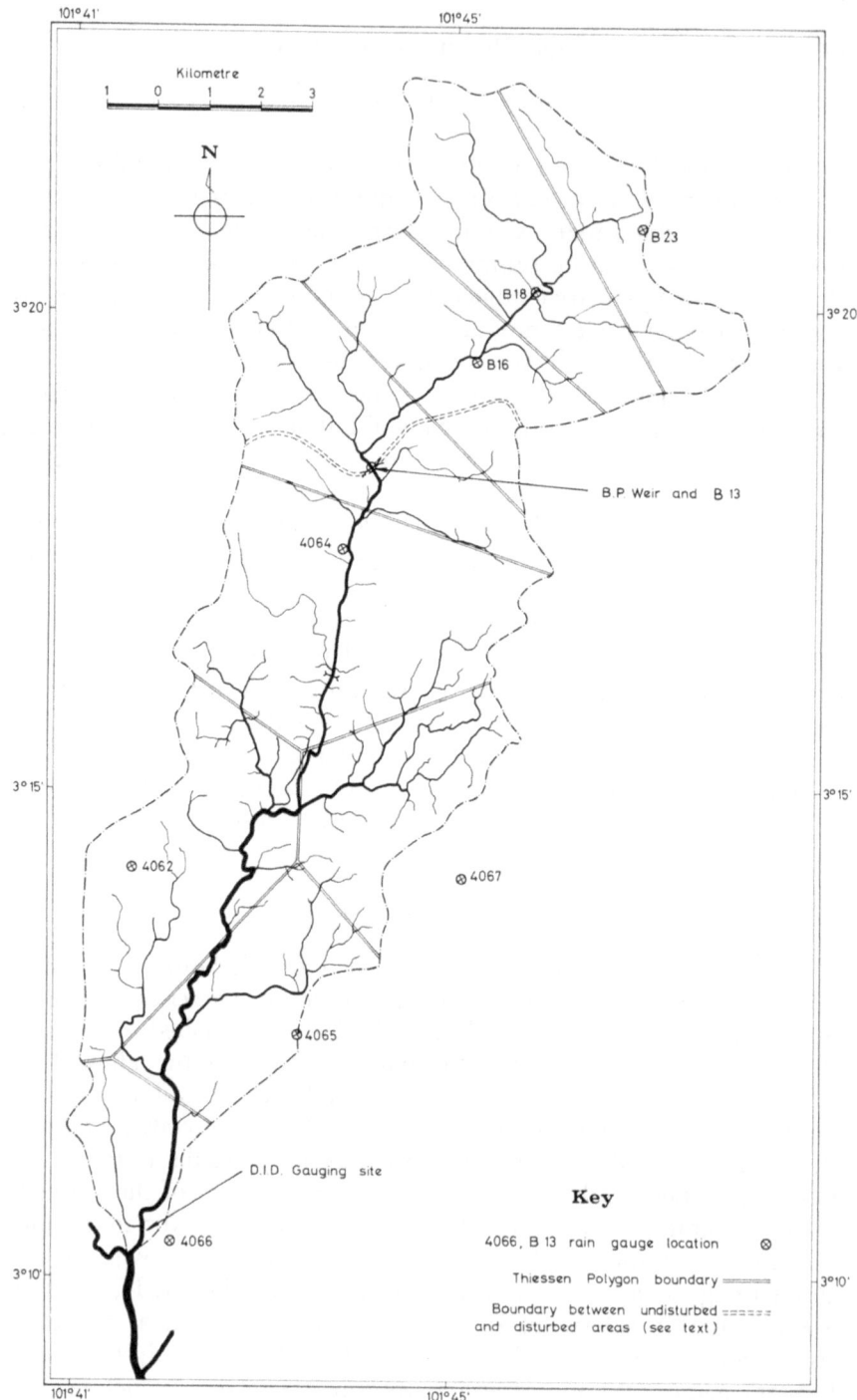

Fig. 6. Distribution of hydrometeorological facilities, catchment divisions and Thiessen polygons.

Table 8. Area of Thiessen polygons

Rain gauge		Area in sq.km
4066		6.35
4065		12.23
4067		13.99
4062		18.28
4064		18.76
B13+		15.97
B16*		12.83
B18		15.19
B23		9.70
	Total Catchment	123.30
	Lower Catchment	78.37
	Upper Catchment	44.93

+ 8.97 in Upper and 7.00 in Lower Catchment
* 11.07 in Upper and 1.76 in Lower Catchment

2.4.3. Precipitation

Precipitation data were gathered from a network of nine rainfall recording stations, some automatic, others manually read daily, in and around the watershed. Long-term records were available for some stations; others were newly-established for the Kuala Lumpur hydrological survey. Analysis of precipitation over the watershed was done by the construction of Thiessen Polygons (THIESSEN 1911); this made allowances for irregularities in gauge spacing by weighting the record of each gauge in proportion to the area of the polygon surrounding that gauge. The polygon represents the area in which are located all points closer to its gauge than to any other. Gauge sites in Fig. 6 are designated according to their Drainage and Irrigation Department numbers (4000 series) and B-prefixed stations are those of Binnie and Partners (Malaysia). Table 8 shows the area covered by each gauge. B16 and B13 were subdivided to give their areal contributions to the Undisturbed Upper Catchment area and Disturbed Lower Catchment area as separate analysis of these two areas was desired. The partial areas for B16 and B13 are also given in the table. Weekly, monthly, annual and study period mean precipitations were estimated for each section and the whole catchment using the proportionality constants.

2.4.4. Discharge

Discharge was continuously gauged at two sites on the river: at Circular Road, Kuala Lumpur, at the extreme southern boundary of the watershed just upstream from the entry of the Batu River, marked D.I.D. on Fig. 6; just below Location D and above E and F at the boundary between the

Disturbed and Undisturbed Catchments, marked B.P.. At both sites, automatic instruments recorded stage levels which were applied against a periodically-updated rating curve to give discharge. Control at the D.I.D. gauging site is only fair, as there is no defined channel and the natural bottom is subject to scour and deposition. The B.P. station has an artificial weir of compound 'V' form with rounded cross-section, one stabilized bank and stable bottom configuration giving good to excellent stage-discharge correlations. Mean daily, mean monthly, maximum instantaneous, and total discharge were extracted from records at these stations for the upper and total catchment. Long-term records were available for the D.I.D. station for comparison.

In addition, crest staff-gauges designed after the types described by Collet (1942) and Stevens (1942) were placed at Stations I, II, IV and V; again, no installation at Station III was possible. These measured the maximum depth for the period preceeding the reading. On each sampling date, maximum and present depths on the staff-gauge and discharge on a standard reach were measured. Gauging was carried out according to Corbett (1945), by integrating multiple depth and velocity measurements across the width of a smoothly profiled section at equal intervals, usually 0.5 m. Velocities were measured with a modified pitot tube system at 0.2 and 0.8 of the depth or 0.6 where depth was less than 20 cm, the reading used being the mean height over a 15 sec period to allow for short-term variations and surges in flow.

2.4.5. Substrate analysis

a. Physical analysis of bottom sediment was carried out according to the recommendations of Cummins (1962, 1966) following standard procedures (for references see Morgans 1956, Subcommittee on Sedimentation 1957, Cummins 1966, Doeglas 1968, Guy 1969) and based on the Wentworth scale (Wentworth 1922). Samples were collected from each sampling station using the benthos sampler to be described fully under the faunal section. Essentially, this square sampler enclosed an area of 0.1 sq. m with a flow-through system that collected materials put into suspension when the bottom was disturbed in a net (mesh size 90 μ used for sediments) that covered the downstream side. The technique used was to select at random an area in a visually-uniform river section, place the sampler firmly in the substrate and then dig out all the sediments to a depth of 10 cm. The net-caught fraction was then added to the gross sample. At all stations, two such samples were taken from each habitat type, riffle and pool, and analysed fully as a representative sample. In sections where large stones and boulders invalidated this type of sampling, e.g. the erosional areas at Stations I, II and III, larger areas, usually 10 × 10 m, were randomly chosen, all large boulders and stones were measured and weighed with a Slater spring balance and sling where feasible, and only the finer sediments were sampled by the above technique. When a boulder was too large to weigh, its weight was calculated from dimensions and density. The samples were subsequently dried at 60 °C for three days and then passed through a series of grids, 128, 64, 32, 16, 8 and 4 mm, to separate the larger sediments. The remainder were dry-sieved on a mechanical shaker for 15 min using standard sieve series down to 53 μ. All fractions were weighed to 0.1 g on a Mettler model K open pan balance. With the sampling technique used, some of the very fine sediments were lost. The 90 μ net rapidly became blocked and retained most of the finer fractions, but not quantitatively. However, the proportion of these clay and silt fractions was very small and the lost part would have had little bearing on the final size distribution, although it may have been important in microdistribution considerations. Results were expressed as percentage of the weights for each *phi* unit, corrected on a volumetric basis to include the proportional contributions of the large rocks and boulders. *Phi* is defined as -$\log_2$ of particle size in mm, and gives whole integers on an arithmetic scale to the WENT-WORTH logarithmic divisions.

b. Interstitial pore space, defined as the area potentially available for biological processes, was measured for the sediment types at each station. A representative sediment sample, about 0.01 cu. m collected as above, was placed, still wet, in a cylindrical vessel, shaken to settle the particles and eliminate air bubbles, and water added until it reached 1 mm above the surface. This was then drained, through 90 μ mesh, under agitation for 5 min to free all non-capillary water and the volume of water measured. Water was then re-added from a burette up to the saturation point and drained and measured as before. This was repeated until variation was less than 5% between three successive volumes. Substrate volume was measured after each reading and a mean calculated.

$$\% \text{ potential living space was expressed as } \frac{\text{mean vol. saturation water}}{\text{mean vol. substrate}} \times 100$$

c. To investigate the extent of bed movement, the lithology of a single size element of the bottom sediments was measured. At progressive intervals from the source, a minimum of 100 pebbles of the −5 to −6 *phi* (32–64 mm) size group were collected randomly. These were characterized in broad lithologic categories and expressed as a percentage at the location. (This subsection was carried out in collaboration with Dr. I. DOUGLAS, visiting lecturer from the Department of Geography, University of Hull, U.K.).

2.4.6. Erosional load

a. Total Residue, Total Suspended Load, Total Dissolved Load and the inorganic and organic subfractions of these were measured each sampling period at each station by methods recommended in A.P.H.A. (1965), with the difference that in addition to drying at 180 °C for organic, ignition at 600 °C for 1 h to obtain an ash-free dry weight (inorganic fraction) was carried out. The low concentrations of carbonates and sulfates in the water (see chemistry section) made losses through volatilization minimal, and these were disregarded. All weighings were made on an Oertling balance, model R20, to 0.1 mg. Water for these analyses was collected from a riffle area of the station into a narrow-necked 4 l. polyethylene bottle by integrating collections from three or more levels of the vertical depth. One litre was filtered through a Pyrex fritted glass filter (porosity 4, approx. 5 μ) and another litre was used unfiltered. These were evaporated to dryness in Vitreosil crucibles, stabilized at 105 °C for 12 h, weighed, heated at 180 °C for 18 h, weighed, ignited at 600 °C for 1 h and weighed. Residue fractions were calculated by differences. The data obtained routinely in this study were supplemented by load figures obtained by DOUGLAS (personal communication) during the course of his extensive studies on erosion in the catchment. These data included measurements for flood crest and rising stage and indicate the maximum loads that probably occurred.

b. At Stations I and II where interference with equipment left in the river for extended periods was minimal, it was possible to obtain an estimate of bed-load under low and moderate flow conditions. Nets, as used for the drift studies and described later, with minimum mesh dimension of 165 μ, were set flush with the bottom and left for 8 × 3 h once during each sampling period. The volume of water passing through the net as a proportion of total discharge in the river at that locality and the ratio of bottom covered by the net to bottom area exposed to eroding current velocities were calculated. By measuring the volume of the inorganic fraction (only particles larger than *phi* 2.5 were taken with the net used), a rough approximation of bed movement and saltating load could be obtained. Unfortunately, this technique could only be carried out when water depth did not exceed the height of the net (30 cm) so that the data are incomplete for high water periods.

2.5. PHYSICAL OBSERVATIONS

2.5.1. Light

Mean monthly radiation data for the lowland plain area are given in Table 9. More detailed records for periods of interest with respect to algal productivity will be discussed in that section. The variations from month to month are largely a function of cloud-cover and are therefore unpredictable. Day-length at the latitude of Kuala Lumpur varies only 20 min over the year so is just a minor factor. Slightly higher means occurred February to April, a hot, dry period in most years (DALE 1959).

Table 9. Mean monthly solar radiation gram cal/cm²/day recorded from Lower Catchment area

October 1968	402
November	473
December	405
January 1969	471
February	503
March	531
April	492
May	476
June	436
July	461
August	439
September	457
October	468
November	390

Table 10. Proportion of visible light reaching water surface

Station	Range (%)	Mean (%)
I	1.6–12.8	4.5
II	2.3–15.0	7.4
III		100
IV		100
V		100

Over a three year period prior to this study, mean irradiance at the R.R.I. Experimental Station was recorded as 420 gram cal/cm²/day, close to the theoretical 435 gram cal/cm²/day, calculated from meteorological data for the locality (see VOLLENWEIDER 1969).

The proportion of visible light reaching the water surface at each station is shown in Table 10. Under cloudy conditions, proportionately more light reached the river under the forest canopy, as diffuse radiation became more important than the direct component. The mean percentage was calculated from a series of readings taken throughout the year at each station and probably represents a reasonable summary of the conditions. Slight variations of foliage cover and sun-altitude with season had little effect on the proportion, but in the steep valleys, the length of time under the shadow of the surrounding hills, which depends on the relative position of the sun, was a minor consideration. At Stations III, IV and V, there was negligible ($< 3\%$) cover over the water so that 100% of the radiation was assumed to reach the surface. At Station II, a mean of 7.4% and at Station I 4.5% of potential radiation reached the water surface.

Table 11. Vertical attenuation (percentage of intensity 10 cm above surface)

Cm below surface	I	II	Stations III	IV	V
0.5	86	94	97	97	95
10	78	86	92	92	89
20	70	80	86	81	72
30	63	71	80	73	58
40	57	60	73	63	45

Vertical attenuation is summarized in Table 11. The increasing loss with depth at the lower stations, particularly V, was a function of turbidity. At all stations during peak floods, very heavy silt loads were carried with almost complete extinction at less than 5 cm. These usually short-lived conditions were not considered in the calculations as in the Western Rain District most heavy showers causing such conditions occur in late afternoon to early evening (DALE 1960) when effects on light are minimal. The rapid initial losses with depth at Stations I and II compared with the other stations may have resulted because the light there was diffuse, with little reflection off the dark banks and vegetation, whereas the lower stations were open to direct insolation.

2.5.2. Temperature

The efficacy of single weekly maximum-minimum temperature readings in characterizing the general conditions in streams has been discussed at length by MACAN (1958a) and EDINGTON (1966), with the conclusion that if such readings are made over a long period they give adequately reliable results. In the comparatively stable climatic conditions of West Malaysia, the data obtained for the Gombak over 60 weeks are undoubtedly valid for most of the sections represented, but slight local variations affecting the biota at the micro-level are probably common. Short-term fluctuations caused by cloudy days and the warm run-off water after rain are lost in these weekly maximum-minimum readings, but as the magnitude of such oscillation is only of the order of one to two Centrigrade degrees, their effects are probably not of any significance.

The weekly maximum-minimum air temperatures for the upper and lower catchment areas are shown in Fig. 7. For the forested hill Stations I and II, the ranges were: maximum 28.0–33.5 °C, minimum 18.0–21.0 °C and for lower areas of the disturbed catchment (Stations III, IV and V): maximum 31.0–34.5 °C, minimum 18.3–23.4 °C. These figures are notable for their small range and lack of seasonality. There was a slight peak of about two degrees indicated for the March-June period. This was more

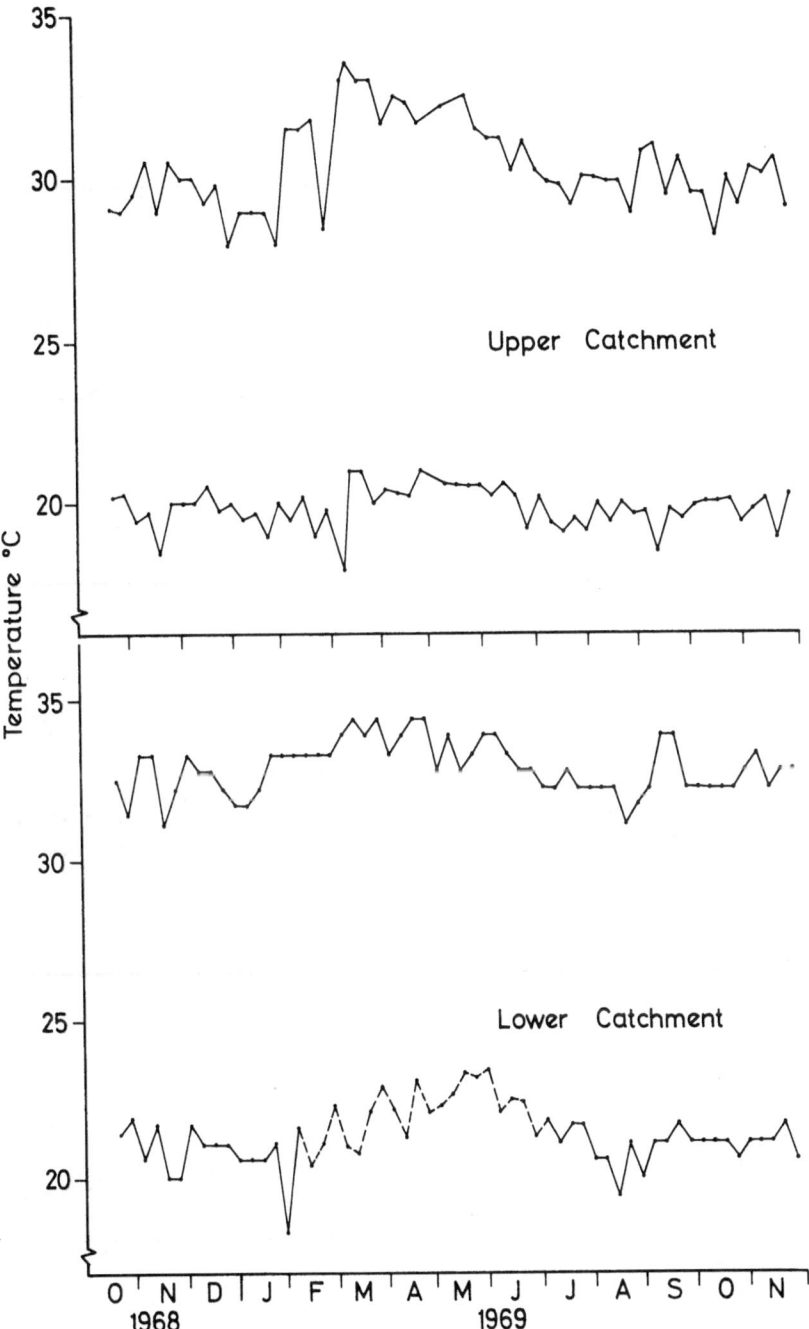

Fig. 7. Weekly maximum-minimum air temperatures in the Upper and Lower Catchments. Note: Broken minimum curve for the Lower Catchment indicates incomplete records from F.R.I.–data obtained from Subang Airport (see text).

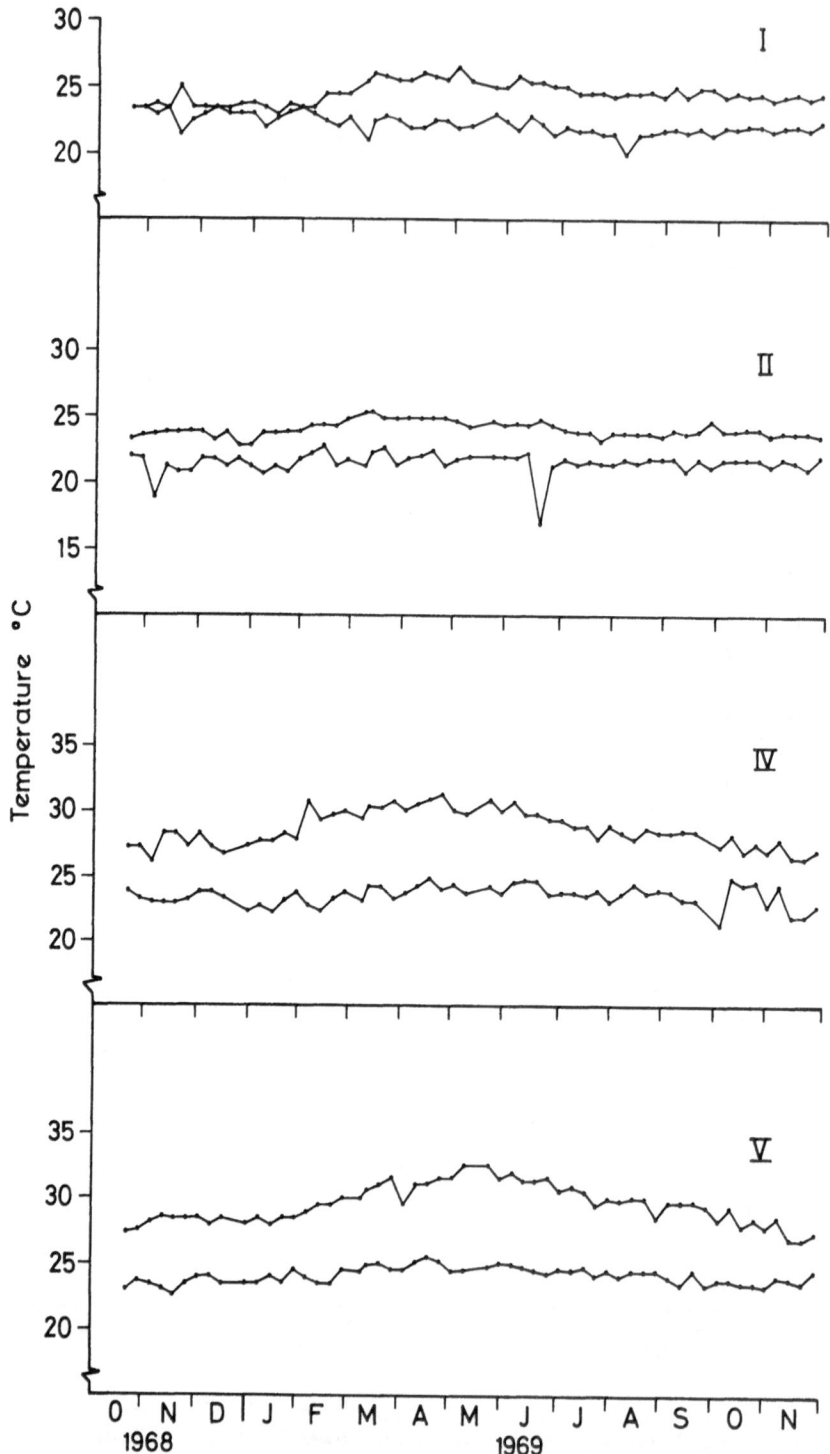

Fig. 8. Weekly maximum-minimum water temperatures at Stations I, II, IV and V.

evident in the upper than the lower catchment where temperature maxima were always high.

Water temperature (Fig. 8) exhibited even less variation at Stations I, II, IV and V where continuous records could be made. The ranges for Station I were: max. 23.0–26.5 °C, min. 20.0–23.5 °C; Station II: max. 22.8–25.5 °C, min. 17.0? or 19.0?–22.8 °C; Station IV: max. 26.2–31.5 °C, min. 21.3–24.8 °C; Station V: max. 26.8–32.5 °C, min. 22.5–25.6 °C, over the period of the investigation. The two extreme minima for Station II, 17.0 °C and 19.0 °C, were probably artefacts of thermometer vibration on the stream bed, especially as there were no corresponding lows at Station I. However, air temperatures in the headwaters at 1000+m may well have been considerably lower than the temperatures recorded at Station II at 220 m. The adiabatic lapse rate (fall in temperature with altitude) in central Malaya is of the order of 1C°/200 m (see BURGESS 1969) so that temperatures in the headwaters may have been as low as 15 °C. Why a plug of cold water, with its temperature unmodified by dilution or friction with the substrate, should descend the valley on two isolated occasions is not explainable. A slight increase in mean temperature for the March-June period was indicated by the water temperature data. This was particularly noticeable at the lower two stations and less pronounced at Station II on the main river under the rain forest canopy. Variation between maximum and mean was less during the major wet period, October-December, at all stations, but particularly evident at Station I. This was probably the result of less variable, higher discharges and cloudy conditions during those months. There was no comparable effect during the March-April wet season. Coefficients of thermal astatism (KAMLER 1965) K = T max./T min. were all low (1.3, 1.5–1.3, 1.5 and 1.4 for Stations I, II, IV and V), emphasizing the temperature stability of the environment.

Diurnal changes in water temperature were slight, as expected. The results from one series of observations taken during normal stage on 28 July 1969 are given in Fig. 9. A slow cooling in late afternoon and throughout the night and a sharp increase at the lower exposed stations as the sun-angle increased in the morning was evident. At Stations I and II shaded both by vegetation and the high hills around the valley, this pattern was not so pronounced. Warming at these stations apparently always began before sunrise, at about 04:00 h, as surface cooling and heat loss from evaporation diminished and the effects of the ground-water temperature began to override the decreasing air-water temperature gradient. WEBER (1959) discussed the isothermal conditions in tropical soils and gave a mean annual temperature of 26 ± 0.3 °C in Panamanian rain forest at 10, 20, 30 cm depth, and a range of 23.5–27.6 °C for African rain forest soils. Ground-water contributions were therefore likely to be at a considerably higher temperature than the nocturnal low of the

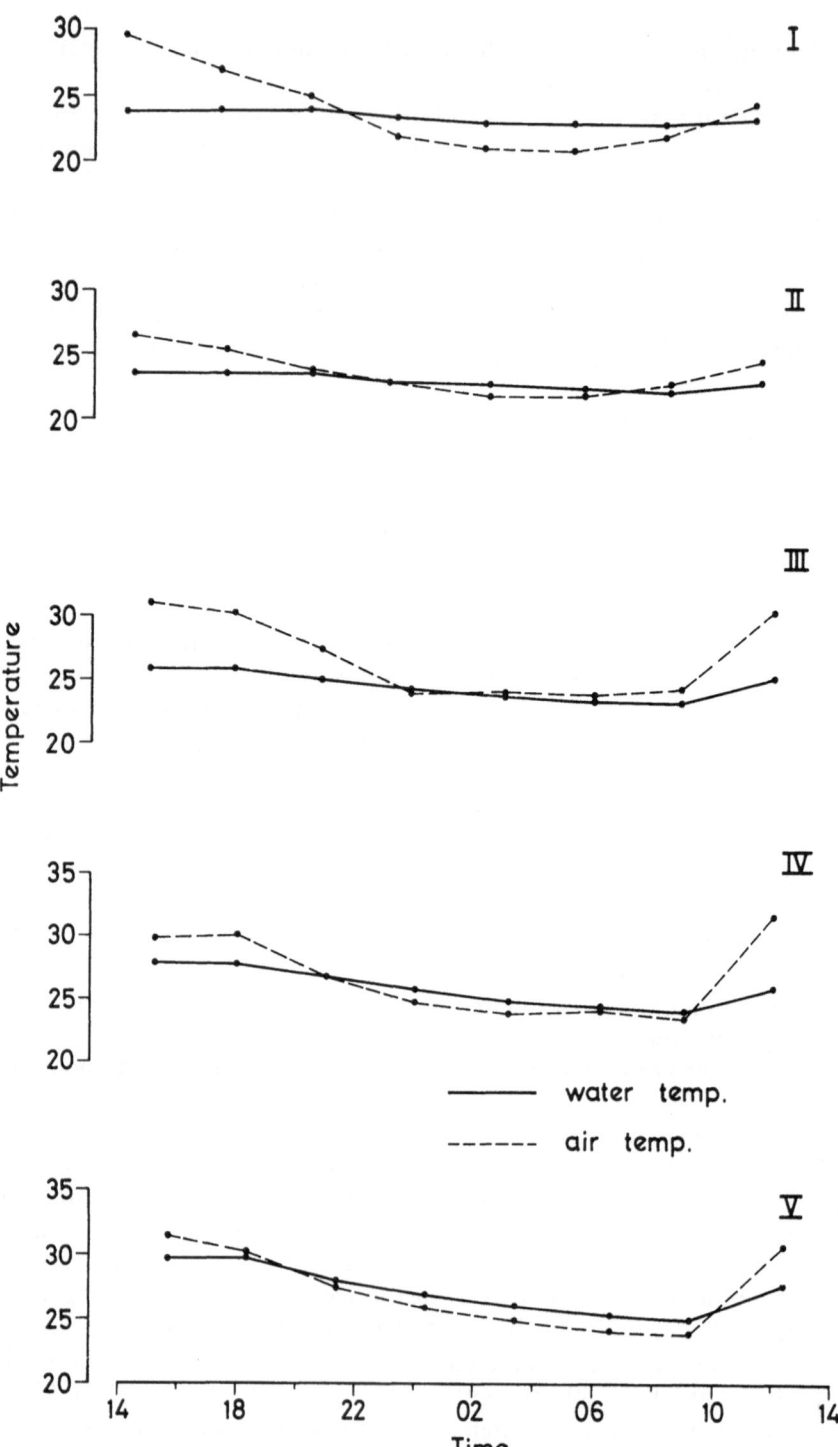

Fig. 9. Diurnal temperature fluctuations at the main stations on 28 July, 1969.

river water and would have had some moderating effect before equilibrating with air temperatures.

Variations in river temperatures during and following rain were slight. The cooling effects of intense rain and increased heat loss to wind in surface layers were quickly dissipated by turbulence. Elevation of the mean temperature of the river by 1–1.5 °C because of surface run-off occurred on several occasions, when the shower was short and intense and followed a day of sunshine. These increases were transitory as losses to the relatively cooler air soon modified their effect. Such changes were not noticeable in the forest areas where surface run-off was more gradual and the ground shaded, but in lowland areas under garden and urban land use, rapid run-off from exposed surfaces was common and increased river temperature frequently occurred.

Vertical gradients (BERG 1943, SLACK 1955) and horizontal variations in temperature over small distances (NEEL 1951, MINCKLEY 1963) were difficult to measure accurately but, as static water was rare in the river, probably neither was common. Slight decreases in temperature, 0.3–0.5 °C at the bottom of large pools in the forested region, more than 1 m deep, were sometimes found. In the lowland reaches, a few deep pools on meander bends, often with overhanging *lalang*, had bottom waters cooler by about 1 °C, but no areas of marked discontinuity or stratified flow were found.

Few long-term temperature records for humid tropical rivers are available. The constancy of temperature conditions in rain forest streams in Amazonia has been described by FITTKAU (1964, 1967) who gives a mean value of 24.5 °C with diurnal and seasonal variations of only about 1 °C. In the forest reaches, the Gombak behaved similarly, although the variation was slightly greater and the mean at Station II lower, probably because of altitude.

Temperature gradients along the profile of a river have been used extensively to classify stream habitats. ILLIES and BOTOSANEANU (1963), KAMLER (1965), WILLIAMS (1966), CHUTTER (1970) and CRISP and LE CREN (1970) have summarized this work. In the Gombak, slight differences in mean temperature between upper and lower stations were obvious, but probably of more significance was the fact that the maximum temperature reached was much higher at Stations IV and V. The maximum-minimum ranges increased at downstream stations, but since these increases were only of the order of five degrees, they were perhaps not as important as in temperate rivers.

The major factors affecting stream temperatures are the warming effects of radiation absorption and the cooling by air-water temperature differences (see RICKER 1934) and for tropical streams in particular, the type of substrate and degree of exposure to direct sunlight (GEIJSKES 1942). The increases in temperature evident at the downstream stations were largely due to increased exposure to radiation. Stations IV and V are almost completely unshaded, and temperature build-up appears proportional to the distance from the forest cover and hence irradiation time. A number of studies have shown woodland to stabilize or decrease stream temperatures (MACAN 1958a, GRAY and EDINGTON 1969) and deforestation is known to increase both maximum readings and fluctuations in temperature, especially in low volume streams (MACAN op. cit., LIKENS et al. 1967). The factors affecting the rates of heat loss and gain are interrelated. The increased surface exposed for evaporation and cooling in riffles and cascades is dependent upon water depth and discharge as also are turbidity, which ELLIS (1936) has shown to increase energy absorption, and water turbulence, which may decrease the heating of bottom sediments by reducing light penetration and the amount of energy absorbed. The water at Stations I and II was always turbulent and, except after large storms, clear, while from Station III down to V, turbulence declined and turbidity rose considerably with increasing silt and clay load. Together these were a definite factor in the warming of the lower river. Wind velocity, often important in temperate latitudes as a factor controlling evaporation rates and disturbing the water surface, is not a significant factor in Malaya where air movements in the forest are negligible and only of low velocity in the lowlands, except during rainstorms.

The higher maximum temperatures at Station I on the tributary, compared to Station II in the same area may be accounted for by the following factors. One kilometre above the sampling location the $3°$ stream trifurcates, increasing the effective channel surface in an area where Route II with its culverts and cleared concession crosses the streams. The water is thus exposed to full insolation for more than 100 m before the tributaries reenter the forest. The sand and stones of the beds of these streams are mostly dark-coloured shales and schists of the sedimentary formation that occupies most of the headwater area. These dark substrates tend to absorb and transfer heat to the water at a higher rate than the white quartz and quartzite sands found elsewhere in the watershed. In an analogous situation, MINCKLEY (1963) reported that bottom materials were 0.1–0.9 °C warmer than the water above them under high insolation.

In summary: water temperatures in the Gombak were seasonally and diurnally stable with small differences caused by rain and prolonged cloudy weather. Mean temperature ranges were narrow for all locations, but showed a longitudinal gradient. The lower stations had higher readings and ranges than the forest stations. Insolation, substrate composition, turbidity, ground-water and rain run-off inflows and heat exchange with the ambient air (both gain and evaporation losses) were important in controlling temperature in the river.

2.5.3. Precipitation

Precipitation in the catchment was characterized by variability of occurrence and intensity. Long-term monthly averages for some of the recording stations and the data for the present survey period are given in Table 12 and show the general features of the rainfall discussed earlier, i.e. increased precipitation in the September-November and May-June inter-monsoonal periods with drier intervening months. From these monthly means and the weekly rainfall totals for each gauge, shown in Fig. 10, the considerable differences between adjacent areas can be seen. The deviation from mean monthly rainfall, defined by a coefficient of variability, is only 13% in the Western Rainfall Region, but in the dry months, particularly July and August, this increases to 50–70% (DALE 1960). The low variability over a large region for long records may be valid, but from gauge to gauge, month to month, the variation was considerably larger than this value of 13%, even for the wet periods. One localized storm of high intensity (>0.3 cm/min has been recorded for Kuala Lumpur (DALE 1960)) from a single unstable cumulonimbus cloud, often covering an area of only a few square km (WATTS 1955), will disrupt the distribution pattern. Although daily totals are obscured by the weekly figures, considerable variation between gauging sites can still be seen. For example, Gauge 4062 for October and November 1969 had aberrantly high readings as a result of four or five local storms. Because of this areal discontinuity in the rainfall, any estimate of total or mean precipitation for the catchment as a whole is difficult without a very dense network of rain gauges (see LINSLEY et al. 1958). Using the Thiessen Polygons, depth and volume of precipitation over the Undisturbed Upper Catchment, Disturbed Lower Catchment and Total Catchment areas were estimated. Table 13 gives these data for each month. The upper catchment had slightly less than proportional rainfall, 34.9% over 36.4% of the area, a reflection of LOCKWOOD's (1967) feelings that rainfall in the foothills is often greater than on the higher slopes. DOUGLAS (1968a) gave figures for a single storm showing 137 mm/24 h at 75 m altitude, 82 mm at 140 m, 67 mm at 220 m, 59 m at 650 m and felt that the maximum distribution of rain in the Gombak catchment was in the foothill region. However, the comparatively greater number of rain-days (Table 14), defined as a 24 h period with more than 0.03 cm precipitation (read as 0.01 inches), at the upper catchment (B series) gauges partially compensated for this and corroborated the empirical field observations that rain in this region was generally more frequent but often less intense than in the lowlands. Many rain-days at these hill gauges were in fact only condensed mist and dew, often amounting to about 0.05 cm, that just came under the definition of a rain-day. This minor component of the precipitation is biologically very important as it helps maintain a high relative humidity. The frequency

Table 12 (a). Average monthly rainfall (up to September 1968)

Gauge	Jan.	Feb.	Mar.	April	May	June	July	Aug.	Sept.	Oct.	Nov.	Dec.	Length of records
4066	14.30	13.25	23.90	26.95	20.25	14.30	12.10	15.90	19.75	28.10	27.10	21.75	51 years
4065	15.25	15.65	27.00	30.80	22.25	16.90	13.40	18.80	23.90	33.00	29.75	22.30	36 years
4067	11.50	10.55	20.45	24.05	25.50	17.80	12.70	19.85	21.40	32.30	28.70	18.60	36 years
4062	13.95	13.35	23.05	27.00	20.85	15.35	13.60	17.85	21.60	29.85	28.10	21.90	56 years
4064	11.70	9.15	18.80	21.35	19.85	14.55	11.95	15.85	31.35	30.40	25.95	18.55	34 years

Table 12 (b). Monthly rainfall over study period (to nearest 0.05 cm)

Gauge	Oct. 1968	Nov.	Dec.	Jan. 1969	Feb.	Mar.	April	May	June	July	Aug.	Sept.	Oct.	Nov.	Total
4066	40.70	14.30	25.00	23.15	15.90	34.80	22.00	39.15	16.95	18.00	24.65	16.15	35.70	25.30	351.75
4065	22.10	17.50	9.65	9.15	9.50	41.45	12.85	33.55	26.75	20.20	26.25	16.40	41.40	20.60	307.35
4067	26.80	10.05	21.90	25.80	11.60	32.85	11.60	21.55	29.55	17.00	29.30	22.50	61.25	19.55	341.30
4062	27.20	18.55	26.55	16.75	10.00	31.65	14.65	27.50	38.20	43.20	50.70	16.70	62.60	61.55	445.80
4064	39.30	17.55	22.75	14.05	7.15	15.05	15.15	25.75	12.20	26.50	26.30	23.95	32.45	24.40	302.55
B13*	37.90	16.85	27.35	17.65	11.00	12.90	11.90	26.85	20.00	24.30	26.50	21.55	34.15	31.15	320.05
B16*	35.85	23.85	19.90	10.30	14.60	14.40	11.10	31.10	25.50	25.95	24.35	25.60	39.80	30.80	333.10
B18*	30.50	17.55	23.50	11.40	15.00	13.70	11.25	26.95	27.30	26.15	21.80	25.50	37.80	31.55	319.95
B23*	32.80	15.40	27.95	11.90	16.90	16.00	21.35	25.65	18.85	28.45	18.60	21.55	34.05	28.70	318.15

* previous records not available

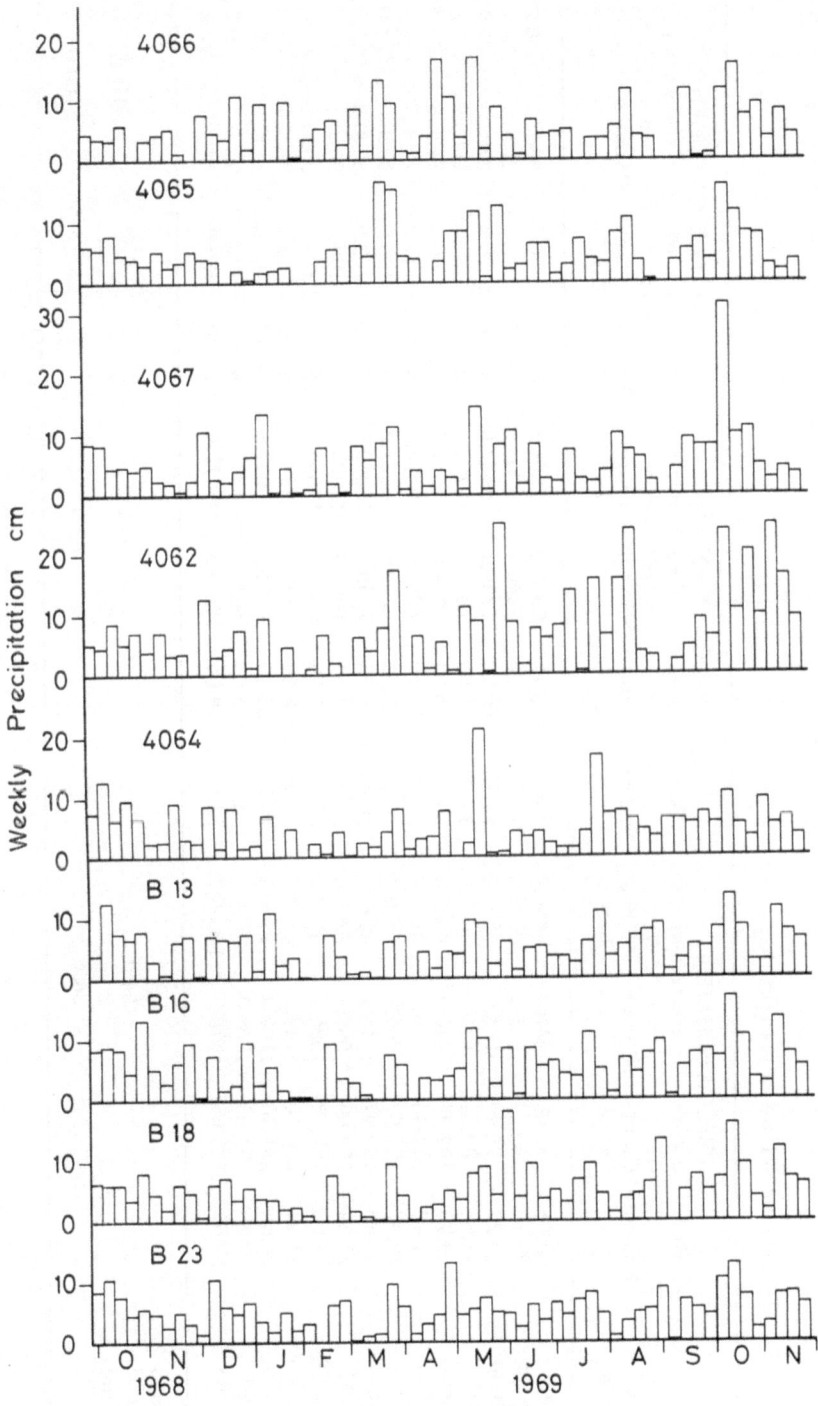

Fig. 10. Weekly precipitation for the nine rain gauges over the study period.

Table 13. Depth and total volume of precipitation

Period ending	UPPER CATCHMENT				LOWER CATCHMENT				TOTAL CATCHMENT			
	Depth on area (cm)		Rain volume m³ × 10⁴		Depth on area (cm)		Rain volume m³ × 10⁴		Depth on area (cm)		Rain volume m³ × 10⁴	
	Monthly	Cumulative	Monthly	Cumulative	Monthly	Cumulative	Monthly	Cumulative	Monthly	Cumulative	Monthly	Cumulative
30–X–68	32.06	32.06	1440.6	1440.6	30.13	30.13	2361.5	2361.5	30.84	30.84	3802.1	3802.1
29–XI–68	20.16	52.22	905.7	2346.3	17.15	47.28	1344.4	3705.9	18.25	49.09	2250.1	6052.2
29–XII–68	23.27	75.49	1045.6	3391.9	22.15	69.43	1736.1	5442.0	22.56	71.65	2781.7	8833.9
31–I–69	13.65	89.14	613.2	4005.1	17.23	86.66	1350.7	6792.7	15.93	87.58	1963.9	10797.8
27–II–69	13.70	102.84	615.5	4620.6	10.27	96.93	804.8	7597.5	11.52	99.10	1420.3	12218.1
29–III–69	11.46	114.30	515.1	5135.7	22.77	119.70	1784.6	9382.1	18.65	117.75	2299.7	14517.8
28–IV–69	13.37	127.67	600.9	5736.6	16.32	136.02	1279.1	10661.2	15.25	133.00	1880.0	16397.8
31–V–69	31.40	159.07	1410.7	7147.3	30.61	166.63	2399.1	13060.3	30.90	163.90	3809.8	20207.6
30–VI–69	23.59	182.66	1059.7	8207.0	25.01	191.64	1960.4	15020.7	24.49	188.39	3020.1	23227.7
30–VII–69	21.96	204.62	986.7	9193.7	25.95	217.59	2033.4	17054.1	24.49	212.88	3020.1	26247.8
29–VIII–69	25.48	230.10	1144.9	10338.6	31.00	248.59	2429.2	19483.3	28.99	241.87	3574.1	29821.9
28–IX–69	24.66	254.76	1107.9	11446.5	16.90	265.49	1324.3	20807.6	19.73	261.60	2432.2	32254.1
31–X–69	40.20	294.96	1806.0	13252.5	50.14	315.63	3929.3	24736.9	46.52	308.12	5735.3	37989.4
30–XI–69	30.67	325.63	1378.2	14630.7	32.42	348.05	2540.5	27277.4	31.78	339.90	3918.7	41908.1
% of total rainfall			34.9				65.1				100.0	
% of total area			36.4				63.6				100.0	

44

Table 14. Number of rain-days in study period of 426 days

Gauge	Oct. 1968	Nov.	Dec.	Jan. 1969	Feb.	Mar.	April	May	June	July	Aug.	Sept.	Oct.	Nov.	Total
4066	20	13	10	10	6	13	11	13	7	8	12	8	17	12	160
4065	18	19	11	7	4	8	6	15	13	8	12	6	18	10	155
4067	16	11	11	8	8	16	12	16	15	12	18	13	21	15	192
4062	16	12	13	11	5	13	12	15	15	13	15	10	20	21	191
4064	17	17	18	5	4	11	13	11	10	12	15	19	21	20	193
B13	20	19	19	13	11	13	17	25	16	22	22	18	23	21	259
B16	21	24	20	16	11	15	13	23	17	27	22	23	31	26	289
B18	26	18	25	15	12	15	17	21	18	20	20	20	28	26	281
B23	26	19	26	15	15	20	20	20	20	21	24	21	28	25	300

Table 15. Maximum number of consecutive rainless days

Gauge	Days	Period
4066	16	5–20 Sept. 1969
4065	18	28 Jan.–14 Feb. 1969
4067	11	24 Nov.–4 Dec. 1968
4062	12	5–16 Sept. 1969
4064	16	13–28 Jan. 1969
B13	9	2–10 Mar. 1969
B16	12	3–14 Apr. 1969
B18	7	7–13 Feb. 1969
B23	6	6–11 Sept. 1969

of rain-days per month indicates the temporal distribution of rainfall, with fewer rain-days in February-March and more in the inter-monsoonal periods. However, light showers or dew made this an unreliable indicator of overall rainfall conditions, particularly in the upper watershed. Total precipitation for the 14 month period 1 October 1968–30 November 1969 was 339.9 cm and for the year 1 October 1968–30 September 1969, 262.3 cm. This annual total was slightly higher than the long-term mean for Kuala Lumpur of 235 cm, but the higher relief over most of the catchment may account for this. From the monthly totals, 1969, particularly July and August, was slightly wetter than the average year. The dates and duration of dry spells (consecutive days without rain) for each gauge are given in Table 15 for 1968/69. These indicate the uniformity of precipitation over a year with 18 as the maximum number of days anywhere without precipitation. In general, the upper rain gauges had shorter dry spells, but this was not particularly valid as a heavy dew could break up a genuine dry spell record.

Short (< 3 h/storm), heavy downpours, often local in extent, are characteristic of most precipitation in Malaya, particularly of convectional rain. This occurs at any season and accounts for the variation between monthly records from year to year and between gauge records for any one storm. Orographic and boundary rain (DALE 1959) may produce persistent rain, but these types occur only in conjunction with the steady winds found at monsoon changes. Even in October 1969 when the number of rain-days was very high, most of the rain was convectional in the Gombak catchment.

In summary: precipitation over the catchment amounted to 340 cm over the 60 weeks, concentrated in two wetter periods of October–December and May–June, but with considerable rain in all months so that rainless periods were short. Rainfall intensity (cm/day) was greater in the lowland and foothill zones where short, heavy downpours were characteristic. In the upper hill areas, precipitation was more frequent, but of lower intensity.

2.5.4. Discharge

Many of the data for this section have been extracted from the records of the Drainage and Irrigation Department and Binnie and Partners (Malaysia). The former hold stage records and stage-discharge rating curves for the gauging station (D.I.D.) at Circular Road, Kuala Lumpur, at the bottom of the catchment, and Binnie and Partners (M) made available daily discharge records for the gauging weir (B.P.) in the upper catchment. The normal water year in Malaysia runs from 1 July to 30 June, but for convenience in this study, the period of minimum flow at the end of the second dry period was used as the initial point and the water year from 1 October 1968 to 28 September 1969 was analysed although data were collected up to 30 November 1969.

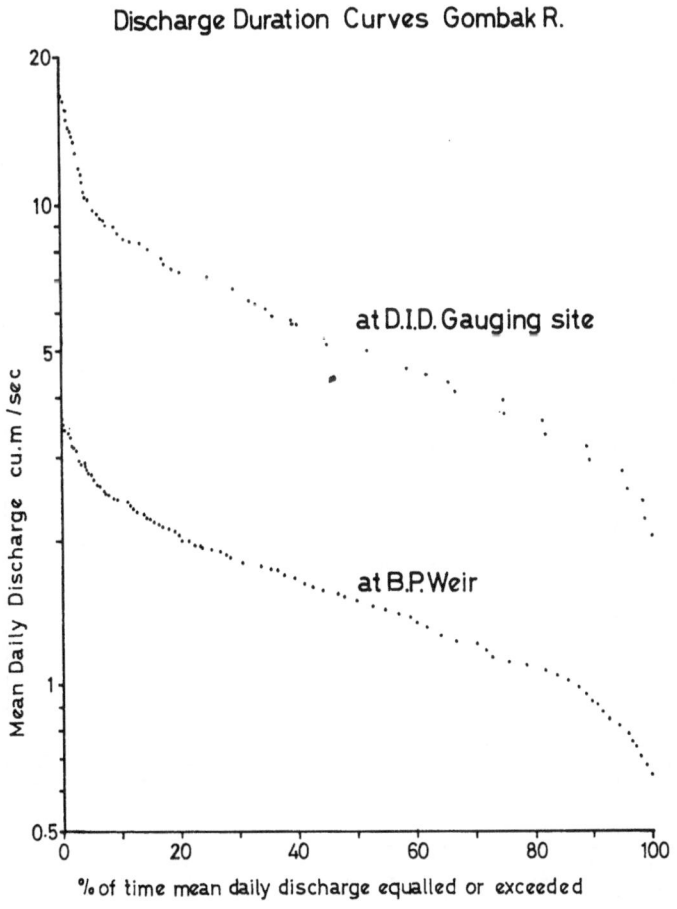

Fig. 11. Flow duration curves for the Gombak River at the B.P. weir and D.I.D. gauging site.

Table 16. Discharge at D.I.D. gauging site (Circular Road, Kuala Lumpur)

Month	Maximum instantaneous discharge cumec					Mean discharge cumec					Mean unit discharge cumec/km²		Extremes		MDD cumec	
	1948/49-1954/55	1955/56-1959/60	1960/61-1967/68	1968	1969	1948/49-1954/55	1955/56-1959/60	1960/61-1967/68	1968	1969	1968	1969	Max.	Min.	Max.	Min.
July	14.11	11.78*	38.23		8.25	3.14	2.49	U		4.28		0.036			7.08	2.80
Aug.	14.96	14.91*	29.17		15.30	2.81	3.13	N		5.65		0.046			8.30	3.93
Sept.	30.35	23.03	43.61		13.03	3.67	3.71	A		4.28		0.035			7.08	2.80
Oct.	39.64	40.07	33.14	16.90	27.40	6.10	7.20	V	5.94	8.98	0.048	0.073	7.78	3.93	16.60	4.60
Nov.	28.35	24.76	44.18	16.20	22.66	6.32	6.38	A	4.96	8.43	0.040	0.068	10.45	2.80	14.10	6.25
Dec.	15.68	32.74	36.82	31.00		4.46	5.80	I	6.70		0.054		16.30	2.80		
Jan.	28.39	18.04	30.30	30.00		3.75	4.10	L	5.07		0.041		11.95	2.06		
Feb.	23.67	18.00	15.72	15.30		3.98	2.94	A	3.46		0.028		7.60	2.80		
Mar.	26.29	10.21	32.00	27.90		4.17	3.15	B	4.11		0.033		14.20	2.06		
April	24.09*	23.03*	21.95	11.90		5.26*	4.83	L	3.77		0.031		8.08	2.06		
May	23.83	21.50*	32.29	27.10		5.22	4.71	E	6.09		0.049		15.00	2.80		
June	10.74	10.91*	38.23	19.90		3.90	2.66		5.60		0.045		16.60	3.15		
Water year mean									5.01		0.041					
Study period mean									5.54		0.045					

* records incomplete

Discharge characteristics of the catchment are summarized in the flow duration curves (Fig. 11) for the undisturbed upper and total catchment zones. These indicate the percentage of time for the study period during which the mean daily discharge in cubic metres/sec (cumec) was equalled or exceeded. The slopes of both curves are moderately flat, indicating considerable ground-water storage capacity, but the even falling-off of the lower end of the curve suggests that the rate of release is uniform with little long-term storage (see JOHNSTONE and CROSS 1949). Storage for the total catchment, top curve, is probably better than for the upper region alone. This might be expected, with the large alluvial plain over capped limestone in the lowland providing a substantial storage area.

Table 16 summarizes the important discharge parameters for the whole catchment. In the Gombak, as in most Malayan rivers, flood crests are of short duration so that maximum instantaneous discharges, listed for the study period and from long-term records, give a distorted impression of high water conditions. The maximum mean daily discharge (MDD) per month is a more realistic indicator and the figure of 16.6 cumec in June and October 1969 is only about half the maximum instantaneous discharge recorded for a number of months. The extreme range between maximum and minimum MDD found for most months reflects the rapid run-off and low retention of the heavy precipitation from single storms. The relatively small difference (factor $\times$ 2 only) between the highest and lowest maximum MDD for the period indicates that flood events occurred at any time in all months of the year as a result of convectional rainfall. This will be discussed further under Hydrologic Cycle. Mean monthly discharges are a good indication of general flow conditions with the lowest values found during the February–March dry period and the higher means in May–June and October–November, particularly in 1969. June, July and August were slightly wetter than average and this partially accounts for their mean cumec being higher than the long-term records. Unfortunately, records for the period 1960–1968 have not yet been published, but from records for the study period in which rainfall was only slightly above average, the mean monthly discharge exceeded the 1948–1960 means for nine of 12 months. This is perhaps a reflection of increased run-off losses and less retention by a watershed that is being used more and more for urban and agricultural purposes. Mean unit discharge in cumec/sq.km had a mean for the study period of 0.045, a minimum in February of 0.028 and a maximum in the very wet October 1969 of 0.073.

Comparable data for the Undisturbed Upper Catchment alone are given in Table 17, but long-term records for comparison are not available. Mean discharge ranged from 2.46 cumec in November 1969 to 0.90 in March, with a mean for the October to September water year of 1.47 cumec. The driest period in this area was later than for the whole catchment whose minimum was in February. Relative variations of MDD were less than for the whole catchment with a 3.63 cumec maximum in October

Table 17. Discharge at B.P. weir

Period ending	Mean discharge cumec	Mean unit discharge cumec/sq.km	Extremes MDD cumec Max.	Min.
30–X–68	2.12	0.047	3.14	1.50
29–XI–68	1.89	0.042	2.92	1.44
29–XII–68	1.82	0.041	2.97	1.33
31–I–69	1.39	0.031	1.93	1.08
27–II–69	1.10	0.024	2.41	0.79
29–III–69	0.90	0.020	1.59	0.65
28–IV–69	1.15	0.026	2.21	0.65
31–V–69	1.60	0.036	3.51	0.99
30–VI–69	1.44	0.032	2.86	1.08
30–VII–69	1.38	0.031	2.44	0.93
29–VIII–69	1.36	0.030	2.44	0.93
28–IX–69	1.48	0.033	2.12	1.02
31–X–69	2.02	0.045	3.63	1.44
30–XI–69	2.47	0.055	3.43	1.76
Water year mean	1.47	0.033		
Study period mean	1.60	0.036		

1969 and a minimum of 0.65 cumec in March and April. The maximum instantaneous discharge of 13.99 cumec occurred on 13 June 1969, a day when the MDD was only 2.66 cumec, giving an indication of the 'flashy' character of floods in the upper river that often peaked and subsided in less than two hours. The average mean unit discharge for the study period was 0.036 cumec/sq.km. This is considerably less than the total catchment value given above and indicates the increased run-off from the lower catchment. DOUGLAS (1968a) reported a greater unit discharge from the upper catchment areas than from the lower for a single storm and attributed this to the steepness of the slopes compared to the more gentle relief of the lower zone. This may have been true for a single storm in which intensity probably varied in the different areas, but over a longer period, rate of run-off was less controlled in the lower catchment.

At the sampling stations, stage, discharge and the maximum stage for the preceeding week were measured on each visit. The stage data are plotted in Fig. 12, with arbitrary base levels set below the lowest depth recorded on the crest gauges. Maxima are not available for Station III. The most striking feature of these readings is that at the main river stations (II, IV and V) flood crests occurred in almost all weeks giving unstable flow conditions. At Station I, the control area where stage was taken was a large pool that damped the effects of flood crests, but relative changes in volume discharged were as great as those in the main

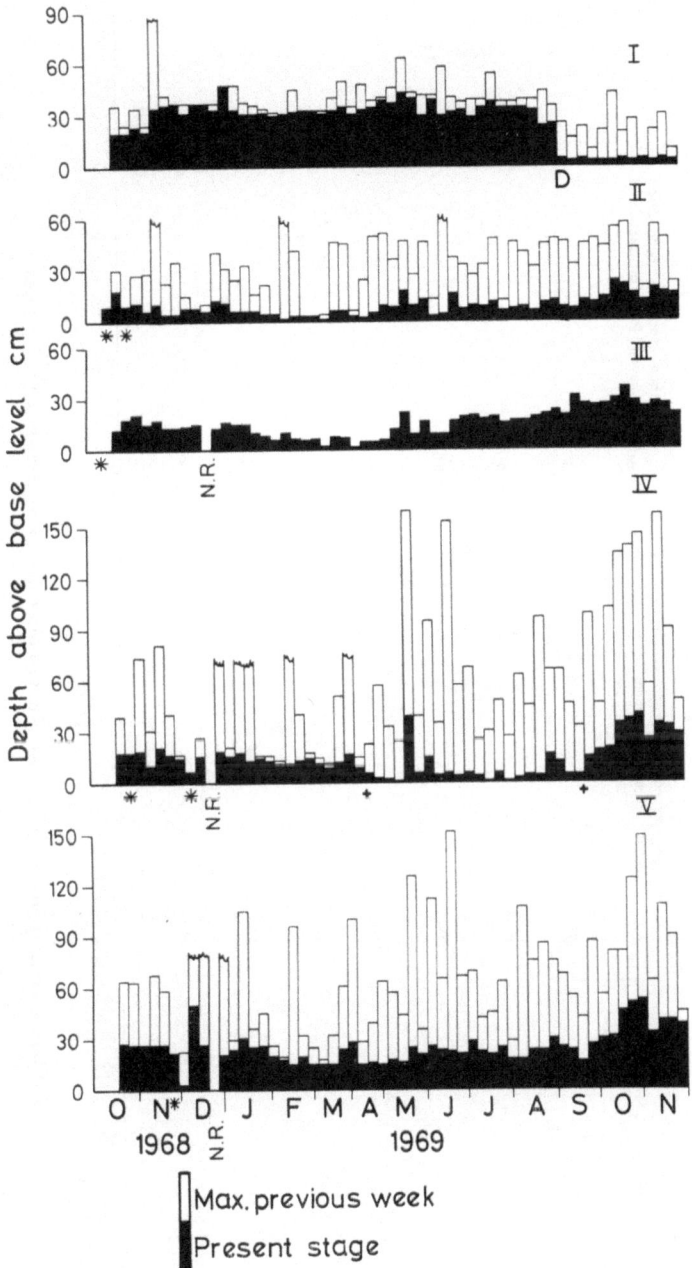

Fig. 12. Stage of river on sampling date and maximum for preceeding week (cm above minimal flow stage). D rupture of bottom, base level altered at Station I: + irrigation ditches above Station IV open between these dates: * no maxima available (Note at Station III): N.R. no record.

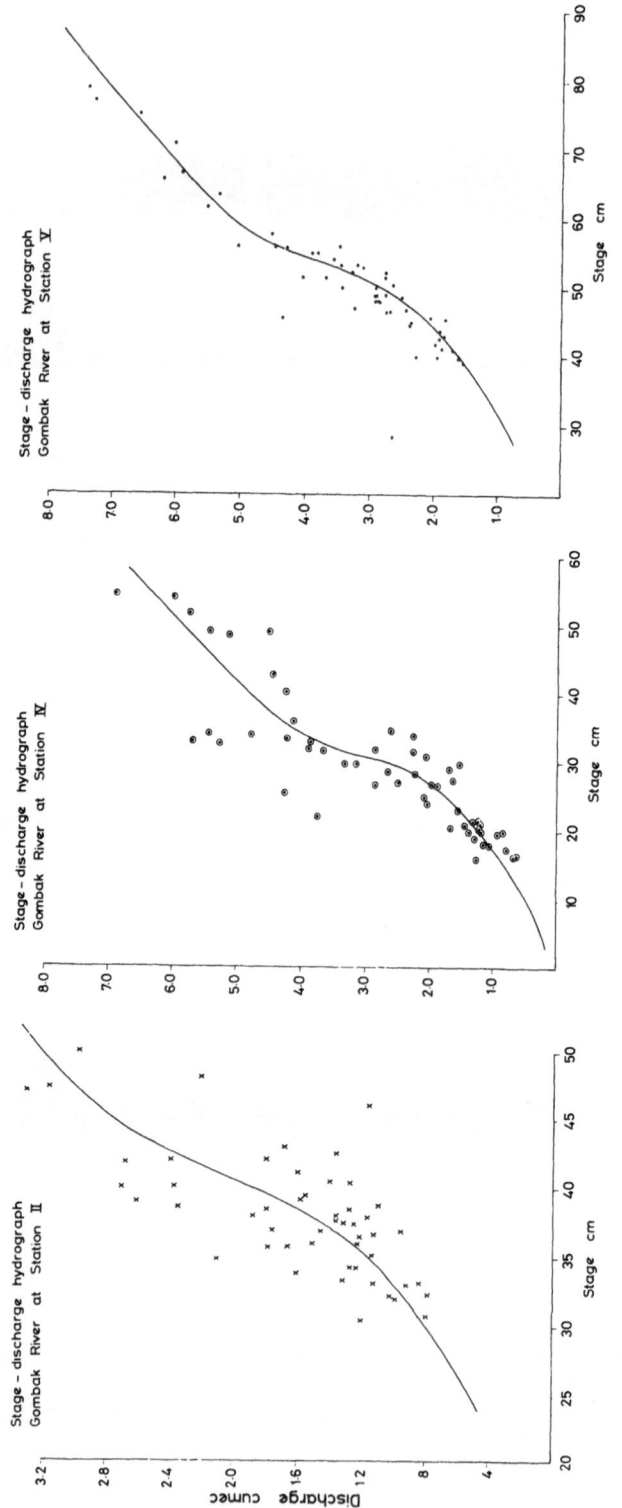

Fig. 13. Stage-discharge hydrographs for the Gombak River at Stations II, IV and V.

river. The dammed area forming this control ruptured in September 1969, dropping the base line 25 cm; hence the discontinuity in the histogram. Immediately above Station IV, two diversion canals extracted irrigation water during the growing season from the end of April to October. This dropped the stage at Station IV to very low levels. Much of this water returned to the river before Station V so that normal levels there were not affected. The volume abstracted was nominally about 0.2 cumec, but off-take volumes at the mouths of these canals were measured on all sampling dates during the irrigation season and were often found to exceed considerably this gazetted volume. Frequently, more than 1.0 cumec was abstracted and on several occasions more than 1.3 cumec which accounted for the drastic drop in water levels at Station IV. Stage-discharge computations were made at all stations but because of highly variable bottom conditions, those at Stations I and III were inconsistent and therefore, not valid. The stage-discharge hydrographs for Stations II, IV and V are given in Fig. 13. The falling-off of the top part of these curves for Stations IV and V was a function of their wide flood plains and the increased sediment loads during large floods that reduced current velocity. At all stations, the wide scatter of points was a result of changing bottom configurations that affected the stage readings.

On 10 April 1969, following a dry period when water levels were near minimum and mean discharges approached base flow, Tributaries A to K were gauged. Their summed discharges (Table 18) accounted for only 54% of the 2.58 cumec discharge at the D.I.D. gauging station. The contributions of A, B, C, D and Station I amounted to 88% of the 0.790 cumec discharge at the B.P. weir. Contributions from other tributaries, particularly the unmeasurable 3° on the NW side below A, probably accounted for most of the remainder in the upper zone, with little direct entry into the channel by ground-water. In the lower catchment, unknown flows from the two large tributaries that feed the tin mining areas and about 0.1 cumec of piped water abstracted from upstream near B.P. entered, but it seems likely that there was also a considerable ground-water contribution from the alluvial plain area. At Station V, just below the entry of the Sg. Blongkong (K), discharge was 1.58 cumec and the tributaries accounted for 89% of this flow.

Depth on area and total flow volume were calculated separately for the Undisturbed Upper, Disturbed Lower and Total Catchments. Table 19 summarizes these data on both a per period and cumulative basis. A March minimum for the Upper and a February minimum for the Lower and Total Catchments were evident, and peak discharge was likewise a month different with the maximum in the Upper delayed until November while the Lower and Total maxima were in October, emphasizing the more direct, rapid run-off in the disturbed area. On an areal basis, the Upper Catchment covering 36.4% of the total area contributed only 28.6% of the discharge. The P.W.D. extracted 2.0 million gallons per

Table 18. Depletion flow on 10 April 1969

Stream	Location	Discharge cumec
Gombak headstream	near A	0.174
Tributary from Gunong		
Bunga Buah	A	0.200
Indian Temple trib.	B	0.068
Tributary	C	0.016
Tributary	Sta. I	0.059
Sg.Pisang	D	0.181
		(Tributary total 0.698)
Main river	B.P.weir	0.790
Sg.Pasir	E	0.065
Sg.Rumput	F	0.084
Sg.Salak	G	0.015
Sg.Pusu	H	0.286
Sg.Semampus	J	0.168
Sg.Blongkong	K	0.087
		(Tributary total 1.403)
Main river	Sta. V	1.580
Main river	D.I.D.	2.580

day (7570 cu.m/day) from the top of the Lower Catchment near the B.P. weir (since increased to 3.0 m.g.d.); this was processed at the Kuala Lumpur waterworks. However, piped water was supplied to the several kampongs and residential estates in the lower zone. Because no accurate estimate of this return contribution could be made, it was assumed to equal the extracted water, i.e. no net loss to the catchment was envisaged. Other water was extracted by the D.I.D. for paddy irrigation at several points, but this all remained within the watershed so the only losses in this regard were from increased evapo-transpiration.

In summary: discharge in the Gombak is highly variable, with little seasonal pattern to flood peaks that may occur at any time, but are concentrated in the inter-monsoonal wet periods. Mean discharge for the study period was 5.54 cumec, maximum MDD was 16.60 cumec and minimum MDD 2.05 cumec for the whole catchment with a mean unit discharge of 0.041 cumec per sq.km. Comparison of sections of the watershed showed that the upper forested zone contributed only 28.6% of the run-off from an area of 36.4%. Mean unit discharge from this zone was 0.033 cumec per sq.km.

The Hydrologic Cycle: Some basic water relationships of the river system emerged from the above sections on precipitation and discharge, and will be discussed from the point of view of the total water budget of the catchment. No attempt will be made to define any of the terrestrial and biological components of the evapo-transpiration $(E + T)$ contribution; this will be left simply as the difference between input and outflow.

Table 19. Depth and total volume of discharge

Period ending	UPPER CATCHMENT				LOWER CATCHMENT				TOTAL CATCHMENT			
	Depth on area (cm)		Discharge m³ × 10⁴		Depth on area (cm)		Discharge m³ × 10⁴		Depth on area (cm)		Discharge m³ × 10⁴	
	Monthly	Cumulative	Monthly	Cumulative	Monthly	Cumulative	Monthly	Cumulative	Monthly	Cumulative	Monthly	Cumulative
30-X-68	12.22	12.22	549.2	549.2	12.67	12.67	993.1	993.1	12.51	12.51	1542.3	1542.3
29-XI-68	10.91	23.13	490.2	1039.4	10.47	23.14	820.8	1813.9	10.63	23.14	1311.0	2853.3
29-XII-68	10.50	33.63	471.7	1511.1	15.41	38.55	1207.9	3021.8	13.62	36.76	1679.6	4532.9
31-I-69	8.81	42.44	395.7	1906.8	14.05	52.60	1101.1	4122.9	12.14	48.90	1496.8	6029.7
27-II-69	5.70	48.14	256.3	2163.1	7.07	59.67	553.8	4676.7	6.57	55.47	810.1	6839.8
29-III-69	5.16	53.30	232.0	2395.1	8.94	68.61	700.6	5377.3	7.56	63.03	932.6	7772.4
28-IV-69	6.67	59.97	299.8	2694.9	9.82	78.43	769.5	6146.8	8.67	71.70	1069.3	8841.7
31-V-69	10.16	70.13	456.7	3151.6	16.30	94.73	1277.1	7423.9	14.06	85.76	1733.8	10575.5
30-VI-69	8.29	78.42	372.5	3524.1	13.77	108.50	1079.1	8503.0	11.77	97.53	1451.6	12027.1
30-VII-69	7.94	86.36	356.9	3881.0	9.67	118.17	757.7	9260.7	9.04	106.57	1114.6	13141.7
29-VIII-69	7.86	94.22	353.1	4234.1	14.03	132.20	1099.8	10360.5	11.78	118.35	1452.9	14594.6
28-IX-69	8.56	102.78	384.5	4618.6	9.11	141.31	714.0	11074.5	8.91	127.26	1098.5	15693.1
31-X-69	12.84	115.62	576.7	5195.3	24.63	165.94	1930.3	13004.8	20.33	147.59	2507.0	18200.1
30-XI-69	14.26	129.88	640.6	5835.9	19.71	185.65	1544.3	14549.1	17.72	165.31	2184.9	20385.0
% of total discharge			28.6				71.4				100.0	
% of total area			36.4				63.6				100.0	

55

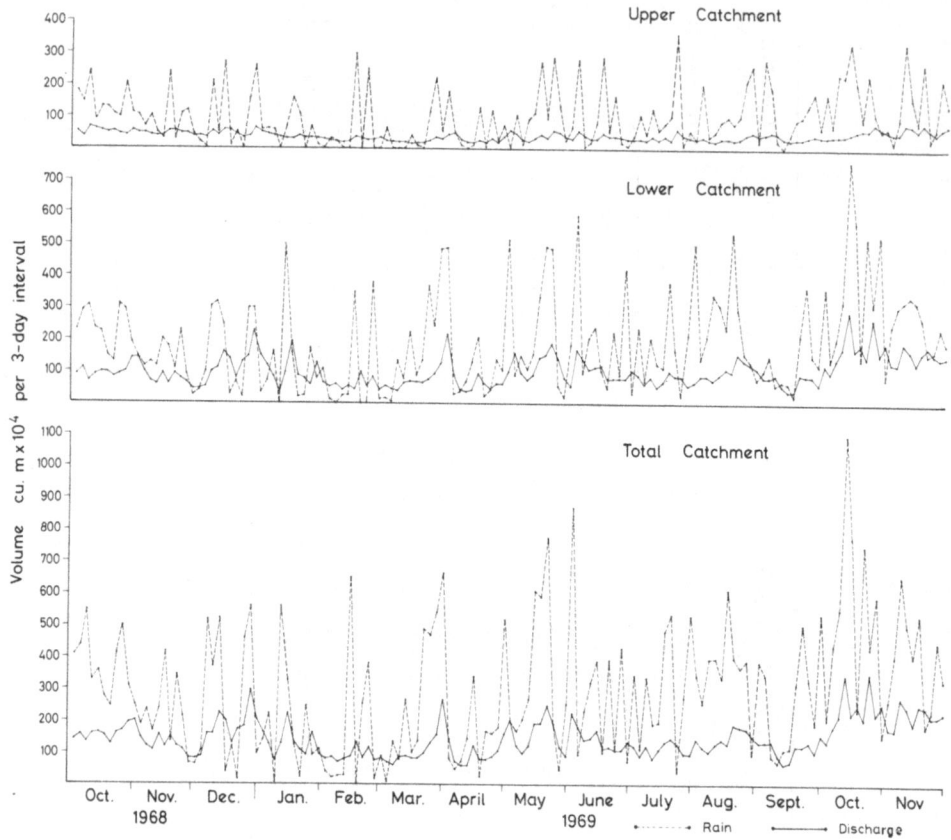

Fig. 14. Three-day discharge hydrograph and three-day total precipitation for Upper, Lower and Total Catchment areas; – – – – – – Rain; ———— Discharge.

Fig. 14, the 3-day discharge hydrograph and 3-day precipitation plotted over the period for the Upper, Lower and Total Catchments, is a summary of the principal hydrologic events. The run-off component of discharge was probably responsible for almost all outflow as discussed earlier and total discharge was, therefore, closely dependent on antecedent rainfall. The directly recruited ground-water component and the difference between abstracted and supplied water (yield) could not be quantified, but were probably insignificant.

Total run-off volume for any storm was dependent on prior catchment conditions. Infiltration and interception loss (base loss) for a steep jungle-clad catchment similar to the Upper Gombak was found to be 1.3–1.8 cm by CHARLTON (1964) but much lower, 0.8 cm, for wet conditions by TAN (1969), i.e. no surface run-off occurred until rainfall exceeded this base loss. No value for a disturbed catchment is available, but an estimate based on proportions of forest, rubber and urban cover for the Lower Catchment is less, 0.5–1.0 cm. Fig. 14 shows the comparatively even flow for the Upper Catchment with moderate peaking following high precipitations and, in contrast, the extreme fluctuations in discharge of the Lower Catchment with each storm. The degree of substrate instability caused by these spates was much greater in the downstream reaches. The frequent, but moderate, flood crests in the torrential reaches had removed most of the very fine substrate materials as they formed, leaving a relatively stable bottom composed of fragments moved only by floods with greater competence. At the lowland stations, the constantly accumulating finer sediments were resuspended and transported by almost any increase in stage (CHEBOTAREV 1962), making biotic conditions in these areas highly variable and transitory. The steep relief of the upper area might have been expected to promote more rapid run-off than the gentle slopes of the lowland area, but this factor appeared to be counteracted by the effects of forest cover on water retention capacity. As discussed earlier, the Lower Catchment is greatly disturbed, with much of the forest cover removed and replaced by monolayer crops, rubber plantations and urban zones, producing many areas of little or no retention. HOOVER (1944) found a run-off increase of about 20%, BOCHKOV (1970) up to 50% and LIKENS et al. (1970) from 39% increase in the first year to 28% in the second year after deforestation. Many other authors (DELFS 1956, RUTTER 1958, BULLARD 1965, BINNS 1969, GRAY and EDINGTON 1969) have discussed the problems of increased run-off from areas where the natural cover has been removed. Their consensus is that removal of evapotranspiration capacity is the primary cause, but that decreases in infiltration rate and porosity which follow the destruction of the root-mat result in more surface flow and are also important factors. HOOVER (1944) commented on the higher maximum flood peaks following removal of tree cover. The role of forest as a regulator of flow peaks and as an agent moderating and increasing discharge minima has been summarized by PENMAN (1963) and BOCHOV (1970). These factors apply to the Gombak and contribute to the discharge instability pattern. TAN (1969) reported that run-off for a gentle-sloped, *Hevea*-covered watershed, similar to the lower Gombak, was up to 50% greater than for a steep-sloped jungle area analogous to the upper Gombak for heavy precipitation in excess of 7 cm, as often occurs in convectional storms. However, it must be reiterated that in the lowlands, rainfall intensity is often greater than in the hills and that consequently pressure on the drainage system is higher.

   Table 20 compares the monthly and cumulative discharge as a percentage of the precipitation. The value of about 40% found for the forested catchment, i.e. with E+T losses of 60%, is higher than the maximum 30% found for Sg. Lui, a D.I.D. reference catchment with forest-clad, steep hills investigated by TAN (1969), but considerably lower

Table 20. Cumulative and monthly discharge as a percentage of precipitation

Period ending	Cumulative % $\frac{\text{Discharge}}{\text{Rain}}$			Monthly % $\frac{\text{Discharge}}{\text{Rain}}$		
	Upper	Lower	Total	Upper	Lower	Total
30–X–68	38.1	42.1	40.6	38.1	42.1	40.6
29–XI–68	44.3	49.0	47.1	54.1	61.1	58.3
29–XII–68	44.6	55.5	51.3	45.1	69.6	60.4
31–I–69	47.6	60.7	55.8	64.5	81.5	76.2
27–II–69	46.8	61.6	56.0	41.6	68.8	57.0
29–III–69	46.6	57.3	53.5	45.0	39.3	40.6
28–IV–69	47.0	57.7	53.9	49.9	60.2	56.9
31–V–69	44.1	56.8	52.3	32.4	53.2	45.5
30–VI–69	42.9	56.6	51.8	35.2	55.0	48.1
30–VII–69	42.2	54.3	50.7	36.2	37.3	36.9
29–VIII–69	41.0	53.2	48.9	30.8	45.3	40.7
28–IX–69*	40.4	53.2	48.7	34.7	53.9	45.2
31–X–69	39.2	52.6	47.9	31.9	49.1	43.7
30–XI–69	39.9	53.3	48.6	46.5	60.8	55.8
			Mean	41.9	55.5	50.4

* end of water year

than the 52% and 63% found for the large Perak River and Kelantan River respectively (Oh 1965) which have partially-cleared watersheds and are subject to frequent flood excesses. The value close to 50% for the catchment as a whole indicates considerable loss of stability from the natural state because of deforestation. Under severe conditions in South Africa, SCHWARTZ (1969) found E+T losses of up to 98% (i.e. percentage discharge over precipitation (% D/R) $\simeq$ 2%) but even so, flash floods still occurred. The cumulative ratios obscure the variations that occurred from period to period. The high % D/R in January, for example, was probably due to increased ground-water contributions from a saturated watershed (see Fig. 14), assuming that E+T losses of precipitation remained the same. Other variations were dependent on the intensity of rain. If the base loss value, dependent on antecedent conditions, was exceeded only rarely or at evenly spaced intervals that allowed stabilization of storage between storms, the D/R ratio remained small, as in July to October 1969. Higher values occurred with frequent heavy rainfall as in the November-December period. Fig. 15 illustrates the relationships between discharge and precipitation for depth on area, with the difference R—D on the histograms representing E+T losses per period. The volume relationships are of course identical.

KENWORTHY (1969) has recently completed a preliminary investigation of the effect of forest cover on a 1° upper Gombak watershed and concluded that E+T unit losses in the forest approached theoretical

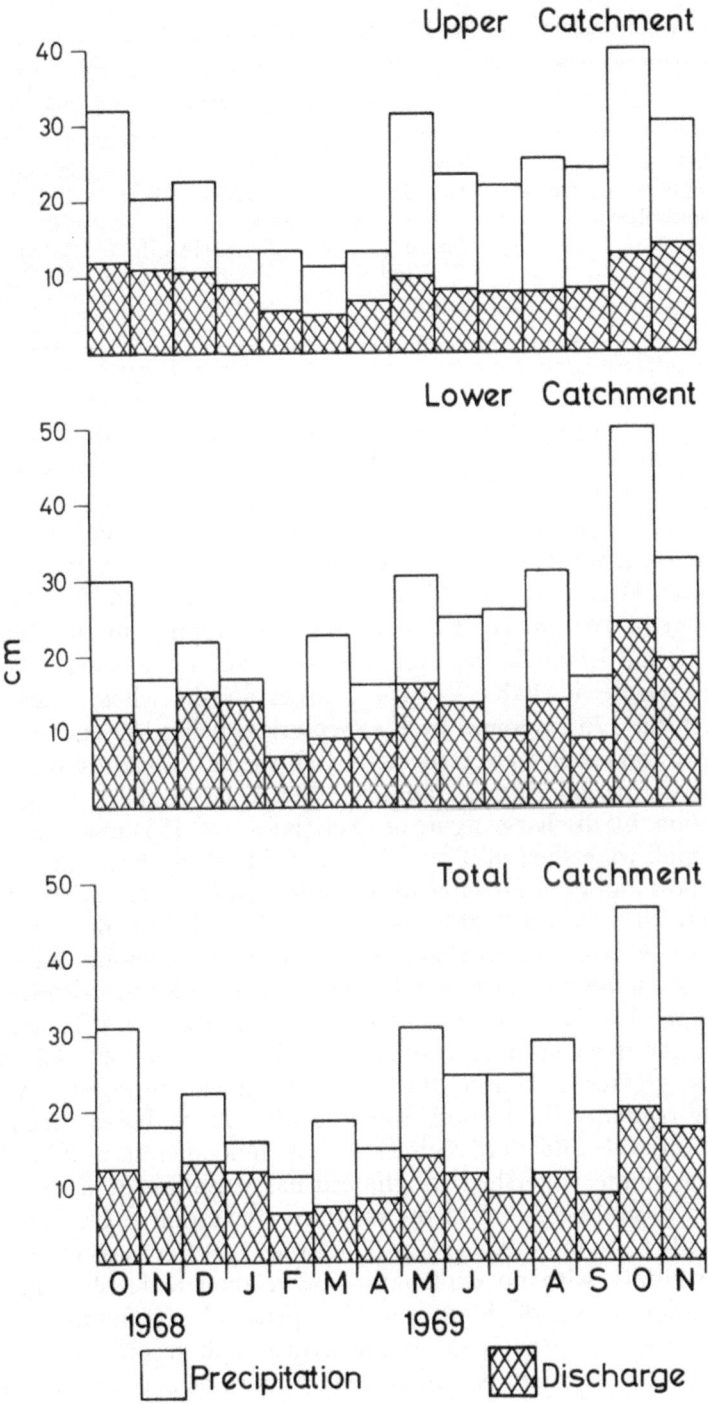

Fig. 15. Depth on area precipitation and discharge per month for the Upper, Lower and Total Catchment areas.

Table 21. Water budget (October 1968–September 1969)

	Upper	Lower	Total Catchment
Rain depth on area cm	254.76	266.49	261.60
Discharge depth on area cm	102.78	141.31	127.26
E_0 (theoretical) cm	197	202	199
$\dfrac{E+T}{E_0}$	$\dfrac{254.76-102.78}{197} = 0.77$	$\dfrac{266.49-141.31}{202} = 0.62$	$\dfrac{261.60-127.26}{199} = 0.68$

values for evaporation from an open water surface (E_0) with $(E+T)/E_0 = 0.89$. The theoretically calculated E_0 of 198.3 cm/yr (WYCHERLEY 1967) and 205.7 cm/yr (CHIA 1967) for Kuala Lumpur were averaged to give 202 cm/yr for the Lower Catchment, and the adjustment for distance from the coast was substracted (KENWORTHY op. cit.) to give an E_0 for the Upper Catchment of 197 cm/yr, with an estimate for the whole catchment of 199 cm/yr. These values were applied to the data from Tables 13 and 19 for the rain year October 1968 to September 1969 and gave the results in Table 21. $E+T$ losses for the whole year did not approach the 175 cm found by KENWORTHY, being 152 and 124 cm for the upper and lower areas respectively. However, his values were derived only for the period August–January when E losses are at a maximum. In addition, his discharge figure of 75 cm is low (cf. 103 cm for this study). In his small watershed of 0.3 sq. km on a side slope, losses through sub-surface flow and ground-water to the valley floor, not considered by him, are likely to be considerable. An $(E+T)/E_0$ of 0.77 is not exceptional (cf. PENMAN 1963) and perhaps water loss from the humid forest is not as high as is sometimes theorized ($(E+T)/E_0 \simeq 1.0$ RUTTER 1967). These data emphasize the effects of different cover types on $E+T$, with losses in the lower area considerably depressed ($(E+T)/E_0 = 0.62$) and propor-tionately greater discharge the result. Removal of forest cover that partially regulates the run-off from the variable Malayan precipitation leads to unstable discharge regimes with high flood crests and little water retention in the watershed. Slight diurnal variation in discharge was noted on several occasions in the tributary at Station I. On clear nights after 21 : 00 h, an increase in stage of between 0.5–1 cm was noted for this small tributary with no explanation other than a decrease in evapo-transpiration demand. KENWORTHY (personal communication) also observed this nocturnal rise on the hydrograph records for the small stream he was studying. No similar effect could be detected on the main river at Station II.

2.5.5. Substrate analysis

a. The results of the physical analyses of erosional (r) and depositional (s) sediments at each station are given in Tables 22 and 23. These data have been plotted as cumulative curves (Fig. 16, 17 and 18) and the median diameter (Md) and the first and third quartiles (Q_1 and Q_3), i.e. 25 and 75 percentiles of the *phi* distribution, calculated. The use of this classification system with the addition of other percentiles where necessary, e.g. 5 and 95, allows precise definition of particle size and homogeneity. The shape of the curves gives an indication of the uniformity of the substrate. When $Q_3 - Q_1 < 0.5$ as in IIIs and approached in IIr and Vs, uniform, well-sorted distribution is indicated. Skewness of size distribution, as indicated by the kurtosis of the curve, is dependent on the dispersal of the quartiles around Md.

At Station I, the erosional and depositional sediments overlapped, with about 35% medium gravel or smaller particles characteristic of pools occurring also in the riffle area (Fig. 16). The pool zones had a bottom composed of well-mixed sand sizes with negative skew ($Md - Q_1 > Q_3 - Md$) and some fine and medium gravels. In the riffle area, uniform proportions of increasing gravel and cobble sizes were found with a few boulders ($phi > -9$). In the forest tributaries, small pool and riffle areas followed each other closely so that great differences in substrate did not develop. These substrates were derived largely from the band of sedimentary rocks forming most of the catchment. The sands were predominantly dark, soft feldspars, shales and mica-schists that fragmented easily, but also included a small proportion of quartz from the disintegration of sandstone. The larger, more resistant gravels were quartzitic, but with some granite and sandstone.

Station II erosion and deposit sediment types were almost completely isolated (Fig. 16). The depositional substrate, IIs, was positively skewed ($Q_3 - Md > Md - Q_1$) with 65% sand, mostly coarse but with 15% medium-fine and very fine sands. The other 35% was very fine, fine and medium gravels. The erosional sediments, IIr, had less than 10% smaller than coarse gravel and were remarkably uniform; 73% was in just two *phi* categories, very coarse gravel and small cobble, only 3% in large cobble category and the remaining 15% distributed over various boulder sizes up to car-sized. These sediments were unevenly distributed in the river, building up behind big boulders and leaving only the larger size groups in areas exposed to strong current. Frequent spates and constant vibration of the sediments by turbulence removed the small particles as they formed by abrasion. Bottom materials *phi* > -3 were largely granitic in origin. Much of the sand was quartz, but some dark mica-schist and feldspars were present as at Station I.

The erosion and deposit substrates at Station III had about 20% common size materials but were visually quite distinct (Fig. 17). The

61

Table 22. Erosional substrates–percentage composition by weight

	Size	I	II	III	IV	V
	512		15.1*			
	256	1.0	3.0			
	128	27.6	44.3	3.1	9.3	
	64	15.9	28.4	37.4	16.3	4.7
> mm	32	22.1	4.0	25.1	33.6	13.0
	16	7.8	2.2	12.9	14.9	25.7
	8	7.5	1.2	6.8	9.1	18.4
	4	5.1	0.7	4.9	6.8	11.8
	2	3.0	0.3	3.3	3.5	6.4
	1000	3.7	0.3	2.7	2.0	7.4
	600	2.8	0.2	1.1	0.9	6.8
	500	0.7	< 0.1	0.2	0.3	1.7
> μ	300	1.1	0.1	0.6	0.9	1.9
	210	0.6	< 0.1	0.8	1.1	1.0
	150	0.4	< 0.1	0.6	0.7	0.6
	75	0.5	< 0.1	0.5	0.6	0.6
	53	0.1	< 0.1	< 0.1	< 0.1	< 0.1
<	53	0.1	< 0.1	< 0.1	< 0.1	< 0.1

* includes contributions, on an areal basis, of all boulders up to 5 m in diameter. The very occasional large boulders or bedrock at other stations were not included.

Table 23. Depositional substrates–percentage composition by weight

	Size	I	II	III	IV	V
	16	15.5	15.4		5.5	0.3
	8	21.2	10.4	0.2	6.6	2.6
> mm	4	15.6	11.2	0.9	18.7	16.8
	2	8.7	11.0	4.3	15.5	24.1
	1000	13.5	20.2	26.5	19.0	31.6
	600	11.4	17.2	49.3	17.5	14.7
	500	3.8	2.6	6.7	4.7	2.0
	300	4.3	4.8	8.1	6.8	2.1
> μ	210	2.3	3.2	2.7	3.6	1.6
	150	1.4	2.0	0.9	1.6	1.8
	75	1.7	1.6	0.4	0.5	2.3
	53	0.4	0.2	< 0.1	< 0.1	0.2
<	53	0.3	0.1	< 0.1	< 0.1	< 0.1

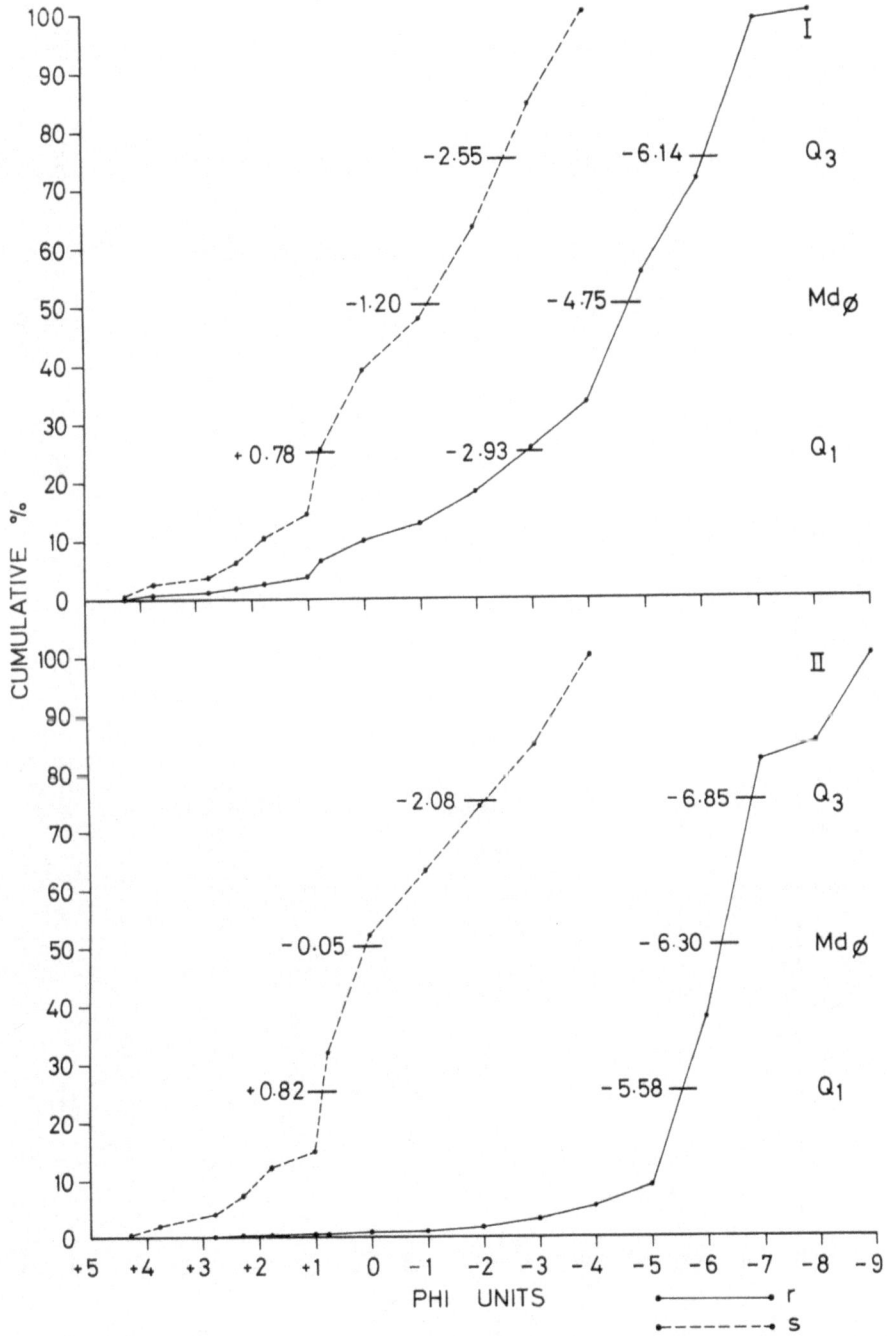

Fig. 16. Cumulative mass curves of depositional (s) and erosional (r) area sediments at Stations I and II.

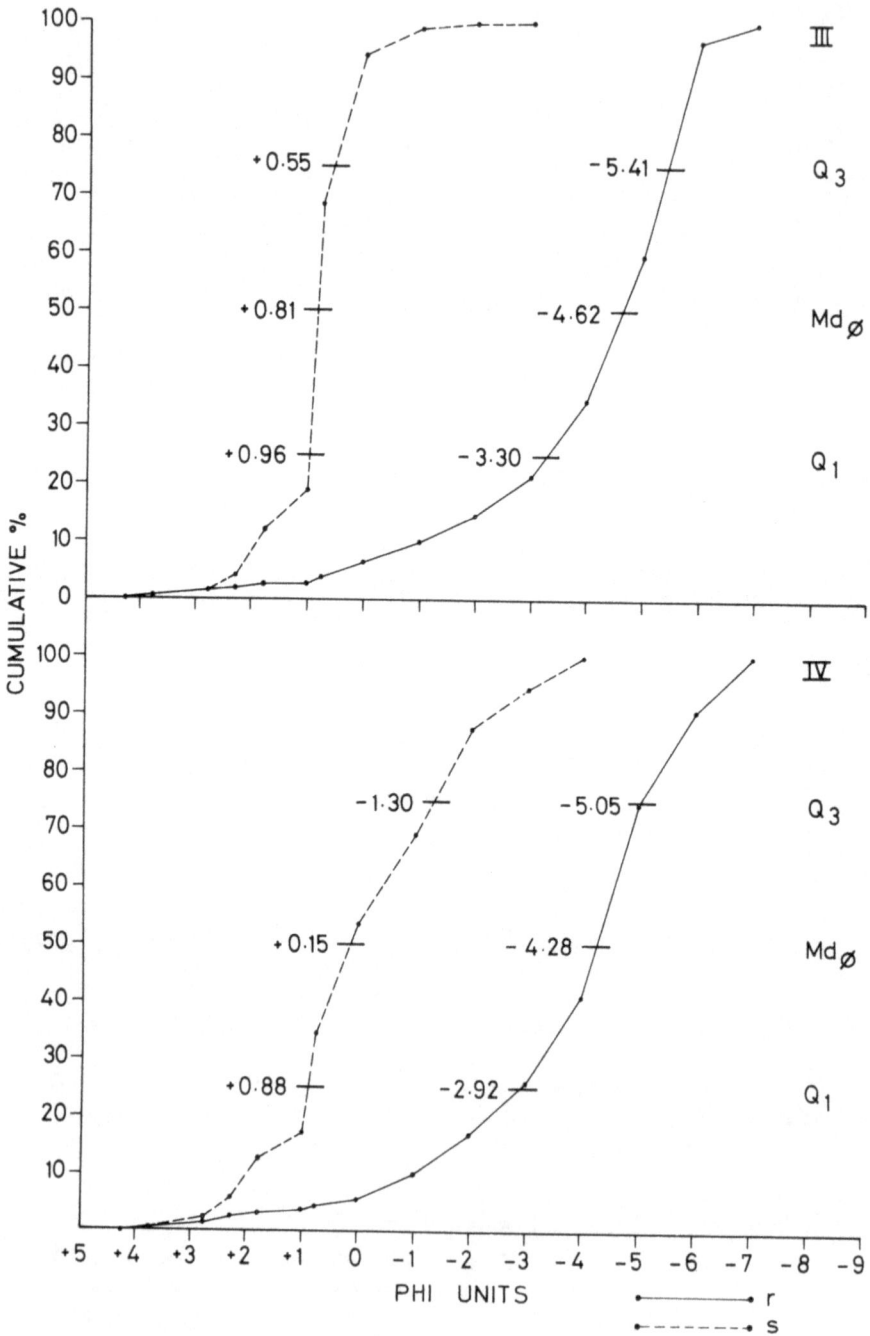

Fig. 17. Cumulative mass curves of depositional (s) and erosional (r) area sediments at Stations III and IV.

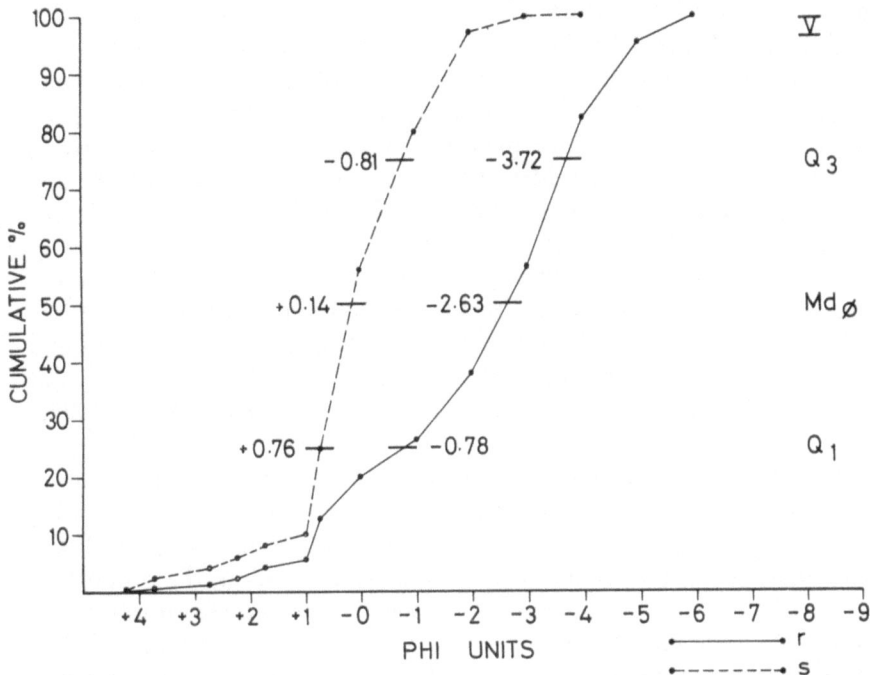

Fig. 18. Cumulative mass curves of depositional (s) and erosional (r) area sediments at Station V.

depositional component was a very well sorted coarse sand with a Q_3 to Q_1 range of only *phi* 0.41 with slight positive skew. Eighty-five percent of the riffle sediments were an evenly mixed series of very fine to very coarse gravels, with a Md of -4.62 and slight negative skew. The flow pattern in this section was generally smooth with uniform velocity so that particles became well-sorted. All sediments exceeding the carrying capacity were left in the erosive zones, with the few large cobbles and boulders and a small proportion of fine material in the interstices. Smaller particles were transported downstream out of the section and deposited where velocity decreased. Because of the evenness of flow, this was a very selective process in this reach with most of the materials finer than *phi*+1 being retained in suspension. The sand material here was entirely quartz, the softer materials having completely weathered. Larger particles were of granite and quartzite.

By Station IV, the river had lost much of its slope and more uniform bottom conditions were the result. The few riffle areas were short and not as well differentiated as at stations further upstream. In addition, erosion of bank deposits and realignment of sand and gravel bars resulted in constant recruitment to the bottom materials. The erosional substrates contained about 40% pool-type sediments deposited between the medium and coarse gravels (*phi*-3 to -5) that dominate this bottom with Md-4.28. Some larger very coarse gravels and small cobbles were also present on the faster areas of the riffles. Depositional sediments were 25% very coarse sand, 30% coarse sand, 20% mixed finer sands and the remainder fine and medium gravels. Md was +0.15 with a strong positive skew (Q_3—Md $\gg$ Md—Q_1). The finer materials were mostly quartz and quartzite but with some flat mica-schist particles; the larger gravels were granite or schist.

The two categories of substrate were not well differentiated at Station V (Fig. 18). Over 80% of the sediment sizes were common to both pools and riffles and the differences were largely of proportion. The erosional bottom was an evenly-graded, well-distributed series from medium sand to very coarse gravel with lesser amounts of fine and very fine sand. The Md of -2.63 and negative skew of this ill-sorted bottom indicated a comparatively small-sized substrate. The pool sediments were a more uniformly textured aggregation. Very coarse and coarse sands (Md+0.14, Q_1 to Q_3 *phi* 1.5) made up 70% of the total with only 10% in all the finer categories and 20% very fine gravel. At this station, the proportion *phi* < +3.5 was definitely underestimated as the silt and clay fraction was lost by the sampling technique. However, the magnitude of this lost portion was probably only 1 to 2% as high water conditions periodically removed the finer sediments. Competence of the river at this point on its profile was very low, and little shifting of particles much larger than *phi* -1 occurred except on very rapidly rising stage after a torrential storm. This accounted for the poor sorting of the riffle sediments. In the depositional zones, only the fine sands and silts dropped after the last flood filled the interstices of the coarser, more stable bottom components. The sand here was all quartzitic.

The apparent break-point in all the curves at or near *phi*+1, between medium and coarse sand, is interesting. Coarse sand was the dominant bottom sediment in the depositional zones at all stations and appeared to represent a natural fragmentation size for materials derived from granite. Some smaller sands originating in other lithologies or resulting from abrasion of the coarse sand occurred, but in smaller proportions.

Fig. 19 summarizes the Q_1MdQ_3 relationship for the sediments, arranges the stations in order of Md and visualizes the skewness of size distribution. The pool deposits in the tributary Station I were larger than at the lower main river stations. IIs, IIIs, IVs, and Vs were quite similar in Md, but the poor sorting and positive skewness of II, IV and V have increased the Md from that of Station III. These depositional substrates appeared to form a similar habitat at all stations, with a uniform bottom type. The erosional bottom substrates in the main river followed the classic pattern with the largest components in the upper zone, IIr, decreasing with slope and erosional age through a progressive series IIIr

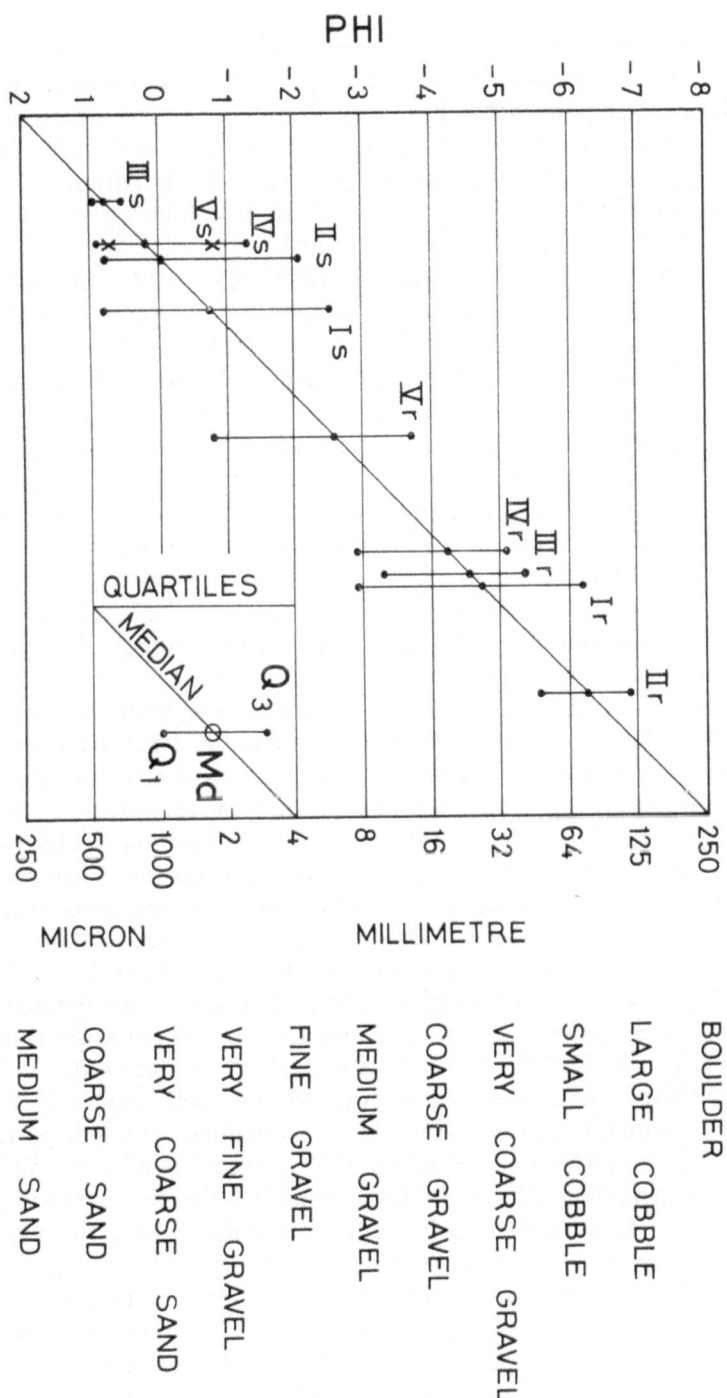

Fig. 19. Q_1MdQ_3 diagram summarizing the structure of the inorganic substrates of the depositional (s) and erosional (r) biotopes at the main stations.

$> \mathrm{IVr} > \mathrm{Vr}$. The sudden decrease in size at Station V was attributed to its location on the alluvial plain at a considerable horizontal distance from the nearest potential parent rock formation. Some limestone out-cropping did occur in the river at Station V but this was a very limited source of substrate. The intermediate position of the Ir tributary bottom type between IIr and IIIr was a function of its comparative youth and the kind of soft parent rock in its watershed.

Since PERCIVAL and WHITEHEAD (1929), the authors of most lotic ecology work have classified the bottom deposits of the watercourses under study. However, their often arbitrary classifications as erosional or depositional (MOON 1939), or with divisions of boulders, rocks, rubble, gravel, sand, mud etc., (BERG 1943, PENNAK and VAN GERPEN 1947, and many others), even if broadly defined as to size, give little indication of the important relationships at the organism-substrate level, particularly with respect to the finer size groups. SCOTT (1958) measured the size of individual substrate units and with RUSHFORTH (1959), SCOTT (1960, 1966) developed the cover fraction $(\mathrm{C_{vf}})$ concept. This considered the ratio of the area of river bed covered by stones to the total area as an index of benthos density, but largely ignored the interstitial areas that are obviously of paramount importance (see later section: Vertical Distribution of the Benthos). KAMLER and RIEDEL (1960) and PASTER-NAK (1968) have sized and counted particles as a substrate analysis technique, ULFSTRAND (1968a) and others have described 'standard areas' occupied by a particular sized-stone and several authors have con-ducted experimental studies on the physiological and ethological responses of various invertebrates to sediment size (see CUMMINS and LAUFF 1969). However, until the recent call by CUMMINS (1966) for some uniform, easily and widely applicable and relevant classification system, little quantifi-cation of the relationship between bottom sediments and benthos had been done. The bottom sediments of the Gombak are suitably defined and separated by the *phi*-Wentworth scheme, and this system appears to be applicable to widely different areas as a means of comparison.

b. Interstitial pore space, analogous to the 'void ratio' of WEBB (1969), is a function of the volume of the individual sediment particles. Theoretical computations of interstitial volume based on spherical particles with various Md are of little biological significance because of the size and shape heterogeneity of most sediments and the variable degree of consolidation. For the sediments from erosional and depositional areas at the main sampling stations, the pore space data are given in Table 24. Pool and riffle sediments are separated by their pore volumes which fall into two groups. Depositional sands had interstitial volumes between 9.0 and 14.5 %, while riffle sediments ranged from 21.5 to 27.0 % space by volume. Fig. 20 shows the plot of pore space on Md *phi* which gives a negative correlation $y = 12.84 - 2.29 \, x$ with $r = -0.46$. The high standard error of estimate (282.62) is a function of variation in

Table 24. Percentage interstitial pore space of the depositional and erosional substrates at each station

Substrate	I	II	III	IV	V
Depositional	13.96	9.08	11.96	14.40	12.12
Erosional	23.22	25.06	22.06	27.00	21.62

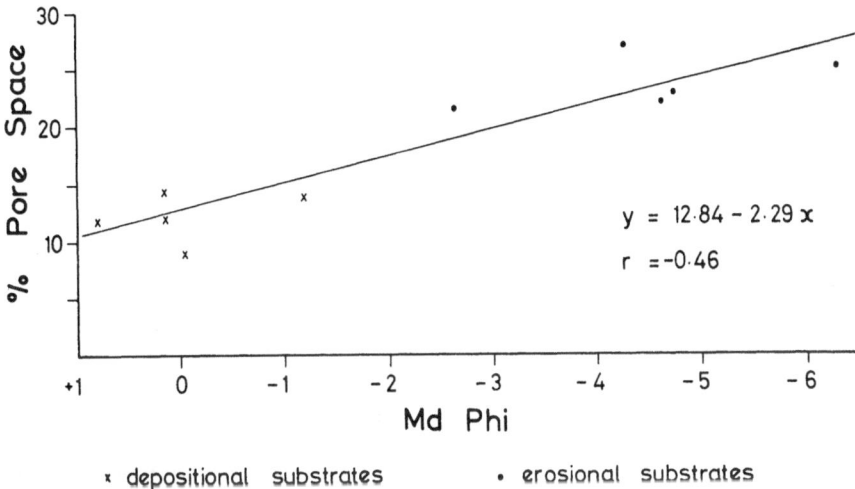

x depositional substrates   • erosional substrates

Fig. 20. Relationship between interstitial pore space and median diameter of substrate particles.

compactness in the sediment and the distribution of odd-sized and shaped particles among the other sediments. The range of particle sizes is particularly important in determining compactness. A small percentage of fine sediments may fill much of the void space and larger angular particles may disrupt the natural packing lattices. Under natural conditions, this sorting is dependent on the vibration by current forces of the sediments as they are deposited. It is likely, therefore, that for any particular area of sediments with a specific size range, the pore volume will vary, within determinable limits, depending on the antecedent flow conditions. Probably, for better definition, a relationship between pore volume and Q_1Q_3 or even higher percentiles would be more realistic than just Md which gives little indication of the distribution of sizes present. This relationship, if generally applicable, gives an estimate of the area available for biological activities, and, coupled to determinations of sedimentary-organic relationships and physical and chemical conditions in the sediment, might be useful as an index of potential productivity in a habitat.

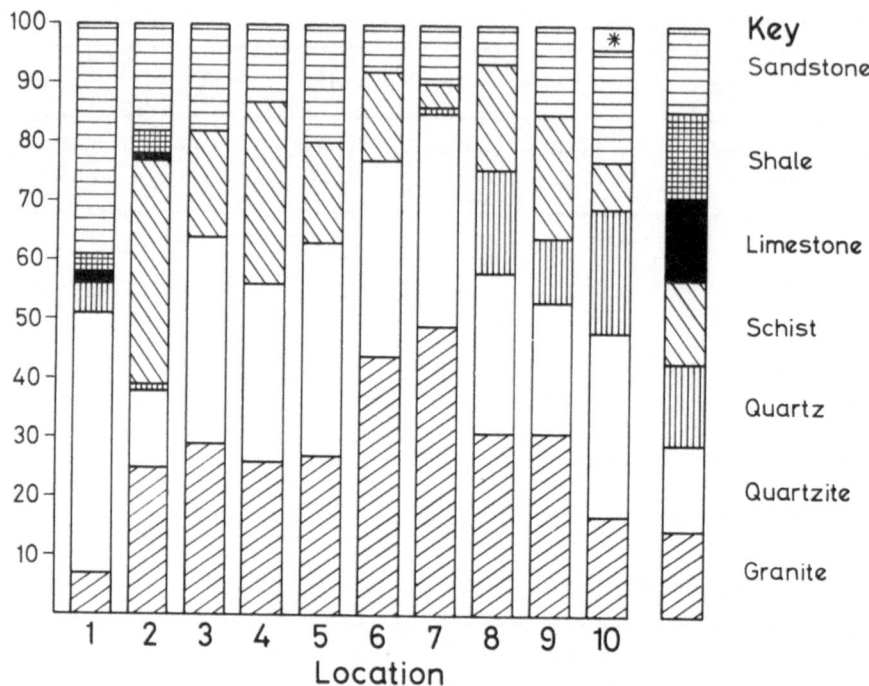

Fig. 21. Percentage lithological composition of the *phi*–5 to –6 sized gravels at various locations on the profile. Locations: 1. Location A; 2. Station II; 3. 0.5 km above D; 4. near E; 5. 2 km above G; 6. near G; 7. below Station III; 8. near J; 9. above Station IV; 10. Station V. * Artefacts–1% concrete, 3% road metalling products.

c. The lithology of the bottom sediments in the size group Md *phi* -5 to -6, collected at various points along the river is illustrated in Fig. 21 and shows some marked differences between localities. The possible petrology of these fragments with respect to their sites of occurrence is an indirect measure of bed-load transport and the resistance of the rock component to erosive and fragmenting forces (DOUGLAS 1968a). The percentages plotted are undoubtedly only a rough estimate of the actual sediment composition. The small number of pebbles sampled and the technique of manual selection of a particular size group for analysis inevitably introduces bias. The areas sampled may have aggregation or depletion of a particular lithology because of local current effects or fragment density and morphology.

At Location A, quartzite (44%) and polycrystalline sandstone (39%) from the local arenaceous and lutaceous sedimentary facies formed the bulk of the sediments with a small (7%) contribution of unweathered granite from the headwater areas (see Geology, Fig. 4). At Station II, granite (25%) was more important as the local supply had increased,

the whole NW catchment being granitic. Schists from the Lower Arenaceous Formation were the dominant rock type (38%) with smaller
percentages of shale, quartz and limestone. Shale, in low quantities, was
found only at these two stations in the vicinity of the lutaceous facies,
and must weather rapidly to sand and clay as it disappeared as soon as
there was no further source of supply. Quartz, in the size range under
study, occurred in low percentages in the headwaters and in higher
percentages in the area of the quartz dykes near and below Location J.
These stones are extremely resistant to both abrasion and chemical
weathering so their high percentage at Station V was a function more of
long distance transport and longevity, compared to other lithologies, than
of local supply. The limestone fragments found in very low percentages
at the upper two locations were probably an unnatural occurrence. When
logging tracks were first built through the jungle, dolomite chips, quarried
from the Batu Caves outcrop in the lower catchment, were often spread
in the soft spots to consolidate the road bed. Several such tracks cross the
upper headwaters and may account for the presence of this limestone
in the bottom sediments. However, there is a small area of calcitic
limestone in the extreme north of the catchment that could also be the
site of origin.

Schist had a peculiar distribution: not present at Location A in the
lutaceous rock region, at least not in the size range sampled; predictably
high through the section below the sedimentary bedrock area, but with a
low proportion near Location D where the 18% contribution was only
about half the percentage at the stations above and below. This may be a
sampling anomaly. Below Location E, schist declined in proportion as it
disintegrates without further supply until the Hawthornden and Dinding
formations are reached at J. Below this location, the percentage was
elevated (20%) for 2–3 km but then declined to 8% by Station V as there
was no continued local source.

The proportion of granite downstream from Station II was relatively
constant through the broad granitic region of the middle catchment
where local supply was abundant. The percentage increase in the vicinity
of Location G and Station III was more a function of disintegration of the
schist and sandstone components than increased supply. Near Location J,
the granite contribution returned to 30% as the sedimentary types increased, and subsequently declined to 17% at Station V with a concomitant increase in weathered sandstone fragments. Granites rapidly
fragment to form component sand-sized particles and the proportion
was therefore almost entirely due to local recruitment.

The percentage of quartzite which, like quartz, is hard and resistant
to wear was dependent on supply of other components. Throughout the
length of the river it was abundant, only decreasing in proportion where
there was a local source of schistic material as at Station II. The high
percentage at Station V was an indication of its relative resistance to

abrasion, longevity in the bed, and the demise of other lithologies as important components.

Of interest were, at Station V, the total absence of limestone derived from the Kuala Lumpur Limestone Formation which is generally 10 m or more below alluvial level and the presence of road-building artefacts, mostly tar-coated dolomite. The latter were not of local origin but had been transported considerable distances. The nearest highway crossing is above Station IV, but the bituminous covering decreases density, thus encouraging transportation, and protects against bed wear.

The river substrates at different localities, as gauged by a single size category, were therefore dependent on continuous recruitment from local lithologies, but may have been considerably modified by the presence of selected alien materials, with increased resistance to abrasion and chemical decomposition, that persist in the river for variable distances downstream from their point of origin.

2.5.6. Erosional load

The amount of material carried downstream, either in the water mass as suspended and dissolved fractions or along the bottom as bed-load, had important effects on the biotic conditions in the river. The indirect effects of turbidity on the flora and the direct implications of silt and sand movements to the fauna and its distribution will be discussed later. In this section, an attempt will be made to summarize the main terrestrial erosive processes and their contributions to the particulate and dissolved load of the river. The theoretical aspects of denudation systems in the humid tropics, their energetics and stability have recently been reviewed by DOUGLAS (1969). The current state of erosion in Malaysia and suggested corrective measures have been described by many authors (Central Electricity Board 1956, POORE 1961, SPEER 1963, OW 1965, Shawinigan Engineering Co. Ltd. 1967) and summarized by DOUGLAS (1970). Studies specifically on the Gombak catchment have been made by DOUGLAS (1967, 1968a, and in preparation). Erosive processes take two forms under tropical forest: sub-surface flow through the weathered soil mantle and surface run-off. The former is more important in the stable undisturbed forest on steep slopes and is responsible for much of the dissolved load derived from the parent lithology and its soil cap. In addition, it carries the finer clay particles down through the interstitial spaces of the larger sand particles, concentrating them as eluvial soils at the base of slopes and increasing their propensity for entrainment by surface flow. Surface run-off is also a factor in the forest and becomes increasingly more important as ground cover is removed and the natural conditions are altered by anthropogenic activities. Under the rain forest canopy, percolation, which theoretically is high for the friable loamy soils found on the granite, is often inhibited by splash effects from leaves which flatten the soil surface and wash clay particles into the pore spaces, partially sealing the surface. Large drops from the foliage also loosen materials and displace them by the splash energy, generally downslope. Water is purported to rush down the trunks of trees forming a bare patch on the lower side where accumulated materials have been washed away (RICHARDS 1967, DOUGLAS 1968a). This was not seen to occur to any great extent in the Gombak forest with the rainfall intensities recorded during the study, so its effects are considered to be minor. Adding to these lateral movements downslope are soil creep and landslipping that result from waterlogged surface layers sliding over less permeable underlying areas. These are especially common in the more clayey soils derived from sedimentary rocks, but they also occur in granitic weathering profiles where core stones cause instability. Such movements result in large quantities of sediment reaching areas of surface and sheet flow which carry them to the watercourse, particularly in the dipterocarp forest where ground cover is limited, and in areas affected by man's activities. Alterations of the plant soil cover, by gardens, plantations or construction, expose ground surfaces directly to heavy rain and hardening agents and generally decrease porosity or, on roads and urbanized areas, increase the velocity and erosive capacity of run-off water (GRANT 1957, MITCHELL 1957). Mining activities, particularly fluvial tin extraction as practised in West Malaysia, and logging with its rough-cut tracks and skid trails through the jungle (TEBO 1955, BERRY 1956, DOUGLAS 1968a) are major sources of sediments, either directly through run-off of washings and surface soils or indirectly through exposure of large surface areas to accelerated erosion after removal of the plant cover.

a. The results of erosion load measurements obtained during a routine sampling program are summarized in Fig. 22, 23 and 24. These component analyses were done at set periods, not on specific rising, peak and falling stages and therefore give an incomplete picture. However, they cover the normal state of the river although including none of the large flood peaks that generally occur in the early morning following late afternoon and evening rain. Data for the first four sampling dates were obtained using A. P. H. A. (1965) methods, i.e. final drying at 180 °C, but this was found to give unsatisfactory organic fractions so subsequent analyses for these were made after ignition at 600 °C. Some error was thereby introduced because of loss of waters of hydration and volatile salts, but this was unavoidable. In Table 25 giving the ranges of sediment loads at the main sampling sites, data for these four dates obtained at 180 °C were not included in the minima and mean calculations. The values for all fractions steadily increased from Station II to V, with no outstanding increments. Station I, draining a watershed composed largely of shales and schists, had a slightly higher total load than the adjacent main river Station II. The relationship between discharge, and hence current velocity, and erosional load is complex, dependent on

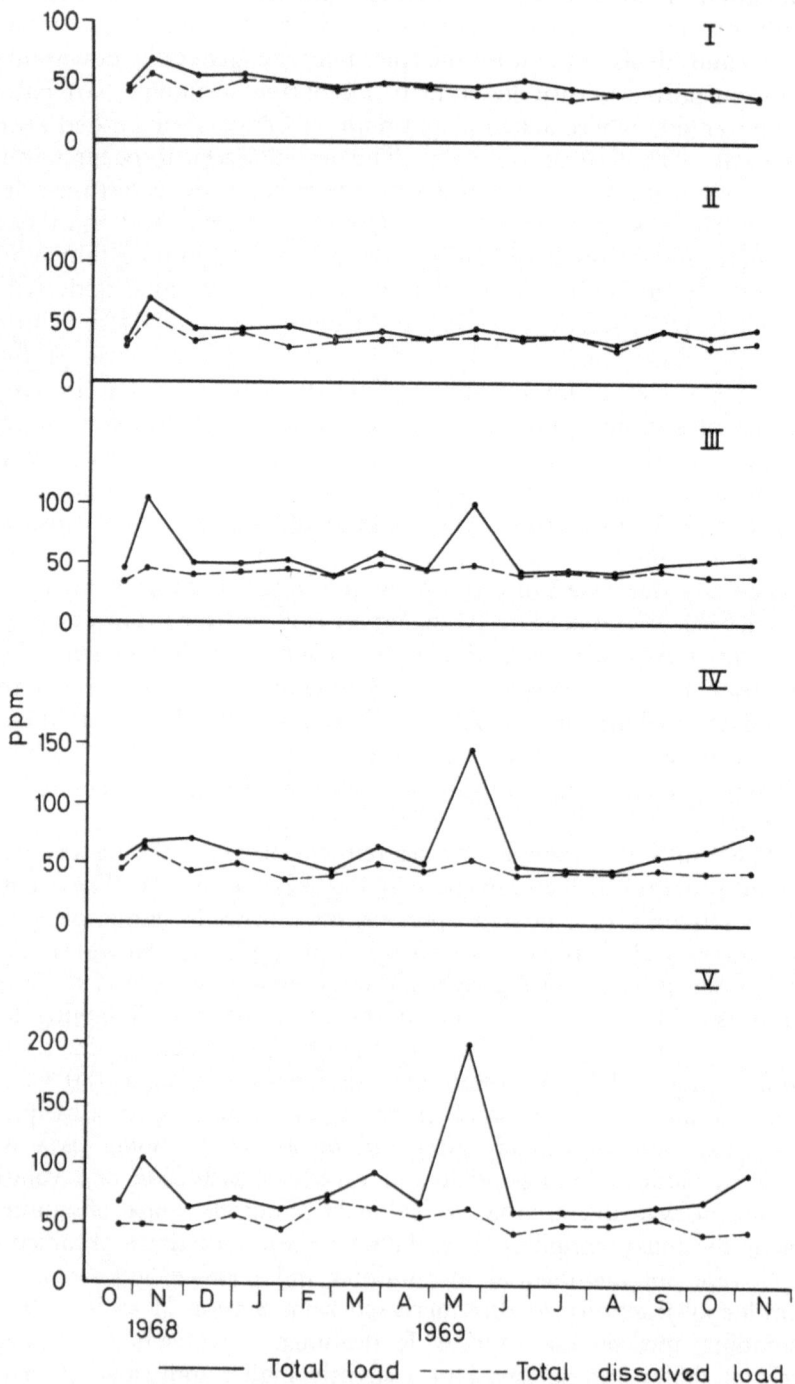

Fig. 22. Total and total dissolved erosional loads in ppm at regular four-weekly intervals at Stations I to V.

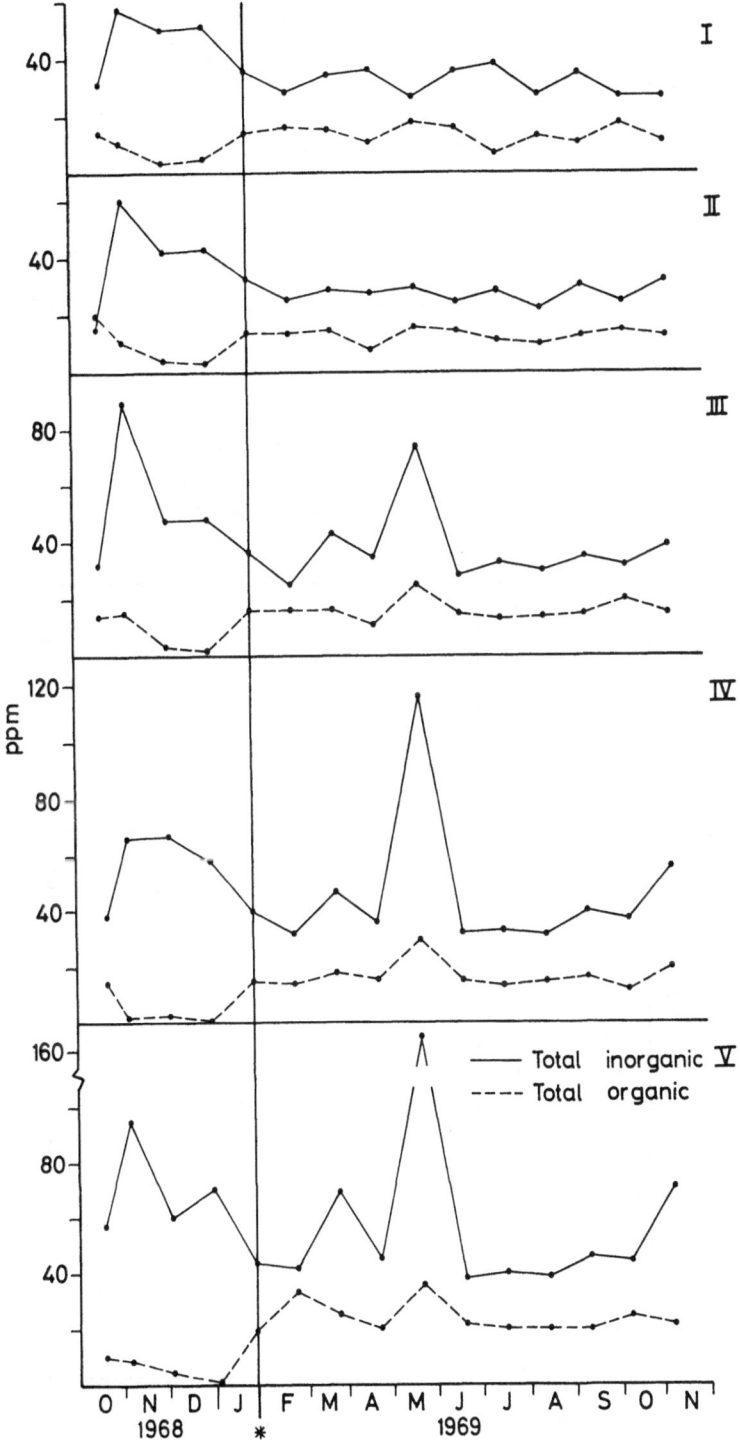

Fig. 23. Total inorganic and total organic erosional loads in ppm at Stations I–V;
* Technique of determining total organic changed (see text).

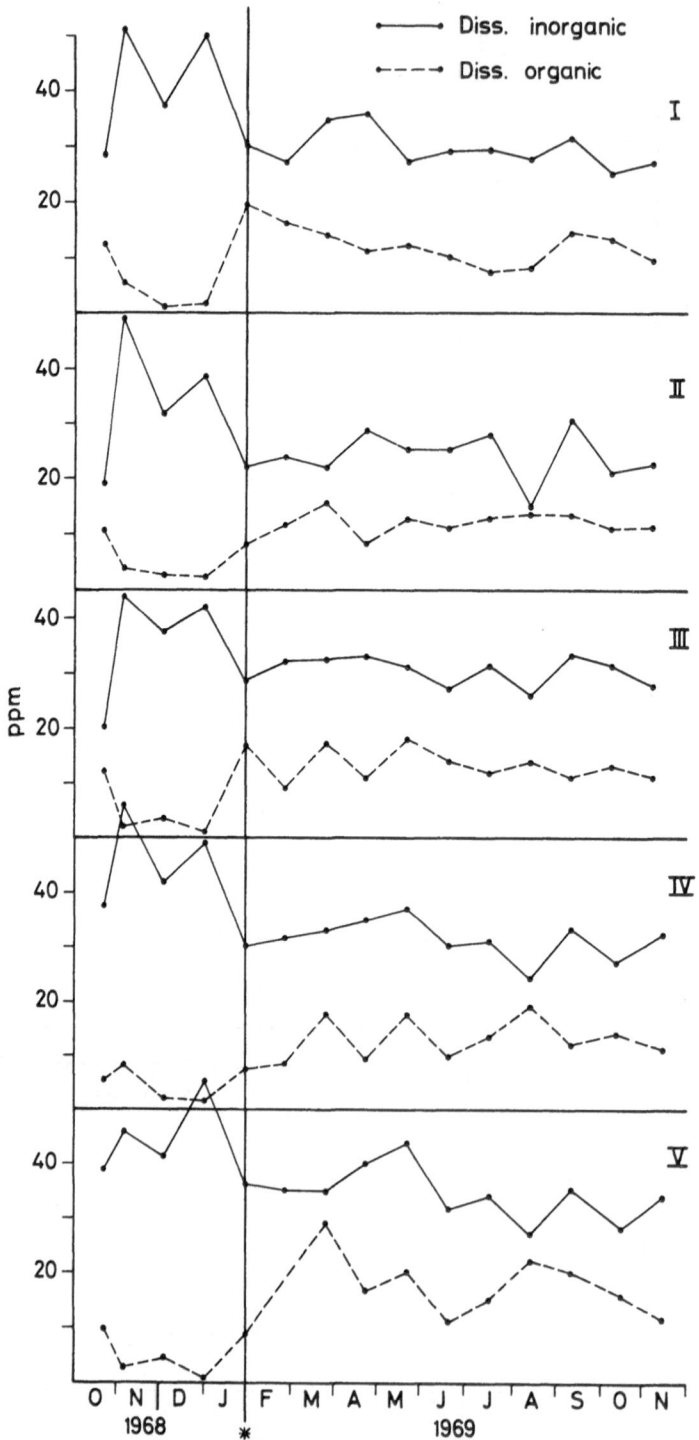

Fig. 24. Dissolved inorganic and dissolved organic loads in ppm at Stations I–V;
* Technique changed (see text).

Table 25. Erosion load at main stations over 15 sampling periods in ppm

| | Total load | | | Total inorganic | | | Total organic | | |
	max.	min.	mean	max.	min.	mean	max.	min.	mean
I	67.3	38.8	46.2	56.9	27.3	32.3	17.8	4.6	14.0
II	71.3	33.7	43.8	60.2	15.7	28.3	20.0	3.3	13.1
III	103.9	41.2	55.9	89.4	25.3	37.4	25.0	1.7	16.1
IV	146.0	45.5	62.9	116.7	30.9	45.4	29.4	1.1	19.1
V	200.5	59.2	80.4	164.8	39.1	58.4	35.7	1.1	23.6

| | Total dissolved | | | Dissolved salts | | | Dissolved organic | | |
	max.	min.	mean	max.	min.	mean	max.	min.	mean
I	55.9	36.7	43.5	50.7	25.0	29.5	19.6	1.5	11.3
II	53.3	28.4	36.8	49.1	14.7	23.9	15.3	2.4	11.7
III	49.5	32.3	42.6	44.0	20.2	30.0	17.9	1.5	13.3
IV	64.3	37.4	45.7	56.0	24.4	31.2	19.0	1.1	12.7
V	70.4	42.7	49.1	54.9	27.0	34.5	35.3	0.9	18.8

gradient, rainfall conditions in the catchment prior to the sampling period and a number of indeterminate factors such as landslips into tributaries and collapse of overhanging banks. An additional complication is the drag effect of high silt load. On several occasions of high stage and turbidity, the current velocity was noticeably reduced by the increased viscosity of the water-sediment slurry. The dependence of suspended and bed-loads on current velocity has been summarized by EINSELE (1960), who delimited velocity ranges and corresponding bottom composition under ideal conditions. In these ranges, currents of < 20 cm/sec will not move fine sand, < 40 cm/sec – coarse sand and fine gravel, < 60 cm/sec – medium and coarse gravel and cobble, < 120 cm/sec – large cobble and small boulders. In the upper Gombak, sustained velocities of any kind were rare. In general, short discontinuous zones of flow existed so that a mosaic of sediment types was built up, with selective erosion of boundary areas. In floods, when bottom stability was ruptured by increased velocity and pressure, all sizes of substrate were eroded. In the lower reaches, more uniform flow patterns were established. At Stations III and IV for example, sustained currents of > 50 cm/sec were common, resulting in a constantly shifting bed and relatively high turbidities. In his study of erosion in the Gombak, DOUGLAS (1968a) found log log correlations between rate of denudation, expressed both as dissolved and suspended load, and discharge. The dissolved load on discharge had regression coefficients for all stations between 0.55 and 0.89 as solute concentration decreased almost linearly with increased discharge. Total dissolved load usually increased with discharge but at a

Table 26. Extreme erosion loads in ppm for selected locations (from DOUGLAS unpublished)

	Total		Dissolved	
	max.	min.	max.	min.
2° near B	54.0	35.3	42.1	7.0
2° near II	116.8	30.0	77.1	10.0
Station II	677.0	10.0	86.0	6.0
	(> 3000)*		(374)	
B.P.Weir	644.0	19.5	67.5	12.2
Trib.E	388.0	44.2	63.8	14.5
	(10664.4)			
D.I.D.site	5342.5	150.0	137.2	17.5

* Results in brackets obtained in early 1970

much slower rate than discharge volume increased. Suspended load increased at all stations much more rapidly than discharge (regression coefficients 1.89 to 2.77). Instantaneous load varied considerably and time-integrated sampling over the duration of a run-off fluctuation cycle would be necessary for an accurate assessment of total load. Under rapidly rising stage, turbidity was often very high, while on falling stage, it dropped below normal flow concentrations. Unfortunately, sampling at this level was not feasible.

The data for samples taken by DOUGLAS (personal communication) at different stages for a number of storms during the study period are shown in Table 26 to give an indication of the extremes. The two 2° tributaries have small catchment areas and are probably free from interference effects. Their loads may be considered as those expected from steep-sloped forest areas and are of similar magnitude to those at Station I. At Station II, a maximum of > 3000 ppm for total and 375 ppm for dissolved load were recorded for separate storms in early 1970, but total suspended solids concentrations between 200 and 400 ppm for large flood crests were more normal. Some gradient has been lost at the B.P. weir, accounting for the observed decrease in suspended load at this point. The loss of dissolved concentration was probably a function of dilution by the large Tributary D that drains a granite watershed and confluences just above the weir.

The small Tributary E illustrates the effects of increased velocity on sediment load. The enormous total load of > 10600 ppm (i.e. more than 1 %) in early 1970, was largely soil picked up in road drainage after a landslip just above the station, at a time when the discharge had increased from a normal 0.1 cumec to about 1 cumec. There was considerable dissolved load (47.9 ppm) at the same flood, but this concentration was below the maximum recorded at this location.

Table 27. Erosion loads for tributaries in ppm (at basal flow on 10 April 1969)

	Total load	Total inorganic	Total organic	Total dissolved	Dissolved salts	Dissolved organic
A	67.1	51.0	16.1	61.2	45.5	15.7
B	51.8	38.3	13.5	43.6	32.0	11.6
C	55.6	37.9	17.7	50.3	33.3	17.0
D	48.9	37.6	11.3	43.7	35.1	8.6
E	68.1	50.5	17.6	59.0	44.8	14.2
F	60.8	44.6	16.2	45.0	30.7	14.3
G	42.0	25.2	16.8	33.2	21.0	12.2
H	56.2	38.4	17.8	41.0	27.6	13.4
J	44.4	31.0	13.4	33.9	24.3	9.6
K	183.2	122.6	60.6	97.5	59.7	37.8

At the D.I.D. gauging site, the river has received two large tributaries from tin mining areas and passed through a residential and industrial suburb. WEIBEL et al. (1964) reviewed studies on urban land run-off and reported solids up to 2000 ppm from an American housing estate and between 1000 and 3000 ppm suspended effluent from Moscow streets. Comparable data for tropical zone cities are not available but are likely to be higher because of smaller paved areas and in some cases unlined drainage ditches. The dissolved load maximum (137 ppm) reflected pollution largely from domestic effluents. However, at high discharge, this component was often diluted (e.g. 17.5 ppm). Rising stage of large storm floods almost always carried between 1000 and 3000 ppm solids, mainly fine clays and sands from the alluvial tin mining areas. These were deposited on banks and vegetation with falling stage and became incorporated in subsequent suspended loads when they were washed into the river by rain or picked up by rising flood waters. As shown by the minimum figure of 150 ppm from DOUGLAS' data, considerable particulate material was entrained even in periods of low flow, as this was almost three times the minimum found at Station V above the tin mining zone.

Under minimum stable flow conditions on 10 April 1969, erosional loads of various tributaries were assessed. The data (Table 27) show concentrations close to the mean for the adjacent part of the main river with perhaps some slight elevation because of the minimum diluting conditions. Tributary K is polluted with domestic and rubber factory effluents and has elevated concentrations of all components. This inflow and the return water from various irrigation ditches are the major causes of the differences between sediment loads at Station IV and V.

Flood crests, as catastrophic events in the hydrologic cycle, carry a large proportion of the total annual sediment load. In semi-arid temperate areas, this may result in 50% of the annual load being carried on

Table 28. Volume of net-measured bed and saltating load at the forest stations

Date	STATION I Measured load cc	Factor*	Estimated load cc	STATION II Measured load cc	Factor	Estimated load cc
5–X–68	128.0	6.11	782.1	112.0	64.26	7197.1
30–X–68	30.6	3.56	108.9	640.0	41.34	26457.6
28–XI–69	222.4	4.01	891.8	960.0	52.06	49977.6
1–I–69	112.3	4.68	525.6	4250.4	55.22	234707.1
24–I–69	48.0	6.21	298.1	296.0	37.76	11177.0
23–II–69	96.0	2.95	283.2	64.2	42.81	2748.4
19–III–69	31.5	1.91	60.2	53.1	43.89	2330.6
17–IV–69	8.2	5.32	43.6	96.2	39.23	3773.9
23–V–69	308.1	9.85	3034.8	1093.4	57.98	63395.3
12–VI–69	90.7	2.52	228.6	226.3	48.19	10905.4
10–VII–69	17.5	5.07	88.7	235.9	36.07	8508.9
7–VIII–69	83.0	2.69	223.3	221.3	39.06	8644.0
3–IX–69	583.7	5.40	3152.0	454.8	51.14	23258.6
2–X–69	54.4	3.52	191.5	150.3	40.00	6012.0
30–X–69	230.5	8.05	1855.5	2241.0	36.67	82177.5

Total			11767.9			541270.9
Estimated 1969 load**			0.31 m³			14.25 m³
Denudation m³/km²			0.07			1.94

* F = (Total discharge/Volume filtered)
** includes Nov. 1968 to complete annual period

only four days (see LEOPOLD et al. 1964). Under climatically more uniform tropical conditions, this is often not so pronounced. For the Gombak, DOUGLAS (1969) estimated that half the load was carried in 24 days, thus demonstrating that in spite of high precipitation levels in the tropics denudation may proceed more slowly than in areas with great rainfall fluctuations. The main cause for this is the protection given the soils by plant cover and the stabilizing effects of roots and leaf litter. The increased loads found for the lower Gombak catchment, particularly at Station V and the D.I.D. station, were mainly the consequence of man's activity in the watershed. LIEBERMAN and HOOVER (1948), RUTTER (1958) and LIKENS et al. (1970) have commented on similar increased erosional loads, both particulate and dissolved, from small catchments after the plant cover was removed. VAN DIJK and EHREN-CRON (1949) described the heavy erosional load of Javanese rivers as a result of intensive cultivation.

Dissolved loads in tropical streams generally are no greater and are often lower (e.g. Amazon) than those of rivers in higher latitudes. This is unexpected in view of the rapid removal rate of silica from tropical rocks, but emphasizes the almost closed cycle for most of the other ions under forest conditions. In the Gombak, dissolved silica concentrations ranged from 10–25 ppm under 'normal' discharge, decreasing to 4–6 ppm at high stage, and constituted the major fraction of the dissolved salts in the river. The mobility of this element will be further discussed in the section on water chemistry.

b. Data for bed-load and saltating load are almost non-existent for Southeast Asian and tropical rivers. The figures obtained in this study are given in Table 28. The projected totals/day, worked with a factor that allowed for channel width and current velocity, may be $\pm$ 10%, but the estimated totals for the annual period November 1968 to October 1969 (0.3 cu.m and 14.3 cu.m for Stations I and II respectively) are undoubtedly an underestimate, as only low and moderate discharges could be monitored. The effects of the large flood crests on bed movement could not be assessed with the apparatus available but the addition of a coarse trap anterior to the fine mesh to prevent large stones, wood etc., from damaging the net might permit measurement in all conditions. Eddy turbulence created at the base of the net, leading to local recruitment of sands and gravel, can be minimized by a flat plate anchored underneath and in front of the net mouth. Mesh size in the nets used selected particles with the Md > phi 2.5 (fine sand) and probably therefore underestimated bed-load by a small factor. The boundary between saltating load and suspended load was, however, difficult to determine where velocity was not uniform.

At the stations in the Middle and Lower Zones, the volumetric capacity of the nets available was insufficient to permit any measurements. Within 30 min, 3–4 l. of sand were collected over a net width of 30 cm. The depositing bottom at these lowland stations was perpetually moving, with the top 0.5 cm of sand continually rolling and bouncing downstream. The magnitude of this bed-load was estimated by recording the amount of washed building sand removed by a one-man operation, about 1 km below Location J. The operator, who extracted only from a long flowing pool 100 × 10 m, filled one truck (4.5 cu.m) per day. If the work year were 300 days, a minimal figure, this meant extraction of 1350 cu.m sand/yr without any deepening of the channel. During low water periods, the area often became almost depleted, but any flood crest restored the half-metre depth of sand and stones. The trap provided by this operation retained only a small fraction of the bed-load, with considerable material moving through the area and on down the river even during modal flow periods. Larger stones, also moving as bed-load, were removed with the sand and sold at irregular intervals, but no definite estimate of the volume of these could be made. The large increase of measured bed-load at this point over that at Station II (× 100 approximately) was an indication of the weathering effects on bed materials and recruitment from the foothill tributaries under rubber and orchard cultivation. The total bed-load here, and further down the river, was undoubtedly much higher

Table 29. Recapitulation of zone characteristics

	UPPER ZONE	MIDDLE ZONE	LOWER ZONE
Zone length km	4.50	5.69	11.99
Channel width m	6.5(4–10)	10.8(7–12)	14.0(10–25)
Channel relief m	1225–90	90–55	55–30
Gradient ‰	78.0	4.7	2.2
Channel structure	alternating riffle-pool with some cascades	short riffles, long slow reaches	slow-flowing reaches, occas. very short riffles
Cover (exposure %)	complete (4–10%)	open (80–100%)	open (95–100%)
Vegetation type	rain forest	rubber, mixed crops	scrub, gardens
Land use	Forest Reserve	smallholdings	rice cultivation, tin mining, residential, light industrial

	Station I	Station II	Station III	Station IV	Station V
Elevation m	233.2	219.5	70.1	50.3	39.6
Stream order	3	4	5	6	6
Km from source	3.98	6.78	15.81	20.37	24.23
Drainage area km²	4.7	27.7	80.0	93.4	102.8
Temp. range max. °C	23.0–26.5	22.8–25.5	31.0	26.2–31.5	26.8–32.5
min. °C	20.0–23.5	19.0–22.8	21.0	21.3–24.8	22.5–25.6
Substrate	stones, gravels sand	boulders, gravels sand, no silt	stones, gravels sand	cobble, gravels, sand, some mud	gravels, sand, mud some bedrock
Q_1MdQ_3 phi values erosional areas	−2.93, −4.75, −6.14	−5.58, −6.30, −6.85	−3.30, −4.62, −6.41	−2.42, −4.28, −5.05	−0.78, −2.63, −3.72
depositional areas	+0.78, −1.20, −2.55	+0.82, −0.05, −2.08	+0.96, +0.81, +0.55	+0.88, +0.15, −1.30	+0.76, +0.14, −0.81
Emergent vegetation	bank areas bonded by Saraca roots		few Graminae in margins	marginal Graminae	Graminae, Colocasia
Submergent veg.	none	none	none	none (Cyperus in tribs only)	isolated Hydrilla plants

than the estimate, especially in view of the loads moved by the exceptional floods. Below Station V and the tin mining area, bed-load was probably less important, as competence is lost with gradient and the dominant erosional load was fine suspended clays and silts. The selective nature of bed-load was evident with respect to the movement of cobbles in which only the resistant quartz fragments survived any distance from their origin, the other lithologies constantly weathering to component sands and dissolved salts. Almost all the bed-load was quartz sand able to withstand chemical and abrasive erosive forces.

In summary: erosion in the Gombak, estimated at 67 cu.m/sq.km/yr suspended load is not excessive in comparison with other Southeast Asian rivers (DOUGLAS 1968a). However, it is three times higher than the yield from the wholly forested watershed of the Sg. Telom in the Cameron Highlands of West Malaysia (Central Electricity Board 1956). Some of this increased load originates from logging tracks and landslips in the forest zone, but most can be ascribed to the Disturbed Lower Catchment where man's activities have increased the exposure of the soils to denudation processes and altered drainage patterns so that run-off rate is higher and retention reduced (see SIOLI and KLINGE 1966).

The chief erosional agents are chemical weathering that causes the breakdown of most lithologies to sand-sized particles with the release of dissolved salts, particularly silica, to the ground-water and the intense rainfall in short storms that loosens and carries large suspended loads to the river. These high sediment loads have already affected channel capacity in the lower courses and more flooding is indicated for the future as denudation will increase in the catchment with development.

2.6. RECAPITULATION

A summary of the physical characteristics of the catchment and specific features at the main sampling stations on the Gombak River are given in Table 29.

3. Chemical environment

Detailed definition of the chemical environment in rivers of the humid tropics is limited, again with the notable exception of the Amazon (KLINGE and OHLE 1964, SIOLI 1964, 1968a, WILLIAMS 1968, FITTKAU 1970, and others). Most studies on lotic systems have included some basic chemical parameters but systematic long-term analyses considering landscape and precipitation effects are rare, exceptions being TEMPLETON et al. (1969) on Panama-Colombia riverine ecosystems and MARLIER (1954) and MALAISSE (1969) on central African rivers. Investigations on Southeast Asian waters are few. RUTTNER (1931) looked at Indonesian waters, KOBAYASHI (1959) carried out a comprehensive hydrochemical survey of the rivers of Thailand, generally 'harder' than those encountered in Malaya, and JOHNSON (1967a, 1968a) summarized the major features of a variety of freshwater habitats of southern Malaya. For the most part, these were acid or very acid with some blackwater, and differ considerably from the rivers of the granite range of the central and northern peninsula. NORRIS and CHARLTON (1962) conducted a year-long survey of the lower part of the Gombak River in the vicinity of Kuala Lumpur and measured the more important parameters relevant to deterioration of water quality as did TAN and PROWSE (in press) in the Malacca River. Except for those in Panama and central and northern Thailand, the other investigators have described rivers extremely poor in nutrient salts, with very low conductivities and often low pH. The Gombak is not as impoverished as some Amazonian streams, but closely approaches in composition the streams of the central African rain forest and the upper Malay peninsula rivers analysed by KOBAYASHI.

The solutes in river waters are derived principally from precipitation and the chemical weathering of rocks and soils of the catchment. The relative concentration of any component at a particular location is dependent to a degree upon the ionic composition and variation in the rain, but also upon what SIOLI (1968a) has aptly termed 'the urine of the landscape'. The chemical composition of this is a function of discharge volume and of the degree of modification imposed by the terrestrial ecosystem before the run-off water enters the river system. The mineralogy and petrology of the physical substrate, biological uptake by plants and release through remineralization of organic materials, the effects of man's effluents, and interactions between these agents are chiefly responsible for determining the final composition of ground and surface waters.

In setting up the program for hydrochemical characterization of the Gombak, the outline of desirable factors for analysis recommended for IBP studies (VOLLENWEIDER 1969) was followed and wherever possible methods at IBP Level II (GOLTERMAN and CLYMO 1969) were used. Analyses were carried out at the main sampling stations every four weeks for 60 weeks from October 1968 to November 1969, more than covering a complete climatic cycle. These were supplemented by rainfall analyses, diurnal samplings, and irregular test series of extra tributary stations as considered necessary.

3.1. METHODS

Rain-water was collected in an acid-washed, brown glass Winchester bottle. A glass funnel with long, narrow stem to minimize evaporation losses, and washed glass-wool filter to eliminate insects, completely blocked the neck of the bottle. Water was collected over seven days in a large clearing in the forest near Station II. The bottle was placed on a 1 m high plinth and was safely clear of any tree-drip contamination.

During the long dry period January to March, insufficient water was collected in any one week period to permit full analyses but a series of samples was obtained for the period April to November, covering the two wet seasons and the drier southwest monsoon period.

River water for analysis was collected in 4 l. acid-washed, polyethylene bottles from turbulent regions at the top of riffle sections. All air was excluded by screw-capping underwater. Conductivity, pH and alkalinity measurements were initially made in the field, but after comparative field and laboratory measurements showed no change in results, the field estimates were abandoned for better instrumentation in the laboratory. Water was collected sequentially from the uppermost Station I downstream, and analysis begun immediately on return to the laboratory, usually within 45 to 90 min. Filtration, when necessary and as prescribed for the individual tests, was carried out immediately and the sample stored at 4 °C until needed. All tests except B.O.D. were completed within 48 h of collection. Water for O_2 and B.O.D. was collected with a siphon to ensure at least $\times 3$ flushing of the bottles from similar smooth-flowing, unbroken reaches at all stations.

Distilled water for preparation of standards and sample dilution was double distilled, the second time from an all-glass still with sulfuric acid and potassium permanganate crystals added.

All determinations were done in duplicate and repeated when precision was less than method specifications. Jenaglas or Pyrex glassware, prewashed in acid and rinsed twice with distilled water, was used.

pH was measured at 25 °C $\pm$ 1° with a Beckman Zeromatic or Expandomatic pH meter accurate to $\pm$ 0.02 pH units.

Specific conductance was measured with a Dionic Water Tester in the field and a Cambridge Resistance Bridge and conductance cell at 25 °C in the laboratory. Instruments were calibrated against a standard KCl dilution series.

Unless stated otherwise, chemical constituents were determined according to methods in A.P.H.A. (1965).

Alkalinity – potentiometric titration to pH 4.5, then to pH 4.2 with 0.01 N HCl

*Sodium, Potassium – E.E.L. filter flame photometer

*Calcium – EDTA titration

*Iron – o-phenanthroline method

*These techniques were discontinued when an atomic absorption spectrophotometer became available.

Sulfate – turbidimetric method

Chloride – titrimetric method using 0.01 N $HgNO_3$

Phosphate-phosphorus – colorimetric $SnCl_2$-molybdate method

Nitrate-nitrogen – colorimetric phenoldisulfonic method

Nitrite-nitrogen – colorimetric naphthyl-sulfanilamide diazotization method after HAGEMAN and FLESKER (1960)

Ammonia-nitrogen – direct Nesslerization

Reactive silicate – The method of KOBAYASHI (1959), designated as a tentative procedure by GOLTERMAN and CLYMO (1969) for waters with high silicate concentration, was used. The yellow silico-molybdate complex was measured with a filter colorimeter at 365 mμ. A stable K_2CrO_4 solution, calibrated against a solution of Na_2SiO_3 determined gravimetrically, was used as a standard.

Calcium, Magnesium, Sodium, Potassium and Iron were determined routinely with a Techtron Atomic Absorption Spectrophotometer, Model Type AA 4. Trace elements, Aluminium, Arsenic, Barium, Cobalt, Chromium, Copper, Lead, Lithium, Manganese, Nickel, Tin, and Zinc, were also tested for periodically using atomic absorption spectrophotometry. In an attempt to detect these elements at low concentrations, one litre filtered samples were concentrated to 50 ml by two methods, slow evaporation at 80 °C and lyophilization (cf. SHAPIRO 1961 and BAKER 1969).

Dissolved Oxygen – The azide modification of the standard iodometric method was used routinely, with the addition of the alum flocculation technique when turbidities were high. Duplicate tests were fixed in the field and 250 ml titrated as soon as possible on return to the laboratory. In calculating per cent saturation, the tables of TRUESDALE et al. (1955) were used with appropriate corrections for altitude.

C.O.D. (Oxygen Absorbed from $KMnO_4$) – acidic digestion at 100 °C for 30 min with N/80 $KMnO_4$ (MACKERETH 1963). Standard conditions and controls were used throughout the study to obtain comparative data. To determine the efficiency of this technique, the K_2CrO_4 oxidation test of MACIOLEK (1962) was used on a series of samples on several dates.

Turbidity – methods for determination of total suspended solids have been given under Erosional Load. Expressed in turbidity units $\rho = (L \times 10^3)/D$ g/m³ (equivalent to ppm) where L = total load in kg/sec and D = discharge in cumec (CHEBOTAREV 1962).

3.2. ANALYTICAL RESULTS AND COMMENTS

Rain: The results of precipitation chemistry in 1969 are given in Table 30. Only data from complete weeks in which no obvious contamination occurred, particularly by bird feces, are listed. The sample on 8 May 1969 may have included some extraneous material as it had elevated sodium and potassium concentrations and C.O.D.. DOUGLAS (1968b) gave a table of pH values for world rainfall and concluded that in general the pH of tropical precipitation was little different from that in temperate areas unpolluted by SO_2. The narrow range for rain collected at Station II, 6.50–6.95 with a mean of 6.66, which is higher than in many places, but still within the normal range, supports this conclusion. However, no partition collection of rain over a single storm period was made so that any extreme, e.g. pH 7.85 as found in Guinea at the beginning

Table 30. Chemical composition of precipitation (mg/l., ratios from meq/l.)

	pH	μmho$_{25}$	Ca	Mg	Na	K	Fe	Na:Ca	Na:Cl
10–IV–69	6.50	14.40	0.88	0.04	0.67	0.26	trace	0.66	7.39
8–V–69	6.92	27.60	0.56	0.06	4.25	0.40	0.07	6.60	18.74
3–VII–69	6.95	19.20	0.48	0.05	2.62	0.25	nd*	4.75	—
28–VIII–69	6.62	11.40	0.38	0.07	0.90	0.00	nd	2.06	138.91
25–IX–69	6.60	9.48	0.20	0.06	0.69	0.18	0.04	3.00	7.10
23–X–69	6.52	5.90	0.25	0.05	0.03	0.00	nd	0.10	0.42
20–XI–69	6.52	5.83	0.10	0.03	0.12	0.10	nd	1.04	—
mean	6.66	13.40	0.41	0.05	1.33	0.17			

	SiO$_2$	HCO$_3$ Alk.	Cl	NO$_3$-N	NH$_3$-N	PO$_4$-P	SO$_4$	C.O.D.
10–IV–69	1.5	4.88	0.14	0.18	0.04	0.000	1.0	—
8–V–69	1.6	11.47	0.35	0.11	0.06	0.000	0.0	4.99
3–VII–69	2.2	8.42	0.00	0.08	0.02	0.005	0.0	1.41
28–VIII–69	1.5	4.51	0.01	0.06	0.05	0.010	0.0	0.82
25–IX–69	1.4	3.05	0.15	0.08	0.03	0.000	0.0	0.00
23–X–69	1.2	1.53	0.11	0.04	0.02	0.000	0.5	0.68
20–XI–69	1.3	1.46	0.00	0.02	0.03	0.010	0.0	0.87
mean	1.5	5.05	0.11	0.08	0.04	0.004	0.2	1.46

* nd = not detectable

of a thunderstorm (TRICART 1965), would have been undetected. Specific conductance was low, but indicated the presence of enough electrolytes to constitute a significant part of the total ionic composition of the river water. GREEN (1970), who reviewed the literature on ion supply via precipitation, found that in the Mato Grosso the conductivity of rain varied considerably. At the beginning of a climatic cycle, conductivities tended to be high because of dust and aerosols. Individual storms resulting from boundary phenomena had decreasing conductivities as the storm progressed, but convectional rainfall (the dominant rainfall type in West Malaysia) gave uniform specific conductances. The apparently higher conductivities for the first three sampling periods may have been influenced by the first factor as they were at the end of the dry season, but for the 8 May 1969 sample contamination must be considered a possibility. GORHAM (1958, 1961), ALLEN et al. (1968) and HENSON and VIBBER (1969), in comprehensive assessments of ion supply to waters, listed dry fallout and air-borne salts, gases and aerosols as the major sources.

The considerable quantities of calcium undoubtedly originated from dust (cf. YAALON 1964). The dolomite and dolomitic limestone quarries with extensive crushing and blasting operations in the Lower Catchment produce a fine dust which may be carried by up-valley breezes in the daytime. Silica was consistently present at about 1.5 mg/l.. This high concentration must originate as air-borne dust, perhaps from high in the atmosphere, but more likely from local supply when land is stripped for plantation development. Dust particles would be easily carried into the upper air by the strong upward convection systems that develop in Malaya during the early afternoon. The use of a glass collecting vessel may be responsible for a small part of this amount, although the solubility at pH 6.6 is low.

Sodium and chloride ions, normally carried from oceanic areas (GORHAM 1958, DOUGLAS 1968b), were inconsistent in their occurrence. Chloride only once approached the expected equivalent concentration with sodium and was not detectable on two occasions.

The Na:Ca ratio varied seasonally from 0.1 to 6.6, about the same range reported from several sources for east Australia (DOUGLAS 1968b). These ratios are low when compared to the theoretical of approximately 22 for sea water and indicate that much of the cation content does not originate at sea. The importance of local convectional thunderstorms with salts supplied largely from terrestrial sources must be emphasized, especially as the catchment is only some 40 km from the sea. The higher ratios obtained during the April-May inter-monsoon period, when there was little prevailing wind, are difficult to explain, particularly as the next wet period gave the minimum values for this ratio.

Other cations, magnesium and potassium, probably originated from dust and smoke (LIKENS et al. 1967). Iron was usually absent; when present its concentration was low, the maximum 0.07 mg/l., however, being about double that found in Brazil by SIOLI (1969a).

Phosphate-phosphorus, not normally found in rain (KEUP 1968), was found in very low concentrations on several occasions. The source of this was either ash originating from the numerous small sawmills in the lower valley which burn their waste products, or from dust from fertilized agricultural areas (cf. GORE 1968). SIOLI (1969a) found 0.001 mg/l. in Amazonia and DUNN, quoted by MALAISSE (1969), has recorded the improbably high concentrations of 0.3 to 1.0 mg/l. from rainwater in Uganda. In both cases, fly-ash from agricultural burning was the likely source.

Ammonia normally originates from electrical activity in storm clouds. The maximum NH_3-N concentration of 0.06 mg/l. is the same as that given by STEWART (1968) as the mean content of European rain. Most analyses showed considerably lower concentrations than the 0.1 mg/l. reported by SIOLI (1969a) and 0.18 mg/l. found in Nigeria by JONES and BROMFIELD (1970). However, ammonia concentrations determined here are undoubtedly an underestimate as with the collecting technique used, no provision could be made for acid fixation immediately after precipitation. Nitrate-nitrogen concentrations ranged from 0.02–0.18

Table 31. Hydrochemistry of the Gombak River main stations

	I			II			III			IV			V		
	max.	min.	mean	max.	min.	mean	max.	min.	mean	max.	min.	mean	max.	min.	mean
ph	7.35	6.65	6.97	7.39	6.70	7.02	7.25	6.70	7.00	7.10	6.30	6.68	6.95	6.23	6.58
μmho$_{25}$	48.00	31.92	38.42	33.60	25.20	29.55	33.60	26.40	29.65	33.60	25.20	28.99	50.40	35.40	41.19
Ca mg/l.	0.80	0.20	0.49	0.61	0.20	0.42	0.50	0.19	0.33	0.40	0.10	0.27	1.00	0.25	0.64
Mg mg/l.	1.15	0.39	0.88	0.74	0.31	0.57	0.48	0.21	0.33	0.35	0.09	0.25	0.78	0.26	0.50
Na mg/l.	3.20	0.70	2.11	2.50	0.60	1.81	3.10	0.80	2.12	3.20	0.90	2.27	3.70	0.90	2.59
K mg/l.	1.70	0.98	1.27	1.90	1.08	1.45	2.90	1.45	2.04	2.80	1.50	2.11	4.00	2.10	2.62
Fe mg/l.	0.27	0.02	0.12	0.36	0.04	0.14	0.49	0.11	0.27	0.76	0.19	0.45	1.45	0.31	0.73
SiO$_2$ mg/l.	14.4	8.8	11.7	11.4	7.8	10.1	14.4	5.6	11.5	14.4	9.0	11.6	13.2	7.2	9.8
HCO$_3$ Alk. mg/l.	20.13	13.54	17.35	16.78	11.47	13.58	16.78	11.41	13.71	16.63	10.43	13.10	22.88	14.76	17.93
Cl mg/l.	0.80	0.09	0.44	0.65	0.13	0.40	0.82	0.20	0.42	1.29	0.19	0.54	1.76	0.47	0.88
NO$_3$-N mg/l.	0.28	0.07	0.12	0.20	0.08	0.13	0.26	0.10	0.15	0.21	0.08	0.15	0.33	0.12	0.20
NO$_2$-N μg/l.	4.0	2.0	2.8	4.5	1.0	2.2	4.5	2.5	3.3	5.0	4.0	4.5	6.5	2.5	5.0
NH$_3$-N mg/l.	0.14	0.03	0.08	0.08	0.00	0.05	0.08	0.03	0.06	0.07	0.03	0.05	0.16	0.00	0.07
PO$_4$-P μg/l.	36	0	18	24	0	10	35	5	19	45	5	25	930	150	414
SO$_4$ mg/l.	4.00	0.00	0.90	2.50	0.00	0.45	3.0	0.00	0.40	3.50	0.00	0.47	1.00	0.00	0.13
C.O.D. mg/l.	3.74	0.05	0.89	3.10	0.53	1.25	5.46	0.70	1.89	4.75	0.99	2.03	5.02	1.40	2.78
CO$_2$ mg/l.			4.9			2.9			1.8			1.4			9.3
O$_2$ mg/l.	7.88	7.04	7.59	8.0	7.34	7.72	7.92	7.25	7.64	7.52	6.86	7.18	6.67	4.78	5.96
B.O.D. mg/l.	0.40	0.08	0.21	0.49	0.05	0.19	1.17	0.18	0.48	0.95	0.40	0.68	6.58	1.51	3.40

Table 32. (a) Partial analyses (weight percent) of Kuala Lumpur Limestone (from GOBBETT 1964)

Number[1]	Description	Residue	Measured Ca	Mg	Calculated CaCO_3	MgCO_3
*1678	Pale pink limestone	1.6	39.7	0.2	99.2	0.5
*1680	Grey dolomitic limestone[2]	2.1	35.3	3.1	88.3	8.7
*3597	Pink dolomitic limestone	1.1	26.9	9.4	67.2	23.5
1705	Grey dolomite	2.1	22.8	12.4	57.0	31.0
	Theoretically pure dolomite				54.3	45.7

[1] Specimen number, Dept of Geology, University of Malaya
[2] The dolomitic limestones and dolomites probably contain appreciable amounts of FeCO_3
* Common types near Station IV in Batu Caves outcrop

Table 32. (b) Analyses (weight percent) of main range lithologies (from ALEXANDER 1968)

Constituent	Quartz mica Schist	Light buff Quartzite	Reddish Quartzite	Grey Quartzite	Medium-coarse grained Granite	Medium-fine grained Granite	Granite Porphyry
SiO_2	66.40	89.57	80.16	80.02	70.61	68.86	70.14
TiO_2	0.60	0.32	0.42	0.44	0.40	0.58	0.64
Al_2O_3	16.47	*5.29	*7.89	*8.72	14.01	13.67	14.58
Fe_2O_3	1.33	0.59	4.16	0.56	0.45	0.30	0.64
FeO	5.20	0.43	0.83	3.45	2.75	3.77	2.97
Cr_2O_3	—	nd	nd	nd	nd	nd	nd
MnO	0.03	trace	0.02	0.01	0.07	0.09	0.11
MgO	1.62	0.42	0.68	1.24	0.90	1.63	0.71
CaO	0.56	0.04	0.20	0.09	1.85	2.40	2.68
BaO	0.05	nd	nd	nd	0.04	0.05	—
Na_2O	0.88	0.24	0.19	1.32	2.75	2.48	2.73
K_2O	2.58	1.28	2.29	0.87	4.93	4.94	4.39
H_2O	4.17	1.77	3.13	2.98	1.10	1.08	0.70
P_2O_5	0.28	?	?	?	0.17	0.11	0.07
CO_2	—	0.00	0.00	0.12	0.09	0.05	—
Cl	0.03	—	—	—	0.01	0.01	—
F	0.01	—	—	—	0.01	0.03	—
C	—	—	—	0.16	—	—	—
S	—	—	—	—	0.03	0.06	—

* includes P_2O_5 if any

Table 33 (a). (Partial soil analyses (weight percent) (from ROE 1953))

Parent rock	Depth of horizon	pH	Nitrogen	Fe$_2$O$_3$ + Al$_2$O$_3$*	CaO*	K$_2$O*	P$_2$O$_5$*	MgO*
Porphyritic granite	0''–30'	5.3	0.0336	27.43	0.1666	0.3384	0.0128	0.0659
Weathered phyllite	0''– 33''	5.0	0.0896	16.58	0.0070	0.9485	0.0341	0.0747
Schist	0''– 5''	4.5	0.1036	11.22	0.0112	0.4555	0.0174	0.2263
	5''–10' 6''	5.3	0.0476	11.48	0.0056	0.9419	0.0151	0.0574
River alluvium	0''– 14''	5.7	0.1260	17.18	0.0560	0.2769	0.0930	0.0679
	14''– 19''	5.1	0.0364	13.13	0.0070	0.2820	0.0976	0.0213
	19''– 79''	5.1	0.0588	26.81	0.0056	0.1480	0.0052	0.0533
	79'–15'	4.6	0.0224	6.80	0.0462	0.0714	0.0081	0.0723

* Hydrochloric acid extracts

Table 33 (b). Properties of a Red Yellow Ultisol on Granite (from Ng 1969)

		Mean values			P(0.1N NaOH) ppm	C.E.C. meq/100g	Base sat %	6N HCL Sol. Fe_2O_3 %
	Clay %	pH	C %	N %				
T	37	4.7	1.61	0.13	75	8.85	12	1.74
S	50	4.8	0.34	0.04	27	5.70	11	2.33

T = Topsoil
S = Subsoil
C.E.C. = cation exchange capacity

mg/l., within the ranges found for other tropical areas by Sioli and Jones and Bromfield, but considerably less than the normal 0.3 and 0.7–4.0 mg/l. found in heavily industrialized Europe and North America, respectively (Stewart op. cit.).

The occurrence of sulfate is usually associated with industrial air pollution (Gambell and Fisher 1964). The values obtained here on two occasions may be extraneous or possibly may have originated from the low-grade diesel fuel burned by log transporters and other trucks on the heavily-travelled East Coast Highway up through the valley.

River: The relative importance of the precipitation contribution to total ionic content of the river water is evident on consideration of Table 31, summarizing the chemical conditions in the Gombak. (Appendix A lists the individual results of the chemical analyses.) On average, about 20% of the dissolved load in the river was potentially present in the rain. This value is much lower than that recorded for many Australian rivers by Douglas (1968b) and is similar to the proportion found in European streams (cf. Rapp 1960). Anderson (1945) theorized that in areas of high run-off, the composition of the river water is more dependent on chemical denudation than in low run-off areas where precipitation is the chief source of dissolved salts. This would seem to hold in the Gombak where run-off was approximately 50% of precipitation. Any detailed assessment of input-output relationships between precipitation and run-off would have to be based on much more extensive element-by-element analyses (cf. Gorham 1961, Likens et al. 1967, 1970) that go beyond the scope of this study. However, preliminary comparisons for the major ions are given at the end of this section.

To facilitate explanation of the changes in water chemistry at various stations, the partial analyses of rock and soil types found in the watershed are given in Tables 32 and 33 and a summary of the hydrochemistry of the principal tributaries on 10 April, during a period of base flow, in Table 34. Discharge on this date was derived almost exclusively from ground-water

Table 34. Chemistry of tributary water at low flow on 10 April 1969 (mg/l.)

	pH	μmho$_{25}$	Ca	Mg	Na	K	Fe	Zn	Mn	Cu
A	6.70	17.64	0.18	0.38	1.38	0.90	0.06	nd	0.006	trace
B	7.10	36.00	0.28	0.54	2.95	2.00	0.18	0.20	0.017	nd
C	6.70	29.14	0.08	0.21	2.99	2.40	0.06	nd	0.006	trace
D	6.65	26.64	0.08	1.23	2.60	1.20	0.15	0.20	0.017	nd
E	7.00	35.04	0.21	0.23	3.11	2.70	0.18	trace	0.011	trace
F	7.05	23.40	0.04	0.97	2.73	1.30	0.17	0.20	0.006	trace
G	6.55	17.28	0.04	0.04	2.01	1.80	0.10	trace	trace	nd
H	6.25	22.92	0.12	0.08	2.05	2.10	0.22	0.20	0.006	nd
J	6.15	20.60	0.08	0.09	2.13	1.90	0.37	0.13	0.017	trace
K	6.20	142.79	2.24	2.69	4.20	5.00	1.23	0.27	0.067	0.04

	SiO$_2$	HCO$_3$ Alk.	Cl	NO$_3$-N	NH$_3$-N	PO$_4$-P	SO$_4$	C.O.D.
A	6.0	6.83	0.33	0.15	0.007	0.010	0.0	1.10
B	15.2	18.42	0.59	0.07	0.000	0.030	0.5	0.24
C	20.0	13.97	0.48	0.21	0.025	0.015	0.0	2.64
D	16.8	12.63	0.34	0.13	0.035	0.010	0.0	1.09
E	20.6	17.63	1.05	0.13	0.007	0.030	0.0	0.89
F	17.4	10.98	0.34	0.13	0.025	0.030	0.0	2.13
G	12.6	7.26	0.27	0.10	0.000	0.030	0.0	0.48
H	11.0	9.51	0.57	0.14	0.020	0.055	0.0	1.58
J	10.8	8.42	0.49	0.16	0.007	0.040	0.0	0.69
K	10.6	53.99	4.40	0.40	0.430	6.750	0.5	6.30

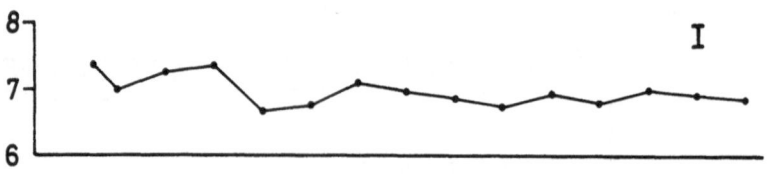

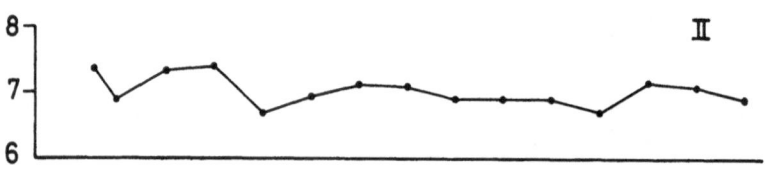

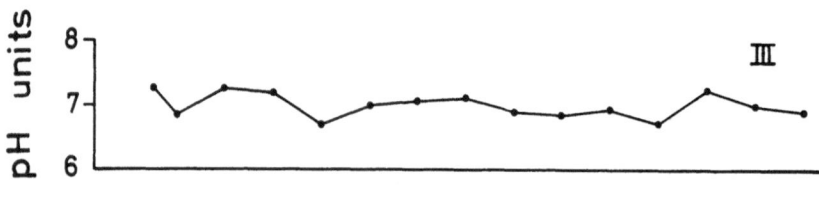

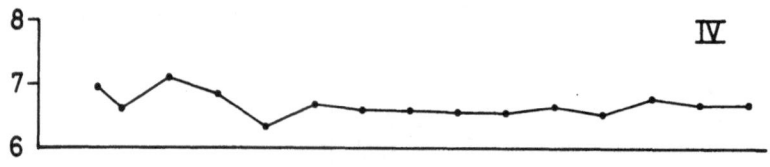

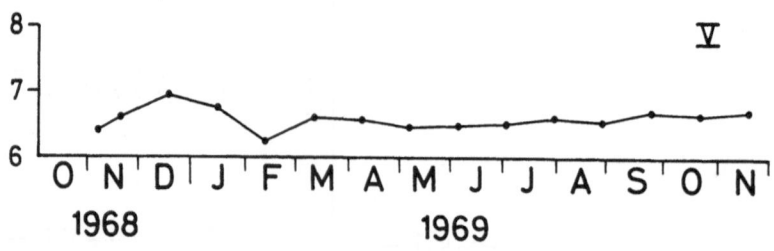

Fig. 25. pH values.

94

sources so that chemical contributions from limited area lithologies could be estimated uniquely, with the usual complicating surface run-off and mixing effects being minimal.

The Gombak river water in the Undisturbed Upper Catchment may be characterized as having relatively low conductivity; pH neutral or slightly acid; low buffering capacity and alkalinity; low concentrations of calcium and nutrient salts, but with high silica concentration; oxygen near saturation. Progressive enrichment (except in silica) occurs downstream with an increase in ions over the limestone formation and elevated nutrient levels below the urban industrial areas. Natural chemical conditions are not as impoverished as those extremes found in Amazonian rain forest streams or the waters of the southern Malaya lowlands, but are very similar to those in the freshwaters of southern Thailand not on limestone formations and to those in Central African rain forest rivers. One notable feature is the 'clear' colour (*sensu* Sioli 1951, 1964) of the water. Little or no humic leachate causing peaty discolouration is visibly present in the water at any time of the year. This is in contrast to conditions in some rain forest rivers, particularly south Malayan lowland waters, in which podsolic soils and very acid-rich humic soil water is the dominant feature of the limnology (Johnson 1968a).

pH: pH values were relatively uniform throughout the length of the river, with little seasonal variation (Fig. 25). The greatest range at any station (IV) was only 0.8 pH units, with no significant diurnal changes as there were very few primary producers. A slight decrease in pH at the lower two stations, from a mean value near 7.00 at I, II and III to 6.68 at IV and 6.58 at V was found and attributed to pollution and build-up of decomposing organic load. The near neutral pH was a surprise as precipitation had a mean pH of 6.66 and soil pHs above both granite and sedimentary rocks lie between 4.5 and 6.0. The primary products of silicate hydrolysis have different solubilities, the bases being much more soluble. These highly mobile products are, therefore, preferentially lost to the drainage water, leaving the acid hydrolysate in the soil. This is responsible for the low pH values found at the surface but increasing toward the bedrock and is, perhaps, a partial explanation for the near neutral pH of the river. However, ground-water seepages and springs at the base of the foothills (at Station III and Location G) had pH values as low as 5.5. so that partial neutralization of such contributions must occur farther downstream, probably by more alkaline ground-water entering the river from the overlain Kuala Lumpur Limestone Formation. The 'typical' granite-derived soil described in Table 33(b) summarizes the situation with acid pH, increasing Fe_2O_3 percentage with depth, very low cation exchange capacity and base saturation percentage, all symptomatic of intense leaching of the electrolytes and alkali silicates.

Alkalinity: Alkalinity is a measure of weak acids and weak acid-salts and for most natural, near neutral waters, this is basically a CO_2-HCO_3^-

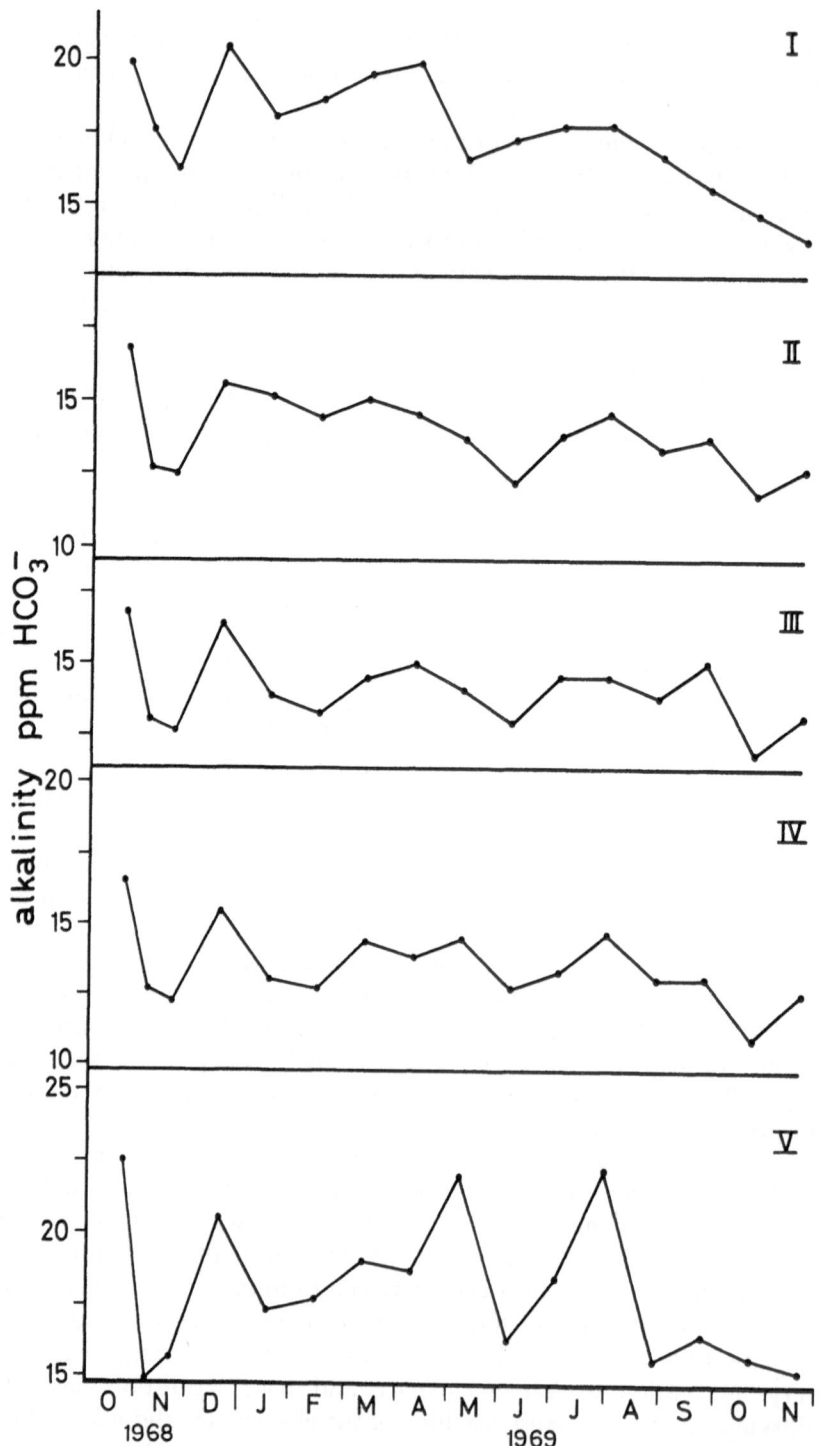

Fig. 26. Titrated alkalinity expressed as ppm HCO₃⁻.

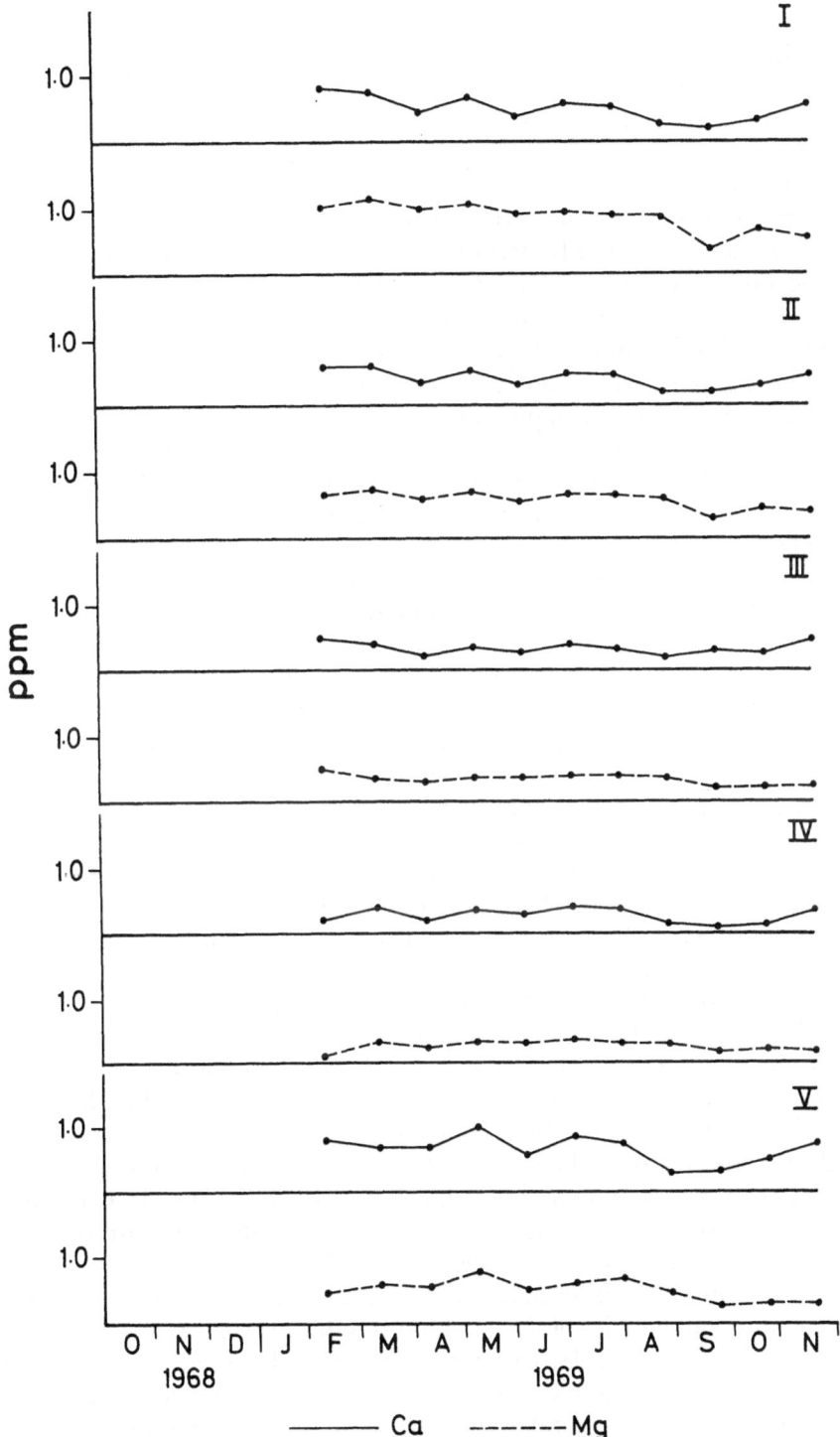

Fig. 27. Calcium and magnesium concentrations in ppm determined by atomic absorption spectrophotometry.

equilibrium system (HUTCHINSON 1957). Many authors conventionally treat alkalinity as a measure of 'hardness', i.e. calcium and magnesium content, but except in waters rich in these elements, there is no justification for this. JOHNSON (1967a) showed that for many local waters, no direct correlation between alkalinity and calcium content exists. Results are generally expressed as meq/l. or ppm HCO_3^- (Fig. 26), although with little reason as the pH buffering system may also be dependent on silica concentrations. This will be discussed further in the section on ionic balance. The data show considerable variation with time, probably as a result of dilution during periods of heavy rain and high flow just prior to the sampling date. Station I, draining non-calcareous sedimentary rocks, had consistently higher alkalinity values than Stations II, III, and IV. Calcium plus magnesium concentration was also higher here, both dissolved and in the parent lutaceous lithology, than at the lower granitic areas, and dissolved silica concentration was very high. The increase in alkalinity at Station V reflected the contribution of ground-water from the limestone formation beneath the alluvium in that area. Total alkalinity values are low when compared to many freshwaters, but are similar to those reported from some areas of the Amazon basin, the Lwiru (MARLIER 1954) and the Sai Buri River (KOBAYASHI 1959) which drain rain forest areas. An organic colloid buffering system could not exist, as reported for various Rio Grande do Sul rivers by KLEERE-KOPER (1955), because organic content and humic colour were very low.

Calcium and Magnesium: Calcium determination by EDTA titration at the low concentrations of these waters proved to be imprecise. The use of the atomic absorption spectrophotometer may underestimate the calcium concentration (cf. LIKENS et al. 1967, who suggested a possible correction factor of $\times 1.6$). This would alter the ionic balance considerably. Fig. 27 shows the spectrophotometrically-determined values for these elements; the unreliable titration results were omitted.

The waters of the upper catchment were extremely poor in calcium and magnesium as are both the sedimentary/metamorphic and granitic rock series and their overlying soils. Magnesium was more abundant than calcium at the upper stations, reflecting their relative concentrations in schists and quartzites (Table 32). At Station I, some contamination of the tributary from the dolomitic fill used as a bed for Route II is possible, giving elevated concentrations over those of the adjacent main river.

Calcium and magnesium concentrations declined to minima at Station IV, indicating that dilution occurs down the granite section of the catchment and that mobility of these ions is probably greater in the sedimentary than in the igneous rocks. Decomposing leaf leachates may also add some calcium to the upper river (cf. THOMAS 1970). Enrichment, particularly of calcium, from the underlying limestone formation occurs between Stations IV and V. A further source of both ions is the precipitation. As seen earlier, considerable concentrations are available from the rain,

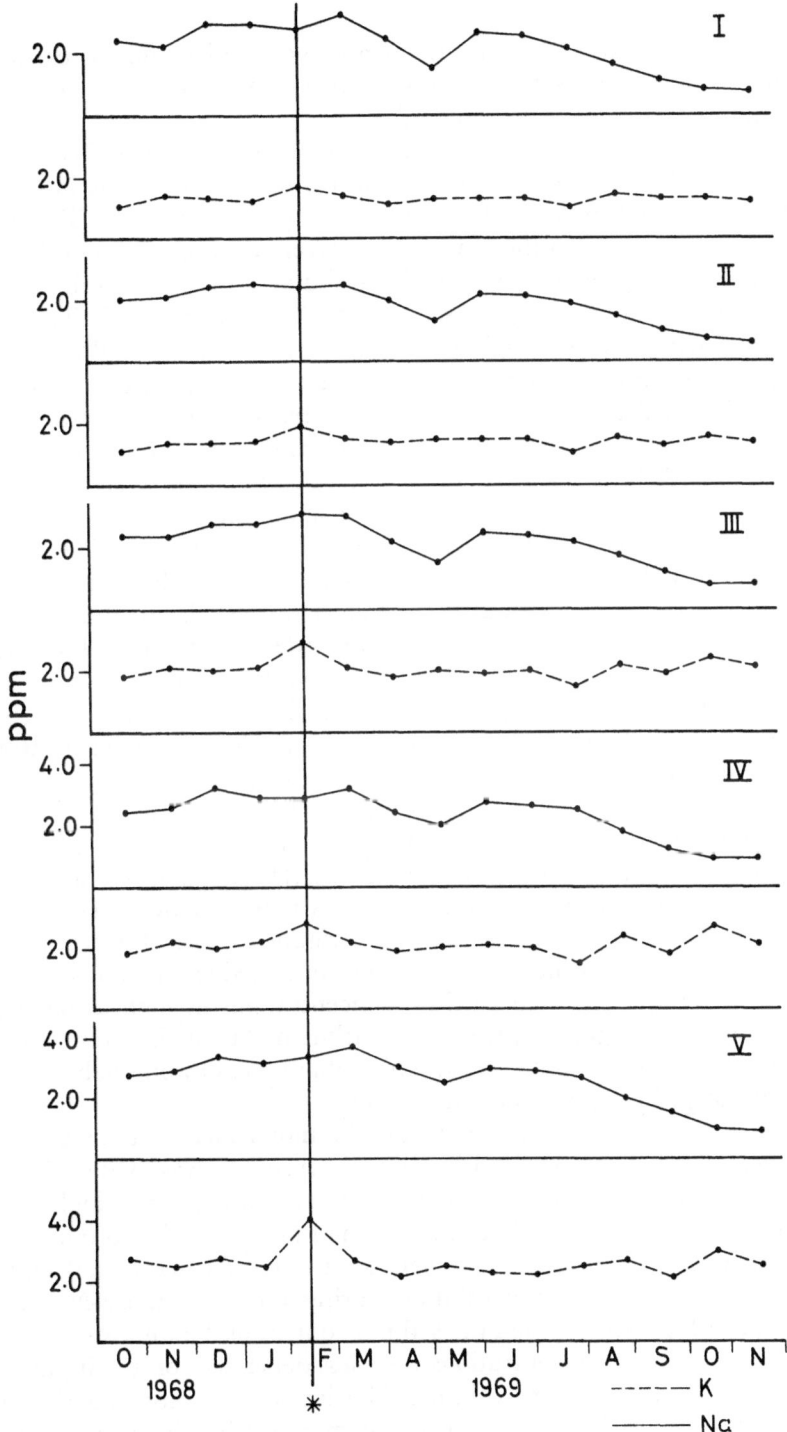

Fig. 28. Sodium and potassium concentrations in ppm; * Change in technique from flame photometer to A.A. Spectrophotometer.

but this source would supply uniformly over the catchment, or deliver higher concentrations to the lower areas, and does not explain the longitudinal discrepancies observed.

The analyses of the tributaries at base flow showed elevated concentrations at A, B, E and K compared with the other locations. K lies on the Kuala Lumpur Limestone Formation (composition given in Table 32), but the soils on this lithology, mostly river alluvia, are poor in both calcium and magnesium (Table 33). Tributaries A and B drain forested watersheds predominantly on sedimentary rocks and are similar to the Station I tributary, although at the watercress farm above B liming is a regular procedure so that B's elevated calcium content must be considered an artefact. E drains a granite catchment, but is probably contaminated by run-off from dolomite-chips used in logging track stabilization. Other tributaries, C, F, G and J draining forest or plantation areas on granite soils, had lower concentrations of both calcium and magnesium.

Under WILLIAMS' (1964) classification (following OHLE 1934), waters with less than 2.5 mg/l. calcium are considered extremely poor. JOHNSON (1967a) found that most southern Malaya waters fell into this category and some had calcium concentrations an order of magnitude lower than the minimum 0.10 mg/l. found for the Gombak. The rivers studied in Thailand by KOBAYASHI generally had much higher calcium concentrations (mean > 20 mg/l.) but the southernmost Malay peninsula stream he sampled had a mean of 1.9 mg/l., in the same range as the Gombak.

The maximum calcium concentration for the river above Station V was only 1.00 mg/l., but in Tributary K on the limestone, 2.24 mg/l. were recorded at low flow, and at the D.I.D. gauging site, a value of 2.79 mg/l. with 1.9 mg/l. magnesium was found. KANAPATHY (1968) recorded 1.55–2.18 mg/l. calcium from irrigation water abstracted in the lower watershed. This however probably contained extraneous materials. GOBBETT (1964) reported calcium concentrations as high as 6.6 mg/l. above the D.I.D. gauging site and attributed this to the ground-water inflow off the limestone bedrock. This value was not approached in this survey except in highly polluted inflows.

Sodium and Potassium: The data for the monovalent cations are summarized in Fig. 28. A general increase downstream was evident in both elements (except that sodium concentration at Station I was higher than at Station II) as a result of accumulation of mineralization products and soil leachates. Slight seasonal changes occurred, indicated by a decline in sodium concentration with dilution during the wet August-November period of 1969 during which the sodium content of the rain was less than 1 mg/l.. Potassium concentrations were less dependent on the input from rainfall and on discharge volume. The latosol soils are poor in these electrolytes as leaching is rapid and comprehensive. The parent rocks contain relatively low concentrations but the granites have higher percentages of both elements than the sedimentary series. This accounted

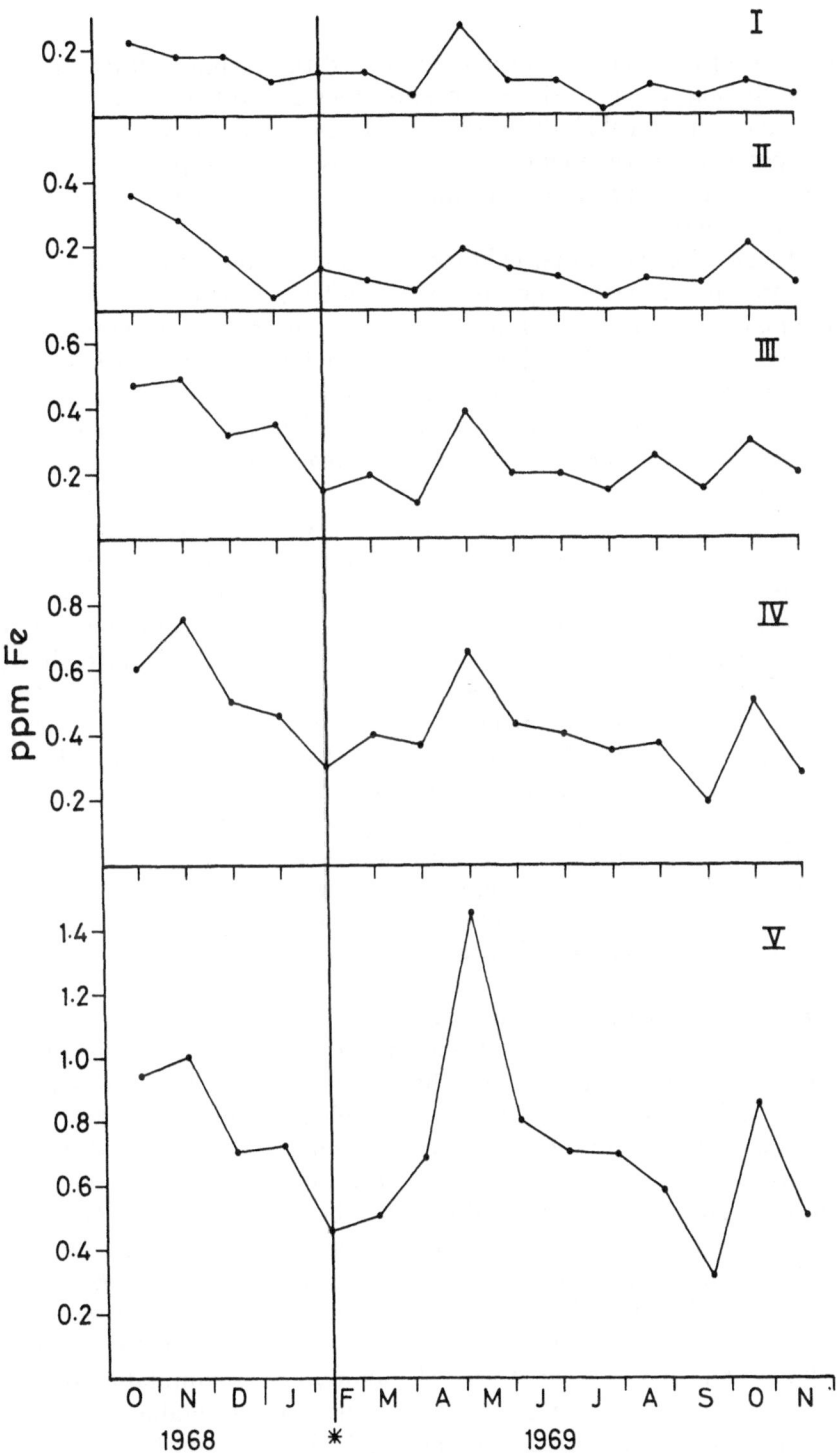

Fig. 29. Total iron content in ppm; * Change in technique from o-phenanthroline method to A.A. Spectrophotometer.

for some of the increase as the river crossed the granite belt downstream from Station II. Potassium concentrations in both igneous and sedimentary/metamorphic lithologies are higher than those of sodium so that potassium enrichment from leachates is also disproportionate and by Station V their mean concentrations were approximately equalized at about 2.6 mg/l..

The data for individual tributaries (Table 34) do not corroborate any increased mobility from granite as few differences can be seen between streams draining the various lithologies and soils. Higher evaporation rate in the main, unshaded river, as compared to the tributaries, is one possible source of enrichment. Also, the mechanism described for Australian rivers by DOUGLAS (1968b) in which calcium and magnesium are removed from the water by ion exchange with the clays and loams, and replaced by sodium and potassium, may affect the Gombak. This would partially account for the decline in calcium and magnesium as discussed previously. Enrichment of sodium and potassium from human activities, particularly gardening which exposes the soils to rapid leaching, but also from small amounts of domestic effluent from the isolated dwellings between Stations III and IV, was possible. Tributary K showed the effects of mild pollution, mostly direct sewage run-off, with 4.2 and 5.0 mg/l. for sodium and potassium respectively. The higher values for electrolytes at Station V were the result of a series of minor inflows of sewage from the kampongs and of the return of irrigation water from the rice-growing areas where the evaporation rate was increased. As found by KOBAYASHI, JOHNSON, KLINGE and OHLE (1964) in Amazonia, sodium is normally the dominant cation in these soft waters, but, as was evident in the lower catchment, potassium may become important when a local supply is available. However, the absolute concentrations of both were still very low.

Iron and Aluminium: The fluctuations in total iron content, shown in Fig. 29, resulted principally from changes in discharge. The highest values occurred during wet, peak flow periods when the soil was waterlogged and wash-out and transport of colloidal and precipitated iron was maximal. During the drier seasons (January-February and August-September), the concentrations were at a minimum. The immobility of iron and aluminium in tropical soils has been discussed by RICHARDS (1957), SIOLI (1968a) and others and unless considerable concentrations of electrolytes and humic colloids are present (SHAPIRO 1966), these elements tend to remain static as sesquioxides in the latosol soils. The data in Table 33 for leached soils when compared to Table 32 for parent rock materials indicate the degree of stability and accumulation of these elements. However, low iron concentrations were detectable throughout the sampling period increasing progressively downstream. In local waterlogged accumulations of leaf detritus and humus, where decomposition may produce anaerobic conditions, iron is reduced and solubilized. In

Table 35. Reactive silicate (as SiO_2 mg/l.) for the Gombak River main stations

	I	II	III	IV	V
14–IV–69	14.4	10.6	12.6	14.0	9.6
8–V–69	8.8	7.8	5.6	10.6	7.2
5–VI–69	10.4	8.0	10.2	9.0	7.2
3–VII–69	12.6	11.0	14.3	14.4	13.2
31–VII–69	13.3	11.4	14.4	12.6	11.0
28–VIII–69	12.3	11.0	12.2	10.6	10.6
25–IX–69	12.2	10.7	13.6	11.4	10.6
23–X–69	10.6	9.6	10.0	9.2	7.9
20–XI–69	11.0	11.0	11.0	12.4	11.0
mean	11.7	10.1	11.5	11.6	9.8

the lower zones particularly, seepages of ground-water directly into the river carried dissolved iron. At the top of riffles where impervious strata reached the surface, beds of precipitated iron hydroxide sometimes formed during low stage conditions. At Station IV, where sandbars were constantly being shifted and some pyritic material was available from the local schist formation, leaf accumulations were often buried and their decomposition created conditions conducive to the mobilization of iron. SHAPIRO (1966) has shown that humates may hold iron in solution under conditions in which it would normally be precipitated, but these compounds are not common except in seepages in the Gombak.

The data for the tributaries substantiate the overall pattern with very low iron concentrations in the extreme headwater streams and small contributions from the other tributaries. At Station V and Tributary K, there was some pollution from workshops and garbage dumps where local reducing conditions mobilized iron.

Aluminium was not detected during the investigation at any station. This may have been due to poor sensitivity (about 1 mg/l.) of the atomic absorption spectrophotometer even with a nitrous oxide flame.

Silica: Reactive silicate, reported here as silicon dioxide mg/l. but determined in an unknown formulation, was the most important single solute in the Gombak River. The high mobility of silica from igneous rock has been described by KOBAYASHI (1960) for Japanese river systems and by DOUGLAS (1968b) for Australian granites. The data for SiO_2 determinations from May 1969 are given in Table 35. No standard was available before this date and the K_2CrO_4 standard of MACKERETH (1963) proved to be inaccurate (cf. Moss 1969). Little longitudinal variation in concentration occurred as supply from granite is constant down to Station IV. A slight decrease at Station V may be attributed to dilution and lack of enriching tributaries running off granite, some diatom

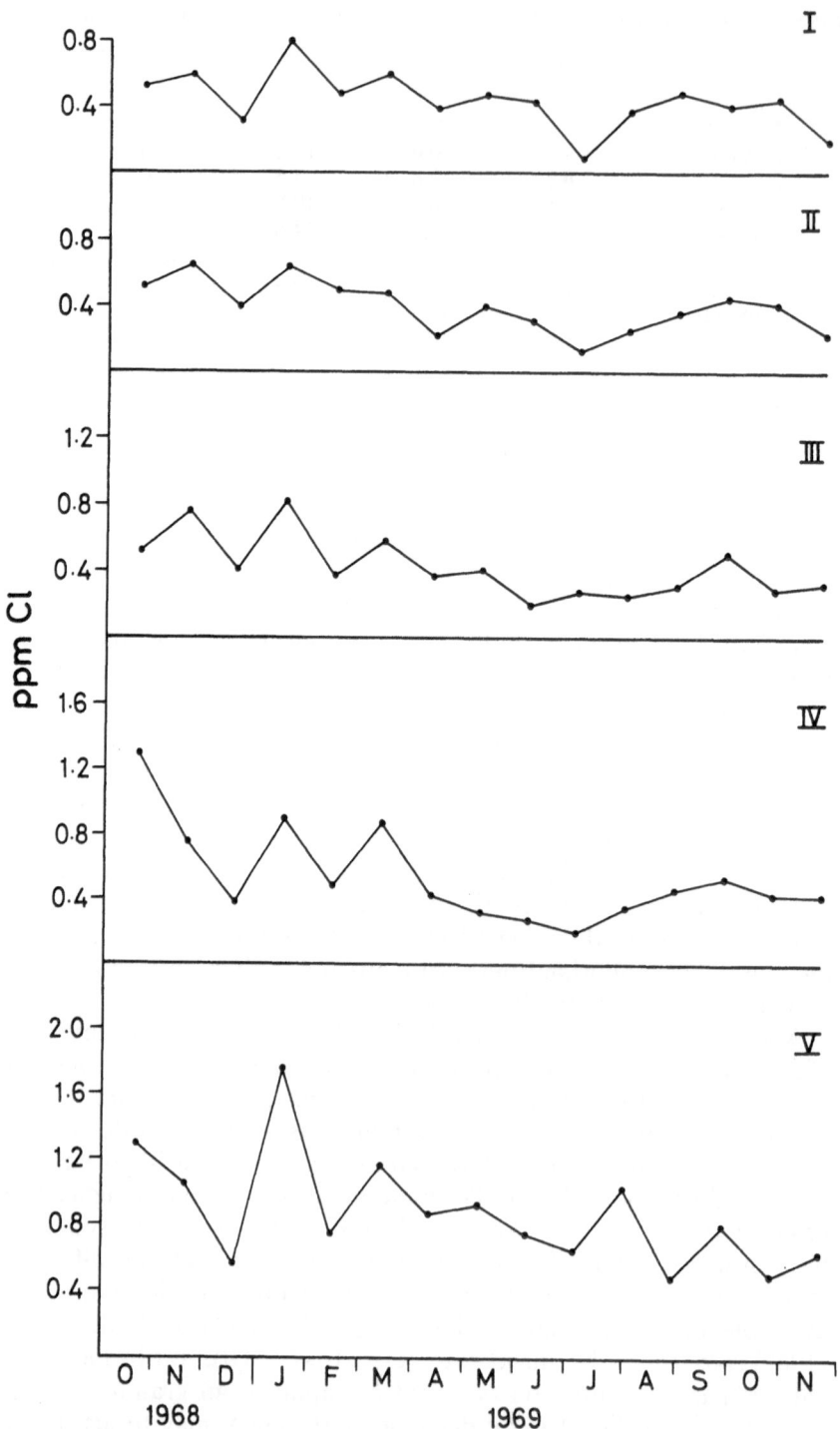

Fig. 30. Chloride ion content in ppm.

growth but probably not enough to seriously affect the concentration of silica (WANG and EVANS 1969), and most important, the use of large volumes of water in the paddy area in which considerable amounts of silica are taken up by the rice plants (see KOBAYASHI 1959 for references). Some seasonal variation was evident in that May, June and October had lower concentrations at all stations than the drier periods; maxima occurred in April at Station I and in July at II, III, IV and V. This is in agreement with the general conclusions made by DOUGLAS (1969) in a summary of the factors governing silica removal in the tropics: that precipitation and other factors affecting run-off control the rate of denudation of silica more than temperature, which was here consistently high.

The SiO_2 values obtained for the tributaries on 10 April show that the highest concentration was derived from the upper, coarse-grained granite watersheds of Tributaries D, E and F, minimum concentration from the quartzite, shale and chert at A and a slightly higher concentration of dissolved silica from the fine-grained granite bedrock of the middle and lower foothill Tributaries G, H and J. Gradient and rate of run-off probably govern this relationship more than density of the bedrock. DOUGLAS (1969) gave a figure, after LIVINGSTONE, for mean dissolved silicon dioxide concentration for Asian rivers as 11.7 mg/l.. The Gombak is, therefore, a typical Asian watercourse with respect to silica. DOUGLAS reported 15 mg/l. as the mean concentration at the mouth of the Gombak, and 24 mg/l. mean content in the headwaters. These data seem high in comparison to the present results, but he used a visual nesslerization technique.

Chloride: Chloride concentrations (Fig. 30) were extremely low considering the proximity of the ocean (cf. values in Australia (DOUGLAS 1968b)). Input from precipitation was insignificant and the content of the parent rocks is low. The data for the tributaries are similar to those for the main river (the high value at E cannot be explained), with slight enrichment downstream, particularly in Tributary K which carried a considerable pollution load and was largely responsible for the increased concentration at Station V. There may be a considerable influx of chloride to the watershed in the treated water piped to the Lower and part of the Middle Zone and used copiously by the inhabitants. Much would be lost by local adsorption but some undoubtedly reaches the river. Analyses at the D.I.D. gauging station gave a range of 5–12 mg/l. and a mean at modal flow of about 7.5 mg/l., indicating the considerable influx of sewage and some industrial effluent between Station V and that point. The variations between sampling dates are attributed to changes in discharge volume. These data, with a mean for the whole river of about 0.9 mg/l., are very similar to those found in both north and south Malaya and are in the range of values found for forest rivers in Amazonia.

Sulfate: In contrast to the waters of southern Malaya in which sulfate is often the dominant anion, reaching concentrations of more than

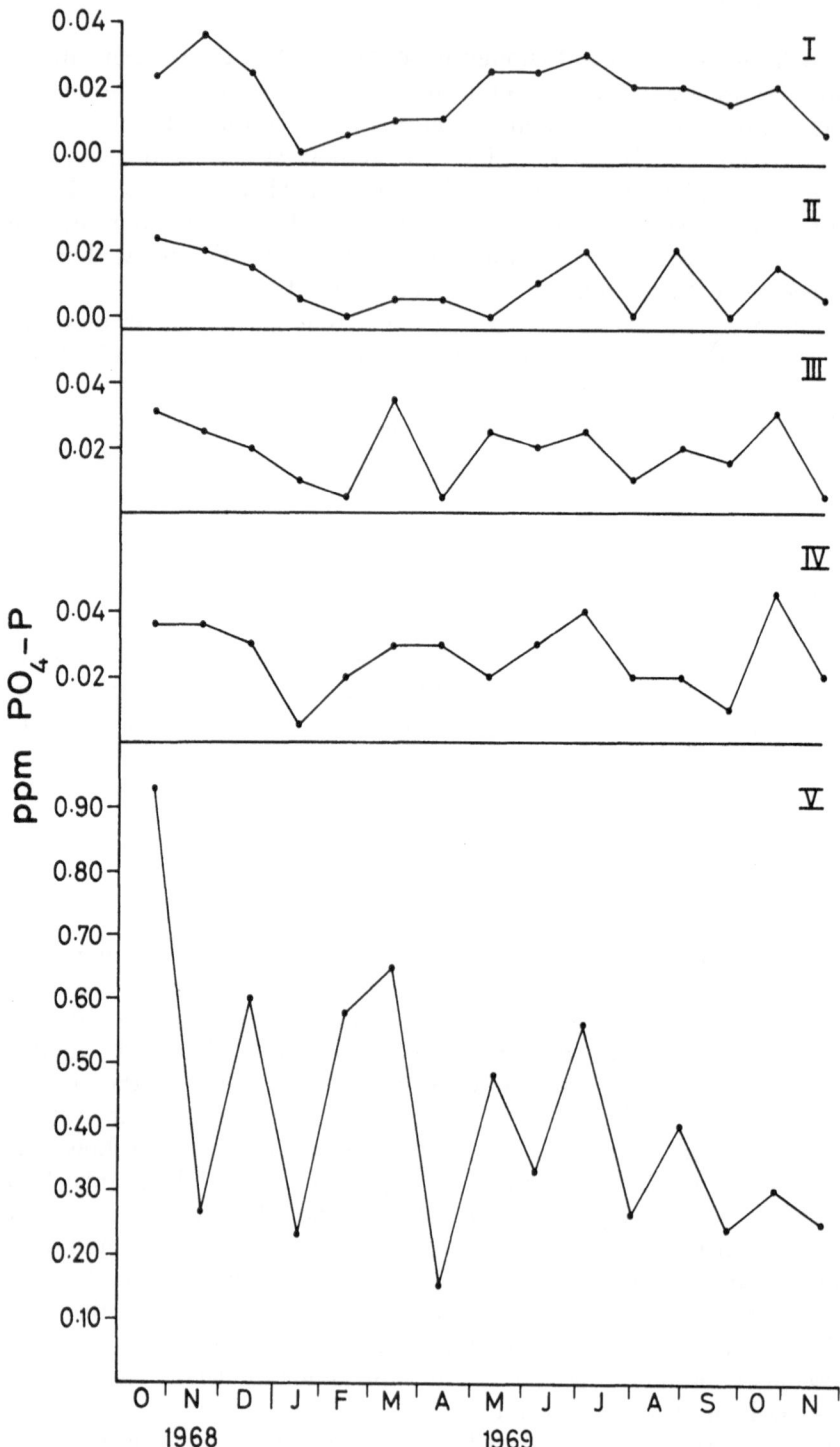

Fig. 31. Concentrations of phosphate- phosphorus in ppm; Note change in scale for Station V.

150 mg/l., but usually in excess of 3 mg/l. in hill-streams (JOHNSON 1967a), the Gombak is poor in sulfate. At all stations, 'not detectable' levels were found on a number of sampling dates and in the analysis of the tributaries (Table 34). Even the normally polluted Tributary K had only 0.5 mg/l. on 10 April. The erratically high maximum values, e.g. Station I, cannot be easily explained as there were only traces of sulfur in the granitic rocks and none in the sedimentary/metamorphic series. Below Station II in an area of soft sands derived in part from the lutaceous shales of that area, black deposits, possibly iron sulfide, formed on the galvanized benthos sampler on several occasions, at depths greater than 10 cm. Quantities of sulfur may therefore be present, either in the sediments as a mineral not analysed as part of the schists and shales complex in Table 32, or more likely as a result of anaerobic decomposition of organic leaf materials. Similar bottom conditions occurred throughout the length of the Station I tributary as a result of the rapid fragmentation of the soft sedimentary rocks into fine particles. The higher sulfate concentrations found there (1.0 mg/l.) may have resulted from sudden changes in discharge that flushed these reduced materials out of the sediments into the oxygen-rich surface waters. Iron concentration was earlier seen to be maximal during peak flow periods so this may account for the presence of both sulfate and iron. Normally, absorption by algae and trailing tree-roots probably held dissolved sulfate to negligible concentrations. These nearly zero concentrations, generally indicative of conditions in the river, are closely analogous to those in southern Thailand, Africa (BEAUCHAMP 1953) and the Amazon basin.

Phosphate – Phosphorus: The phosphate-phosphorus concentration levels found for the Gombak (Fig. 31) were extremely low except at Station V where pollution had added considerable amounts. The availability of phosphorus in the ultisol soils and rocks is very low (Tables 33 and 32), less even than in Amazonia, so that the small mean concentrations (< 0.025 mg/l. in unpolluted areas) are predictable. These diminutive reserves of phosphorus in the soils and waters of a forest ecosystem that reportedly supports the highest productivity of any forest indicate the efficiency of the closed recycling system. The equilibrium between soluble phosphate and that bound to suspended organic particles, critical at about 10 µg/l. (KEUP 1968, after GESSNER 1960), may have been operative in the Gombak, although zero concentrations were periodically recorded. It is more likely that most free phosphate was quickly taken up by the roots and the algal crop. KOBAYASHI (1959) usually found 0.00 or occasionally 0.01 mg/l. in his analyses of Malay peninsula rivers in Thailand. The temporal variations in concentration seen on the graph for Stations I to IV were most likely related to dilution, but the wild fluctuations at Station V were the result of pollution. Much mineralization of organic material occurred in stagnant drains and ditches during drier weather and these enriched waters were flushed into the main river

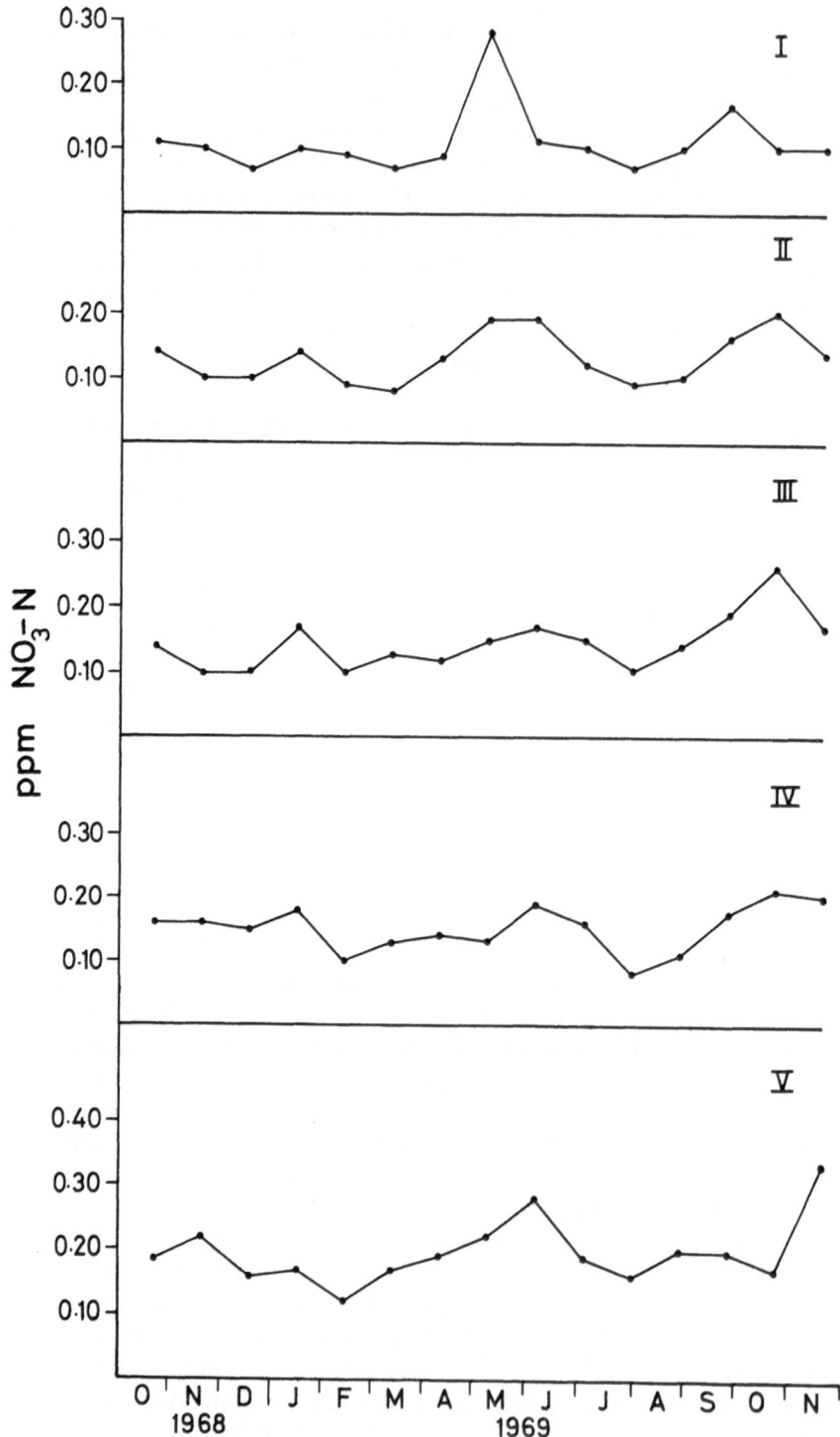

Fig. 32. Nitrate-nitrogen concentrations in ppm.

with every storm. In addition, in the Lower Zone the river was used for bathing and laundry and moderate contributions of soap and detergents high in phosphate were added throughout the lower section. These may have contaminated the water samples even though collection was usually made during the mid-afternoon siesta. The high phosphate-phosphorous concentration in the lower tributary K indicated the degree of pollution that may occur from unregulated riparian urbanization.

Nitrogen Compounds: The dissolved nitrogen compounds in water do not originate from bedrock materials under the wet conditions of tropical rain forest; rather they result from biological processes that also determine their formulation. In contrast to Amazonia (KLINGE and OHLE 1964), the Gombak was relatively rich in dissolved nitrogenous compounds. Nitrate-nitrogen concentrations ranged between 0.07 and 0.30 mg/l. at all stations with little seasonal variation (Fig. 32). The values for the tributaries were within this range, except for K which showed some enrichment. Nitrite-nitrogen was present in very low concentrations (0.001–0.003 mg/l.) with a progressive increase to 0.005 mg/l. at Station V as a result of pollution and local areas of reducing conditions. Ammonia-nitrogen was surprisingly present in considerable concentrations even in the forest sections. Mean concentrations were less than 0.08 mg/l. but values twice as high occurred as maxima. At Tributary K, a pollutional value of 0.4 mg/l. was found.

The source of nitrogenous compounds was probably not the soil directly, although nitrogen is present in all soil types. Of interest is the observation that the mean total nitrogen values for top- and sub-soil levels of the general granite-derived ultisol (Table 33) are almost identical to those given by KLINGE and OHLE (op. cit.) for a brown loam soil under forest in Amazonia. Some of the nitrate and probably most of the ammonia were derived from the rain, presumably from intense electrical activity, but the remainder likely had botanical origins. In the rain forest, considerable development of blue-green algae was evident on the soil and tree-trunks and these are undoubtedly associated with nitrogen-fixing bacteria. In addition, Leguminosae, notably *Saraca* which often has its roots in the water, are common in the Lowland Dipterocarp forest. These floristic elements must continually contribute considerable concentrations of nitrate-nitrogen by 'trickling-fertilization (Trauf-dungung)' (KLINGE and OHLE op. cit.), but especially during periods of heavy rain when sheet run-off occurs. Other possible sources are direct leaching from plants (cf. NYE 1961) or from fallen leaves in the river which give up some nitrogen as leachate immediately after entering the water (HYNES and KAUSHIK 1969).

Trace Elements: Atomic absorption spectrophotometer analysis results for trace elements are given in Appendix A. The only elements detected with any regularity were manganese and zinc. Manganese, when present, was always in low concentrations (< 0.018 mg/l.-trace), except at

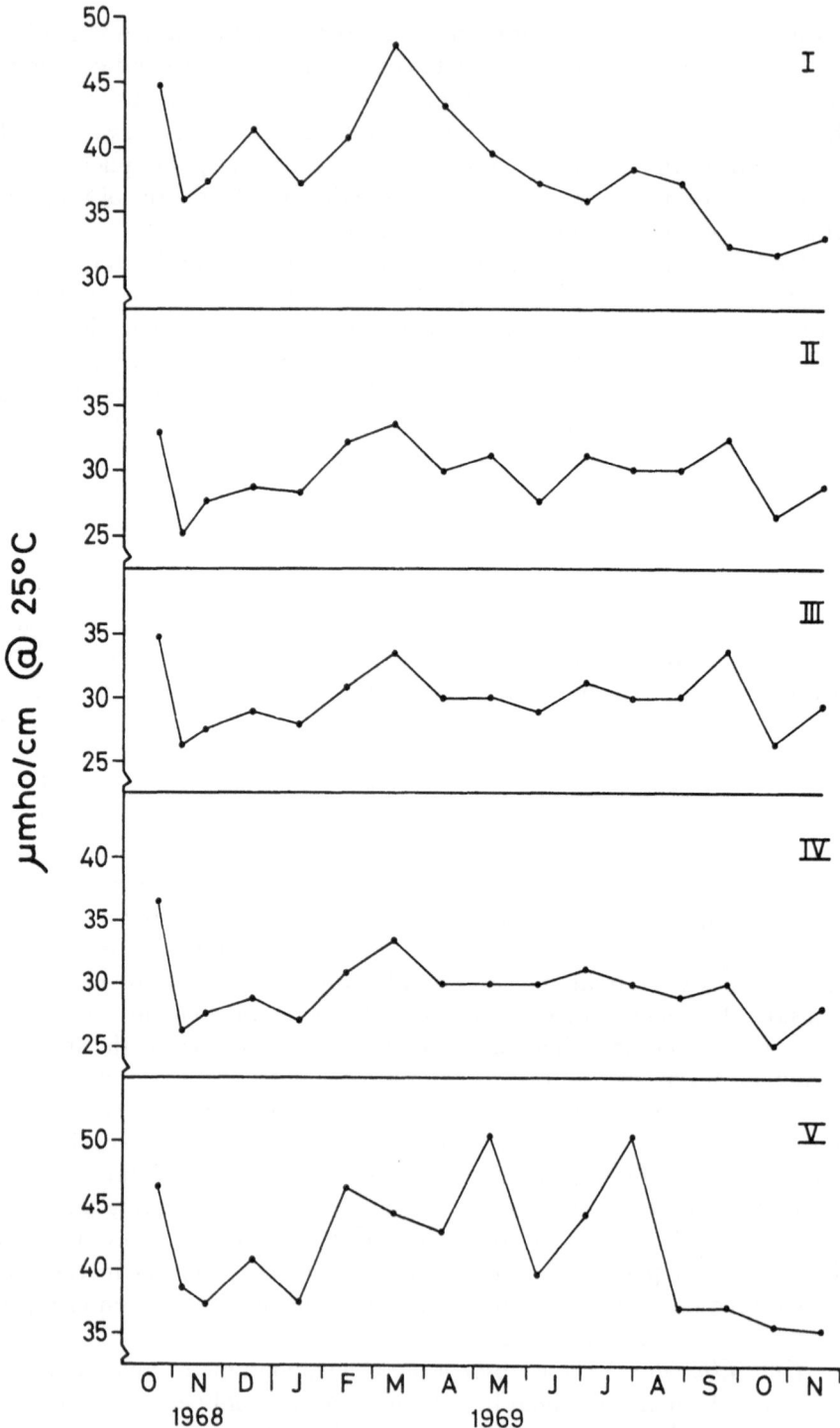

Fig. 33. Specific conductance in μmho/cm at 25°C.

Station V where a maximum of about 0.080 mg/l. was recorded on one occasion. The tributaries, even at base flow, had very low to trace concentrations only. Zinc was often not detectable; at other times values of trace to 0.250 mg/l. at Stations I to IV and to almost 1 mg/l. from industrial effluents and garbage run-off at Station V were found. Copper was very occasionally present at Station V in trace (< 0.01 mg/l.) amounts and in one sample at Stations III and IV also. Copper compounds are used as a fungicide in *Hevea* plantations (WYCHERLEY 1967) and heavy precipitation immediately after spraying is a possible source for this element and would account for its irregular occurrence.

Concentration ($\times 20$) failed to detect any aluminium, arsenic, barium, cobalt, lead, lithium, nickel or tin, in either the heat-concentrated or lyophilized samples. Traces of copper appeared at Stations I and II and of chromium at Stations IV and V in the heat-concentrated samples, probably the result of metal workshop pollution. Aluminium and tin might be expected to be present in the water, but with the system available, neither was detectable even after concentration. Arsenic contamination from the popular practice of applying sodium arsenite as a weedkiller in rubber plantations was also expected, but most of the 20–25 lb/acre/year applied is bound by the soil and accumulated to high concentrations (JONES and HATCH 1937). The solubility of As_2O_3 from the soil into ground-waters is dependent on high electrolyte concentrations (GREAVES 1934) and is therefore limited in the Malayan context of poor soils. Run-off from heavy precipitation immediately after spraying probably carries locally high concentrations of short duration.

Specific conductance: Conductivity data, indicated in Fig. 33, summarize the ionic condition of the river at the various stations. Stations II, III and IV had almost identical specific conductance readings throughout the study. Station I was constantly higher as a result of its greater ionic load. The maximum and mean for Station V were higher than at the other stations, reflecting the influx of industrial wastes and salts from mineralized organic accumulations. Tributaries B and E had values similar to those at Station I, the higher ion concentrations coming from the watercress farm and the highway surface respectively. K had a very high electrolyte content from various effluents. The other tributaries had lower conductivities than the main river.

Seasonal differences at each station were caused mainly by variations in the ionic composition of the precipitation and the diluting effects of large volumes of rain. The peak at Stations I, II, III and IV at the end of the February-March dry period and the generally lower values in November and December were a function of this. SLACK and FELTZ (1968) recorded higher alkalinity and conductivity and a drop in pH as a result of increased leachate from fallen leaves. At the end of February and beginning of March, much of the *Hevea* and some forest trees were replacing their foliage, creating a large leaf load in the river at a time of

Table 36. Summary of mean ionic conditions in the river

	I	II	III	IV	V
Fe^{+3} μeq/l.	6.45	7.52	14.50	24.17	39.22
Ca^{+2} μeq/l.	24.45	20.96	16.47	13.47	31.94
Mg^{+2} μeq/l.	72.39	46.89	27.15	20.57	41.13
Na^+ μeq/l.	91.79	78.74	92.22	98.75	112.67
K^+ μeq/l.	32.47	37.08	52.16	53.95	66.99
NH_4^+ μeq/l.	5.70	3.56	4.28	3.56	4.99
H^+ μeq/l.	0.107	0.095	0.100	0.209	0.263
Total cations μeq/l.	233.357	194.845	206.880	214.679	297.202
Cl^- μeq/l.	12.41	11.28	11.85	15.20	24.82
PO_4^{-3} μeq/l.	0.57	0.32	0.60	0.79	13.08
SO_4^{-2} μeq/l.	18.75	9.37	8.33	9.79	2.71
NO_3^- μeq/l.	8.57	9.29	10.71	10.71	14.29
NO_2^- μeq/l.	0.20	0.16	0.24	0.32	0.36
OH^- μeq/l.	0.093	0.105	0.100	0.048	0.038
Titrated alkalinity μeq/l.	284.40	222.60	224.80	214.80	294.00
Calculated $H_3SiO_4^-$ alk. μeq/l.*	103.364	110.050	141.920	156.351	111.235
Calculated HCO_3^- alk. μeq/l.	89.400	54.270	33.130	21.470	130.670
'True' alkalinity μeq/l.	192.764	164.320	175.050	177.821	241.905
Silicon mg/l. (Ionic state variable)*	5.46	4.71	5.37	5.41	4.57
Ca + Mg 'hardness' μeq/l.	96.84	67.85	43.62	34.04	73.07
Na + K μeq/l.	124.26	115.82	144.38	152.70	179.66
Ratio Ca + Mg : Na + K	0.78	0.59	0.30	0.22	0.41
Ca : Mg	0.34	0.45	0.61	0.66	0.78
Na : K	2.8	2.1	1.8	1.8	1.7
Na : Cl	7.4	7.0	7.8	6.5	4.5
μmho/cm_{25}	38.42	29.55	29.65	28.99	41.19
pH	6.97	7.02	7.00	6.68	6.58

* see text

low flow. The considerable fluctuations at Station V resulted from the flushing out of drains and ditches with individual storms and were dependent on rainfall conditions immediately prior to the sampling date. Unfortunately, neither KOBAYASHI nor JOHNSON recorded specific conductance in their surveys, but the data here indicate similar conductivities to those found in Amazonia, if perhaps a little higher. FITTKAU (1970) found low 10–15 μ S_{20} in the headwaters of the Xingu and KLINGE and OHLE (1964) gave readings between 10 and 45 μ S_{20} for a variety of water types. These conductivities were largely attributed to the equivalent conductance of free hydrogen ions at low pH values (GESSNER 1960, and see BERG 1961 for the Congo basin), while the values obtained in the Gombak reflected ionic composition, dominated by silica and bicarbonate ion and with little contribution from free acid radicals.

Ionic Balance: The overall ionic conditions in the river are summarized in Table 36 which gives the mean concentrations of anions and cations in μeq/l. for each station. These should balance, within ± 5%, after the equivalents of the species determining pH are added, if all ions have been determined. However, with the alkalinity values as found by titration, there was an apparent anionic excess at all stations on the Gombak that was incompatible with the pH status. JOHNSON (1967a) reported a similar phenomenon in some of the sulfatic south Malayan waters he investigated. Normally, in acid waters a cation excess would be detected, dominated by hydrogen ions released when mono- and divalent ions are adsorbed by humic complexes (cf. KLINGE and OHLE 1964). However, as discussed by GOLTERMAN (1967), the structure and chemical nature of the humic acids is unknown so that this cation displacement mechanism is only an attractive theory.

Except for a very small contribution from manganese and perhaps other trace metals, the cation contribution can be assumed to be complete. The apparent anion excess is probably an artefact, almost certainly due to the alkalinity data. Much of the alkalinity will originate from the usual CO_2–HCO_3^- system, but a contribution from organic or silicon compounds may also be important. No attempt was made to assess the buffer capacity of the dissolved and colloidal humic complexes, but, as indicated elsewhere, their influence on hydrochemical conditions in the Gombak was probably small. A second possibility is that conditions of the standard HCl titration method artificially elevate the silica contribution. According to GOLTERMAN and CLYMO (1969), the titration may include NH_4^+ and $H_3SiO_4^-$ ions with the HCO_3^- under particular conditions. The high pK (> 9) of ammonium ion ensures that at the pH of titration involved here all is present as NH_4^+ and so will not contribute to the basic system, and as the pK $Si(OH)_4$ = 9.5 this species also ought to have little effect on the acidimetric titration (STUMM and MORGAN 1970). SiO_2 resembles CO_2 in its chemical properties but is less strongly dissociated (RUTTNER 1963), so that in the presence of CO_2 and HCO_3^- it occurs mostly as free silicic acid, H_2SiO_3 (H_4SiO_4), either dissolved or in colloidal form. However, it is the alkali silicates that are leached preferentially from the soils and rocks and some will remain in that form in solution in an equilibrium $SiO_2 + 2H_2O$ (H_4SiO_4) $\rightleftharpoons$ $H_3SiO_4^- + H^+$ at the pH of the Gombak. If, during the titration with HCl, this equilibrium is pushed to the left, the alkalinity equivalence determined will be too high. The proportion of the total silica concentration in each ionic species in this equilibrium will depend on the condition of the CO_2–HCO_3^- equilibrium and its association state; i.e. the extent of the over-titration is dependent on the ratio of free CO_2 to free SiO_2 in the system. The 'true' total alkalinity, therefore, is dependent on both the CO_2–HCO_3^- system and the ionized state of the Si at the naturally-occurring pH before the titration was begun. For example: at Station I, the mean anion excess, using the titrated alkalinity data, is 324.993 — 233.357 = 91.636 μeq/l., assuming all other ion contributions are correct and complete. If all the 5.46 mg/l. Si had been reactive with HCl, that would be equivalent to 195 μeq/l. of anion. However, only 91.636 μeq/l. of the total silicon concentration was excess, i.e. originally in acid or colloidal form, leaving the other 103.364 μeq/l. as the alkalinity contribution from the silica. The true HCO_3^- alkalinity is, therefore, 233.357 — (103.364 + other anions) = 89.400 μeq/l. and the true total alkalinity is 103.364 + 89.400 = 192.764 μeq/l.

i.e. [Total Si] — [Anion excess] = [Si Alkalinity]
[cations] — [Si Alk. + other anions] = [HCO_3^- Alk.]
[True Alk.] = [Si Alk.] + [HCO_3^- Alk.]

If there had been no anion excess in the original balance sheet, either all the titrated alkalinity would have been due to HCO_3^-, without any contribution from the silica concentration, or all the silica would have been titrated as alkali silicate ion. In this situation, the actual state could be checked by a CO_2 determination and the calculation of the HCO_3^- concentration from the theoretical pH-CO_2 relationship.

This mechanism is suggested to perhaps explain some of the anomalies found by JOHNSON (1967a), although in his blackwater sulfatic areas, podsolization restricts the mobility of silicates and ionic conditions are dominated by the strong acid ion species. This back calculation method, if tenable, might be usefully applied to other silica-rich waters with apparent ionic imbalance i.e. the surface waters of hot, humid regions draining igneous watersheds. The alkalinities for the tributaries given earlier would also be affected by this overestimate but have not been recalculated. The silica fraction in rain originates largely as SiO_2 dust and would probably remain in acid form at the pH of precipitation.

From the relationship between total alkalinity and pH (GOLTERMAN and CLYMO 1969), the mean CO_2 at each station can be estimated: Station I 110.75 μeq/l.; Station II 66.21 μeq/l.; Station III 40.75 μeq/l.; Station IV 31.99 μeq/l.; Station V 211.70 μeq/l.; i.e. 4.9, 2.9, 1.8, 1.4 and 9.3 mg/l. CO_2 respectively. Because of the errors inherent in the silica determination ($\pm$ 5%) the CO_2 concentration calculations are accurate only within these limits. CO_2 concentrations are naturally related to HCO_3^- alkalinity, being highest at Stations I and V and lower at the other poorly buffered stations. The absence of any aquatic vegetation and presence of only slight algal growth meant that little photosynthetic modification of the pH-CO_2 status occurred and that this relationship was primarily determined by physical-chemical controls. At Station V, an increase in CO_2 concentration from organic decomposition processes was noted, but as will be seen in the section on oxygen, these conditions were not severe.

The ratio [Ca + Mg] : [Na + K] was discussed by JOHNSON (1967a) for the soft waters of Southern Malaya and Singapore and rejected as an index of trophicity as used by ZAFAR (1959). However, this ratio does provide information on the parent formations of the watershed and the

114

extent of their influence on the waters draining them. Hydrochemical dependence on the 'landscape' (SIOLI 1968a) for the Amazon area with its large lithologic blocks has been well delimited (SIOLI et al. 1969). On a smaller scale, these interrelationships can be seen for the Gombak. Some salient points have been described already in the discussion of individual elements in the waters of particular tributaries (e.g. calcium and magnesium). Overall conditions in the river are considered here. 'Hardness', [Ca + Mg], always less than 0.1 meq/l., decreased while [Na + K] increased passing downstream from Stations I to V as already discussed, and the ratio between them dropped from 0.78 to 0.22 (Table 36). The mean ratios found by DOUGLAS (1968b) for Australian waters were 0.7 from metamorphic and 0.5 (range 0.2–1.5) from granitic lithologies, with higher ratios, up to 1.9, from pure sedimentary formations. The data obtained for the Gombak are typical of metamorphic headwaters grading into a granite area (cf. also CORBEL (1964) for similar ratios from Ivory Coast rivers). The extreme calcium poverty of these waters is further emphasized by the Ca:Mg ratio, which gradually increased from 0.34 to 0.78. This magnesium excess is also common in Amazonia (KLINGE and OHLE 1964) and may be partially explained by the preferential uptake of bicarbonate-bound calcium over magnesium salts by humus colloids in the soil, leading to a relative surplus of magnesium. In the metamorphic facies magnesium concentration is much greater than calcium concentration, but calcium is dominant in granites so the downstream increase in the ratio was predictable. The overall decline and deficiency of 'hardness' has been discussed previously. Both ions increased in abundance at Station V, but calcium more than magnesium as the Batu Caves limestones, although dolomitic, do not approach the theoretical composition of dolomite (Table 32). As the total sodium plus potassium concentration increased, the Na:K ratio progressively declined down the river from 2.8 at Station I to 1.7 at Station V. Sodium, with respect to potassium, is comparatively more abundant in granites than in schists and quartzites, but a higher leachate from plants and greater mobility in the soil accounted for the relative increase of potassium in the surface waters and the decline in the ratio.

The Na:Cl ratios were always much greater than any recorded by DOUGLAS or estimated from the data of KOBAYASHI or JOHNSON and were roughly comparable to the ratio found in the rain as suggested by ERIKSSON (1956). The whole catchment is poor in chloride because the rocks contain only very small amounts and the input via precipitation, largely of the convectional type, is low compared to that imported by off-sea rain near the coast in Queensland, southern Thailand and Singapore. The gradual decrease in the ratio downstream is attributed to increased chloride from human and animal sources.

Ion Budgets: Partial budgets for the principal dissolved cations, Na, K, Ca, Mg and for NO_3-N, Cl and SiO_2-Si were calculated for the 8-month

Table 37. Ion budgets for the Undisturbed Upper Catchment (kg/sq. km for 8-month period, April–November 1969)

Element	Input	Output	Net loss or gain
Ca	805.7	283.8	+ 521.9
Mg	110.3	399.2	— 288.9
Na	3334.3	977.8	+2356.5
K	356.1	1084.7	— 728.6
NO_3-N	155.4	144.3	+ 11.1
Cl	229.1	226.5	+ 2.6
SiO_2-Si	1507.5	3534.1	—2026.6

period, April–November 1969, for which chemical analyses on rainfall were available. Although not a complete year, this period covers the main seasons. Only the Upper Catchment could be assessed as the lower inhabited area receives significant, but undefinable amounts of ion in the form of wastes from the processing of allochthonous products, rubber, food, etc.. The polluting effects of the population would obscure any input-output relationships caused by land use and alteration of the cover on the catchment. Input was calculated from the product of the ionic concentration of the rain and the volume of the precipitation over the appropriate period, output from the product of the volume of discharge at the B.P. weir for the corresponding period and the ionic concentration as determined at Station II. There was probably little difference in river water quality between these sites as both are in the undisturbed forest reserve, only about 2 km apart. The data are given in Table 37. No estimate of bicarbonate was possible for reasons already given and the presence or concentrations of the other ions was too variable to justify extrapolation from the chemical analyses at four-week intervals. The relatively large gains for Ca and Na (5.2 and 23.6 kg/ha/8 months respectively) are difficult to explain in the light of the net losses found by LIKENS et al. (1970) for Hubbard Brook. The calcium originates as dust and is retained by the otherwise calcium-poor watershed, presumably becoming incorporated into the forest biomass. Much of the sodium entered the watershed during the second and third periods (May, June, July) when precipitation concentrations were high (4.25 and 2.62 ppm) and the largest outputs, still smaller than inputs, occurred during the third and fourth periods. If these two anomalously high periods are disregarded, a net input is still indicated. Overall, this implies continuous uptake and absorption of sodium by the soils and the plant cover. Mg and K had small net losses of 2.9 and 7.3 kg/ha/8 months respectively. The watershed gained in small amounts both anions calculated, Cl 0.03 and NO_3-N 0.11 kg/ha/8 months. JOHNSON et al. (1969) suggested that the

116

mean chloride concentration in stream water should be approximately equal to mean precipitation concentration after adjustment for evaporation, and this would seem to apply in the Gombak. The large net loss of SiO_2-Si, much of it probably as $H_3SiO_4^-$ as discussed earlier, is the direct result of leaching from the geologic substrata.

Enrichment: Chemical enrichment of the river as a result of anthropogenic activities was evident, particularly in the Lower Zone. Under natural forest conditions, dilution and increased ionic input from tributaries would be expected to balance, leaving the river with an almost constant chemical composition. However, as seen, chloride, phosphate, nitrate, nitrite, iron, sodium and potassium concentrations increased with distance from the forest edge, with only calcium, magnesium and sulfate decreasing. Most of these substances showed only modest gains to Station IV and then a larger increment at Station V. A number of studies (see SIOLI 1969b, LIKENS et al. 1970 for review) have shown that agricultural use of forest land, particularly by drastic cut and burn methods, leads to severe losses of nutrient compounds by increasing the rate of leaching. This breakdown of the conservative plant-soil recycling system of the climax tropical rain forest in an area of low nutrient reserves in the substrate and abundant, very dilute rainfall may have serious repercussions. The increase in suspensoids as the soil loses its stabilizing cover and the resulting fluctuations in temperature and discharge pattern have already been discussed. Much of the increase in the Middle and Lower Gombak Zones was attributed to the removal of the natural vegetation for plantation, orchard and casual farming. Associated with this agricultural land use is the increasing application of fertilizers and chemical control agents. The rice-growing areas are enriched both chemically and with manure so that the return water from irrigation carries part of these nutrients as well as normal solutes concentrated by soil desalination and high evaporation. Copper and arsenic compounds used as fungicides and weed-killers and various chemical insecticides are also added.

The increases in nutrients between Stations IV and V were attributed to the effluents of a progressively larger riparian population. The transitory phosphate load from washing detergents has been described, but a much larger contribution of this and nitrates, chlorides and electrolytes had its origin in the mineralization of human wastes, both in side ditches and the main river. The chemical enrichment at Station V was low compared to conventional pollution standards, in reality only 'fertilization', but considerable when compared to nutrient levels in the river above this point.

The magnitude of the dissolved chemical load carried by some inflows was evident from the data for Tributary K. Below Station V, another small stream, initially running through a housing estate, then draining a derelict area, enters the river (the lowest tributary from the east on the map). This had the following composition on 25 September 1969 in ppm:

titrated alkalinity as HCO_3 210.3; Ca 14.2; Mg 7.2; Na 9.3; K 14.2; Fe 0.9; Mn 1.7; Cl 8.2; PO_4–P 13.8; SO_4 155? C.O.D. 33.6; pH 6.87; conductivity at 25 °C 976.7 µmho/cm (interferences prevented determination of NO_3 and NH_4). A number of inflows in the extreme Lower Zone draining densely-populated areas had similar grey-black colour (FeS), highly malodorous water and probably similar chemical composition. Such pollution, even though in comparatively small volumes (at low discharge on that date approximately 0.1 cumec into approximately 3 cumec in the main river) severely affected the chemical quality of the river. Biological changes were immediately apparent and will be discussed later.

NORRIS and CHARLTON (1962), in their pollution survey of the lower river, routinely sampled just below the D.I.D. gauging site and obtained data as follows: C.O.D. 1.7–19.0, mean 5.6 ppm; B.O.D. 1.4–9.6, mean 4.7 ppm; nitrate less than 0.2 ppm and ammonia up to 1.3 ppm. Unfortunately, no analyses for sulfate or phosphate were made and their data for chloride appears unreliable; therefore, no significant comparison was possible. Several 'spot' determinations at this site indicated equivalent or lower concentrations for roughly comparable discharge. Specific conductance was considerably elevated (44.5–136, mean 95 µmho/cm) with the highest values at low discharge when tributary inflow contributions were at a maximum. Comparison with NORRIS and CHARLTON's data for a site located adjacent to Station V indicated some deterioration in water quality since 1958/59. The mean B.O.D. concentration increased from 2.0 to 3.4 ppm and C.O.D. from 2.1 to 2.8 ppm with much higher maxima than previously recorded. However, the dissolved nutritional load had not increased significantly, so that enrichment was by undigested organic wastes and factory effluents rich in reduced salts.

Oxygen: The dissolved oxygen conditions in the river are summarized in Table 31 and expressed as per cent saturation in Fig. 34. The rough, boulder-strewn bed and comparatively low organic load ensured near saturation levels at Stations I, II, III and IV. As turbulence was lost and temperature increased downriver, a slight decrease in total dissolved oxygen was found. At the forest stations, no appreciable algal growth develops and oxygen relationships are controlled by physical factors (see also SIOLI 1967a). After the river becomes phototrophic with the removal of the tree cover, some effects of the seasonally rich periphyton growth are evident. At Station III, for example, during the stable flow period February-March when a good coating of diatoms developed in the riffles, slightly elevated per cent saturation levels, relative to Station II, were obtained, but these were still below 100%. Station IV showed a minor reduction in mean oxygen content, but the full effects of the increasing organic load at this point were moderated by considerable algal development in riffle areas that restored some of the oxygen loss. At Station V, the mean oxygen concentration was less than 6 mg/l.,

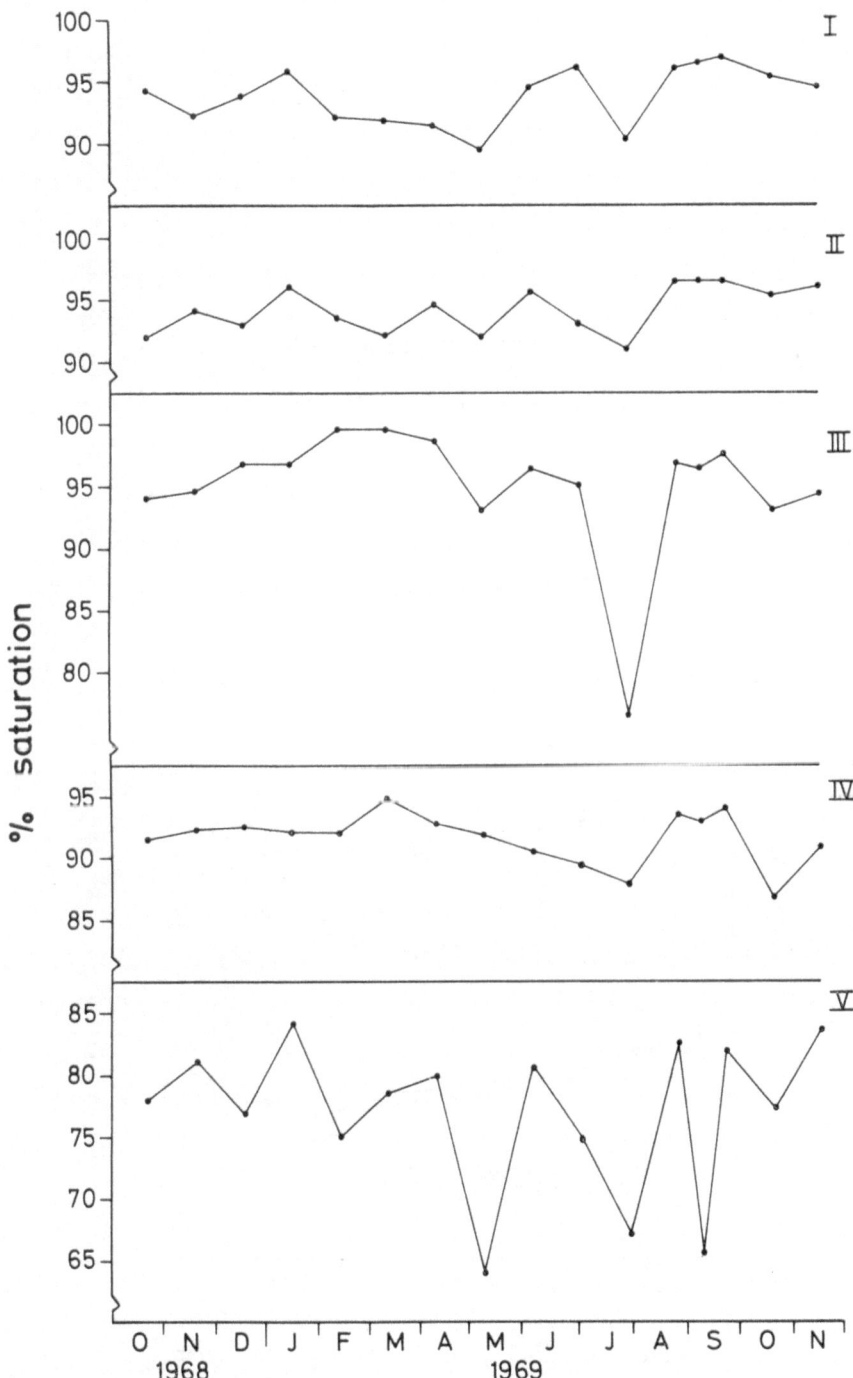

Fig. 34. Dissolved oxygen as percent saturation (spot readings taken on four-weekly sampling dates).

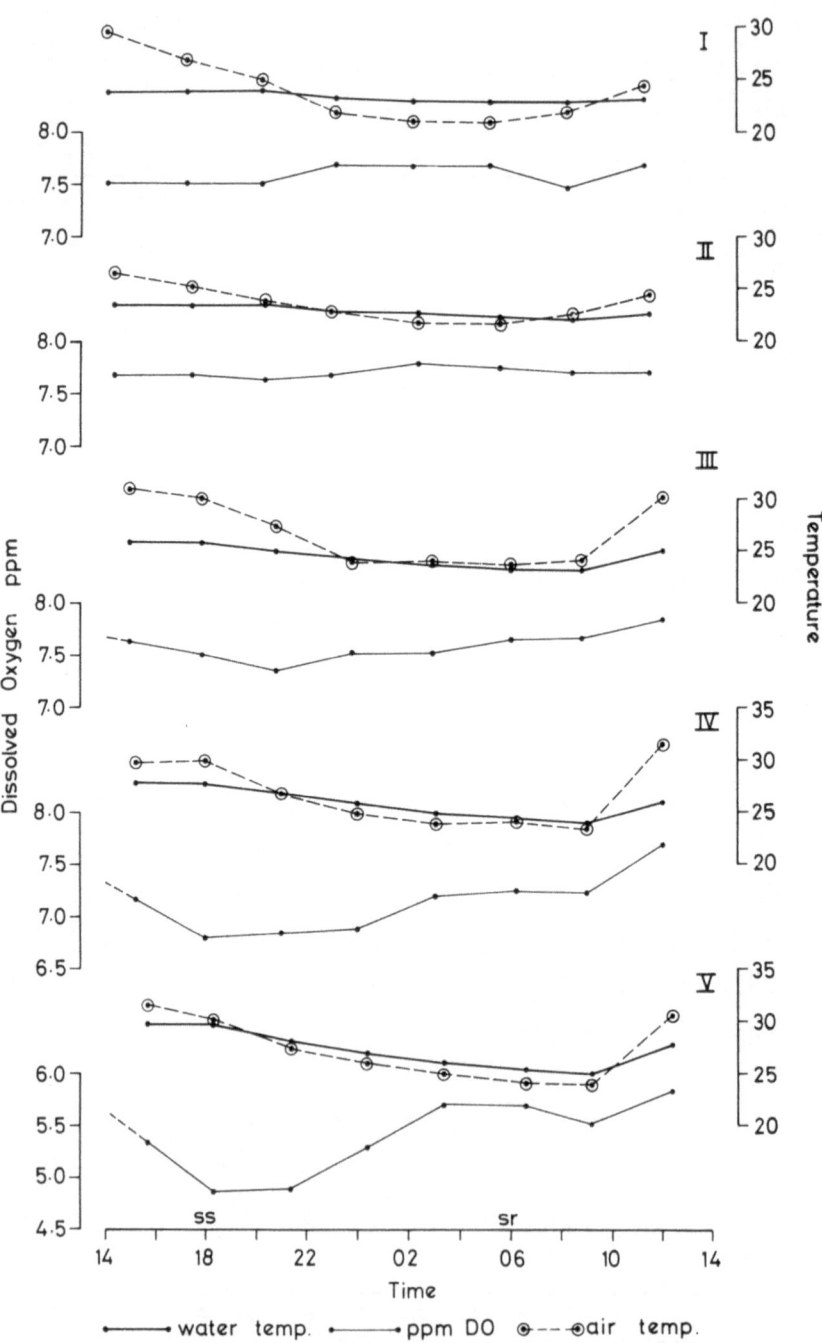

Fig. 35. Diurnal oxygen curves for Stations I–V on 28 July 1969 at modal discharge and with no interference from precipitation (oxygen concentrations in mg/l.).

giving about 78 % saturation at the high mean temperature at that station. The increasingly heavy organic load, reduction in current velocity and reaerating turbulence coupled with little algal growth, because of heavy silt load and low light penetration, were responsible for this decreased level. Periodic sampling at a few typical sites at each station, as was necessitated by the nature of this survey, is of limited use in assessing the gaseous conditions of a river as the extreme depletion concentrations are unlikely to be sampled. Variations between months are really irrelevant; the differences seen were the result of short-term (often of only a few hours duration) conditions immediately antecedent to the test period. The severe fluctuations at Stations V resulted from local effluent inflows (caused by spates introducing large loads of reduced materials that accumulate in the side ditches) that consumed oxygen. The reduction in oxygen content at all stations on 31 July 1969 came during a period of low flows. A short, sudden storm washed out many pockets of decomposing materials, accumulated leaves, frass, etc., radically increased the organic load and caused low oxygen concentrations right down the river (cf. SCHNELLER 1955, SLACK 1955, MINCKLEY 1963).

Diurnal profiles were made on a number of occasions. The results from 28 July 1969, at low flow with no rain or other interferences, are recorded in Fig. 35. Note: the sampling times are not synchronous; the differences were the travel time between successive stations. The paucity of the algal flora at Stations I and II was indicated by the absence of a respiratory load at night. The oxygen concentration in fact rose at both stations in response to the slight decrease in water temperature. At Stations III and IV, the algal mat on the riffles produced the classical effect (BUTCHER et al. 1930, ODUM 1956) of causing a decrease in oxygen concentration immediately after sunset. At both stations however, increased gas absorption and transfer capacity with the nocturnal decrease in temperature overrode the respiratory requirements and the oxygen content began to rise after 01:00 h. This rapid reaeration rate was attributed to the high surface to volume ratio and turbulence of the river (EDWARDS et al. 1961). Station V, with its considerable organic load, showed a moderate drop in oxygen content after sunset and rapid recovery with temperature fall, as at III and IV. Such a recovery is remarkable as the flow pattern at this station is relatively smooth, the bottom normally has a large fauna of macro-invertebrates (worms particularly) and a considerable hetero-trophic flora must be present in the organic sediments. However, as discussed by EDWARDS (1962) and EDWARDS and ROLLEY (1965), per-haps no correlation exists between the organic content and oxygen con-sumption of the sediments. The shape of the depression curves at Stations IV and V indicates the possible presence of a pollutant, with high oxygen demand, for a limited time between 14:00 and 18:00 h. After recovery from this, the river assumes a normal oxygen-temperature dependent relationship. This pattern (see SCHMASSMAN 1951) could result, for example, from flushing of rubber factory settling tanks at the end of the day's work. There are no pollution inflows above Station IV. A further possibility is an effluent from the rice-fields with a high oxygen demand, but this would require further investigation. VAAS and SACHLAN (1956) reported a similar occurrence in the polluted Tjibunut River in Bandung, Java, where oxygen tension was at a maximum at night when contamination was minimal and decreased following the early morning input of sewage and garbage. An oxygen deficit was never found anywhere on the main river, although several polluted tributaries, described earlier, were probably anaerobic. Oxygen analysis was not made at these locations because of the large concentrations of interfering substances, particularly organic matter and reduced sulfur.

An attempt to determine whether stratification occurred in the deeper pools was made using a micro-level syringe sampler after FOX and WINGFIELD (1938) and ERIKSEN (1963), but this was not completely successful. In some deeper pools in the forest zone, with thick layers of decomposing leaves, a decrease of 0.3 mg/l. was found on several occasions in the water immediately above the bottom. This may have represented a real decrease as the temperature was 0.5° lower at the bottom, but the imprecision of the method precluded the drawing of any definite conclusions. Readings taken at the sediment-water interface in slow-flowing reaches at Station V were 0.6–1.4 mg/l. lower than the dissolved oxygen concentration at the surface; this was attributed to high oxygen consumption in the organic-rich muds by the dense oligochaete and microconsumer fauna. Most 'pool' areas were not stagnant, having slow to moderate currents running continuously, so that conditions were not conducive to the establishment of oxygen gradients except in layers immediately adjacent to the bottom and behind obstructions where deadwater occurred. In flood-pools or the isolated backwater areas, oxygen relationships would have been entirely different, depending on temperature, algal development and the nature of the bottom debris. These habitats were not included in the present study except as potential sources of plankton.

Chemical Oxygen Demand: The $KMnO_4$ acid digestion under the standard conditions employed in this study (100°C for 30 min) does not measure organic carbon quantitatively (SLATER 1954), but provides a useful comparative parameter. The value obtained includes the immediate chemical demand of reduced compounds and the amount of oxygen absorbed by a certain fraction of the organic material present. These data are given in Fig. 36. Mean and extreme values increased sequentially with station (see Table 31) and were similar to those obtained by NORRIS and CHARLTON (1962) for the Gombak in 1958/59. Differences between stations reflect progressive fragmentation and, hence, oxidizable surface area of the leaf and frass organic load, and the addition to the river of waste products from the urban areas near the lower stations. These

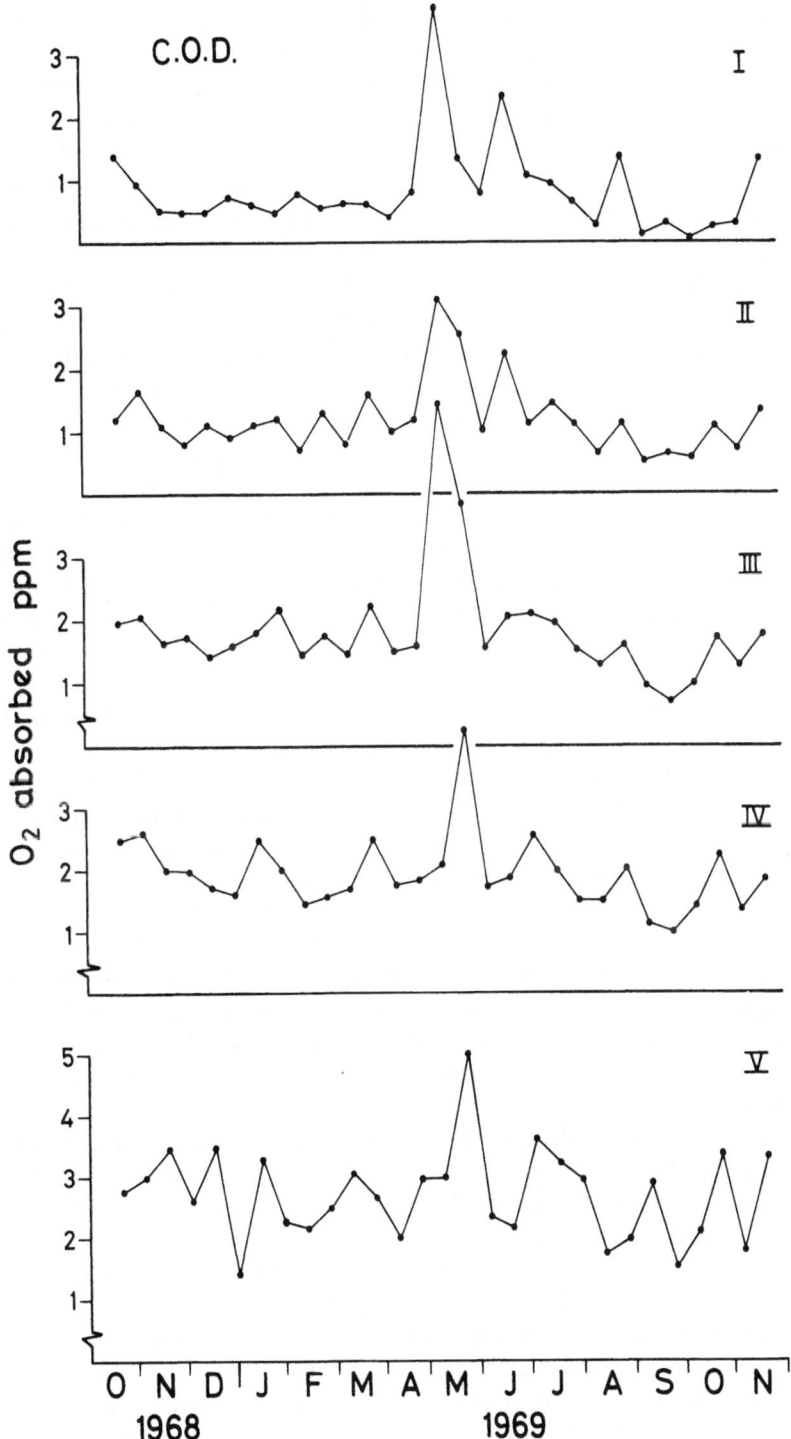

Fig. 36. C.O.D. expressed as ppm oxygen absorbed from KMnO₄.

wastes contain considerable oxygen demand (WEIBEL et al. 1964), as does the run-off from agriculturally-disturbed land; this load normally was a factor only on rising stages with sheet run-off and ditch flushing. The spot-sampling method that had to be followed here rarely sampled these 'optimal' conditions so that the absolute C.O.D.s, in spite of increased volume, were probably considerably higher than any value obtained. The variations between dates are almost certainly the result of discharge-dilution factors. The peak in May (interestingly occurring one sampling period earlier, 8 May, at Stations I, II and III than at the lower two stations, 23 May, indicating a succession of the causative products) came during a sustained high discharge period. HARRISON and ELSWORTH (1958) associated high absorbed oxygen values with turbidity and load of soil organics and this probably applies in the Gombak also. By most standards, these C.O.D. levels are low, disputing the assumption that tropical forest streams carry very large organic loads.

JUDAY and BIRGE (1932) estimated $KMnO_4$ oxidation efficiency at about 40 %. K_2CrO_4 oxidation of duplicate series of samples on three occasions, following the technique of MACIOLEK (1962), gave variable efficiencies at different stations at different times. This variability probably reflected the nature of the inorganic rather than the organic fractions responsible for the oxygen demand. For example, on 3 July 1969 at modal flow, the $KMnO_4$ average values for Stations I to V were 1.06, 1.12, 2.07, 2.56, 3.60 mg/l. O_2 absorbed, while the corresponding dichromate figures were 6.11, 5.30, 7.33, 13.52, 19.70, i.e. with efficiencies of 17.3, 21.1, 28.2, 18.9, 18.3 %. If the mean $KMnO_4$ values from Table 31 are modified using these conversion factors (and adding 15 % to accommodate MACIOLEK's 85 % efficiency rating to his technique) the results are still considerably lower than the 'total organic' determined by ignition (Table 25). This indicates that a part of the organic fraction is very resistant to oxidation and therefore perhaps unavailable to the microconsumers as a short-term decomposition substrate. Despite these deficiencies, the permanganate C.O.D. test, under rigidly standarized reaction conditions, gives reasonably comparable values if the load is chemically homogeneous.

The normal procedure of using filtered and unfiltered water samples was not applicable. KJENSMO (1965) discussed the fallibility of filtration in oxygen demand determinations and gave results showing increases after filtration because of organic materials originating in the filter pads. To avoid this problem, sintered glass filters (pore size approx. 5μ) were used after acid-washing and thorough rinsing in distilled water. The results after filtration through these were often higher than from unfiltered water, so the procedure was abandoned. The most probable explanation of this is that much of the organic matter exists in large, loosely-associated colloidal masses or as assemblages of bacteria and fungi around an organic particulate core. On filtration, these fragment, increasing the surface area exposed to oxidation, and giving what amounts to false oxygen demand; under natural conditions only the outside surfaces would be actively taking up oxygen.

Turbidity: As normally defined in ppm SiO_2, turbidity is largely a measure of inorganic load only. Where a considerable organic particulate fraction is present, as at the lower stations, this should be included in the estimate if possible. Some ratio of volume of load to discharge volume would be the ideal parameter to determine, but this was impractical. The total load in g/cu.m, already reported under erosional load (Table 25), included the organics, but underestimated their importance on a volume basis. WANG and BRABEC (1969), in a discussion of conditions in the Illinois River, related turbidity directly to the concentrations of particulate forms of silicon, phosphorus and iron and suggested that a stoichiometric relationship $[Fe] = [Si] + [P]$ normally exists. This does not appear to hold for the Gombak. The iron, or even iron plus aluminium, particulate fraction in the soils, which constitutes the largest source of turbidity, is small in relation to the high silica fraction and the contribution of the phosphorus complexes is vanishingly small. As discussed earlier, normal turbidities are low in the forest zone and increase with abusive land use downstream. At elevated flood discharges the turbidity may reach extremely high values for short periods, depending on the nature of the area drained, e.g. the more than 10,000 ppm in Tributary E, eroding off the highway.

4. Organic environment

The critical role of allochthonous detritus, variously defined by a number of authors, but perhaps most adequately by STRICKLAND (1960) as 'non-reproductive organic matter', in the trophic structure of aquatic ecosystems has become increasingly evident in recent investigations. ODUM (1957a, b), HYNES (1961, 1963) and SZCZEPANSKI (1965) have noted the importance of accumulation and decomposition of leaves from riparian vegetation and LEVANIDOV (1949), NELSON and SCOTT (1962) and MINCKLEY (1963) reviewed the position of detritus as food for aquatic invertebrates. The importance of terrestrial photosynthesis as a source of organic materials, particularly in the context of tropical rain forest streams where primary production is negligible (as in the forest zone of the Gombak) has been discussed by HULOT (1951) for Congo rivers and more fully by FITTKAU (1964). In Amazonia, and probably in most acid waters, little decomposition of leaf material occurs so that flowers, pollen and fruits become the major source of food in these rivers. The direct nutritional contribution of imported organics, especially leaves, is a matter of conjecture as many nutrients are translocated by the plant during senescence and leaf litter, particularly in the tropics, is low in nutrients (KLINGE 1968). However, DARNELL (1961) and others subsequently (see KAUSHIK and HYNES 1968) have demonstrated that the remaining materials, and particularly the protein component of leaves, serve as a substrate for bacterial and protozoan microconsumers which constitute a major food source for invertebrates (MINSHALL 1967, 1968, HYNES and KAUSHIK 1969). Several authors (e.g. MANN 1964, 1965) have described and assessed direct utilization of plant detritus by some fish. The first order relationship between faunal distribution and plant detritus in streams has been thoroughly analysed by EGGLISHAW (1964, 1965, 1968, 1969), BUSCEMI (1966) and MACKAY (1969), the latter describing the role of leaf accumulation 'packets' as both substrate and nutrient source. The dynamics of utilization of leaf materials in lotic habitats has received less attention. MATHEWS (1967), EGGLISHAW (1968), KAUSHIK and HYNES (op. cit.) and MATHEWS and KOWAL-CZEWSKI (1969) have assessed the effects of water quality on the decomposition of leaf detritus with the general conclusion that nutrient levels, particularly nitrogen content, are critical to the efficiency of break-down and utilization processes. Leachates from leaves contribute considerable concentrations of inorganic and organic materials to the water (10–25 % of initial dry weight in the first three days). These dissolved fractions are particularly important in the impoverished waters found in tropical streams and there is growing evidence that these nutrients are rapidly incorporated, either *de novo* through physical phenomena or by microbial flocculation activity, into particulate organic components that are a major source of food for filter feeders as well as the microconsumers (SEKI et al. 1969, HYNES 1970).

In an attempt to assess the significance of detritus in the Gombak, both as a food source and as a substrate, the input of allochthonous materials, relative concentrations of dissolved and suspended organic load and the proportion of organic material in the bottom sediments were measured.

4.1. METHODS

4.1.1. Organic content of the water

a. Dissolved and total organic load were determined as part of the erosional load and methods were described in that section.

b. In place of the standard Biochemical Oxygen Demand test (5-day at 20 °C) a 3-day at 27 °C incubation period was used as discussed by ANONYMOUS (1953). Facilities for refrigeration to 20 °C were not available and the 27 °C approximated more closely to the conditions in the river (NORRIS and CHARLTON (1962) reported comparable values from the two techniques). To measure the approximate magnitude of total usable carbonaceous load (PHELPS 1944, STRAŠKRABOVÁ-PROKEŠOVÁ 1966), 24-day at 27 °C incubations of a series of tests from each station were made, with interim readings at 3, 6 and 12 days. At Stations III, IV and V, initial aeration and dilution up to ×6 with water taken from Station II (using appropriate controls) were necessary.

c. An estimate of gross organic supply to the river was made using the drift nets as described for bed-load (see Erosional load, b). At Stations I and II, monthly 8 × 3 h samples were taken that collected drifting organic components > 165 μ. Fruit, sticks, leaves, insect cuttings and frass and faunal components were assessed; only the larger branches and logs could not be measured. The percentage of inorganic debris was determined by ignition of representative samples at 600 °C after drying at 60 °C for three days. No regular sampling could be made at or below Station III because of human interference with the nets, various activities in the river and rapid blocking of the nets by bed-load. Occasional samples of short duration were taken at the lower stations to assess the drifting organic load.

4.1.2. Sedimentary organics

The volume and biomass of gross organic material present in the bottom sediments were measured at 4-week intervals using the benthos sampler (used earlier for substrate collection and described in the faunal section) with a net of pore size 165 μ. Two (31.5 × 31.5 × 10) cu.cm samples were collected and pooled from eroding (r) and depositional (s) areas separately at each station. The organic fraction was repeatedly elutriated to eliminate large inorganic particles and the wet volume and dry weight after three days at 60 °C were measured. The ashable component for representative samples was found after ignition at 600 °C for one hour.

4.1.3. Calorimetry studies

Caloric determinations were made on the organic materials collected in 1 and 2 above. The fractions were stored over a silica gel desiccant, run through a culinary meat mincer and then fine-milled twice in a Culatti ricemill to ensure uniform particle size and relative homogeneity. Samples of approximately 0.5 g were burnt in a Gallenkamp Ballistic Bomb Calorimeter and mean kg cal/g found from a minimum of six replicate determinations on each fraction.

4.2. RESULTS AND COMMENTS

4.2.1. Organic content of the water

a. The dissolved and total organic load data are summarized in Table 25. The dissolved fraction at all stations in the Gombak constituted the

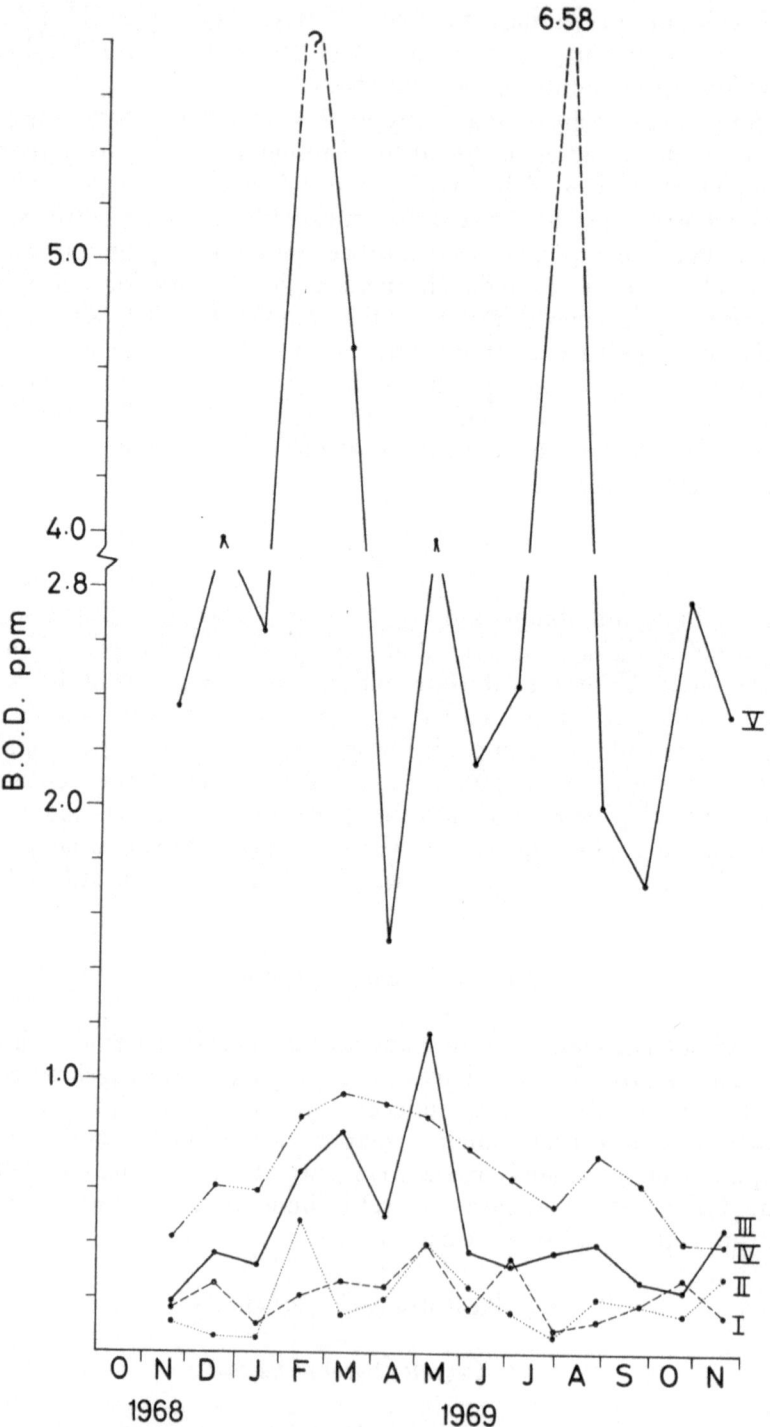

Fig. 37. B.O.D. expressed as ppm for Stations I–V.

major part of the total load, but the latter as measured may be an underestimate. In most published reports this is the case, as the technique of collecting a small water sample for analysis cannot include the occasional larger suspended fragments (leaves, twigs, fruits, etc.) that are, by mass, an important component of the load. These large fragments are dealt with later. BAALSRUD and HENRICKSON (1964) have discussed at length this and other problems involved in sampling suspended matter in lotic habitats.

Stations I to IV were similar in their mean dissolved fraction indicating little recruitment or break-down after the river leaves the forest. The total organics loads at Stations I and II were almost identical and of the same order as the 0.01 ± 0.01 g/l. found by SIOLI and KLINGE (1962) for forest locations in the Amazon basin. The slightly higher value at I was probably a function of the closed canopy supplying proportionately more insect frass directly. The increase in the suspended component at Stations III and IV resulted from continuous recruitment of agricultural rubbish along the river. A considerable part of this at Station IV was semi-burnt, hence less dense, woody materials from the clearing and cultivation of riparian gardens. Below Station IV, irrigation return water from the paddy area and inflow of various urban effluents added an average 6 ppm to the dissolved and 4 ppm to the total load.

Seasonal variations in organic load were not pronounced (Fig. 23 and 24). The dissolved component was reasonably uniform except at Station V where its concentration tended to vary inversely with discharge. The inflow contributions from side channels were always significant, so that during high mean discharge months, October–November, April–May, concentrations were minimal, with maxima occurring during February–March and August–September when dilution was low. Total organic content showed little fluctuation as its main sources, the fall of leaves in the forest and agricultural clearing, were more or less continuous processes. The first four points on these curves were obtained at 180 °C (not ignited) and so probably represent underestimates, while the subsequent determinations, treated at 600 °C for one hour, may be slightly too large because of votalization of inorganic components.

b. The B.O.D. test results are plotted in Fig. 37. Stations I and II had low values throughout the year, less than 0.5 ppm. Stations III and IV generally had higher oxygen-consumed readings than the forest stations with peak loads, still less than 1.2 ppm, in the 'dry' seasons. Station V showed relatively high fluctuating B.O.D., typical of sewage contamination in a variable-flow river. KLEIN (1957), HUET (1958) and HYNES (1960), *inter alia*, have discussed the ecological significance of the 5-day B.O.D. test and quote consensus values of 1–2 ppm or less for clean water, 4–7 ppm for slight to moderate pollution and more than 8 ppm for severe pollution. Much higher values are often reported in rivers with mixed effluent, but were not found in the Gombak at the stations sampled. The

B.O.D. at the D.I.D. gauging site was slightly higher than at Station V, mean 4.7 ppm and maximum approximately 10 ppm (NORRIS and CHARLTON 1962), but heavy pollution from oils, tin tailings and other agents made routine testing at that site difficult. On two dates, 13 February and 31 July, there was a potential oxygen 'debit' (dissolved oxygen less than 3-day B.O.D.) at Station V, but the conditions causing this rarely persisted for more than two or three days; reaeration was relatively efficient so that anaerobic conditions did not ensue for the entire water mass. Local deficits, particularly in the bottom sediments, probably occurred under these conditions, but the lowest recorded dissolved oxygen was 3.4 ppm.

Raw sewage generally has a B.O.D. of approximately 200 ppm (WIEBEL et al. 1964), but industrial effluents have variable demand, depending on their susceptibility to microbial degradation. Much of the oxygen demand of the urban run-off was normally satisfied in the side drains before it entered the river. Dense Cyanophyta, and occasionally *Cladophora* mats, choked most ditches, indicating rapid break-down and mineralization. During periods of frequent rain, flushing of the drains before oxidation was complete dumped considerable quantities of wastes in the river. Most of this was sufficiently diluted to prevent serious deoxygenation, so that high B.O.D. was usually associated with low main river discharges and short local rain storms that carried wastes into the system. In addition, the oxygen demand of the run-off from house compounds and streets that enters the river directly must be met fully by the river. WIEBEL et al. (op. cit.) have recorded a range of 100–285 ppm for such run-off water in temperate latitude cities; with the more primitive waste disposal system of the Kuala Lumpur suburbs, this figure is likely to be much higher.

The differences in quality between the 'natural' organic load of the river above Station IV and the man-made effluents near Station V are obvious when the B.O.D. values and the total load obtained by ignition are compared. The wood, leaf and soil components (mostly lignified) and various humic colloids are stable biologically and exert little immediate oxygen demand as decomposition is protracted. The urban waste contributions, on the other hand, are readily available as bacterial substrates.

Certain short-comings of the test may have affected the results. At the upper stations, the microorganisms needed to break down the organic load may not have been present in sufficient numbers to effect a useful test; there may have been insufficient nutrients in the water to enable development of this flora (phosphate could very well be limiting at Stations I, II and III). Contaminants in the wastes near Station V may have adversely affected the activity and sensitivity of the B.O.D. test. Fungicides, weedkillers and insecticides from rubber plantations and paddy fields may have accounted for some of the variations in the B.O.D.

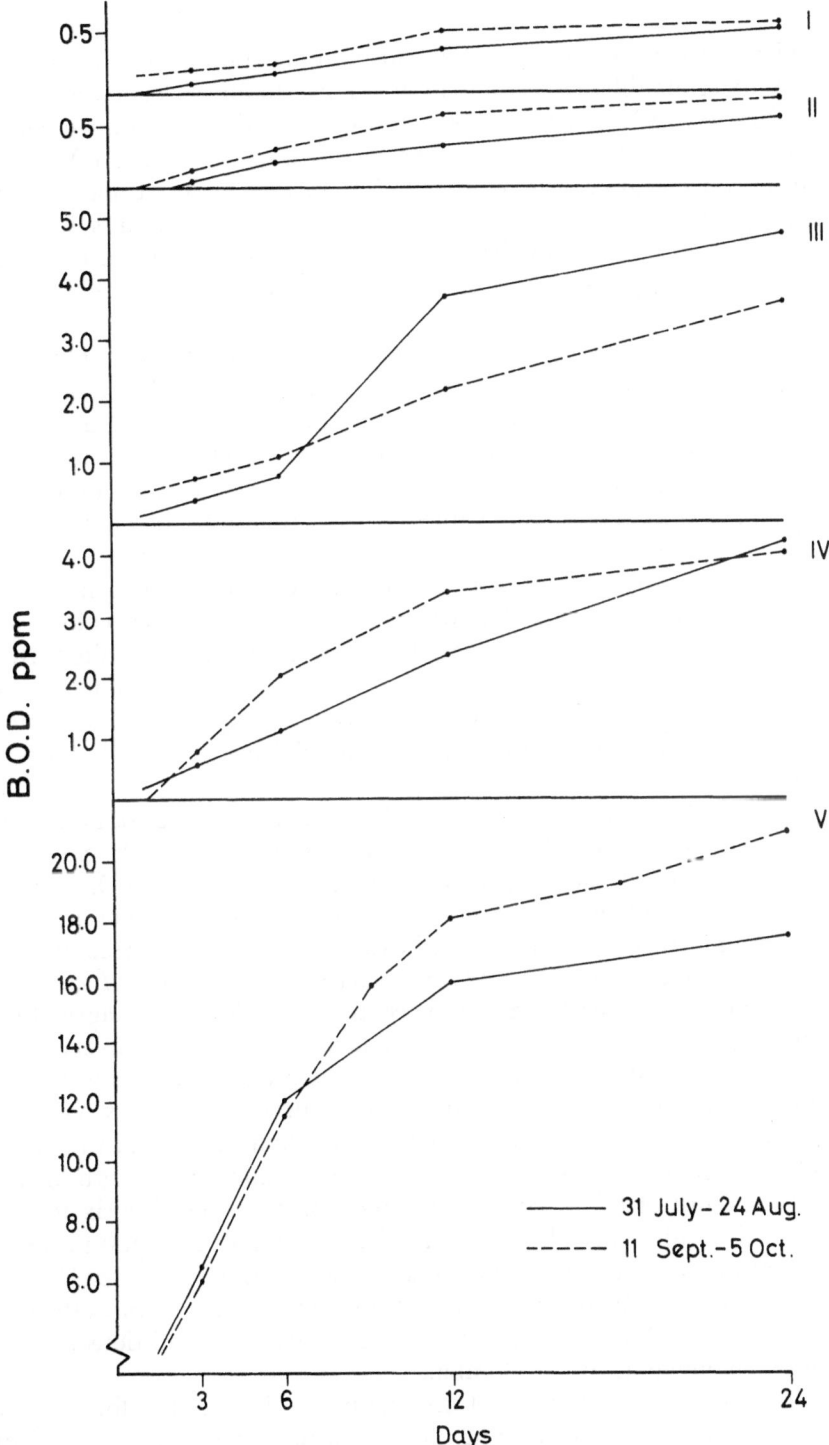

Fig. 38. 24-day B.O.D. test series run on 31 July–24 August and 11 September–
5 October 1969.

results; however, the time and extent of application of such compounds was indeterminable. Heavy metals are known to inhibit bacterial activity (cf. BUTCHER 1946, HYNES 1960) and copper and chromium were found as traces at the lower two stations on occasion. At Station V, where nitrate concentration was relatively high, further underestimation of the load may have resulted from reduction of nitrates as the dissolved concentration dropped to low levels. Sulfate ion, a similar potential oxygen donor, was not present in sufficient concentration to be considered important.

This test was measuring a different component of the organic load from that measured by the C.O.D. test discussed earlier. The relationship between the loads measured by the two methods has been used as an index of pollution by many workers (see HYNES 1960). Generally, waters with values of the ratio B.O.D./C.O.D. less than one are considered to be unpolluted and those with values greater than this to contain some organic contamination. From the composited data in Table 31 and the individual readings listed in Appendix A, the ratios at Stations I to IV were invariably less than 0.5 while at Station V, higher values were found, usually approximately equal to or greater than one. WISNIEWSKI (1958), working with homogeneous algal organic loads, found a linear relationship between 5-day B.O.D. and volatile suspended solids, but no similar relationship was found for the Gombak because of the biologically unreactive fractions in the solids.

The 'total' B.O.D. tests run over 24 days (Fig. 38) substantiated the conclusion that two different components were present. At Stations I and II, where all the load was forest products, low B.O.D. equal to only about 4% of the total organic load was found. The rate of oxygen consumption was slow with no indication of any rapidly-decomposable materials. At Stations III and IV, where small volumes of anthropic waste enter the river by direct defecation, in addition to agricultural products, considerably more of the total load (almost 25%) was accounted for over the 24-day incubation period and included a minor rapidly-utilized fraction. From the shape of the curve, the 24-day values at Station V (17–21 ppm) approached the real total B.O.D. of the sample and amounted to more than 75% of the organic load determined by ignition. Most of this decomposition took place in the first six to eight days with a subsequent levelling-off of the oxygen consumption rate. No attempt was made to determine the theoretical Ultimate Oxygen Demand of GAMESON and WHEATLAND (1958) discussed by LENHARD (1965), as this has little relevance in a system where chemical conditions are constantly changing as in the Gombak.

There are many criticisms of prolonging the B.O.D. test for extended periods, the most serious being the exhaustion of particular nutrients, the potential build-up of poisonous excretion products, particularly ammonia and methane, and a stabilization of the bacterial population

Table 38. Organic drift load (net mesh 165 μ)

Date	STATION I Leaf fraction > 1 cm (cc)	Detritus fraction < 1 cm (cc)	Total cc/day	STATION II Leaf fraction > 1 cm (cc)	Detritus fraction < 1 cm (cc)	Total cc/day
5–X–68	1103.1	3565.2	4668.3	22105.9	19544.8	41650.7
30–X–68	271.7	922.4	1194.1	12289.0	24761.2	37050.2
28–XI–68	597.1	2675.5	3272.6	23748.9	45209.1	68958.0
1–I–69	762.2	3472.0	4234.2	26893.8	60627.3	87521.1
24–I–69	1132.7	2627.0	3759.7	9883.5	25254.0	35137.5
23–II–69	1169.5	2336.6	3506.1	7683.6	32279.4	39963.0
19–III–69	1365.1	605.8	1970.9	23140.0	9221.1	32361.1
17–IV–69	846.6	1037.0	1883.6	24533.7	11954.6	36488.3
23–V–69	1485.0	6431.7	7916.7	18753.1	59339.9	78093.0
12–VI–69	670.7	1608.0	2278.7	13109.0	33456.4	46565.4
10–VII–69	1061.6	1435.1	2496.7	12071.1	20123.0	32194.1
7–VIII–69	655.9	927.0	1582.9	12341.5	19631.4	31972.9
3–IX–69	859.3	3234.4	4093.7	20365.2	52593.1	72958.3
2–X–69	822.2	2145.3	2967.5	30923.2	26775.3	57698.5
30–X–69	1595.6	5346.5	6942.1	14808.6	33562.8	48371.4
Σ 15 days	14398.3	38369.5	52767.8	272650.1	474333.4	746983.5
Estimated m³/yr	0.35	0.93	1.28	6.63	11.54	18.17
m³/km²/yr	0.07	0.20	0.27	0.24	0.42	0.66

Table 39. Percentage composition of drift load

Station	Leaf fraction > 1 cm	Detritus fraction < 1 cm
I	27.3	72.7
II	36.5	63.5
III	28.8	71.2
IV	25.2	74.8
V	3.3	96.7

at a sub-maximum level so that the rate of decomposition drops before the organic substrate is exhausted. These tests did indicate that much of the unusable load (about 12 ppm) at Stations I and II, which presumably was carried on down the river, was altered with time in the water and became biologically reactive by Station V, where abundant bacteria and increased nutrients were available.

c. Drift load of allochthonous organic materials at Stations I and II, given in Table 38, includes all fractions greater than 165 μ, except the very large fruit, logs, etc. that pass down the river periodically, but contribute comparatively little to the total organic load. The differences between sampling dates were caused largely by discharge variations as, within limits, the load was dependent on volume. In the Station I tributary, the large fragments probably all originated as direct leaf-fall into the stream bed as sheet flow was rarely competent to erode big leaves. Much of the frass and other small particles were transported downslope into the stream channel. The main river carried the organic contributions from the tributaries, supplemented by a large component of leaves that fall directly onto its surface. Under low flows, this load was almost entirely from contemporaneous recruitment, but as stage rose, the accumulations of leaves, detritus and flowers behind boulders and caught in the bankside roots were freed and added to the load. The percentage composition of the drifting organics is shown in Table 39. The 'leaf fraction' included fruits, flowers and twigs separated by hand, while the less than 1 cm component was the nondescript homogeneous remainder. The coarse component was most abundant at the forest stations but had nearly disappeared by Station V. The leaf load at this bottom station was almost exclusively bamboo and *Acacia* leaves from local sources, the forest-derived material having disintegrated or become waterlogged. This sharp decline in the large fragment load was probably a function of the length of time in the water, hence exposure to decomposition, and the decreased velocity that permitted settling-out in the deeper meander pools in the lower zone. The proportion of drifting detritus increased downstream both as a result of disintegration of bigger fragments and from large increments brought in off the land in irrigation and drain inflows.

4.2.2. Sedimentary organics

The weights of gross organic material present in the substrate at the various stations on monthly sampling dates are given in Table 40. These include all particles greater than 165 μ that were extracted from 2×0.01 m³ in each gross substrate type, and have been corrected for the inorganic fraction adhering after repeated washings. In the Gombak, algal growth, although locally important on the tops of riffle stones at Stations III and IV, was not extensive enough to produce much organic load. Mosses did not occur in the aquatic habitat and aquatic cryptogams were absent so that there was no significant autotrophic contribution to the bottom organics. Within the limitations of the standardized technique (e.g. mesh size and completeness of collection), these data compare the allochthonous organic material present in the substrate. The loss of the fine materials smaller than 165 μ was a serious omission as in the pools this fine fraction was deposited with the silts and formed a rich habitat for certain components of the benthic fauna. Various methods were tried to quantitatively sample the entire organic fraction, but the technique of BUSCEMI (1966) and other coring devices were not operable except in shallow depositional areas. This shortcoming might best have been solved with a freezing sampler (cf. EFFORD 1960), but facilities for this were not available.

The erratic differences in organic sediment content between months in both types of substrate were the result of local discharge variations prior to the sampling date. Even small spates removed much of the surface organic material from riffle areas, and large floods with strong eddy turbulence and pressure often denuded these erosional areas to below the 10 cm sampling depth. Pool areas, conversely, accumulated materials during small floods, but tended to become barren after the severe storms which completely changed their bottom layers. During low and modal discharge, especially if maintained for extended periods, considerable aggregation of organic fragments of all sizes occurred in slow-water areas, often as repeated layers of leaves-detritus-sand. On riffles, leaf fragments worked their way into the interstices and formed the nucleus of detritus-silt pockets between and underneath the stones.

The nature of the bottom organics changed downriver. At the forest stations, many whole, unaltered leaves, leaf cuttings and coarse detritus were commonly present. At Station III, skeletonization was generally advanced, probably through autolysis and microbial activity rather than invertebrate feeding. Some of the bottom material here was derived from local sources as the riparian vegetation fronting orchards and gardens was controlled and the cuttings were dumped in the river. On occasion, much of the organic fraction both here and at Stations IV was water-logged charred wood from land clearance. Particles at the lower Stations IV and V were generally much smaller; few leaves, other than bamboo, were found, but increasing proportions of alien particles (rubber processing

Table 40. Amounts of organic materials in the bottom substrates (ash-free dry weight, g per 0.02 cu.m)

Depositional areas	I	II	III	IV	V
19–X–68	7.02	3.39	12.60	10.67	2.39
14–XI–68	18.49	5.71	2.35	20.94	6.08
12–XII–68	6.86	1.54	28.63	35.58	5.23*
9–I–69	19.77	2.62	2.16	2.94	3.40
6–II–69	9.58	1.93	8.19	8.41	6.53
8–III–69	5.17	2.79	8.63	5.36	3.73
3–IV–69	10.91	7.58	5.64	3.87	3.43
1–V–69	6.61	2.40	2.11	8.25	6.91
29–V–69	5.58	5.67	2.16	5.52	4.44
26–VI–69	3.33	3.39	5.10	15.78	5.34
24–VII–69	4.87	1.67	3.92	16.14	4.96
21–VIII–69	2.25	2.10	6.23	4.28	4.63
18–IX–69	4.46	3.95	6.08	5.52	9.30
16–X–69	1.38	2.10	3.78	6.76	5.86
13–XI–69	4.66	2.02	4.31	4.33	2.87
mean	7.40	3.26	6.79	10.29	5.01
g/m² × 10 cm deep	37.00	16.30	33.95	51.45	25.05

Erosional areas	I	II	III	IV	V
19–X–68	3.83	3.92	11.77	4.15	4.13
14–XI–68	3.33	4.02	14.36	8.00	3.74
12–XII–68	2.58	3.42	17.35	7.51	2.75*
9–I–69	2.34	3.82	3.26	4.58	2.02
6–II–69	2.44	2.18	17.55	4.09	3.54
8–III–69	1.54	1.93	9.51	7.02	4.87
3–IV–69	7.41	6.34	2.99	8.31	2.51
1–V–69	4.03	4.61	5.65	4.64	4.18
29–V–69	4.18	4.22	8.11	6.11	7.23
26–VI–69	2.73	6.24	3.72	7.64	3.98
24–VII–69	3.58	6.10	6.85	5.19	5.66
21–VIII–69	2.24	5.75	5.32	4.46	6.24
18–IX–69	4.03	5.31	10.44	8.98	6.49
16–X–69	3.98	8.18	11.50	14.78	5.07
13–XI–69	3.83	2.38	12.83	4.70	5.66
mean	3.47	4.56	9.41	6.68	4.54
g/m² × 10 cm deep	17.35	22.80	47.05	33.40	22.70

* samples taken 19–XII because of high water on 12–XII

wastes, sawdust, cloth and paper remains) were evident, mixed with the fine detritus.

As seen in the section on driftload above, the potential supply of allochthonous organic materials to the sediments from this source was fairly constant throughout the year with slight seasonal variations in leaf-fall having little effect. The state of the bottom organics was, therefore, a factor of the time they had been exposed to decomposition, hence position down the river and, closely allied to this, the stability of the inorganic sediments in which they were trapped. The mean data for the riffle areas illustrate this. At the forest stations where velocities were considerable, about 20 g/m² organic fraction was found. At Station III with a semi-stabilized riffle area, this was more than doubled to 47 g/m², mostly small particles lodged deep in the interstitial spaces. The subsequent declines at IV and V were a result of the advanced level of physical break-up of the fragments of forest origin and the partially-decomposed state of most of the locally-derived organics.

There were no correlations between organic sediment content and either the interstitial pore space or the Md of the substrate. EGGLISHAW (1969) found an increase of between 0.2 and 0.5 g/m² for each 1 cm increase in stone length. However, the size of the surface interstices, as opposed to the total available pore volume throughout the substrate, affects the stability of the surface organics in the current. This may account for some of the decline in content downstream from Station III as the size of the riffle stones decreased at Stations IV and V, and the proportion of fine inorganics, that occupy the same spaces, increased.

Comparative data are scarce (see BUSCEMI 1966, MINSHALL 1967). NELSON and SCOTT (1962) collected 200–3000 mg/500 cm² from the Middle Oconee River riffles; EGGLISHAW (1968) obtained 2.9–11.4 g/m² from riffles in various Scottish streams using a net mesh of 670 μ and similar amounts using 200 μ nets (EGGLISHAW 1969); ULFSTRAND (1968a) found considerably higher quantities (mean about 90 g/m²) of coarse detritus filtered through 145 μ mesh in Lapland rivers. No data are yet available on the sedimentary organics of Asian tropical rivers.

4.2.3. Calorimetry studies

To obtain an estimate of the energy passing through the aquatic system from allochthonous sources, the kgcal/g of organic material were determined for the various organic fractions. These data are given in Table 41. The values obtained at Stations I and II for recently fallen leaves from a large number of species, 4.98 and 4.69 kgcal/g organic, compare well with the caloric content of 5.05–5.25 kgcal/g obtained for willow, oak and sycamore leaves by MATHEWS and KOWALCZEWSKI (1969). The energy content of the detritus fractions at Stations II to IV was not significantly different from that of the parent leaf materials,

Table 41. Mean caloric content of drifting and sedimentary organics

Drifting organics		kgcal/g material	percentage organic	kgcal/g organic
I	leaves	4.63	93.06	4.98
II	leaves	4.38	93.48	4.69
I	detritus	3.55	86.83	4.09
II	detritus	3.81	82.65	4.61
III	detritus	3.63	71.79	5.06
IV	detritus	3.35	72.34	4.63
V	detritus	3.55	66.46	5.34
Sedimentary organics				
Depositional areas	I	2.88	51.22	5.62
	II	0.84	42.93	1.96
	III	2.04	49.03	4.16
	IV	2.38	51.57	4.62
	V	1.39	37.33	3.72
Erosional areas	I	1.51	49.72	3.04
	II	2.14	49.59	4.32
	III	2.44	66.48	3.67
	IV	2.50	61.08	4.09
	V	2.01	49.18	4.09

indicating little change in composition with decay. The low value at Station I cannot be explained unless some unique chemical condition in this tributary accelerated the leaching of the soluble organics from the detritus particles. The apparent energy increase at Station V is in agreement with the findings of KAUSHIK and HYNES (1968) and MATHEWS and KOWALCZEWSKI (op. cit.) who found increases in both organic content and caloric content after an extended period in the river. The first authors associated this increase, particularly in protein content, with the development of a microflora, predominantly of hyphomycetes, and suggested that this was dependent on the nutrient quality of the water. In a subsequent paper (HYNES and KAUSHIK 1969), the nitrogen content of the water was indicated as the likely factor controlling the protein increase. At Station V, high nutrient concentrations were available and may have been responsible for accelerated saprobe growth and activity that increased the mean energy level. EGGLISHAW (1968) found that the decomposition rate of leaves and detritus was a function of calcium concentration and at Station V, this was elevated. However, the concentrations involved in the Gombak are much lower than the ranges discussed by him.

The energy content of the sedimentary organic fractions, with the

Table 42. Estimated total organic drift load > 165 μ

Station	kg/yr	Area above station km²	kg/km²/yr
I	232.3	4.7	49.4
II	3718.9	27.7	134.3
III	16415.1	80.0	205.3
IV	24060.9	93.4	257.6
V	30482.0	102.8	296.5

exception of the anomalous high and low values for depositional areas at Stations I and II respectively, was variable but generally lower than the content of the drifting organic materials. This decrease resulted from the advanced state of decay and mineralization of many of the bottom organic constituents. The upper, newly-deposited layers may have had energies equivalent to those of the source materials, but the deeper, buried components depressed the mean. No consistent difference between the erosional and depositional areas was obvious. The organic materials present in the sediments contained considerable potential energy for utilization by the benthos throughout the year.

Synthesis: To obtain a first approximation of the movements of organic loads in the river, estimates of annual total export past the various stations were made. Although samples were taken on only 15 occasions, these were considered to cover all except the severe flood conditions and extrapolation to a year was made on a time/volume basis. This may have resulted in an underestimate as often large loads of coarse fragments were carried on a rising stage. Counteracting this were the very low loads carried on falling stage. Other isolated events such as bank collapse or heavy flowering and fruiting of a single or group of riparian trees may have added considerably to the overall load, but without constant monitoring these could not be evaluated.

Total organic drift estimates are given in Table 42. The rate/unit area yield for Stations I and II draining similar forested watersheds might be expected to be the same. The greater load in the main river was probably the result of larger leaf contributions from bankside trees and, more important, the higher velocities that kept the channel cleared of accumulated organic materials. In the tributary, much of the input of leaves and detritus decomposed *in situ*, without being washed out. The progressive denudation/unit area downstream reflected the cleared state of the middle and lower watershed, with large volumes of surface run-off. LIKENS et al. (1970) found a decrease in organic outflow after deforestation but their watershed was small and plant regeneration well-controlled by defoliatives. Here, growth of cultivars or secondary jungle was rapid and production and decay of short-lived species was considerable.

Table 43. Estimated total organic load from evaporation and ignition (g/day)

Date	Station I	Station II	Station III	Station IV	Station V
22–X–68	156.9	3801.6	2575.4	5716.6	3886.4
8–XI–68	108.6	2594.1	3821.0	1034.9	2722.7
5–XII–68	48.8	619.0	477.7	732.5	1012.7
2–I–69	52.2	684.8	328.0	470.8	429.4
30–I–69	114.7	1848.7	2233.9	4324.3	4540.4
27–II–69	119.9	1257.5	1594.6	2721.0	6671.7
27–III–69	103.1	1490.1	1889.2	4522.3	6075.2
24–IV–69	91.4	823.4	898.0	1344.5	3391.5
23–V–69	418.3	4431.2	9686.8	15200.0	26431.5
19–VI–69	138.4	1709.7	1934.7	1718.5	5086.6
17–VII–69	61.6	1373.1	1426.5	1450.1	4290.1
14–VIII–69	144.0	1139.8	1266.5	1058.2	5065.9
11–IX–69	127.4	1299.0	1745.3	1706.7	4307.0
9–X–69	220.4	1722.0	3882.3	2795.0	7375.5
6–XI–69	202.8	2060.6	3206.7	7291.2	8348.4
Σ15 days (g)	2108.5	26904.6	36966.6	52086.6	89545.0

Table 44. Total exported organic material

Station	kg/yr	kgcal/yr	kg/km²/yr
I	283.6	12.3×10^5	50.3
II	4373.6	20.3×10^6	157.9
III	17314.6	87.6×10^6	216.5
IV	25328.3	11.7×10^7	271.2
V	32660.9	17.4×10^7	317.7

Table 43 lists the total organic loads (T.O.L.) from evaporation and ignition of water samples, worked up to daily total load. The small contributions to these T.O.L. values of particles larger than 165 μ (i.e. the overlap with drift load) was ignored, as the nature of the water sampling method precluded collection of leaf and twig fragments and, as seen in Table 25 previously, the dissolved load (particles smaller than about 5 μ) was always the major constituent.

The T.O.L. values, estimated on an annual basis, were added to the drift loads to give the total exported organic matter (Table 44). At modal flows, this represented the excess over the holding capacity of the bottom, but the frequent floods cleared all materials from the upper sediment layers destroying the balance. The total weight passing Station V, approximately 32.5 metric tons, was a moderate load for a river draining a high productivity forest and cultivated area, but is consistent with the findings of CUMMINS and coworkers (1966 and subsequent personal communication) that more than 80% of the particulate matter that enters a stream is processed there. The value at Station II, about 160 kg/km²/yr was surprisingly low in light of the litter fall of 5–7 tons/ha/yr (about 6×10^5 kg/km²/yr) reported by BRAY and GORHAM (1964) and WANNER (1970) for dipterocarp forest. However, much of this litter was rapidly decomposed and recycled and therefore inaccessible to erosion. The exported load was probably derived only from a narrow band of riparian vegetation with a few wind-blown additions from adjacent areas. Total energy loss from the watershed above Station V (this included certain extraneous waste contributions from materials imported and processed in the lower valley) was estimated at about 17×10^7 kgcal/yr. The loss from the forest above Station II, 20×10^6 kgcal/yr, may have represented a considerable part of the non-recycled energy in the otherwise closed system.

III. THE ALGAE

1. Composition of the Flora

The phytoflora of the Gombak River is limited in mass and productive potential, but relatively rich in diversity. In the upper forest zones, the development of an extensive algal community in precluded by shading of the stream, the instability of all but the largest substrate fragments and the nutritive deficiency of the water. Those algae that do occur have a markedly discontinuous distribution, often being found only on large stones in the more open parts of the river and on root-masses trailing in the water. Lower down the river out of the forest, where light is potentially available and nutrient supply increases somewhat, the effects of silt and an eroding substrate prevent extensive autochthonous production. High turbidities hinder light penetration, but at the shallow depths of the Gombak, this is not as important a factor as the molar action of the silt-sand load that continually removes or smothers whatever periphyton develops. However, in these reaches, wherever a stable substrate eroded clean of depositing materials is found, substantial epilithic and epiphytic growth occurs and during modal periods, a considerable transient epipelic community often develops. The phytoplanktonic element of the flora is absent, but as a result of current action, continuous recruitment to a tychoplankton occurs and provides a source of colonizers for denuded substrates. Flood-pools and small depressions in stream-side boulders constitute a special habitat in which nutrients may become more concentrated through evaporation. These develop a very different flora from that of the adjacent river and during floods contribute alien species to the main stream. This pattern is similar to that found in the Amazon forest creeks described in detail by FITTKAU (1964, 1967) and summarized by SIOLI (1968b). However, there are no slow-water, near-lacustrine shore lagoons or large backwaters, even in the lower Gombak, to allow development of any true phytoplankton as found in many tropical rivers (BLACHE 1951 – Cambodian waters; HOLDEN and GREEN 1960 – Sokota River; RZÓSKA 1961 and others – Nile; SIOLI 1967a, 1968b – Amazonian rivers; VENKATESWARLU 1969 – MOOSI and summary for Indian rivers).

Few previous studies have been made on the algae of Malayan streams. HIRANO (1967) surveyed a number of Southeast Asian waterbodies and briefly collected from a small river in Templer Park, about 20 km west of the Gombak watershed. RATNASABAPATHY (1971) has described the associations found in streams of Gunong Jerai, an isolated 1200 m peak in the NW corner of West Malaysia. Several ponds and hot springs adjacent to the lower Gombak were studied by BISWAS (1929). SANDS (1934) and JOHNSON (1970) have investigated the algae of Malayan rice-fields. PROWSE, in a series of papers (1957, 1958, 1962a, b), taxonomically surveyed various algal groups from a multitude of habitats and with RATNASABAPATHY (1970) compiled a species list for the lakes of the Taiping region of the main range.

In their reviews of the ecology of river algae, BUTCHER (1932), BLUM (1956a, 1957, 1960) and ROUND (1964) have discussed the problems of sampling. In this study, an attempt was made to define qualitatively the algal assemblage in terms of subcommunities, based on the substrate type, to minimize intra-site variations. In the forest zone where periphyton development was generally dispersed in small, isolated patches, the problems were particularly acute. The main divisions discussed by ROUND (op. cit.) were used, i.e. epilithic, epipelic, epiphytic, and a fourth, 'culture', added to cover the tychoplanktonic element. Conventional plankton sampling by nets or centrifuge was inapplicable in the Gombak because of the particulate organic load and the scarcity of planktonic forms. A crude culture technique was used therefore to determine the floristic composition of the plankton.

No frequency/diversity assessment of the natural assemblage was feasible because of the uneven mosaic growth pattern and severe spate and erosion factors that made any enumeration of individuals of an unstable community meaningless without large numbers of quantitative samples beyond the scope of the present preliminary survey.

1.1. METHODS

Periphyton was collected from all stations at least monthly from October 1968 to May 1970. Because of the patchy distribution in the forest section, supplementary samples were taken at locations above and below Station II in the Upper Zone of the main river and in several hill tributaries similar to the Station I stream. The findings from these were included with those of Stations II and I respectively. Surveys in a number of the lowland tributaries added few new species to the flora. These have been considered with the collections from the adjacent main river with the notable species additions discussed separately.

Epilithic forms were collected from 10–15 rocks and boulders of mixed sizes and from a cross-section of current conditions at each site on each sampling occasion. Obvious algal aggregates and scrapings from all faces of the stones were sampled and preserved in the field in 5 % formalin. At the lower stations, scrapings were also made of the epiflora on erosion-control pilings and other emergent structures. Wherever possible, only truly aquatic forms were collected; however, because of the rapid and considerable variation of stage, some aerial species and lichens were occasionally included.

Epipelic samples were collected whenever bank and bottom conditions were stable for a few days. Several vials of mud were collected and examined fresh where possible, before preservation. Larger samples were collected periodically and allowed to stand overnight before examination of the upper layers for motile species.

The category of epiphytes was here restricted to those forms found on trailing bankside vegetation and roots, particularly the *Saraca* and liana assemblages. Small pieces of leaves, roots, etc. were taken fresh to the laboratory for examination and then preserved, usually in F.A.A..

Separation of planktonic cells from the mass of suspended organic materials was impossible. Four-litre water samples were taken in clean semi-opaque polyethylene bottles and allowed to stand lightly-capped, in indirect light for two to four weeks without the addition of any nutrients. The forms that developed on the walls or in the water were then sampled. To supplement these data, collections were made from several overflow and rock-pools which were flushed at high water and presumably received their propagules from the planktonic species. These pools provided natural culture conditions, but were subject to enrichment by evaporation and from the leachates of leaves. Because of the standing water conditions that eliminated most of the reophilous forms and the organic enrichment that occurred between flushings, these areas often developed a unique assemblage of species. Neither method measured the true tychoplankton because of the stagnant growth conditions, but an interesting, although relatively unimportant, part of the flora was revealed.

Other problems inherent in any sampling program were the result of the growth habits of the algae. Mud and root samples were often only arbitrarily separated in the Lower Zone where muddy banks are common. Some epiphytes, particularly diatoms on filamentous algae, were inevitably recorded as coming from the rock or epipelic community of the host filament. At the lower stations, no attempt was made to sample the flora of irrigation and drainage ditches or paddy fields as these were not considered part of the river system because of their gross nutrient enrichment. However, casual species washed in from these areas were probably present in some samples, particularly of the epipelic flora.

Most of the algal species were identified in conjunction with Mr. M. RATNASABAPATHY, University of Malaya, who is conducting floral surveys in various freshwater habitats in Malaya. His assistance is gratefully acknowledged.

1.2. RECORDS AND COMMENTS ON DISTRIBUTION

Appendix B lists taxa recorded from the Gombak River system; 89 genera and 194 taxa were found in this investigation, but this is undoubtedly an underestimate of the total flora. A number of the smaller diatoms, several Oscillatoriaceae and many Euglenophyta at the lower

Table 45. Number of algal taxa recorded from each station

	I	II	III	IV	V	Whole river
Cyanophyta	9	18	9	7	10	26
Rhodophyta	5	5	1(1)	1	1	7
Cryptophyta	2	3	2	1	2	4
Bacillariophyta	23	35(4)	32	27	33	67
Chrysophyta	2	3(1)	3(1)	1	2	4
Pyrrophyta	—	1	—	—	—	1
Euglenophyta	5	20(12)	2(1)	19(1)	6	35
Chlorophyta	12	13(2)	11	30(5)	21*	47
Charophyta	—	—	—	3	1	3
Total	58	98 (19)	60 (3)	89 (6)	76	194

* includes 3 spp. found only in Tributary K just above Station V. Numbers in parentheses are species found only in 'culture'. Station I includes records from other headwater tributaries.

stations have probably been missed. Table 45 gives a summary of the number of species in each division found at each station or its environs, using the taxonomic system of SILVA (1962). Only Station II had more than 50% of the species, but included in its 98 taxa were 19 species found exclusively in pool areas. Station III had three species and Station IV six species collected only from backwaters. Station I and other hill tributaries had 58 recorded species, eight of which came uniquely from Tributary B draining the watercress farm and whose presence may have resulted from allochthonous introduction and culturing in the artificially-enriched water. The total at Station V included three chlorophytes found only in the polluted Tributary K, just above its confluence with the main river.

In Table 46, the occurrence of each taxa per community per station is given. Most notable was the dominance of the epilithic flora of Bacillariophyta, Cyanophyta and Rhodophyta at Stations I and II with some uncommon or occasionally found filamentous Chlorophyta and desmids. The diatoms of the forested reaches and Station III (probably constantly recruited from above) included many genera common throughout the river, with the most numerous and frequent being the *Gomphonema – Pinnularia – Achnanthes – Eunotia* spp. group. *Melosira* spp., *Hydrosera whampoensis*, *Fragilaria* spp. and *Synedra* spp. were much less common, occurring almost exclusively in the shaded area and not lower downstream. The commonly found *Surirella* spp. (except *S. tenera*) seemed to prefer the lower Stations IV and V with increased nutrients and no

Table 46. Distribution of algal taxa in subcommunities at each station (+ = present; T = present in upper tribs other than I; B, H, K, = present in or near trib. B, H, or K only; P = present in rock-pools or backwaters only; ? = probably present.

	EPILITHIC					EPIPELIC					EPIPHYTIC					CULTURE				
	I	II	III	IV	V	I	II	III	IV	V	I	II	III	IV	V	I	II	III	IV	V
CYANOPHYTA																				
Aphanothece stagnina	B																			T
Merismopedia ? glauca	B				+															
Unknown Chroococcaceae	B	+															+	+		
Entophysalis sp.	T	+										+				+				
Chamaesiphon sp.		+	+																	
Stichosiphon sansabaricus		+																		
Unknown Chamaesiphonales		+																		
Lyngbya ? allorgei	+	+										+								
Lyngbya sp.		+						+	+			+	+					+		
Plectonema sp.		+			+						+	+								
Microleucus sp.	+	+								+										
Oscillatoria 4 spp.						2		3	4	4										
Schizothrix sp.								+												
Unidentified filamentous sp.	+	+			+		+	+	+	+						+		+	+	
Anabaena sp.																+	+	+	+	
Nostoc sp.		+																		
Scytonema sp.		+										+								
Tolypothrix tenuis		+		H																
Tolypothrix sp.		+																		
Calothrix sp.	+	+														+				
Stigonema ocellatum	B	+																		
RHODOPHYTA																				
Compsopogon ? coeruleus		+			+															
Audouinella sp.	T	+										+								
Batrachospermum sp.a																P	P			

146

Batrachospermum sp.b

Unknown Batrachospermaceae

Unid. cf. chantransia of
Lemanea

Caloglossa leprieurii forma
typica

CRYPTOPHYTA

Chilomonas paramecium

Chilomonas sp.

Cryptomonas 2 spp.

BACILLARIOPHYTA

Melosira roseana

M.ruetineri

M.italica

Hydrosera whampoensis

Fragilaria vaucheriae

Fragilaria sp.

Synedra ulna var
amphirhynchus

S. ? ulna

Synedra sp.

Eunotia ? monodon

E.pectinalis

E.robusta

Eunotia 2 spp.

Achnanthes brevipes

A.crenulata

A.inflata

A.lanceolata

Achnanthes sp.

Cocconeis ? thumensis

147

Table 46. continued

	EPILITHIC					EPIPELIC					EPIPHYTIC					CULTURE				
	I	II	III	IV	V	I	II	III	IV	V	I	II	III	IV	V	I	II	III	IV	V
Cocconeis sp.	+	+	+	+	+							+		+	+					
Cymbella javanica	B																			
C.sumatrensis			+	+	+			+												
C.lanceolata	+	+	+	+	+		+	+	+	+			+	+						
C.tumida	T		+	H	+		+	+	+	+			+	+					P	
C.ventricosa	T	+	+	+	+		+	+	+	+	+		+	+	+		P	+	P	
Cymbella sp.			+													P	+	+	P	+
Gomphonema gracile											+	+	+	+	+					
G.longiceps	+			+																
G.parvulum																				
G.subventricosum	T	+	+	+	+	+		+	+	+	+	+	+	+		+	P	+	P	
Gomphonema sp.												+	+		+		P	+		
Frustulia javanica	+				+															
F.rhomboides				+	+															
F.saxonica	+																			
Diploneis ? ovalis																				
Pleurosigma sp.	T	+	+	+	+			+	+						+				P	
Stauroneis sp.										+					+					
Neidium sp.	+								+	+										
Navicula amphibola																				
N. cancellata																				
N. elegans									+											
N.feurborni	B																			
N. confervacea	B							+												
N. ? pupula																				
Navicula sp.	+	+	+	+	+	+	+	+	+	+				+			+	+	P	+ +
Pinnularia biceps var minor																P		+	P	+ +
P.viridis																			+	+

148

P. gibba	+		+	+	+	+		P P + +
P. legumen	+	+	+	+	+	+	+ +	P P +
P. microstauron	+							P
P. ? trigonocephala	T							+
Hantzschia sp.		+ +	+ +					
Nitzschia palea				+	+			P
N. sigma	B							
Nitzschia sp.	+	+		+				
Surirella tenuissima		+	+					
S. angusticostata	+		+					
S. capronii	+ +	+ ?	+					
S. biserata			+	+	+ + +		
S. muelleri								
S. robusta					+			
S. robusta var splendida					+			
S. tenera	T + +	+ ?	+ +	+ + +	+	+	+ + +	+ + P
Epithemia gibberula	+		+	+				+
E.gibberula var producta				+				

CHRYSOPHYTA

Ochromonas sp.	T +	?	+	+			+ +	+ +
Mallomonas sp.a	T	?	+	+				+ + P
Mallomonas sp.b					+			P

Bumilleria sp.

PYRROPHYTA

Gymnodinium sp.

EUGLENOPHYTA

Euglena agilis
E. elongata
E. fusca
E. mutabilis

149

Table 46. continued

	EPILITHIC					EPIPELIC					EPIPHYTIC					CULTURE				
	I	II	III	IV	V	I	II	III	IV	V	I	II	III	IV	V	I	II	III	IV	V
E. ? vivida	B															P				
Euglena sp.									+	+				+		P				
Lepocinclis ovum									+	+						P			P	
L. salina									+							P			P	
Phacus ? onyx				+					+	+										
P. platalea									+	+										
P. stokesii				+					+							P	P			
Phacus sp.																	P	P		
Strombomonas ? australica																P				
Trachelomonas curta																P	P			
T. dubia																P				
T. hispida									+	+										
T. ? oblonga									+							P				
T. ? similis									+							P				
T. volvocina									+	+										
T. volzii var *cylindracea*									+											
Astasia sp.	?	+																		
Entosiphon ovatum	B	+			+		+				+					+	+			
Heteronema leptosmum												+				+				
H. polymorphum																			P	
Heteronema sp.						+														
Notosolenus stenochismos																P				
Notosolenus sp.			+																	
Peranema cuneatum	T																			
P.trichophorum				+																
Peranema 3 spp.	1T				1		1		3		1					1,P				
Petalomonas heptaptera																P				
P.mediocanellata									+							P				

Taxon	Occurrence markers
Petalomonas sp.	+
CHLOROPHYTA	
Chlamydomonas sp.	P, +, ++, +
Sphaerellopsis sp.	+, +
Palmella sp.	+, +
Asterococcus sp.	+
Pediastrum sp.	+, ++, +
Dictyosphaerum pulchellum	+
Dimorphococcus sp.	++, ++, +, +
Coelastrum sp.	+, +, ++, +, +
Ankistrodesmus falcatus	+
A. falcatus var spirilliformis	+
Scenedesmus quadricauda	+, +, +
Scenedesmus sp.	P, +, K
Ulothrix subtillissima	P, +, K
Ulothrix tenuissima	+
Uronema sp.	B
Cylindrocapsa ? geminella	+, +, +, T
C. ? conferta	+
Schizomeris leibleinii	K
Stigeoclonium sp.	P, +, ++, +
Cladophora sp.	?
Rhizoclonium ? heiroglyphicum	P, +, +, ++, T
Oedocladium sp.	P, ++, +, ++, T
Oedogonium sp.	P, +, +, B
Mougeotia/Debarya sp.	P, +, +, ++, ++, +, T
Spirogyra sp.a	P, +, +
Spirogyra sp.b	+, +
Actinotaenium sp.	+
Cosmarium ? obsoletum	+
C. pseudoconnatum	P

Table 46. continued

	EPILITHIC					EPIPELIC					EPIPHYTIC					CULTURE				
	I	II	III	IV	V	I	II	III	IV	V	I	II	III	IV	V	I	II	III	IV	V
C. granatum	+																			
C. ? pseudogranatum	B	+	+	+				+	+	+									P	
Closterium ehrenbergii			+	+	+			+	+										P	
C. libellula				+	+						+								P	
C. ? rostratum								+	+	+										
C. acutum								+	+											
C. moniliferum				+	+				+											
Closterium sp.	+	?	+	+	+	+	+	+												
Euastrum ? spinulosum																			P	
E. binale var *brevius*									+										P	
Micrasterias crux-melitensis																				
M. foliacea									+											
Penium sp.			+																	
Pleurotaenium ovatum									+	+										
P. cylindricum	B																			
Pleurotaenium 2 spp.									2											
Staurastrum sp.									+											
CHAROPHYTA																				
Chara sp.									H											
Nitella acuminata									H											
Nitella sp.									H	K										

shading. The cyanophyte flora was surprisingly rich in the Upper Zone with *Tolypothrix*, *Nostoc* and hormogonia of various *Lyngbya* spp. dominating the macroscopic assemblages. Notably absent throughout the river was the encrusting genus *Phormidium* which is a prominent member of the algae in waters richer in calcium (FRITSCH 1929, MINCKLEY 1963). However, at Stations I, II and III *Chamaesiphon*, the first component of FRITSCH's succession before *Phormidium*, was present in small colonies. Apart from a ubiquitous, but unidentified, filamentous form, there were few records of cyanophytes in the stone flora of the lower river. The Rhodophyta, with the exception of single records for *Compsopogon* at V and *Audouinella* at IV, were restricted to the forest stream. None was ever common in the river. The *Batrachospermum* spp. were rare and attained substantial size only in the flood-pool areas. A notable record was an unidentified Batrachospermaceae found quite abundantly, but uniquely, in Tributary B, above the watercress farm. This genus has a macroscopic (up to 15 cm), narrow, cylindrical, lax, reddish-brown thallus with occasional short branching. The cells of the axis and branches are similar and are not differentiated into nodes and internodes. The appearance is reminiscent of *Tuomeya*, but other features do not correspond. Another unidentified species with a flat purple thallus, resembling the chantransia stage of *Lemanea*, was regularly collected in epilithic samples from Stations I and II. At Location B, in addition to the above, were found several species recorded elsewhere only in the lower polluted reaches, e.g. a *Nitzschia* sp. and an unidentified *Euglena* sp.. The only collections of *Uronema* and *Merismopedia* were made at this location and *Mougeotia/Debarya* was found here and in the next tributary as an attached form. The species found solely at these high altitude sites and not in other tributaries are indicated in Table 46 as B.

The epipelic flora at the upper stations was scarce. None was recorded from the stream at Station I where no true mud bottom existed even in the pools. At Station II, only two species of *Oscillatoria*, the most common diatoms and two euglenoids were found. A single, probably incidental, record of the chantransia-like species was also made here. Station III added a third *Oscillatoria* sp., a *Schizothrix* and *Anabaena* spp., more diatom spp. and a few desmids. Station IV, with its extensive muddy areas and some natural organic enrichment, showed a rich development of Euglenophyta and desmids in areas where current effects allowed mud banks to stabilize. The presence of *Nitzschia* spp. in this habitat indicated some pollutional enrichment (BLUM 1957). ROUND (1964) suggested that flagellates and Cyanophyta generally increase in sediments rich in organic matter and that the epipelic of flowing systems usually contains many motile forms; this was generally corroborated in the flora of the Gombak. The *Chara* and *Nitella* spp. recorded for this station were from tributaries only, *Chara* rarely from Tributary H and *Nitella* very abundantly in the small tributaries returning irrigation water to the river. *Nitella* was reported from the main river in the vicinity of the D.I.D. gauging site by NORRIS and CHARLTON (1962) but none was seen in this survey. Conditions were not optimal for either genus. VAIDYA (1967), in a review of the ecology of Charophyta, placed *Nitella* in acid oligo-meso and *Chara* in neutral-basic meso-poly hard waters. The calcium and magnesium concentrations found for the Gombak were much below the 10–20 ppm usually considered minimal for the growth of these genera. Various unicelled Chlorophyta, *Oedogonium* and *Spirogyra* spp., were occasionally present after long stable flow periods, both here and at Station V. At this latter station only a reduced desmid flora and few euglenoids were found, these having been replaced in the rich muds by heterotrophic ciliates and flagellates. Lack of light because of high turbidities and/or the presence of toxic substances may have eliminated the Euglenales.

The root epiphyte community was generally species-poor, containing only a few diatoms, notably *Achnanthes* and *Cymbella* spp., at all stations. In the Upper Zone, a true association, repeatedly found and of considerable extent in local areas, comprised *Lyngbya* spp., *Plectonema*, the flat thalli of *Caloglossa leprieurii* and the centric diatom *Hydrosera whampoensis* growing in the masses of *Saraca* roots lining the banks. In areas where the forest had been cleared from the banks and various Graminae with trailing suckers and *Colocasia esculenta* flourished in the increased light, the *Caloglossa* dropped out, but the *Lyngbya* and *Hydrosera* were common, often forming grey-black sheets along the banks, with the bright yellow macroscopic colonies of the diatom visible near the attached end of the *Lyngbya* filaments.

The 'culture' of the river water collected at modal flow revealed a tychoplankton poor in both numbers and species. Some Chroococcaceae, *Nostoc* and *Anabaena* spp. developed and the more common *Eunotia* spp. and assorted other diatoms were found at various stations. Stations I, II and III developed *Epithemia gibberula;* a single record of an unidentified *Hantzschia* sp. was made at III and Station V grew a considerable enrichment flora of naviculoid diatoms. *Entosiphon ovatum* at Stations I and II and a *Peranema* sp. at II were the only euglenoids grown. Station V water carried propagules of *Scenedesmus*, a *Ulothrix* and *Ankistrodesmus*, species not found at other stations and perhaps derived from the ricefield areas. Various unicell and palmelloid Chlorophyta grew in cultures from Stations I and II and *Mougeotia/Debarya* from III. Chrysophyta: *Ochromonas* sp. and *Mallomonas* spp. were taken at Stations II and III, a *Mallomonas* sp. from IV and the single Pyrrophyta, a *Gymnodinium* sp., was found from Station II, from whose water an almost pure culture developed. In contrast, the specialized flora of the flood-pools and backwaters was more extensive, particularly at Station II. The sampled pools in the forest received organic leaf-leachate enrichment, accounting for the large numbers and considerable diversity of Euglenophyta. Some diatoms, notably the *Eunotia* and *Achnanthes* spp., were abundant and the filamentous greens, *Cladophora, Rhizoclonium, Oedogonium* and *Spirogyra* showed some development in the waters with higher cation content resulting from concentration by evaporation between floods.

Nowhere on the river was periphytic growth extensive. In the forest, epilithic growth was sparse and grazing by the considerable larval population of bufonids, pelobatids, and *Amolops larutensis* helped to keep the flora in a state of constant renewal (COSTA and BALASUBRAMANIAN 1965, DICKMAN 1968). Algal abundance has also been shown to be limited in this way by other benthic faunal elements by BROOK (1955) and DOUGLAS (1958) – various insects and protozoa, and MINCKLEY (1963) – *Goniobasis*. In pools where a few filamentous forms were found, their growth was always stunted. *Batrachospermum* developed thalli only 3–5 cm long and the Chlorophyta often had filaments less than 5 cm in length. In regions of high insolation and flow, as, for example, in exposed cascades in a small tributary above Location B, *Spirogyra* spp., *Oedogonium* and *Rhizoclonium* developed profusely, forming 2–3 mm thick mats over the rocks. High available light in the shallow water and the steep diffusion gradient for nutrients as a result of sheet flow, the 'physiological enrichment' of RUTTNER (1963), discussed at length by WHITFORD (1960) and WHITFORD and SCHUMACHER (1964) for these genera, accounted for this local proliferation. The inherent current demand for many species, especially diatoms (see McINTIRE 1966 for summary), coupled with high insolation conditions, was rarely satisfied in the Gombak because of high turbidity and increasing depth down the river. Occasionally at Stations III and IV *Spirogyra* developed small mats on muddy banks, but these were carried away or stranded with the next change in stage and never established a permanent aggregation. BLUM (1957) in the Saline River and MINCKLEY (1963) in Doe Run noted that *Cladophora* was confined to unshaded areas, but the deleterious effects of siltation and mechanical detachment on the production of large colonies, also described by these authors, were probably more restrictive in the Gombak. Even in the lower reaches with better light and nutrient conditions, tufts of this alga rarely exceeded 1 cm because of periodic abrasion by silt-laden floods. The characteristic epiphytic diatom community described by MINCKLEY (op. cit.) and discussed by MARGALEF (1960) as the 'Cladophoretum' complex, was largely intact in the Gombak with *Cymbella, Cocconeis*, stalked *Gomphonema* spp. and various Myxophyceae often attached to the *Cladophora* filaments. At Stations III and IV, a slimy, encrusting 1–3 mm thick layer, mostly diatoms of various species, but with a few desmids, formed over the stones in the riffles during low water periods when transparency was at a maximum. This layer was destroyed by the molar action of the shifting bed-load as soon as a flood occurred. Such production was limited to a small area, that of riffles in the Middle Zone and upper part of the Lower Zone. Below Station IV and before Station V, the infrequency of shallow riffles, the high turbidity and smothering by large flat mica flakes and finer silts all but eliminated this flora. At Station V, very little growth, even on the riffles, was found in comparison with that at Stations III and IV. In the drainage ditches of the Lower Zone, however, extensive Cyanophyta and Chlorophyta mats developed, often up to 1 cm thick.

Of particular interest was the specificity shown by the epilithic forms for certain lithologies. The soft, constantly eroded schists and limestones never had any epiflora. Quartz and fine-grained quartzite were colonized only by the Chroococcaceae and palmelloid green algae. In the forest, diatom growth was almost exclusively restricted to the darker shales, conglomerates and granites, while the Cyanophyta were more catholic in their choice of substrate. This may indicate that the algae in this ion-poor environment were deriving some of their nutrient requirements from the weathering rocks, as the granites contain comparatively higher concentrations than the metamorphic series. On the surface of the limestone bedrock outcrop in the river bed at Station V, a dense growth of desmids and the coccoid *Merismopedia* developed between floods. There was, of course, no development of any marl-depositing community at the very low calcium concentrations of the river.

The development of a subaerial flora was severely restricted in most areas by the variation in water level. A few encrusting forms, *Trentepohlia*, *Fischerella* and *Mesotaenium* were found on bankside and emergent boulders, along with various lichens, but they did not survive prolonged submergence. Diatoms, particularly the epiphytic *Gomphonema*, *Synedra*, *Eunotia* and *Cocconeis* spp. that are reportedly unable to withstand desiccation (DOUGLAS 1958), were restricted almost exclusively to the areas below modal water depth.

In summarizing the numbers of taxa found in each assemblage (Table 47), the problem of casuals becomes acute. In a river subject to sporadic wash-out, many forms will drift downstream and may re-establish themselves temporarily before adverse conditions eliminate them. Records of single cells were not considered adequate to place an alga in a sub-community, but this criterion probably did not eliminate all the extraneous forms. The epilithic community was pre-eminent in the forest zone as 56 out of a total of 58 taxa recorded at Station I and 69 of the 79 river-only taxa at Station II occurred on rock substrates. The flora at Stations III and IV was dominated by epipelic forms, particularly at IV where numerous euglenophytes and desmids occurred. The root epiphytic community was surprisingly poor at all stations, but as expected, more species were found in the forest zone where potential substrates were readily available. The composition of the tychoplankton, pool and backwater flora has already been discussed.

The only satisfactory way to define communities, species distribution and succession adequately would be to quantify relationships by a suitable frequency analysis (cf. PATRICK et al. 1954) that would indicate

Table 47. Number of taxa found in each algal assemblage

	Epilithic					Epipelic					Epiphytic					Culture				
	I	II	III	IV	V	I	II	III	IV	V	I	II	III	IV	V	I	II	III	IV	V
Cyanophyta	8	16	2	1	3	—	2	7	6	5	1	4	—	—	—	2	2	4	—	3
Rhodophyta	4	4	—	1	1	—	1	—	—	—	1	2	—	—	—	—	1	1	—	—
Cryptophyta	2	3	—	—	2	—	—	2	1	1	—	—	—	—	1	—	3 (3)	(1)	1 (1)	—
Bacillariophyta	23	30	22	16	25	—	7	17	19	11	7	7	2	4	5	2	15 (14)	8 (1)	10 (6)	8
Chrysophyta	2	1	1	1	1	—	1	1	1	—	—	—	—	—	1	—	3 (1)	3 (1)	1	—
Pyrrophyta	—	—	—	—	—	—	—	—	—	—	—	—	—	—	—	—	1 (1)	—	—	—
Euglenophyta	5	6	—	1	5	—	2	1	17	2	2	2	—	1	—	1	16 (14)	1 (1)	3 (3)	—
Chlorophyta	12	9	6	8	12	—	1	8	21	10*	1	1	1	—	—	—	8 (6)	4 (1)	6 (6)	5
Charophyta	—	—	—	—	—	—	—	—	3	1	—	—	—	—	—	—	—	—	—	—
Total	56	69	31	28	49	—	14	36	68	30	12	16	3	5	7	5	49 (39)	21 (5)	21 (16)	16

* includes 3 spp. found only in Trib.K just above Station V.
Numbers in parentheses are species included in total but found in 'pool-cultures' only.

Table 48. Numbers of common algal taxa and coefficients of association between stations

	EPILITHIC						EPIPELIC			
	I	II	III	IV	V		I	II	III	IV	V
I	56	36	18	16	21	I	—	—	—	—	—
II	0.58	69	21	19	65	II	0.00	14	8	8	6
III	0.41	0.42	31	18	17	III	0.00	0.32	36	20	16
IV	0.38	0.39	0.61	28	19	IV	0.00	0.20	0.38	68	22
V	0.40	0.42	0.43	0.49	49	V	0.00	0.27	0.48	0.45	30

	EPIPHYTIC				
	I	II	III	IV	V
I	12	9	1	1	1
II	0.64	16	1	2	1
III	0.13	0.11	3	2	—
IV	0.12	0.19	0.50	5	1
V	0.11	0.09	0.00	0.17	7

	CULTURE						TOTAL				
	I	II	III	IV	V		I	II	III	IV	V
I	5	3	2	—	—	I	58	41	33	24	26
		(1)							(1)		
II	0.11	49	14	8	4	II	0.53	98	35	30	29
		(39)	(10)	(7)	(3)		/0.60/	(19)	(5)	(3)	(1)
III	0.16	0.40	21	6	4	III	0.56	0.44	60	32	33
			(5)	(3)	(1)		/0.56/	/0.44/	(3)	(1)	(2)
IV	0.00	0.23	0.29	21	4)	IV	0.33	0.32	0.43	89	44
				(16)	(2)		/0.34/	/0.33/	/0.44/	(6)	(3)
V	0.00	0.12	0.22	0.22	16	V	0.39	0.33	0.49	0.53	76
							/0.36/	/0.47/	/0.52/		

Numbers in ( ) are species included in total but found in pool-cultures only
Numbers in / / are coefficients calculated with the pool-culture taxa omitted

dominance and eliminate the casual or extraneous species of erratic occurrence. Because of the unstable habitats and the tendency, particularly in the upper river, for isolated colonies of various small assemblages of species to develop, no measure of species diversity without a very large number of samples would be worthwhile. However, an attempt was made to show the similarities between the floras at the different stations and in the designated subcommunity types. Delimitation of communities has been thoroughly reviewed by KONTKANEN (1957) among many others (see references in ROBACK et al. 1969, BARNARD 1970) and a host of indices of similarity and coefficients of association have been proposed (e.g. JACCARD 1902, SØRENSEN 1948, MORISITA 1959, MOUNTFORD 1962). The SØRENSEN index was used here.

I = coefficient of association between A and B

$I = 2j/(a+b)$ where j = number of spp. common to A and B

a = number of spp. present at A
b = number of spp. present at B

The presence/absence data available were subject to errors discussed above and use of a more refined index did not seem justified. This index gave equal weight to each species, irrespective of either absolute or relative abundance and therefore gave no indication of succession or dominance. In addition, it did not take into account the absence of a species from one assemblage which may be of equal significance to its presence in another aggregation.

In Table 48, the numbers of common taxa and the coefficients of association between stations for the various subcommunities are listed. The usual procedure in analysing association coefficients, cluster analysis, was not carried out for two reasons: 1. introduction of distortion by the unweighted pair-groups method commonly used to define clusters; 2. more serious, the production by clustering of hierarchical aggregations irrespective of the original relationships in the matrix of association coefficients, i.e. stations or subcommunities could be clustered into a hierarchy that did not exist, or was meaningless, in nature. The delimitation of such entities, that disregard actual spatial arrangement, was unjustified in a river where sequential relationships were more important than overall or absolute similarities. In most cases the matrices of association coefficients were readily interpreted and special factors and exceptions peculiar to a specific relationship could be noted and rationalized. Cluster analysis does not give this flexibility.

The epilithic spp. aggregation was reasonably uniform, with particular affinities, as expected, between Stations I and II (forest) and III and IV (unshaded, unpolluted), with a considerable number of common species between Stations IV and V and II and III in areas where overlap was predictable. The epipelic floras of the upper and lower river had little in common, but some association existed among the lower three stations. Root epiphytes showed a similar pattern to the stone flora with strong aggregation of the Stations I and II species, generally on the same *Saraca*-root habitat. Stations III and IV with comparable trailing shrubs and grasses as substrates also had allied microfloras. However, the small numbers of species recorded at the lower stations make these comparisons dubious. The 'culture' subcommunities showed little association on total species content, but comparison of only the numbers of species recorded in pools and backwaters generated a coefficient of association of 0.40 between Stations II and III, but less than 0.31 for the other combinations. The coefficients between stations (calculated on total taxa irrespective of subcommunity) were always more than 0.3, indicating a group of recurring, widely tolerant species. When the 'pool-only' species were excluded from the analysis, Stations I and II and I and III had considerable affinity (I = 0.60, 0.56), mostly of the rock encrusting forms.

Table 49. Numbers of common algal taxa and coefficients of association between subcommunities

Station I	S	M	R	C		Station II	S	M	R	C
S	56	—	10	3		S	69	9	14	25(19)
M	$\overline{0.00}$	—	—	—		M	$0.\overline{22}$	14	3	5(3)
R	0.29	0.00	12	2		R	0.33	$0.\overline{20}$	16	7(4)
C	0.10	0.00	$0.\overline{24}$	5		C	0.42	0.16	$0.\overline{22}$	49(39)

Station III	S	M	R	C		Station IV	S	M	R	C
S	31	11	1	8(2)		S	28	15	4	10(6)
M	$0.\overline{33}$	36	2	9(2)		M	$0.\overline{31}$	68	5	13(9)
R	0.06	$0.\overline{10}$	3	2		R	0.24	$0.\overline{11}$	5	2(2)
C	0.31	0.32	$0.\overline{17}$	21(5)		C	0.41	0.29	$0.\overline{15}$	21(16)

Station V	S	M	R	C		All Stations	S	M	R	C
S	49	13	6	6		S	117	42	27	48(25)
M	$0.\overline{33}$	30	1	2		M	$\overline{0.40}$	95	17	35(17)
R	0.21	$0.\overline{05}$	7	2		R	0.38	$0.\overline{28}$	27	19(9)
C	0.18	0.09	$0.\overline{17}$	16		C	0.49	0.20	$0.\overline{36}$	78(42)

S = Epilithic; M = Epipelic; R = Epiphytic; C = 'Culture'; Numbers in ( ) are species included in total but found in pool-cultures only.

Stations II–III and III–IV–V, particularly the lower two stations with an index of 0.53, had a number of floral elements in common.

Few affinities existed between the species aggregations of the sub-communities for each station (Table 49), as most taxa, with the exception of the common diatom spp., showed fidelity to a particular association. The aggregation of species found as root epiphytes was almost always well represented in the epilithic, e.g. 10 of 12, 14 of 16, 4 of 5 and 6 of 7 taxa found on roots common with the stone flora at Stations I, II, IV and V respectively, but the low total number of species in the epiphytic community resulted in relatively low association coefficients. At all stations, the epipelic constellation had few common elements with either the lithic or phytic group; this agreed with the contention of ROUND (1964) that the mud flora was generally a distinct assemblage of species. The 'culture' subcommunity showed some affinity to the epilithic community from which it was probably derived, except at Station I where it was poor in taxa and at Station V where the tychoplankton was derived from the communities upstream that had different species composition. These relationships are emphasized in the association table for 'All Stations' in which all the 27 epiphytic taxa are indicated as common with the rock flora and the 'culture'-epilithic index is 0.49.

Table 50. Numbers of common algal taxa and coefficients of association between stations for the prominent algal Divisions

CYANOPHYTA

	I	II	III	IV	V
I	9	7	3	—	2
II	0.52	18	5	4	5
III	0.33	0.37	9	5	5
IV	0.00	0.32	0.63	7	5
V	0.21	0.36	0.53	0.59	10

BACILLARIOPHYTA

	I	II	III	IV	V
I	23	18	18	13	12
II	0.62	35	19	16	16
III	0.65	0.57	32	17	14
IV	0.52	0.52	0.58	27	17
V	0.43	0.47	0.51	0.57	33

EUGLENOPHYTA

	I	II	III	IV	V
I	5	3	1	3	3
II	0.24	20	2	6	2
III	0.29	0.18	2	1	1
IV	0.25	0.31	0.10	19	2
V	0.55	0.15	0.25	0.16	6

CHLOROPHYTA

	I	II	III	IV	V
I	12	6	8	6	7
II	0.48	13	5	6	7
III	0.70	0.42	11	7	7
IV	0.29	0.28	0.34	30	14
V	0.42	0.41	0.44	0.55	21

The total numbers of common species and coefficients of association between stations for the four prominent algal Divisions, irrespective of subcommunity, are given in Table 50. There appeared to be two assemblages of Cyanophyta, one at the Upper Zone stations, the other for the open, lower river (best developed at Stations III and IV) with only a few common species between the two groups. This separation was essentially that of epilithic from epipelic blue-greens. The diatom association coefficients were all fairly high showing the uniformity of the flora. Sequential associations with adjacent stations showing the most affinity were the rule, except for I–III which was similar to I–II, reflecting the wash-out and drift of upper river types and the suitability of conditions at Station III for recruitment and successful colonization. Little affinity existed between the Euglenophyta at any of the stations, accentuated by the low total numbers of species found at Stations I, III and IV. Relationships within the Chlorophyta were masked by the sporadic occurrence of the filamentous genera in localized areas uncharacteristic of their general habitat, e.g. the records for *Spirogyra* spp. and *Oedogonium* sp. high in the headwater tributary, where exposure in a cascade provided suitable conditions over a localized area. Analogous development of mats of *Spirogyra* in torrent tributaries of the Great Berg River was reported by HARRISON and ELSWORTH (1958). These records were responsible for the high index (0.70) between the headwater tributaries and Station III, for example. A number of common species were found between all stations, but the true abundance relationship (rare individuals in the Upper Zone and many plants in the lower reaches) was lost in this analysis method. Stations IV and V shared an aggregation of desmids and some filamentous species.

Other microflora/fauna: In association with the algae, considerable numbers of microconsumers were found. A list of the more prominent and recurring forms is given below. No attempt was made to identify any bacteria, fungi, the majority of the Protozoa or the Rotifera.

At all stations – fungi and bacteria

> – numerous Ciliophora, *Anthophysa vegetans* (O. F. Müller) Stein, other Zooflagellata (especially rich in types at Stations IV and V)
> – various amoebae; *Arcella* spp. and other Rhizopoda
> – Rotifera and Nematoda (not numerous)

Station II – *Salpingoeca* sp.; unidentified Heliozoan; *Euglypha* sp., *Centropyxis* type, *Difflugia* type; *Vorticella* sp., *Paramecium* type; *Chaetonotus* sp.

Station III – *Actinosphaerium* sp.; *Euglypha* sp., *Centropyxis* type; *Trachelius* sp.

Station IV – *Centropyxis* type; *Vorticella* sp.; *Stylonychia* sp., *Epistylis* sp., *Paramecium* type

Station V – *Centropyxis* type; *Vorticella* sp., *Stylonychia* sp., *Epistylis* sp., *Paramecium* type, *Acineta* sp.; *Chaetonotus* sp.; *Hydra* sp.

A notable omission from the list (not really a microform) was the Porifera. The ionic poverty of the water probably excluded this phylum from the river, and the severe spates and high turbidities would also mediate against its occurrence.

Filamentous bacteria occurred in several specialized habitats. Hynes (1960), Cooke (1961), Ormerod et al. (1966) and Phaup (1968) have reviewed the ecology of these forms and their relationships with enriched conditions in the water. In the ochraceous seepages, especially at Station IV, extensive flocs of colloidal iron hydroxide occurred. These were composed largely of a sheathed bacterium, probably a *Leptothrix* sp., that deposited the oxidized iron in its sheath. In the main river at Station V, the typical *Sphaerotilus* growth-form developed on trailing vegetation and pilings during periods of low, stable flow. In Tributary K, long aggregations were always present as an epiflora in close association with the *Schizomeris* sp. found only in that location. The growth of both these bacteria (Phaup maintains that they are only forms of a single genus) is promoted by enrichment of nutrient salts and is usually taken to be indicative of mild to moderate pollution. There is some evidence that *Sphaerotilus* is autotrophic (see Phaup) but the common condition is that of a heterotroph utilizing a ready supply of inorganic salts. Large beds of bacteria attached to submerged objects, as often seen in rivers, never developed, not even at the lowest station near the D.I.D. gauging site, because of the unstable substrate and frequent spates. *Beggiatoa* occurred in abundance in a small flood-pool at Station II, but there was no apparent hydrogen sulfide available. However, decomposing leaves may have produced reduced sulfur locally in the sediments of the pool.

2. Primary production studies

Production studies on the periphyton of lotic systems have become numerous in the last 15 years, particularly in north temperate streams and rivers. GUMTOW (1955), ODUM (1956, 1957a, b), McCONNELL and SIGLER (1959), CLAUS (1961), KOBAYASI (1961), HOHN and HELLERMAN (1963), DUFFER and DORRIS (1966), BESCH et al. (1967), CUSHING (1967), KING and BALL (1967), WRIGHT and MILLS (1967), BEERS and NEUHOLD (1968), BESCH and HOFMANN (1968), STOCKNER (1968a), BACKHAUS (1969), FLEMER (1970), McFARLAND and WEBER (1970) and others have investigated various aspects of this problem. In the tropics few studies have been made, as the emphasis has been on the plankton of the larger rivers (see VENKATESWARLU 1969).

In temperate and mid-latitude rivers, the physical and chemical environment is subject to cyclical changes of sufficient magnitude to result in periods of varying periphytic productivity and seasonal community succession. Temperature, light availability, tree cover and extent of auto-shading are seasonally different and nutrient supply varies with the availability of leaf decomposition products, agricultural activity that releases enrichment elements and seasonal meteorological effects that control the rate of addition of these products to the river system and degree of dilution. These factors often produce an annual sequence of a period with low algal biomass and high productivity, followed by stable, high standing crop, moderate production conditions and then a period of community senescence. In the humid tropics, many of the seasonally variable environmental effects may be muted. This is particularly true in central Malaya west of the main range where monsoonal influences tend to be minimal. Thus, apart from fluctuations in water level, which are generally transitory, the aquatic environment is almost in a steady state. Production rate might, therefore, be expected to be relatively low and biomass constant, with few of the oscillations of the temperate or monsoonal river flora.

Sampling the biomass and determining production rates of attached algae have been continuing problems with a number of ingenious, but often inadequate in some respect, techniques suggested. SLÁDEČKOVÁ (1962) has made a comprehensive review of these techniques and other authors (COOKE 1956, LUND and TALLING 1957, CASTENHOLZ 1961, SLÁDEČEK and SLÁDEČKOVÁ 1963a, 1964, WHITFORD and SCHUMACHER 1963, WETZEL 1964, ROUND 1965, HOHN 1966) have summarized the mass of previous work and critically assessed the results and shortcomings. Removal of natural populations for frequency and biomass determination by coring type devices that sample a known area (DOUGLAS 1958, CUMMINS 1966, ULFSTRAND 1968a) or by collodion film (MARGALEF 1949) has proved useful in some cases, but the problem of sampling discontinuously distributed floras has limited the utility of such techniques. A method widely used by many of the previously mentioned authors has been the provision of artificial substrates for colonization by the periphyton. As shown by BUTCHER (1932), PATRICK and co-workers (see HOHN 1966) and others, the periphytic community that develops on glass slides suspended in a water mass generally has a similar, but not identical, species composition to that found on natural benthic and macrophytic substrates. However, under specific conditions of current and temperature, some substrates are more efficient in obtaining representative populations, e.g. styrofoam and stone often collect species not able to attach to smoother glass or polyethylene surfaces (HOHN and HELLERMAN 1963). A number of field instruments such as the Catherwood Diatometer (PATRICK 1949, PATRICK et al. 1954) have been used to enumerate the periphytic community and to determine variations in composition caused by pollution (PATRICK 1963 and others). A large number of materials other than glass have been tried as substrates with varying success (see review by WETZEL and WESTLAKE in VOLLENWEIDER 1969). The quantity of algae growing on such substrates, the rate of increase and the chlorophyll *a* content are the commonly measured factors leading to production estimates. The position of the substrate is known to determine to a large extent the quantity of attached algae (cf. NEWCOMBE 1949, 1950, SLÁDEČKOVÁ 1962, PIECZYNSKA 1968). DUMONT (1969) has shown that the orientation and depth of the substrate are particularly important as slight differences in current may radically alter both the supply of nutrients and the accessibility to planktonic 'seed' cells. However, in flowing river systems, depth and hence light availability would appear to be more important since nutrient renewal and gas exchange by diffusion enrichment is fairly rapid and algal propagules and displaced cells are likely to be available throughout the water mass in a turbulent river. In many studies the substrates have been suspended in convenient positions irrespective of current or light attenuation and have therefore recorded only the rate of accumulation and production of the planktonic element under surface conditions. In determining the natural benthic periphytic production, the position and attitude of the substrate is critical. CASTENHOLZ (1961) found vertical slides collected less organic material than horizontally-placed substrates in a ratio 1:6–12, depending on season. This agreed with NEWCOMBE's (1950) data and indicated that large amounts of detritus were accumulated with the algae, making biomass determinations erroneous. Duration of submergence was discussed fully by SLÁDEČKOVÁ (1962) and most workers have used two to six-week periods, depending on trophic conditions. Sloughing-off of thick growths and peeling when the lower layer of cells dies result when the optimal time needed to develop a stable or climax community is exceeded. As discussed fully by WATERS (1961a), SLÁDEČEK and SLÁDEČKOVÁ (1964), CUSHING (1967) and WETZEL (in VOLLENWEIDER 1969), the rate of accumulation of biomass on artificial substrates is a net production rate. Primary and secondary growth, consumption, decomposition, and accrual and loss of materials to the current occur throughout the exposure period and are involved in determining the ultimately measured biomass. VOLLENWEIDER (1969) felt that net production calculated from changes in biomass tended to be underestimated under natural conditions where grazing and current effects were severe.

The difficulties in obtaining the true periphytic contribution to volume or mass of the growths removed from slides or other substrates because of the accretion of fine detritus or heterotrophic organisms have led to the development of the indirect estimation of production through quantitative pigment analysis. This has been widely used in plankton and laboratory stream studies, but not extensively for the epiflora of natural stream habitats. YOUNT (1956) determined chlorophyll content of slide-grown periphyton in Silver Springs as did ODUM (1957a) who also analysed the 'aufwuchs' community of *Sagittaria* leaves. McCONNELL and SIGLER (1959), WATERS (1961a) and FLEMER (1970) estimated pigment from the periphyton of stone substrates, the last two authors providing concrete surfaces and removing them after various colonization periods. KURASAWA (1959) used tile substrates and GRZENDA and BREHMER (1960), plexiglass panels. Other pigment analysis studies have been made by MARGALEF (1960), KOBAYASI (1961), SLÁDEČEK and SLÁDEČKOVÁ (1963a) who used glass slides and WETZEL (1964), EATON and MOSS (1966), BROCK (1967), BROCK and BROCK (1967) and MOSS (1967a, b, 1968). Several difficulties are inherent in the method. Both living and ametabolic cells are extracted so that the accumulated pigments do not necessarily reflect the producing biomass and the pigment degradation products interfere with the determination of the photo-synthetically-active pigments. Extraction with the standard techniques has variable efficiency, depending on the algae and the type of substrate. In addition, the relationships between pigment content, usually expressed as chlorophyll *a*, biomass and production capability are not constant, again depending on the species involved, the physiological state of the community, and physical conditions, particularly light (KALFF 1969, WETZEL and WESTLAKE in VOLLENWEIDER 1969). Therefore, as discussed by WETZEL (1963), estimates of producer biomass based on pigment content must be interpreted with considerable caution.

Primary production measurements based on oxygen, carbon dioxide, pH or specific conductance changes with time have limited application in the flowing water habitat except under certain restricted conditions. Community metabolism measurement in the field on non-isolated sections of a river has been tried by a number of workers but overall changes in O_2 or CO_2 balance are generally non-existent in lotic waters where turbulence and diffusion currents maintain concentrations in equilibrium with the air. The analysis of diurnal curves and photosynthesis/respiration ratios has been used with varying success by ODUM (1956, 1957a, b), STOCKNER (1968a) and others (see OWENS in VOLLENWEIDER 1969 for review), but the determination of diffusion rates and exchange coefficients, magnitude of CO_2-rich ground-water inflow and loss of O_2 by bubbles are serious problems. The upstream-downstream method used by some workers (EDWARDS and OWENS 1962, WRIGHT and MILLS 1967) has similar limitations, especially in waters with low production such as the Gombak, where temperature-induced changes are more important than photosynthetic/respiration effects.

Measurements on isolated communities, by the common *in situ* incubation techniques have little application in shallow, fast-flowing lotic systems where many of the epiflora are dependent on the current for nutritional enrichment and metabolic waste disposal (MCINTIRE 1966). Enclosure in bottles, 'plastic sausages', etc. creates an artificial situation without diffusion exchange and in which the heterotrophs are provided with an increased surface area for colonization (VERDUIN 1969). However, some workers (ODUM 1957a, b, KOBAYASI 1961, KEMMERER and NEUHOLD 1969) have used these techniques with varying success. The limitations and advantages of the various techniques have been thoroughly reviewed by a number of authors in VOLLENWEIDER (1969). Of the three possible methods of measuring production, changes in biomass, natural community metabolism and *in situ* isolated periphytic production, only the first had any utility in the Gombak where standing crop was generally low and physical factors rather than photosynthesis/respiration determined the chemical conditions.

2.1. METHODS

CASTENHOLZ (1960) and SLÁDEČEK and SLÁDEČKOVÁ (1964) found that Cyanophyta were under-represented on glass slides; HOHN (1966) indicated that extruded polyethylene obtained more representative species composition than glass; NEAL et al. (1967) described Chlorophyta as slow colonizers of glass and polyethylene. In view of these results and the empirical observation that most of the Chlorophyta were epilithic in the Gombak, three types of substrate were used in this study; a. standard soft glass microscope slides; b. 2 mm thick polyethylene sheet cut to the same size as glass slides; c. asbestos-cement fire wall compound in thin sheets cut to 75 mm × 25 mm.

Ten glass, ten polyethylene and five asbestos-cement slides were exposed at each station for periods varying from four to seven weeks depending on river conditions. A galvanized wire cage with mesh size about 1 cm that prevented most tampering and obstruction by leaves, but allowed light, current and small benthic fauna to enter was used to hold the slides. The glass slides were placed in racks and held in a vertical position parallel to the current with about 1.5 cm between slides. Wires

166

strung across the cage held the asbestos-cement and polyethylene plates (through holes near the upper edges) in a similar orientation, with 1.5 cm between them maintained by rubber distance pieces. The cages were sited in mid-riffle areas at the level of the bottom sediments where potential colonizers would be abundant, in a position where light exposure and current conditions were judged to be average for the section. No maintenance other than periodic removal of leaves from the outside of the cage was made. Prior to placement in the field, the substrates were washed thoroughly and extracted in 90% aqueous acetone. The glass and polyethylene slides were dried and weighed to 0.1 mg on an Oertling model R20 balance. In the very soft water of the Gombak, the asbestos-cement plates eroded so no estimate of biomass accumulation on them was possible.

On removal at the end of the colonization period, the slides were gently rinsed to remove surface adhering particles and then allowed to air-dry at 27 °C in the dark over silica-gel desiccant for 72 h. SLÁDEČEK and SLÁDEČKOVÁ (1963b) found little difference between air-dried and oven-dried weights (8.8 to 8.0% of wet weight) so that drying under the conditions described was admissible. All glass and polyethylene slides were weighed for biomass estimation. The polyethylene, asbestos-cement and five glass slides were kept for pigment extraction. The other five glass slides were placed in Vitreosil crucibles and ignited in a muffle furnace at 600 °C for one hour to determine comparative ash-free dry weights. When, as happened on several occasions, the cage was partially buried by shifting sediments, the level of this was usually visible on the slides. The atrophic area was scraped prior to weighing and extraction and an appropriate correction to the area available for colonization was made. Additional glass and polyethylene slides for microscopical study were added when needed. On removal, these were placed in river water in screw-capped Coplin jars in the field for fresh examination.

The slides for pigment extraction were placed in screw-capped polyethylene Coplin jars (1 per substrate per station) and pre-chilled 90% aqueous acetone (analytical grade) with one drop NaOH added was poured in to cover the slides (about 40 ml). The jars were capped and placed in the dark at 2–4 °C for 20 h for extraction. No ultrasonic vibrator apparatus was available and scraping and grinding was not feasible, particularly with the asbestos-cement slides. The efficiency of this extraction was unknown (as low as 50% is reported for some algal types, STRICKLAND and PARSONS 1968), but a subsequent treatment with 90% methanol as solvent yielded an almost undetectable concentration of pigment. The extracts were filtered through fritted glass filters to remove suspended materials, made up to 50 ml in volumetric flasks and read against a reagent blank in a Beckman DU spectrophotometer using 1 cm quartz cuvettes. Readings were taken at 665, 645, 630, 510, 480 and 430 mμ and a blank at 750 mμ subtracted as a measure of background absorption. Chlorophyll *a* was calculated using the equation $11.9 \times D_{665}$ = mgChl.*a* of TALLING and DRIVER (1963). The 'trichromatic' equations of RICHARDS with THOMPSON (1952) as modified by STRICKLAND and PARSONS (1960) were also used, with reservations discussed later, to estimate the concentrations of the chlorophylls and carotenoids.

In an attempt to estimate comparative gross primary production rates, duplicate polyethylene slides from each station were brought back to the laboratory in river water after the colonization period, sliced longitudinally into halves and placed in Winkler bottles with river water. The bottles were incubated at 23.5 °C in a plant culture room, one set under about 3500 lux and the other in the dark.

The records have several gaps, notably at Station III for the periods ending 27 March 1969, 23 May 1969 and 27 November 1969. On a number of occasions vandals destroyed the apparatus at Station III and twice, after severe flooding, one or more of the cages were buried in sediment so that the whole set was discarded. The fourth and fifth sets had to be extracted together because of instrumentation difficulties. The slides from 27 March were stored frozen, in the dark, but the very low Chl. *a* concentrations found may have resulted from pigment degradation during storage (STRICKLAND and PARSONS 1968).

2.2. BIOMASS

The biomass/sq. m expressed as mg ash-free dry weight for the glass and polyethylene substrates are given in Table 51. These represent the differences between accrual of adherent organic materials, periphytic and various heterotrophic propagules, primary and secondary production by these components and attrition by current action, grazing by invertebrates, decomposition and catabolic losses, i.e. net 'production' (increase in organic content) of the attached community over the exposure period. These data exclude a variable inorganic fraction that was an integral part of the assemblage, particularly large at the lower stations where silt loads were heavy. Mean ash contents over the experimental periods were Station I 7.9%, Station II 5.0%, Station III 22.8%, Station IV 57.2%, Station V 32.4%. These materials must have contributed to the shading of the algal community. In almost all samples, the amount of organic material on the polyethylene substrates was greater than that on the smooth glass slides. The exceptions, notably at Station V, were attributed to the large population of *Hydra* (often 10–15/sq.cm) that grew preferentially on the glass. These weights are expressed as rates mg/sq.m/ day in Table 52 and give an indication of net production rate of the whole community on the different substrates. Station II, with fast current and low nutrients, supported a very poor community, considerably less than the adjacent tributary. At Station I, the slides were exposed at the top of a flowing pool adjacent to a riffle as the water depth

Table 51. Accumulated ash-free dry weight in mg/sq.m on glass (G) and polyethylene (P) substrates

Period ending	STATION I			STATION II			STATION III			STATION IV			STATION V		
	No.of days	G	P	No.of days	G	P	No.of days	G	P	No.of days	G	P	No.of days	G	P
5–XII–68	48	1084.8	1651.2	48	91.2	3072.0	47*	3445.1	4638.9	47*	752.0	1640.3	45*	4266.0	7749.0
9–I–69	32	2694.4	4214.4	32	796.8	1491.2	32	3168.0	5372.8	28*	400.4	814.8	32	2832.0	3017.6
6–II–69	28	1234.8	2651.6	28	243.6	669.2	28	2486.4	3878.0	28	800.8	747.6	28	3488.8	2357.6
27–III–69	42	3990.0	4565.4	42	197.4	676.2	—	—	—	42	684.6	945.0	42	1562.4	1688.4
23–V–69	43	1535.1	2936.9	43	1388.9	2326.3	—	—	—	43	1358.8	1857.6	43	2850.9	3422.8
24–VII–69	28	716.8	1565.2	28	nd	nd	28	3245.2	4337.2	28	495.6	1022.0	28	2254.0	2198.0
3–IX–69	27	1301.4	3496.5	27	291.6	823.5	27	3890.7	2386.8	27	353.7	899.1	27	4077.0	4136.4
9–X–69	28	789.6	658.0	28	277.2	890.4	28	2399.6	2934.4	28	1772.4	2816.8	28	2100.0	4578.0
27–XI–69	35	1253.0	1869.0	35	609.0	1585.5	—	—	—	35	280.0	966.0	35	2145.5	2429.0

* Number of days varied because of flooding which prevented establishment of substrates on same date.

Table 52. Rate of accumulation of ash-free dry weight in mg/sq.m/day on glass (G) and polyethylene (P) substrates

Period ending[+]	STATION I		STATION II		STATION III		STATION IV		STATION V	
	G	P	G	P	G	P	G	P	G	P
5–XII–69	22.6	34.4	1.9	64.0	73.3	98.7	16.0	34.9	94.8	172.2
9–I–69	84.2	131.7	24.9	46.6	99.0	167.9	14.3	29.1	88.5	94.3
6–II–69	44.1	94.7	9.7	23.9	88.8	138.5	28.6	26.7	124.6	84.2
27–III–69	95.0	108.7	4.7	16.1	—	—	16.3	22.5	37.2	40.2
23–V–69	35.7	68.3	32.3	54.1	—	—	31.6	43.2	66.3	79.6
24–VII–69	25.6	55.9	nd	nd	115.9	154.9	17.7	36.5	80.5	78.5
3–IX–69	48.2	129.5	10.8	30.5	144.1	88.4	13.1	33.3	151.0	153.2
9–X–69	28.2	23.5	9.9	31.8	85.7	104.8	63.3	100.6	75.0	163.5
27–XI–69	35.8	53.4	17.4	45.3	—	—	8.0	27.6	61.3	69.4
mean[1]	46.6	77.8	12.3	34.7	101.1	125.5	23.2	39.4	86.6	103.9
	±17.3	± 26.7	± 7.1	±13.3	± 20.7	± 29.8	±11.2	± 15.9	± 22.7	± 31.2
s.d.[1]	25.9	40.0	10.7	19.9	25.4	36.5	16.8	23.8	34.1	46.9

[+] See Table 51 for length of exposure period
[1] Weighted to compensate for differences in exposure period

over the riffle was too variable to ensure continuous submergence. This may have led to increased accumulation of both propagules and particulate organic material in the quieter flow conditions. Station III had the greatest accrual rate. The river at this point carried only an average load of very fine materials, was shallow with complete exposure and little light attenuation and had moderate currents even at high stage so that losses through physical erosion were minimal. Station IV suffered severely from both siltation (see the ash content above) and scouring currents during high flow periods. The moderate organic accumulation at Station V was composed almost entirely of heterotrophs utilizing the increased available nutrients from the mild pollution. The large differences between sampling periods, apparently independent of time (compare December 1968 and January 1969), reflected in the large standard deviations for the mean values, were the result of discharge effects rather than variation in available nutrients or light. Changes in the 'aufwuchs' tended to be spasmodic, from a denuded condition after a flood to a semi-stabilized community, rather than the seasonally regulated periodic fluctuations found in most streams studied by others to date.

If the assumption made here, that after four to six weeks the community reached a stable level if it was not violently disrupted, were valid, the values in Table 51 would represent an approximation of the standing crop. There was no indication of a particular time period for greatest accrual because of the intermittent spates, but the maximum biomass data were obtained after 28–35 days at most stations. The data from the six-week exposure (December 1968) showed less organic material and empirical evidence from an abandoned series, in June 1969, indicated sloughing-off from the glass slides at Station IV after seven weeks, i.e. post-maximum development.

2.3. PHYTOPIGMENTS

The accumulated Chl. *a* concentrations in mg/sq. m, corresponding to the biomass data of Table 51, are given in Table 53. The low readings for the two sets extracted together on 23 May were perhaps caused by the storage of one set or, more likely, were the result of an extended rainy period with accompanying high silt loads that shaded and smothered algal growth even at the upper stations. The rates of Chl. *a* accumulation in µg/sq.m/day are given in Table 54. The mean values for each substrate at each station show clearly that algal development was greatest at Station III. Station IV had considerable growth also, but the large inorganic silt load found with the periphyton reduced production below the Station III level, although physical and chemical conditions were nearly identical. HAMILTON (1961) reported little effect of sand-pit washings on the encrusting and filamentous algae, but inhibition of growth was obvious in the Gombak. Station V, with high turbidity,

Table 53. Accumulated Chl. *a* in mg/sq.m on asbestos-cement (A), glass (G) and polyethylene (P) substrates

Period ending[+]	STATION I			STATION II			STATION III			STATION IV			STATION V		
	A	G	P	A	G	P	A	G	P	A	G	P	A	G	P
5-X-68	0.96	0.50	0.25	6.93	0.75	0.50	41.22	3.01	1.51	6.21	5.52	5.52	2.87	1.00	2.01
9-I-69	0.00	1.26	0.00	0.00	0.00	0.00	34.65	22.59	21.34	0.00	0.00	0.00	0.00	0.00	0.00
6-II-69	1.67	0.00	0.94	7.17	0.00	0.00	37.04	11.30	20.71	19.59	6.28	8.85	2.87	0.00	0.00
23-V-69*	1.08	0.97	0.85	1.55	0.28	0.19	—	—	—	2.03	0.85	0.94	1.31	0.19	0.28
24-VII-69	2.03	1.21	2.52	2.76	0.00	0.46	21.91	16.84	24.41	8.71	11.98	28.99	3.63	0.75	1.91
3-IX-69	1.45	0.42	0.69	3.77	0.17	0.46	34.10	10.78	10.60	3.92	0.42	0.99	5.08	0.50	1.86
9-X-69	1.89	0.00	0.48	5.51	0.28	1.14	41.21	32.87	41.57	59.06	39.48	42.50	10.59	1.40	2.67
27-XI-69	2.32	0.25	0.42	9.58	0.25	0.85	—	—	—	10.88	7.12	0.37	8.56	0.42	1.27

* Two sets extracted together; [+] See Table 51 for length of exposure period.

Table 54. Rate of Chl. *a* accumulation in μg/sq.m/day on asbestos-cement (A), glass (G) and polyethylene (P) substrates

Period ending[+]	STATION I			STATION II			STATION III			STATION IV			STATION V		
	A	G	P	A	G	P	A	G	P	A	G	P	A	G	P
5-X-68	20	10	5	144	16	10	877	64	32	132	117	117	64	22	45
9-I-69	0	39	0	0	0	0	1083	706	667	0	0	0	0	0	0
6-II-69	60	0	34	256	0	0	1323	404	740	700	224	316	103	0	0
23-V-69*	13	11	10	18	3	2	—	—	—	24	10	11	15	2	3
24-VII-69	73	43	90	99	0	16	783	601	872	311	428	1035	130	27	68
3-IX-69	54	16	26	140	6	17	1263	399	393	145	16	37	188	19	69
9-X-69	68	0	17	197	10	41	1472	1174	1485	2109	1410	1518	378	50	95
27-XI-69	66	7	12	274	7	24	—	—	—	311	11	203	245	36	12
mean¹	51 ±18	21 ±14	28 ±22	161 ±68	8 ±4	18 ±12	1134 ±218	558 ±304	698 ±398	533 ±550	317 ±382	462 ±440	160 ±92	22 ±14	53 ±26
s.d.¹	24	18	29	90	5	13	268	373	488	728	506	582	122	16	32

* Two sets extracted together; [+] See Table 51 for length of exposure period; [1] Weighted to compensate for differences in exposure period.

Table 55. Results of split-plot analysis of variance of rates of Chl.a accumulation on different substrates at the five stations

		Significance level[a]
Differences between stations		* * *
Differences between substrates		* *
Station-substrate interactions		* *

Inter-station differences (2-tailed test)

Station	I	II	III	IV	V
I	—	NS	* * *	*	*
II		—	* * *	*	NS
III			—	+	* * *
IV				—	*

Inter-substrate differences

	G	P	A
G	—	NS	* *
P		—	*

[a] Levels of significance are indicated by asterisks: * P < 0.05; ** P < 0.01 *** P < 0.001; NS, not significant (P > 0.05); (+ = significant at P < 0.1, not significant at P < 0.05)

elevated nutrient content and numerous microconsumers, had a very limited phytoautotrophic community, substantiating the contention of BACKHAUS (1969) that biomass cannot be used to indicate eutrophication. Standard deviations were considerable for the reasons mentioned in connection with biomass and also as a result of shading by accreted silt that built up under stable or falling stage conditions. No seasonal differences paralleling physical conditions were evident.

Differences between stations and substrates were obvious, but to better define these relationships a split-plot analysis of variance (COCHRAN and COX 1959) was carried out with the results as indicated in Table 55. This analysis was made on the readings of the five complete series only, i.e. the incomplete data from 9 January 1969, 23 May 1969, and 27 November 1969 were omitted. The rates of Chl. a production at the two forest stations were not significantly different from each other, but were different from the rates of the middle and lower river. The 'not significant' result between Stations II and V reflects the severe conditions for periphytic development at the lower station. Stations III and IV in the high available irradiation zone were significantly different from the other stations. The data at Station III were highly significantly different from those at other stations except IV. The inter-substrate differences were less marked, but still significant. As expected, the asbestos-cement plates, which approximated natural rocks in texture, developed significantly more periphyton than either the smooth glass or polyethylene type

Table 56. Sample pigment concentrations derived from the trichromatic equations of RICHARDS with THOMPSON (1952).

STATION	Glass slides					Polyethylene slides					Asbestos-cement slides				
	I	II	III	IV	V	I	II	III	IV	V	I	II	III	IV	V
26 June–24 July 1969															
AFDW mg/m²	716.8	nd	3245.2	495.6	2254.0	1565.2	nd	4337.2	1022.0	2198.0					
Chl. a mg/m²	1.41	—	20.69	14.74	0.91	3.04	0.55	29.91	35.85	2.32	2.39	3.29	26.99	10.72	4.44
Chl. b mg/m²	0.35	—	0.53	0.38	0.33	0.36	0.12	1.41	19.03	0.26	2.55	0.32	—	—	—
Chl. c m-SPU/m²	3.50	—	26.26	17.12	3.46	5.26	1.06	39.33	35.18	3.15	5.98	5.85	25.50	9.07	5.63
Ast. carot. m-SPU/m²	0.76	0.81	2.37	—	0.77	0.71	0.31	3.67	2.78	0.47	1.03	1.25	3.04	1.65	1.37
Non-ast. carot. m-SPU/m²	0.17	—	5.28	7.62	—	0.51	—	6.64	7.15	0.44	0.55	0.55	8.07	2.14	0.88
7 August–3 September 1969															
AFDW mg/m²	1301.4	291.6	3890.7	353.7	4077.0	3496.5	823.5	2386.8	899.1	4137.4					
Chl. a mg/m²	0.49	0.20	13.20	—	0.62	0.81	0.55	13.03	1.17	2.06	1.67	4.61	41.93	4.80	6.20
Chl. b mg/²	0.16	0.04	1.04	3.56	0.03	0.23	0.12	0.66	0.31	0.22	0.44	—	—	0.02	0.02
Chl. c m-SPU/m²	1.25	0.39	15.42	19.32	0.60	1.85	1.06	14.27	2.56	1.66	5.61	5.14	41.32	3.67	6.68
Ast. carot. m-SPU/m²	0.42	0.23	1.10	0.30	0.22	0.49	0.25	1.07	0.43	0.19	1.03	0.93	4.28	0.75	1.15
Non-ast. carot. m-SPU/m²	—	—	2.91	—	0.11	—	0.00	2.90	0.04	0.51	0.20	1.20	12.44	1.39	1.71
11 September–9 October 1969															
AFDW mg/m²	789.6	277.2	2399.6	1772.4	2100.0	658.0	890.4	2934.4	2816.8	4578.0					
Chl. a mg/m²	—	0.33	36.54	48.76	1.55	0.56	1.37	50.87	52.55	3.20	2.24	6.76	50.14	72.81	13.00
Chl. b mg/m²	—	0.13	2.30	0.80	0.89	0.27	0.30	3.41	0.16	0.50	0.40	—	—	0.02	—
Chl. c m-SPU/m²	—	0.97	46.02	48.53	7.50	0.56	4.32	65.39	53.32	6.43	3.85	6.23	137.20	59.99	10.64
Ast. carot. m-SPU/m²	0.02	0.21	2.92	3.41	1.53	0.35	0.47	2.85	3.87	1.12	0.88	1.03	4.78	6.14	1.71
Non-ast. carot. m-SPU/m²	0.09	0.09	4.68	10.00	—	—	0.01	8.51	11.30	0.30	0.36	1.62	13.63	20.95	3.23

of slides. There may have been slight enrichment from this substrate, particularly of calcium, but its effects could not be assessed. No appreciable difference was evident between the Chl.*a* accumulation rate on glass and on polyethylene although the latter usually had slightly higher concentrations.

The most natural periphytic assemblage probably developed on the asbestos-cement plates, but to evaluate biomass relationships, the comparative weight-Chl.*a* data derived from the glass and polyethylene slides and the ash-free dry weights only detectable from the glass substrates were necessary, because of the difficulties of weighing the asbestoscement.

The concentrations of the various phytopigments derived using the trichromatic equations are listed in Table 56 for the 26 June–24 July, 7 August–3 September and 11 September–9 October periods which were representative. (Note: Chl. *c* and the carotenoids are given in thousandths of a Specified Pigment Unit (m-SPU) as defined by RICHARDS with THOMPSON (1952)). The Chl. *a* values obtained here are similar to those from the TALLING and DRIVER equation given in Table 53, but almost always higher as discussed by TALLING and DRIVER (1963), VOLLENWEIDER (1969) and others. To avoid this overestimation, several authors have suggested abbreviated expressions for determining Chl. *a*: STRICKLAND and PARSONS (1960) recommended 15.0 D_{665}; ODUM et al. (1958) and BROCK (1967) used 14.3 D_{665}; MOSS (1967b) found a factor of 11.0. The TALLING and DRIVER constant 11.9 was used as it appeared to represent the best empirical absorption coefficient for Chl. *a* in acetone. On the supplied substrates subject to constant current effects, there was probably little accumulation of dead or moribund periphyton, and hence of phaeophytins or other pigment degradation products that would invalidate the use of the RICHARDS with THOMPSON method (HUMPHREY 1963). The presence of these fractions shifts the maximum absorption from 430 to 410 mμ (Moss 1967a). The extracts from Gombak algae always had the ratio $D_{430-435}$: D_{410} greater than one, indicating that the concentration of degradation products was low and could be ignored (cf. BROCK and BROCK 1967), and that differentiation between Chl. *a* and phaeophytin *a* concentrations by the acidification techniques of LORENZEN (1967) and Moss (1967a, b) was not necessary.

As already noted, the epiflora at all stations, but particularly those in the forest, included many Cyanophyta. In addition, phycoerythrincontaining Rhodophyta were often present in the encrusting epilithos. The pigments from these groups absorb green-yellow light and transmit in the blue and red part of the spectrum respectively (ROUND 1965), thus interfering with chlorophyll absorption in an indeterminate way depending on their concentration. Therefore, the warning by STRICKLAND and PARSONS (1960) repeated by STRICKLAND (1963) of the invalidity of the trichromatic equations when phycobillins are present makes the use of data derived from them suspect. The significant feature revealed by these data was the relatively high concentration of Chl. *c* in almost all the extracts. Allowing for the m-SPU to be greater than 1 mg (STRICKLAND and PARSONS 1968) and bearing in mind the findings of HUMPHREY (1963) that Chl. *c* was overestimated by 20–50% by these equations, significant concentrations of Chl. *c* would still be present in most extracts. This was expected as Bacillariophyceae were the largest periphytic component by numbers and these contain considerable concentrations of chlorophyll *c* (WESTLAKE in VOLLENWEIDER 1969).

The animal carotenoid pigments were also present in surprisingly high concentrations compared to the non-astacin, largely diatom, carotenoids. The equations gave obviously erroneous results, often zero or less than zero, for these non-astacin pigments even though much of the growth was epiphytic diatoms!

No relationship between Chl. *a* and ash-free dry weight was evident because of the often large and indeterminable heterotroph and organic detritus contribution to the organic weight. A number of authors, working in lakes particularly, have been able to obtain significant correlations between these parameters (WETZEL 1963, SLÁDEČEK and SLÁDEČKOVÁ 1964, CUSHING 1967). In a review of literature to 1964, SLÁDEČEK and SLÁDEČKOVÁ found a range of 0.09–2.4% and compromised by assuming Chl. *a* to be 1.3% of ash-free dry weight. ROUND (1965) extended this range to 0.01–6% under exceptional metabolic conditions. The mean percentages calculated from total Chl. *a* and total accumulated biomass at each station for the glass and polyethylene substrates respectively were: Station I 0.032, 0.026; Station II 0.044, 0.031; Station III 0.523, 0.510; Station IV 0.941, 0.811; Station V 0.017, 0.032. With the exception of Station IV where heterotrophic growth was minimal, these are low. At the upper two stations, a higher percentage might have been expected with increased production of chlorophyll by the subdued-light adapted forms (ROUND 1965) than at the fully exposed, high light intensity lower three stations where photo-oxidation and inhibition probably occurred (see EDMONDSON 1956, KALFF 1969). As a result of the reduced intensities and spectrally-shifted light reaching the forest stream, greater concentrations of accessory pigments may be produced to take advantage of these conditions; this would partially account for the success of Myxophyceae and Rhodophyceae and the relative scarcity of Chlorophyceae under the jungle. However, there was no evidence of this from this study, but it might warrant further investigation. For the purpose of estimating possible production relationships, the percentage Chl. *a* at Station IV was assumed to be indicative of the actual pigment content of the heterogeneous flora of the Gombak and a rounded figure of 1% Chl. *a* by weight was adopted. From this and the mean rates of Chl. *a* accumulation, a biomass figure for periphyton alone, free of other organics and consumer biomass, could be calculated and the algal component on the asbestos-cement substrates quantified.

In Table 57, the important parameters for net production estimation have been listed. The mean organic and mean Chl. *a* content were calculated assuming that at the end of each of the colonization periods a stable equilibrated community was established, representative of standing crop (biomass) as discussed earlier. From these data Table 58, showing projected net annual production and the turnover ratio, was calculated. Values for the three substrates at each station for primary production and for the glass and polyethylene slides for the entire organic assemblage irrespective of source are listed. The total ash-free dry weight production is a measure of the potential food available to the stone-scraping macro-invertebrates living on the riffles.

The last entry in Table 57 gives the net efficiencies of conversion into organic matter of the light reaching the substrates. The mean daily rate of Chl. *a* accumulation (net) was converted to biomass using the 1% by weight factor and expressed as calories at the rate of 4.9 kg cal/g (HAR-GRAVE 1969). This was then calculated as a percentage of the mean visible radiation energy reaching the substrates, as given in Table 59, assuming that 50% was photosynthetically available to the algae (STRICK-LAND 1968). SZEICZ (1966) gave a range, 38–51% mean 41% for cloudy-sunny conditions, but the more commonly used 50% figure was used here. Also evident in that table are the effects of shading by the forest at Stations I and II and by turbidity at Station V. The slight variances between Stations III and IV were caused by differences in light attenuation at the depths of the slides.

In the discussion of standing crop (biomass), net production and turn-over ratio, the optimal substrate only will be considered, i.e. asbestos-cement for periphyton and polyethylene for total organic. The asbestos-cement slides probably did not accumulate much more total organic than the polyethylene slides, except for the enhanced algal growth which was only a comparatively small part of the total. Few studies on small rivers are available with which to compare the Gombak data, and none on tropical streams. Table 60 compares the mean values of Chl. *a* and biomass found in other lotic locations with those from the Gombak; for wider comparison see the tables in Moss (1968). The forest stations had very low Chl. *a* content, paralleling the laboratory streams of McINTIRE et al. under low illumination. Stations III and IV had values in the same range as the rivers of KOBAYASI, WATERS and CUSHING, and two to four times the 6–9 mg Chl. *a*/sq. m found in epipelic communities on the Niger by EATON (see MOSS 1968). The reduced periphyton at Station V was largely the result of suspended silt shading the substrates. Despite relatively enriched nutrient content, particularly of phosphate, nitrate and calcium, algal growth appeared to be limited by the reduced light. An extensive microconsumer fauna may also have helped to keep the

Table 57. Total weight, mean biomass and rate of accumulation of (a) ash-free dry weight (AFDW), and (b) Chl. *a* on artificial substrates. (c) Net efficiency of energy conversion

	STATION I			STATION II			STATION III			STATION IV			STATION V		
(a) AFDW	A	G	P	A	G	P	A	G	P	A	G	P	A	G	P
Total accumulated mg/sq.m	—	14599.9	23608.2	—	3895.7	11534.3	—	18635.0	23548.1	—	6898.3	11709.2	—	25576.6	31576.8
Mean biomass mg/sq.m	—	1622.2	2623.1	—	432.9	1281.6	—	3105.0	3924.7	—	766.5	1301.0	—	2841.8	3508.5
Mean accumulation rate mg/sq.m/day	—	46.6	77.8	—	12.3	34.7	—	101.1	125.5	—	23.2	39.5	—	86.6	103.9
(b) Chl. *a*	A	G	P	A	G,	P	A	G	P	A	G	P	A	G	P
Total accumulated mg/sq.m	11.40	4.61	6.15	37.23	1.73	3.60	210.13	97.39	120.14	110.40	64.90	94.91	34.91	4.26	10.00
Mean content mg/sq.m	1.27	0.51	0.68	4.14	0.19	0.40	35.02	16.23	20.02	12.27	7.21	10.55	3.88	0.47	1.11
Mean accumulation rate mg/sq.m/day	0.051	0.021	0.028	0.161	0.008	0.018	1.134	0.558	0.698	0.533	0.317	0.462	0.160	0.022	0.053
(c) Efficiency	A	G	P	A	G	P	A	G	P	A	G	P	A	G	P
%	0.032	0.013	0.016	0.055	0.003	0.006	0.027	0.013	0.016	0.012	0.007	0.011	0.005	0.001	0.002

Table 58. Annual production P(mg/sq. m) and turnover ratio TR($P/\bar{B}$ mean biomass).

	Periphyton*			Total organics		
	P	$\bar{B}$	TR	P	$\bar{B}$	TR
Station I						
A	1861.5	127	14.7			
G	766.5	51	15.0	17009.0	1622.2	10.5
P	1022.0	68	15.0	28397.0	2623.1	10.8
Station II						
A	5876.5	414	14.2			
G	292.0	19	15.4	4489.5	432.9	10.4
P	657.0	40	16.4	12665.5	1281.6	9.9
Station III						
A	41391.0	3502	11.8			
G	20367.0	1623	12.5	36901.5	3105.8	11.9
P	25477.0	2002	12.7	45807.5	3924.7	11.7
Station IV						
A	19454.5	1227	15.9			
G	11570.5	721	16.0	8468.0	766.5	11.0
P	16863.0	1055	16.0	14381.0	1301.0	11.1
Station V						
A	5840.0	388	15.1			
G	803.0	47	17.1	31609.0	2841.8	11.1
P	1934.5	111	17.4	37923.5	3508.5	10.8

* Calculated from Chl. *a* content assuming Chl. *a* = 1% of periphyton ash-free dry weight

standing crop small at this station. Data for the Gombak are not really comparable to those from the chemically much richer Mission River of ODUM et al. Total biomass was similar to values reported by KOBAYASI and CUSHING, being more dependent on current velocity, and perhaps grazing pressure, than primary production. The large accumulation in the artificial streams of MCINTIRE et al. reflected the slow current conditions. Mean calculated periphyton standing crops for the Gombak were small, 127, 414, 3502, 1227, 388 mg/sq.m, at the five stations. Net production is usually considered to be about half of gross production (WESTLAKE in VOLLENWEIDER 1969), but this approximation would apply only to the autotrophic algal component and so would increase the total organic production by just a small amount.

The net production rates were very modest compared to the 12300 mg/sq.m/day of Silver Springs (ODUM 1957a), but at Station III on the asbestos-cement slides, the value of 1.13 was not far from the 1.32 and 1.57 Chl. *a* mg/sq.m/day of WATERS and CUSHING respectively. It must be reiterated that this production occurs only on the stable riffle areas; little or no primary production is possible in the shifting sand areas so that production per mean unit area of the river would be much lower.

The turnover ratio, production/mean biomass ('Aktivitätskoeffizient')

Table 59. Total visible light energy (gcal/cm²) available at the water surface and at the depths of the artificial substrates

Period ending	Station I		Station II		Station III		Station IV		Station V	
	Surface	Slides	Surface	Slides	Surface	Slides	Surface	Slides	Surface	Slides
5–X–68	1011	708	1670	1378	22015	19611	22015	19831	20983	13679
9–I–69	615	480	1018	839	13741	12240	9971	8982	13741	8958
6–II–69	627	489	1036	854	14006	12476	14006	12616	14006	9130
27–III–69	1017	793	1676	1382	—	—	22642	20396	22642	14760
23–V–69	925	722	1527	1259	—	—	20644	18596	20644	13458
24–VII–69	569	444	938	773	12687	11301	12687	11428	12687	8271
3–IX–69	543	424	899	741	12142	10816	12142	10938	12142	7915
9–X–69	613	386	1002	827	13555	12075	13555	12210	13555	8836
27–XI–69	661	417	1031	892	—	—	14602	13153	14602	9519
Total days of exposure	311		311		190		306		308	
Total light available gcal/cm²/period	4863		8945		78519		128150		94526	
Mean light available gcal/cm²/day	15.64		28.76		413.26		418.79		306.90	

Table 60. Comparison of values of Chl. *a* and biomass (ash-free dry weight) found for the Gombak and other rivers

Reference	Location	Substrate and conditions	Chl. *a* mg/m²	Biomass g/m²
YOUNT (1956)	Silver Springs, Fla.	Glass slides – 16 days	34.2	—
		– 28 days	80.0	—
ODUM (1957a)	Silver Springs, Fla.	*Sagittaria* blades	(2950)*	188
ODUM et al. (1958)	Mission River, Texas	Cyanophyte mat (polluted)	1960	—
McCONNELL and SIGLER (1959)	Logan River, Utah	Rocks	300 –1400	25
KOBAYASI (1961b)	Arakawa R., Japan	Rocks	30 – 70	2.5–7.0
WATERS (1961a)**	Valley Creek, Minn.	Concrete – 14 days	18.5	—
McINTIRE et al. (1964)	Artificial streams	Rocks – 2000-6000 lux	0.5– 0.8	187
McINTIRE (1966)	Artificial streams	Rocks	230 – 380	—
CUSHING (1967)	Columbia R., Wash.	Glass slides – 14 days	22	4.2
Present study	Gombak R.	Asbestos-cement, glass and polyethylene slides – 27–48 days		
	Station I	shaded	1.3	1.6
	II	shaded	4.1	1.3
	III	open	35.0	3.9
	IV	some silting	12.3	1.3
	V	slightly polluted, turbid	3.8	3.5

* included pigment of *Sagittaria* – not comparable; ** calculated by CUSHING (1967).

is a measure of the mean specific rate of renewal of the community. Total ash-free weight was renewed at all stations at approximately the same rate, once every 35 days, but this has little significance as much of it is simply deposited detritus. The microconsumers must recycle at a much higher rate. The number of renewals for the periphyton component alone varied from 15 at Stations I and II (once every 24 days) to 12.5 (29 days) at Station III, 16 (23 days) at IV and 17 (22 days) at V. These rates justify the length of time the slides were exposed, which provided adequate time, after an initial period for 'seeding-on' of propagules, for growth and development of a stable biomass, if physical conditions did not interfere. The following comparative values listed by SLÁDEČEK and SLÁDEČKOVÁ (1964) were based on total biomass rather than on periphyton alone, but in lenitic habitats the greatest part of the accumulated organics is likely to be of algal origin: Silver Springs (ODUM 1957a, b) had a turnover time of 14.4 days; the lakes investigated by CASTENHOLZ (1960) in Washington 22 days; by NEWCOMBE (1950) in Michigan 32 days; for the Sedlice Reservoir, Czechoslovakia (SLÁDEČEK and SLÁDEČKOVÁ) almost 60 days. Of interest was the consistently smaller ratio of periphyton turnover found for the asbestos-cement than for the other substrates. This perhaps reflected a greater stability and resistance to erosion that required less frequent renewal.

Gross Primary Production: The estimates obtained for gross primary production by the light and dark oxygen technique suffered from most of the diffusion and microorganism colonization shortcomings of the test discussed by VERDUIN (1969). In addition, the serious difficulties inherent in confining a lotic flora in a static habitat for an extended period, in which the normal nutrient-uptake/photosynthetic regime must be disrupted, were a major factor. On several occasions at Stations I and II, oxygen concentrations in the light bottles were anomalously lower than those produced in the dark, perhaps as a result of photo-oxidation in the strong light of the test conditions compared to the intensities in the forest stream or from photokinetic effects on the microorganisms as discussed by TALLING (in VOLLENWEIDER 1969). The data for the communities grown on polyethylene slides during the period ending 9 October are given as an example:

Stations I and II: Heterotrophic changes in light bottle greater than respiratory changes in dark as discussed above.

Station III:

317.3 mg O_2/g AFDW/24h equivalent to 933 mg O_2/g Chl. *a*/h

Station IV:

210.0 mg O_2/g AFDW/24h equivalent to 579 mg O_2/g Chl. *a*/h

Station V:

30.0 mg O_2/g AFDW/24h equivalent to 2155 mg O_2/g Chl. *a*/h

These production rates are of the same order as the 0.5–0.7 mgO_2/mg Chl. *a*/h found by McCONNELL and SIGLER (1959) in the Logan River.

The total 'aufwuchs' community of the forest and polluted stations (I, II and V) was probably heterotrophic under most conditions. HARGRAVE (1969), in a study on lake sediments, found that the microheterotrophs constituting only 4% of biomass were responsible for 67% of total community respiration, so that where algal standing crop was small, as at these stations, net production was negative, i.e. the respiratory requirements of the bacterial and protozoan population were greater than the oxygen produced. Moderate positive production occurred only at Stations III and IV. The gross production rate at enriched Station V may have been high, but net production was low or negative and grazing or erosion kept the biomass at a low level as seen.

Efficiencies: The efficiencies of net production for the Gombak are extremely low when compared to the average 4–5.2% gross production efficiencies of various Florida springs (ODUM 1957a, b). KING and BALL (1967), in the warm water, slightly polluted Red Cedar River, found a net efficiency of conversion of light at the substrate of 0.298% and NELSON and SCOTT (1962) found 0.28% efficiency by *Podestemum*, values about ten times the 0.027% found on the optimal asbestos-cement substrates at Station III. The low Gombak values (Table 57) emphasize the nutrient poverty of the river and the harsh physical conditions not conducive to algal development. Enrichment by current effects which provide a constantly renewed supply of nutrients (ODUM 1956, RUTTNER 1963) may in many situations compensate for poor water quality (see review by MCINTIRE 1966) but appeared to be ineffective here. Predictably, the highest efficiencies occurred at the forest stations (0.03–0.05%) in subdued illumination where inhibition did not occur, and decreased as total radiation increased at the Middle and Lower Zone stations (see DUFFER and DORRIS 1966). However, these efficiency estimates are based on the assumption of 1% Chl. *a* content in the periphyton. If this is erroneous, or if other pigments make unusually large accessory contributions to photosynthesis, especially under the forest, these efficiencies would be meaningless.

Pigment Ratios: MARGALEF (1960, 1961a, b, 1968) has developed the theory that the pigment diversity expressed as D_{430}/D_{665} may be used as an indicator of the stage in algal succession, as the ratio measures biochemical diversity, and hence indirectly species diversity, in a population. GARCIA (1970) and others (see KORMONDY 1969 for review) have used this scheme with indifferent success to delimit successional stages and ecological conditions.

The concept of 'climax' in respect to algal communities has been reviewed by BLUM (1956b), WHITFORD and SCHUMACHER (1963) and MARGALEF with the consensus opinion that a stable species assemblage, constituting a climax community in the successional sense, may develop ephemerally in a short season and then decline, perhaps to reappear at a subsequent time with the same composition. In frequently disturbed

Table 61. Values of the ratio D_{430}/D_{665} for selected sets of pigment extractions

Period ending	5–XII–68			3–IX–69		
	A	G	P	A	G	P
Station I	3.75	6.00	12.00	4.10	5.00	3.83
II	2.41	4.00	4.00	3.35	4.50	4.00
III	2.50	14.38	3.50	3.00	3.12	2.99
IV	2.42	1.18	3.41	3.30	4.80	3.62
V	2.67	3.50	3.13	3.34	4.50	3.36

habitats, as at Stations III, IV and V, no such assemblage can develop. The diversity depends on the success of the first colonizers rather than on an ordered succession of species adapted to the changing environmental conditions. The pigment index should, therefore, decrease as the water is modified downstream, parallel to decreases in algal species diversity which have been found by PATRICK et al. (1954), FJERDINGSTAD (1964), KULLBERG (1968), STOCKNER (1968b) and many others to result with various pollutants. However, the value of the ratio may also depend on the presence of pigment degradation products with maximum absorbence at 410 mμ, so that a high ratio in a community may be a function of senility rather than species composition. The ratio defines a general relationship between carotenoids and chlorophyll *a* and has been suggested as an indicator of various nutrient levels and hence enrichment (WILIIM and LONG 1969). MARGALEF (1961b) stated that if nutrients increase, Chl. *a* should increase at a faster rate than the other pigments. In a community, this should represent a change in species composition and diversity of the kind reported by NEEL (1953) in which a diatom flora was replaced by *Cladophora* in response to increased nitrate and phosphate levels.

Because of the problems of clumping and sparse erratic periphytic growth, no valid numerical assessment of algal diversity was possible in this study. The ratio D_{430}/D_{665} was calculated to see if there were any correlation between it and total numbers of species at each station or if any linear succession were obvious. The complete data for two colonization periods are given in Table 61; the other dates were either incomplete because of loss of slides or their results were similar to those presented. The index generally had highest values at Stations I and II, emphasizing the dominance of diatoms in the epilithic flora. The lowest values were usually found at Stations III and IV and this would be consistent with the increased development of Chlorophyta, particularly Desmidaceae at these stations. The high value of 14 on glass at Station III must have represented an almost pure diatom growth. Similar high values were found on other dates at Stations I and II and by WINNER (1969). The increase in the ratios at Station V was a result of the considerable contribution of astacin carotenoids from Rotifera and other consumers, rather than a floral change. If the number of species present in only the epilithic flora are considered, the succession diversity pattern and index values are parallel. From Table 47, the numbers of species were 56, 69, 31, 28, 49 at successive Stations I–V, which agrees more or less with the changes in index. However, without quantitative diversity data indicating dominance relationships, such a comparison has little relevance.

The values of the index may also indicate the degree of stability in the community they represent. The low values found here were generally from Stations III and IV where the molar action of eroding bedload during spates resulted in a periodically renewed flora; MARGALEF (1968) found values around two for young cultures. At the forest stations, erosion losses were less severe without the sand load in the water, and grazing by invertebrates (BROOK 1955, DOUGLAS 1958) would be constant but not intense enough to cause continuous rejuvenation. DICKMAN (1968) described tadpoles browsing selectively on the filamentous green algae rather than diatoms and this would also tend to increase the D_{430}/D_{665} ratio. MARGALEF found values of 3–6 for old, stabilized cultures, perhaps analogous to these communities. However, COOKE (1967) reported that ratios did not reflect microcosm maturity, as following an initial increase in the value of the ratio with age, a decline occurred, returning the index to a value not appreciably greater than that of a young system. The algae grown on asbestos-cement substrates tended to have lower index values than those on the other slide types, indicating perhaps that the smooth glass and polyethylene surfaces were not accumulating the full range of species, particularly Chlorophyta. From the microscopical examination of glass and polyethylene slides, the number of species found was always less than the total known from that station. However, the flora present invariably included the most common epilithic and epiphytic species and was considered to be representative of the whole assemblage even if the less frequently found taxa were not present.

IV. THE INVERTEBRATE FAUNA

The invertebrate community of lotic ecosystems is, with some specific exceptions, a remarkably conservative assemblage of types that recur in similar biotopes regardless of geographical location. Similar environmental niches (physical/chemical/biotic) harbour analogous taxa, often of the same familial or generic group, wherever such habitats are found (see BOTOSANEANU 1960, ILLIES 1961b, HYNES 1970). The fauna of the upper Gombak River, in its essential features, has few differences from and considerable affinities with the rheophile rain forest assemblages of Africa, Ceylon and tropical America, while the lower river shares many features with lowland 'potamon' reaches of rivers throughout the world. Structurally, the community parallels that found in more temperate streams with the notable difference that species diversity in many groups is greater and the population of any particular species comparatively smaller. Asynchronous and non-seasonal life cycles combine to keep population densities low and this, with the variety of relatively constant, but not necessarily stable, microhabitats available, permits multiple congeneric combinations or groups of unrelated invertebrates apparently exploiting the same feeding niches to occur. Under these conditions of reduced density-dependent interaction, some species are able to exist in a biotope and even successfully complete their life cycles, under marginally suitable conditions. In seasonal streams, with temporal fluctuations in resources and demographic pressures, such taxa would be eliminated.

To date, low latitude streams, particularly those of the humid tropics, have been neglected. Only a limited number of investigations are recorded in the literature (see Introduction), most of which have been qualitative in scope rather than oriented towards assessment of standing stock and population dynamics. The object of the present work was to delineate, as far as the taxonomic position would allow, the composition, distribution and extent of the invertebrate fauna of the Gombak.

The deficiency in knowledge of the systematics of most groups is the major obstacle to studies on the lotic fauna in this region as in most areas outside Europe (cf. WIGGINS 1966), and an investigation of this scope is perhaps premature in Malaya until better definition of the fauna is possible. The Odonata, Hemiptera and Mollusca have been studied systematically but the remaining Orders are almost unworked. Desultory, isolated collecting, particularly of adult forms, has placed specific names, notably of the Diptera and Trichoptera, in the literature, but almost without exception no work has been done on the immature stages of the lotic amphibiotic insects. Comparison of the fauna of peninsular Malaya with records from Java and the lesser Sunda Islands is of little value as many of the species are different, e.g. only one of 44 Ephemeroptera types had been prerecorded when checked against the original 'Sunda' paratypes. What is available, therefore, are fragmentary, unrelated, sometimes poorly described adult records of limited use. Breeding-out to obtain comparable identifiable specimens under prevailing conditions was difficult, mainly because of fungal contamination, and attempts using conventional techniques were unsuccessful.

Identifications of material were made by the following whose considerable efforts on often incomplete collections containing many undescribed taxa are gratefully acknowledged: Mr. I. BALL, University of Waterloo-Tricladida; Dr. A. J. BERRY, University of Malaya – Mollusca; Dr. E. L. BOUSFIELD, National Museum of Canada – Amphipoda; Dr. R. O. BRINKHURST, University of Toronto – Oligochaeta (Microdriles); Dr. D. M. DAVIES and Mrs. H. GYÖRKÖS, McMaster University – Simuliidae; Dr. C. H. FERNANDO and Dr. L. CHENG, University of Waterloo – Hemiptera, adult Coleoptera, *Labronema* sp.; Dr. H. E. HINTON, University of Bristol – Dryopoidea; Dr. B. JAMIESON, University of Queensland – Oligochaeta (Megadriles); Dr. D. S. JOHNSON, University of Singapore – Palaemonidae; Dr. T. KAWAI, Nara Women's University – Plecoptera; Dr. I. MÜLLER-LIEBENAU, Max-Planck-Instituts für Limnologie – Baetidae; Dr. W. L. PETERS, Florida A. and M. University – Ephemeroptera; Dr. G. B. WIGGINS and Mr. YAMAMOTO, Royal Ontario Museum – Trichoptera; Dr. P. ZWICK, Max-Planck-Instituts für Limnologie – Blepharoceridae, Psychodidae. Determined specimens were used in further identification and interpretation of the faunal relationships and the responsibility for any errors in this rests wholly with the writer. Odonata were identified against a reference collection made by Dr. J. I. FURTADO, University of Malaya, and confirmed by him.

Under these nomenclatural conditions, no attempt at zoogeographical discussion can be nor will be justified until much wider studies of the distribution and ecology of the Sundanian fauna have been made.

1. Composition of the Benthos

1.1. METHODS

The main biotopes at each station were sampled quantitatively at four-week intervals using the techniques described below. These were supplemented by casual collections from most of the tributaries and catches of adults made by light- and passive trapping.

The depositional(s) and erosional(r) substrates were sampled with a box-device based on the principal used by NEILL (1938) in his cylindrical sampler, but with modifications to suit local conditions. This sampler

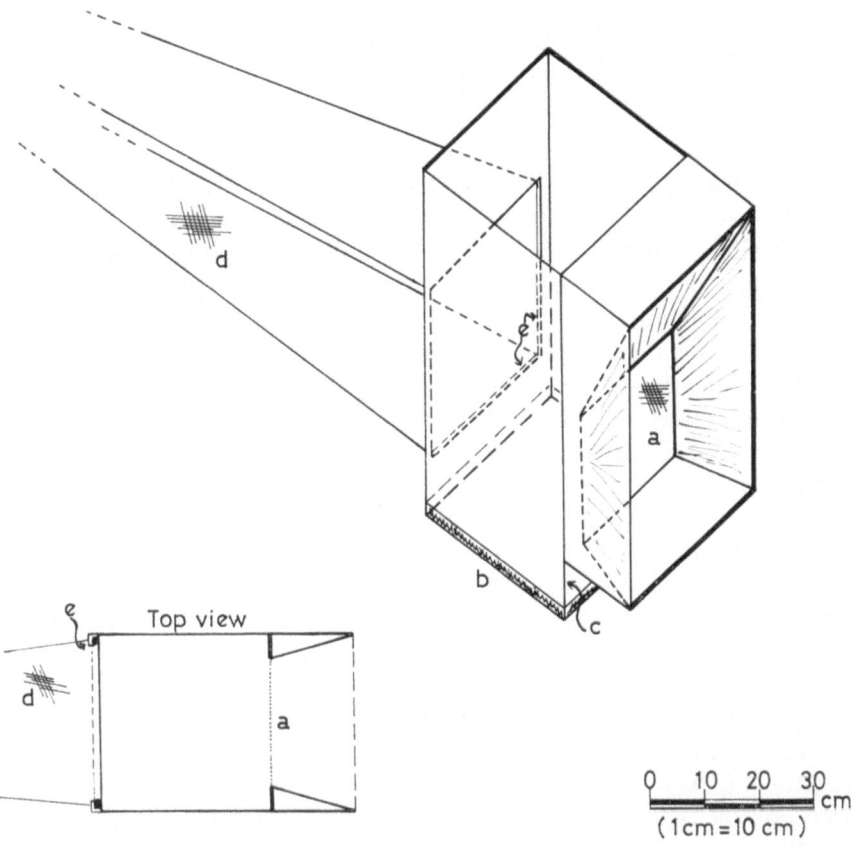

Fig. 39. Diagram of benthos box-sampler used in erosional and depositional biotopes; a. #60 mesh copper screen (22.5 × 22.5 cm), b. reinforced cutting edge, c. pre-set depth, d. 165 μ mesh collecting net (28.5 × 28.5 × 200 cm), e. slides to hold net-frame.

(Fig. 39) was an open-bottomed box constructed of heavy gauge (about 2 mm) galvanized sheet metal, enclosing an area of 1000 sq. cm. The front was fitted with a current scoop and #60 mesh copper screen to prevent entry of drifting fauna and other organics, and the back with a detachable, 2 m-long tapering net bag made of 165 μ terylene mesh. The bottom edges of the sampler were reinforced for cutting and a flange at 10 cm pre-set the penetration depth. In practice, the box was rapidly inserted into the substrate from downstream with the screened front facing into the current. The larger enclosed stones were picked out by hand and scrubbed in the mouth of the net; the remaining gravels and sands were then vigorously disturbed with a flat, spoon-like digger of 10 cm blade length. The enclosed area was systematically worked by inserting the digger and violently lifting the substrate into the current

189

stream. This action was continued for a timed 3-min period per sample and ensured thorough disruption of the bottom. In slow currents in pooled areas, the suspended contents of the box were flushed into the net by forcing water through the copper screen. The total contents of the net, both organic and inorganic were washed into polyethylene bags and preserved with formaldehyde. At each station, four samples were taken on each date using the sampler in a) a slow, usually deep, pooled area, b) a flowing pool or shallow depositing reach, c) a slow to moderate velocity riffle, and d) a rushing, usually shallow, riffle area; a plus b, and c plus d were pooled to make composite depositional and erosional area samples for each location.

The root-bank-leaf-packet biotope was sampled with a coarse meshed (8/cm), triangular-shaped net held vertically against the current while the substrate was violently disturbed by hand or foot (see HYNES 1961, MORGAN and EGGLISHAW 1965). No attempt was made to sample a defined area as variation in age of leaf-drifts and in the extent of root-mats was considerable. In marginal areas where the current was reduced, particularly at the lower stations, the net was swept back and forth through the bankside submergents and accumulations of debris. These samples were preserved as before. Special habitats not effectively sampled by the above methods, such as torrential cascades, thin water films and the surfaces of large boulders, were qualitatively searched by hand. The biotopes sampled by coarse net were occasionally checked with a 90 μ mesh net to obtain the smaller forms. The large volumes of debris and detritus caught in this fine net were examined unpreserved in enamelled trays in the field and the separated invertebrates preserved in 70% ethanol.

Ancillary separation of the invertebrates from the inorganic sediments was carried out in the laboratory by repeated (usually ×4) flotation in saturated calcium chloride solution and elutriation through a 165 μ or finer sieve. Leaf fragments and other larger organic particles were removed and the remainder of the sample preserved in 70% ethanol. The sand and gravel residue was hand-picked for molluscs and cased Trichoptera. In most samples, the volume of material was too great to count in its entirety; when this occurred, subsampling, using a modification of the technique of ALLANSON and KERRICK (1961), as used in BISHOP and HYNES (1969a, b), was carried out prior to identification, enumeration and sizing. No allowance was made for shrinkage in the ethanol, up to 20% after 15 months as reported by LEONARD (1939) and others.

Subsequent to counting, the organisms were placed in trays in convenient categories, dried at $25 \pm 2\,^{\circ}$C over desiccant for 18 h and weighed to 0.1 mg on an Oertling R20 balance. Cases of caddisfly and lepidopteran larvae were removed, but molluscs were weighed intact. No compensation for loss of alcohol soluble solids (cf. LONGHURST 1959) was made.

190

Adult records were obtained for the upper catchment from light-trap catches made 3–5 days a week in a trap located 6 m above the river between Stations I and II. These were supplemented by occasional trapping at all stations and at several upper tributaries using a portable inverted-cone light-trap. No sustained trapping was possible at the lower stations because of human tampering with any permanent installation. Sweep net collections from riparian vegetation were made each sampling date in 1968 but proved singularly fruitless because of the general scarcity of adult stages and so were discontinued in 1969.

Discussion of the sampling procedures: Obtaining representative samples is perhaps the most critical problem in lotic ecology and until recently has been largely ignored. Further discussion of quantitative aspects will be made in the section on vertical distribution. MACAN (1958b), ALBRECHT (1959), HYNES (1961, 1970), CUMMINS (1966), SCHWOERBEL (1967), *inter alia*, have summarized and assessed the commonly used techniques and in general found them deficient in one or more aspects. Alternative methods and apparatuses have been continually suggested as each worker attempts to find an instrument suitable for his local river, but almost all have to compromise on efficiency and frequency of sampling, area covered or selectivity of the invertebrates that can be obtained. Even if a sampler were operationally 100% effective, the insurmountable problem of substrate homogeneity and non-random microdistribution of the fauna in it (GAUFIN et al. 1956, ALLEN 1959, MACAN 1961b and others; see HYNES 1970) would still thwart accurate sampling. MOTTLEY et al. (1939) calculated an expectable error of at least 30% using conventional sampling techniques and the classic paper of NEEDHAM and USINGER (1956) showed that to obtain statistically reliable results from a standardized sampler on a visually uniform riffle, an impractically large number of samples were required (see also more recent confirmatory studies by RADFORD and HARTLAND-ROWE (1971) and CHUTTER (1972)). Such innate variability in faunal distribution makes any operable quantitative study a first order estimation only, particularly in a river such as the Gombak where in the upper reaches the substrates are indefinitely inter-mingled in a complex mosaic of types which cannot be objectively separated satisfactorily prior to sampling and in which the interstitial and/or surface areas suitable for faunal occupation vary greatly over short distances. However, most analytical studies have shown that a limited number of standardized replicates, sampling a feasible substrate area will give a reasonable assessment of faunistic composition and, within acceptable limits, an estimate of the dominant relationships of a fauna. NEEDHAM and USINGER (op. cit.) found that two or three samples would usually obtain representatives of all the principal faunal elements and CHUTTER and NOBLE (1966) reported 3 sq. ft. (about 0.3 sq. m) of stony riffle as satisfactory for normal purposes, particularly if two or more samples were pooled.

A little-investigated aspect, but one that further depreciates most sampling in 'uniform' locations, is the seasonal or life stage movements of certain faunal elements in the stream bed. HARKER (1953), LILLE-HAMMER (1966), ULFSTRAND (1968a), BISHOP and HYNES (1969a) and HENDRICKS et al. (1969) have described lateral movement and aggregation of late instar nymphs in bankside locations, and many insects may have nursery areas in deep mid-stream gravels, so that temporal changes in habitat may result in incomplete or biased sampling of which the investigator is unaware.

The chief drawback to cylinder – or box – type samplers is the difficulty in introducing them into the substrate without loss of fauna underneath the edges or through deflected flow scouring the bottom as the apparatus is lowered. In the stony reaches of Stations I and II, both problems were acute but were of only minor concern in the riffle areas of the lower river where substrate size and current velocity were reduced. To minimize such losses, the box was dropped into place in the visually selected riffle or pool area and immediately worked into the substrate to its full 10 cm depth. If, as often happened on the riffles, a good seal was not made with the bottom, the procedure was repeated in a new location until satisfactory samples could be obtained. This often frustrating procedure was, however, considered more reliable than the normally used Surber type net or shovel samples (see MACAN 1958b). Both of these involve techniques that disturb poorly defined areas and entrain in the water column a large proportion of the fauna, a considerable part of which would be lost in the highly turbulent flow characteristic of the Gombak. 'Kick' samples, using the technique of HYNES (1961), recovered only an estimated 50% of the fauna.

A further major consideration in any study must be the size of mesh used, both in the original sampling and in the ancillary processing steps. This may have considerable bearing on the composition of the catch. The selectivity of various apertures for particular size groups, both large and small, is well documented and JÓNASSON (1955, 1958), REISH (1959), HYNES (1961), TANAKA (1967), MUNDIE (1971) and others have described the spectacular increases in numbers recovered when mesh cross-sectional area was decreased by ostensibly modest amounts. However, clogging and loss of large forms by current deflection and slow flow-through are equally serious problems, particularly in estimating biomass relationships where the loss of any macro-component is grave. A compromise within the limitations and objectives of the study was necessary. The mesh-size of the sampler was smaller than that used by MINSHALL (1968) and the South African freshwater school (cf. HARRISON and ELSWORTH 1958, CHUTTER 1970), but not as fine as the silt or plankton netting employed by other workers primarily interested in obtaining all instars for life history studies. Losses from the large meshed bank-net (600 μ) were considerable as MACAN (1957) reported that if nymphs are physically able to escape through a mesh, a large proportion will, but as these samples were supplemented by casual catches, the problem was not considered serious.

Sampling of certain taxa remained an insuperable problem. The Decapoda, both prawns and crabs, were not adequately recovered by any attempted technique. Perhaps only tagging/recovering with all its attendant problems in a torrent stream would enable enumeration of the prawn population. The crabs in their cryptic habitats in the rocks and roots pose an even more difficult problem. Even if they maintain territories (cf. BOVBJERG 1959), any estimation based on areal extrapolation from trapping records as done by TURNBULL-KEMP (1960) would be useless because of irregular distribution of suitable 'home' burrow sites and bait selectivity by different aged individuals. The fauna of the large boulders and torrential thin-water habitats was not quantifiable. The finding that aggregations of caddis and lepidopteran larvae, Simuliidae, Blepharoceridae, Psychodidae and certain Ephemeroptera occurred in some locations but not in other empirically identical sites precluded any estimation of the overall density of these groups. The large assemblage of species restricted to or found preferentially in the *Saraca*-root biotope could be partially assessed for a particular root-mass, but current exposure, degree of siltation, and age of the habitat varied over small distances and resulted in concomitant changes in the fauna. Collection of forms with conspicuous shelters, such as the Lepidoptera, was too biased to yield an accurate population density.

Statistically defensible evaluation of the techniques used was not possible. On a qualitative basis, a cumulative species area plot might have indicated suitable limits to the area necessary for adequate sampling, but the unknown number of species, of which a large proportion were inseparable from broad, invariably present categories or species complexes (e.g. Chironomini, *Baetis* spp.), and many other infrequent species with discontinuous distribution made such a plot impractical. Assessment of significance levels between numerical data was also meaningless because of variations in food supply, current, substrate and biotic conditions between 'duplicate' sites and from one date to another. However, with the standardized procedure always carried out by the same operator, the technique, at least, was reproducible. As will be seen from the data, the dominant species or taxonomic groups at a station were rarely absent from any sample so that an estimated 40–70% of the total number of species were present in the composite sample from any date and this, as discussed by ULFSTRAND (1968a), probably represents adequate sampling.

Presentation of faunal data from a periodic series of samples necessitates the decision as to which data has relevance in respect to the objectives of the study and which can be ignored. To classify the status of a species in a community, some criteria must be applied that select species which really belong in the biotope and eliminate casual migrants. This is particularly important in a stream system where longitudinal differences in physical environment are not marked and survival after displacement by current is a likely proposition. Many authors have subjectively categorized faunas as 'common', 'present', 'rare', etc.; others have quantified these on a numbers per sq. m basis. In various studies, species constituting less than a certain percentage of the total in a number of successive samples have been disregarded in the discussion of the community (see HYNES 1961). Other investigators have classified taxa as 'significant' or 'non-significant' (HARRISON and ELSWORTH 1958) or 'dominant', 'sub-dominant', 'incidental' and 'accidental' (ULFSTRAND 1968a), usually based on an arbitrarily selected percentage of the total catch. Because of the manipulative problems, catch selectivity and distributional errors inherent in most sampling methods, a straight dominance index based on percentage caught is misleading. A rank of constancy of occurrence, perhaps coupled to absolute rather than proportional density, provides a more rational criterion for classification. The disruptive effects of sampling a mass hatching or large aggregation of one species are confined to the index for that species only, not the whole fauna, if a percentage transformation is avoided (see CHUTTER 1970).

In the following sections, a bipartite, four digit index is used that includes indices for both constancy and absolute density. The initial letter, A–O, indicates the number of occurrences of the taxon out of 15 sampling dates, r and s samples considered separately; the second part is the log to the base 10, to two decimal places, of the sum of the total catch per sq. m from all sampling dates, i.e. $\log_{10} \sum_{1=1-15} N_1$ where N_1 is the number of individuals of a taxon in r or s sample taken each date. This degree of definition is all that is warranted by the collecting technique. For the bank-root samples, no consideration of catch density is justifiable so that only an occurrence index is listed. Records obtained in casual or special habitat samples are indicated as such in the tables by an asterisk. A taxon was regarded as 'dominant' if it had an occurrence index J–O or density > 4.00; 'constant' C–I or density $2.00-3.99$; 'significant', infrequently recorded but determined by field observations and subjective evaluation of the taxon's habitat and catchability to be a permanent member of the community; 'incidental' for single records or low numbers found on only one or two occasions. The last was not considered part of the community, but as a stray, if it occurred downstream of an area with 'significant' presence of that species; if such a record constituted an upstream extension of the range of a species it was accepted.

1.2. COMMUNITY STRUCTURE

The 'preferred location' or microhabitat of greatest occurrence for a species is, at the exogenous level, a function of complex interactions of the physical conditions in the biotope. The more generalized parameters of temperature, dissolved oxygen concentration and current will be discussed later. The configuration of the immediate substrate of occupation, both as a refuge and more critically as a source of food, is often the paramount factor governing distribution. The chief faunal elements comprehensively listed in Appendix C will be discussed with particular reference to this.

Three categories of food were immediately evident: 1) autochthonous periphytic algae, particularly the epilithos of the riffles, relatively scarce in the upper reaches but important wherever sufficient light was available; richer periphyton was present at Stations III and IV, but declined with adverse sediment conditions at Station V: 2) allochthonous particulate matter originating as terrestrial detritus, and the dependent population of microconsumers and degraders that may be more important than the leaf products themselves; this food resource was ubiquitously available from the seston, in substrate interstices or in accumulations in areas of reduced current: 3) animal foods whose distribution was secondarily controlled by the occurrence of 1 and 2. No detailed study of food niches was undertaken, partly because of lack of time but also because specific investigations of that nature are of little relevance until the nomenclature of the organisms is organized. In addition, as the research of a number of workers on various orders (e.g. LEVANIDOV 1949, BROWN, 1960, 1961, BRINKHURST and CHUA 1969) has shown, food ingested may not necessarily be utilized, so that without data on digestive efficiencies enumeration of gut contents as done in many studies may be a fallacious exercise (cf. MECOM and CUMMINS 1964). A few animals were markedly stenophagous, particularly the carnivores, but the general impression for most groups was that the utilized food sources were broad-based and indiscriminate. Euryphagy on the detritus-leaf-algae complex of foods and/or alteration of diet from one resource component to another as dictated by availability was probably the common condition among the herbivores, and even some nominal carnivores, particularly in the forest reaches.

1.2.1. Distribution of the principal taxa

Tables 62 to 72 indicate the relationships between fauna and biotope, and longitudinal distribution using the index, as described, where possible. Where several species or groups of taxa could not be separated satisfactorily, the frequency for the higher classification is given and the presence or absence at a location indicated. Adult record localities are appended where available.

Table 62

	Station I			Station II			Station III			Station IV			Station V		
	r	s	b	r	s	b	r	s	b	r	s	b	r	s	b
a. COELENTERATA															
Hydra sp.	A1.30		*							A1.90		*	B2.00	B2.20	*
b. TURBELLARIA (Tricladida)															
Dugesia lindbergi } *Dugesia* sp. }	E2.00	C1.40	H	E1.74	B1.40	I	D1.70		D		A0.70	C	D2.83	B2.39	H
c. NEMATODA															
Labronema sp.	+			+	+		+	+		+	+		+	×	
Diplogasterid type															
Others							+			+			+		
Total Nematoda	E2.27	E2.56		E2.15	E2.00		O3.65	K2.91	A	L2.97	H2.43	A	K3.21	D2.51	
d. GORIIDA															
Gordius sp.															
Paragordius type } *Beatogordius* ? type }	(+ in drift only +)						A0.70		A						
e. MOLLUSCA															
Pelecypoda															
Pisidium (*Neopisidium*) *javanum*							Trib. H only								
Other lamellibranchs										B0.30					
Gastropoda (Prosobranchia)															
Siamopaludina martensi															A
Pila scutata												A			A
Brotia costula			C						A	D1.30	D2.23	M	D2.18	H2.10	O
Thiara scabra								B1.48		B1.40	D1.93		F2.61	K3.19	M
Melanoides tuberculata			A			*	A0.70	A0.70	A			L			O
Melanoides var A												D			B
Melanoides var B												B			F
Total *Melanoides*			A				A0.70	A0.70	A	A0.70	F1.65	L	J2.22	J2.74	O
Gastropoda (Pulmonata)															
Lymnaea rubiginosa										A0.70		A	A0.70	A1.00	D
Gyraulus convexiusculus												E	B1.40	D2.29	N
Indoplanorbis exustus			A			A		A1.30	A		A0.70	A		A1.60	N
Ferrissia javana			*			*			*			*			*

Tables 62–72. Distribution and frequency of the fauna in the biotopes of the main sampling stations. (Station I includes records from headwater torrent streams)

Biotope classification	{	r = erosional
		s = depositional
		b = bank area and leaf-drifts

* = casual records (only listed when none was available from routine sampling)
(×) = single record
× = rare
+ = present
++ = common
+++ = abundant
(?) = presence probable from adult or distribution records

Index (e.g. F2.63): letter = number of occurrences out of 15; A = 1, B = 2, ... O = 15; digits = $\log_{10} \Sigma N_i$ where N_i is the number of individuals in each four-weekly sample.

Coelenterata (Table 62a)

Hydra sp. was found on rare occasions at Station I attached to stones in pool areas and at Station IV on stones in bank areas. At Station V it occurred in locally high numbers on trailing vegetation, pilings and bankside solid substrates in deep slow-flowing reaches. Food was presumed to be protozoa derived from decomposing forest litter and other organic materials at the enriched lower stations.

Turbellaria (Table 62b)

Two spp. of *Dugesia*, one identified as *D. lindbergi*, occurred in low numbers at all stations. Reproduction was almost exclusively by fission as sexual material was rarely collected, perhaps because there was no extrinsic stress stimulus for sexual reproduction. In the *Saraca* roots of the forest zone these triclads were a constant component, while in the open river most specimens were found attached to rocks or the larger organic substrates. These forms are normally carnivorous (DITTMAR 1955, HYNES 1961, MACAN 1962a, b) and the probable diet was small mayfly nymphs in the *Saraca* and oligochaetes and/or chironomid larvae in the lower zones.

Nematoda (Table 62c)

The number of types identified were not representative of the contributions of this group. The diplogasterid type appeared to be restricted to the bottom substrates of the forest and was a minor component of the fauna. The *Labronema* sp. that was common in all samples from the stabilized riffles at Station III and regularly taken, but in smaller numbers, at the two lower stations was a distinctive member of the community. At Station III, the large numbers of plant nematodes collected in specially-taken sand and mud cores probably originated in the riparian gardens.

Gordiida (Table 62d)

Adult gordids were a regular component of the drift at Stations I and II (see later section), but were only rarely collected in benthos samples. Three distinct species were present: a *Gordius*, a *Paragordius* and a *Beatogordius*, all recorded from tettigonid hosts. Whether Gryllidae and tridactylids were also parasitized is unknown. On two occasions specimens were collected at Station III, but never below that point; these could have drifted down from the forest as distribution appeared to be limited to that zone.

Mollusca (Table 62e)

The molluscan fauna of the upper watershed was extremely poor, probably limited by availability of calcium, suitable and stable substrates and food. After only slight physico-chemical modification of the biotope in the lower river, both diversity and population density increased considerably.

Sphaeriidae: *Pisidium (Neopisidium) javanum* had an inexplicable distribution. It was present in moderate numbers in the extreme headwaters of Tributary H and nowhere else in the catchment. Calcium was undetectable at that location. The bottom substrate was quartz sand and fine organic silt.

Unionidae: Two specimens of an unidentified lamellibranch were collected on separate occasions in depositional areas at Station IV, but as these were both very young (< 1 mm), they may have recently detached from a fish host, infected in a mining pool or ditch, moving upstream. Although apparently suitable substrate conditions were available throughout the non-forested river, no clams were found, even at and below Station V where calcium content was enriched from the underlying lithology. No fragmented valves were found in sand drifts or trash accumulations as would occur with a resident population. *Pseudodon* sp. is common in large rivers of the eastward-flowing drainages of the main range, several *Corbicula* spp. inhabit smaller rheophilic habitats and *Contradens* and *Rectidens* occur in lenitic waters of the alluvial plain area (BERRY 1963), but either the suspended silt load or periodic instability of the bottom may have prevented colonization of the river.

Viviparidae: *Siamopaludina martensi* (Frnfld), separated from the more usually described *Bellamya javanica* (Von dem Büsch) following BRANDT (1968), was the largest gastropod in the river but was taken at incidental frequency only at Stations IV and V on marginal trailing vegetation.

Ampullaridae: *Pila scutata* was equally rare, being exclusive to Station V.

Thiaridae: The thiarids were the dominant snail group, occurring throughout the river, but both *Brotia* and *Melanoides* are highly variable genera and the taxonomic situation in the whole family is confused so that the number of taxa could not be determined. *Brotia costula* occurred in occasional casual samples at Stations I, II and III and in the larger upper tributaries. The specimens recovered were always large, 2–5 cm, and may have represented an isolated, barely viable population that bred elsewhere in the river. At Stations IV and V, this was a constant species but only thinly dispersed over all biotopes. A very distinctive colour variant of this species present in the river had many features similar to *B. testudinaria* (Von dem Büsch), particularly as young snails. However, since in some older specimens these patterns intergraded with the *B. costula* markings and chiefly because the distributions were the same, this variant was considered to be a polytypic *B. costula*. *Thiara scabra*, absent from the forest zone, occurred as an incidental species at Stations III and IV, but was a dominant in the enriched river at and below Station V. At the D.I.D. site, this prosobranch was abundant, living in the emergent grasses at the water's edge. *Melanoides tuberculata* was found with similar distribution to *Brotia*, but was more abundant at Station IV than that species. Two other distinct types of this genus, renowned for its variability, were distinguished. Both varieties A and B were more robust than *M. tuberculata* and had distinctive and apparently constant colour patterns; no intermediates were found. More important, they had limited distributions, never occurring above Station IV and not being found in the river much below Station V. Overall population density for the genus was low at Station IV but at V considerable numbers of young were taken in the depositional sediments and adults in the marginal areas.

Lymnaeidae: The common pulmonate in paddy fields and irrigation ditches in the lower catchment was *Lymnaea rubiginosa*, but it occurred very rarely in the river where ionic concentration was lower, erosive spates frequent and suitable egg-laying surfaces few. One specimen only was found at Station IV and fewer than 10 in casual samples at Station V; these were obviously incidental records from the adjacent water bodies. NORRIS and CHARLTON (1962) recorded *L. luteola* and *L. pinguis* from the main river below Station V, but the taxonomic status of these species is doubtful. In the brief survey at the D.I.D. site, no lymnaeids were found, but may have been present in unsampled habitats.

Planorbidae: *Gyraulus convexiusculus* and *Indoplanorbis exustus* were both lower river species reaching significant density only at Station V and for a short distance below in marginal vegetation. Occasional records of the second species were made at Stations III and IV but never in the forest. *Gyraulus*, on the other hand, was collected as single specimens at Station II and in the Station I tributary. These may have been aberrant drift records as discussed later.

Ancylidae: *Ferrissia javana* was ubiquitous but at very low population density. Occasional individuals were collected from *Saraca* roots and

Table 63

	Station I			Station II			Station III			Station IV			Station V		
	r	s	b	r	s	b	r	s	b	r	s	b	r	s	b
ANNELIDA															
Oligochaeta															
Branchiodrilus semperi							×	×		+	+		+	+	+
Allonais inaequalis										+	+		+	+	
Pristina proboscoidea													+	+	+
Other *Pristina* sp.				+			+	+		+	+		+	+	(?)
Chaetogaster sp.	+														
Aulophorus sp.	×	+		×	×		×	+	×	+	+		+	+	+
Stylaria sp. nr *fossularis*	×	+		×	+		×	+	×		+		+	+	
Other Naididae															
Haplotaxis sp.										×	+		+	×	+
Phreodrilus sp.													+		
Limnodrilus hoffmeisteri													+	+	+
Limnodrilus silvani													×	+	×
Bothrioneurum sp.													×	×	
Branchiura sowerbyi	×			×			×			×			×		
Undetermined Enchytraeidae															
Eukerria kukenthali	×			×						×			+		
Phertima sp(p).										×			×		
Ocnerodrilus occidentalis										×					
Pontoscolex corethrurus													+	+	
Hirudinea													×	×	
Glossiphonia weberi	×			×			×			+			×	×	
Batrachobdella reticulata															
Helobdella sp. nr *nociva*															
Helobdella sp.	×			×			(?)			(?)					
Placobdella ? *inleana* }										+			+		
Placobdella sp. 1										×			+		
Placobdella sp. 2										×			+		
Barbronia weberi }															
Barbronia sp. ? var of *weberi*													+	+	
Herpobdelloidea lateroculata													+	+	
Myxobella ? *annandalei*	×												+	+	

On *Notochelys platynota* at Station IV . . . .

	Station I			Station II			Station III			Station IV			Station V		
	r	s	b	r	s	b	r	s	b	r	s	b	r	s	b
Total Oligochaeta	O3.77	L3.31	D	N3.50	M3.39	D	O4.10	O3.79	G	N4.20	O4.16	E	O5.56	O5.54	O
Total Hirudinea		*			*			B		B1.93	A1.30	C	M3.33	J2.29	O

leaves in small tributaries, from diatom-covered rocks in the current and from trailing vegetation, but never more than two specimens were collected at any site in spite of diligent searching. The preferred food was attached algae (see GELDIAY 1956). Calcium concentration is known not to have much effect on the distribution of this family with its conchiolin shell (FITTKAU 1964, SIOLI 1964, MAITLAND 1965).

A further species, *Digoniostoma pulchellum* (Benson) (Bulimidae), was common in ditches and ponds in the alluvial area over limestone but was not collected in the river. Bilharziasis does not at present occur in West Malaysia, but is known from the slightly richer waters in Thailand (HARINASUTA and KRUATRACHUE 1962). Some areas, particularly where pollution is enriching the naturally poor waters, may in future provide suitable habitats for the host snails.

Tributary B and the watercress farm in the upper forest constituted a restricted area rich in gastropods as a result of liming of the irrigated terraces and fertilization that produced large blooms of filamentous algae and diatoms. *Ferrissia, Brotia, Melanoides, Lymnaea* and *Gyraulus* all occurred in abundance, both in the farm area and for a short distance downstream under the influence of its run-off. Some of the incidental records for various species in the forest river may have resulted from wash-out from this nursery area. This hypothesis was reinforced by the negative observance of any young gastropods, except *Ferrissia*, in the upper reaches. However, *Brotia*, at least, was found higher up in the watershed and elsewhere in similar jungle streams and must have some physiological capacity to extract considerable amounts of calcium from its detritus and leaf diet to build its very heavy and sculptured shell.

Annelida (Table 63)

Oligochaeta: The species recorded undoubtedly constitute an incomplete record, particularly in the Naididae. A number of other species are known from Malaya in the genera *Nais, Allonais, Dero* and *Aulophorus* (STEPHENSON 1931, NAIDU 1965) and unidentifiable types were found at all stations. Many more megadriles are probably casually associated with watercourses. No attempt was made to identify all specimens so the commonness-rarity listing for each species was subjective, based only on subsamples. The total numbers counted may also be suspect because of fragmentation of the fragile tubificids during sampling and the occurrence of multiple fission products in the Naididae. KENNEDY (1966) has shown that *Limnodrilus hoffmeisteri* cannot reproduce asexually as in some Tubificidae (BRINKHURST 1964), but that sexual stages are normally present continuously in a population, particularly at temperatures exceeding 15 °C. Only the laterally-gilled 'tails' of the most common *Branchiodrilus* and *Branchiura* were counted to minimize overestimation of population density. The worm fauna was a constant component of the community at all stations, becoming a diverse and dominant element at the lower three stations. *Chaetogaster* and *Stylaria* were present throughout the river and, with various unidentified naidids, a rare *Aulophorus* sp. recorded at Station II and *Pheretima* sp(p). constituted the forest-river fauna. At Station III, the large naidids *Allonais inaequalis* and *Branchiodrilus semperi*, a *Haplotaxis* sp. and *Eukerria kukenthali* were occasionally found with the two omnipresent species given above. At Station IV, *Limnodrilus* was sometimes present in small numbers with a *Phreodrilus* sp., *Pontoscolex* and *Ocnerodrilus* also recorded. Station V with its enriched substrate supported a large worm fauna with *Branchiodrilus*, *Limnodrilus* and *Branchiura* found in clumps mixed with the ubiquitous naidid species. Population density differences between erosional and depositional substrates were never pronounced even in the upper river, so the physical nature and porosity of the sediments was apparently subsidiary to food availability and other biotic factors in determining occurrence. Spatial isolation of species was not apparent, although *Branchiura sowerbyi* occurred almost exclusively in sheltered bank areas. Most tubificids are indiscriminate particulate feeders utilizing different components of the organic substrate either directly or through a symbiotic microflora and/or fauna (BRINKHURST and KENNEDY 1965); this was presumed to be the case here. At the D.I.D. gauging site with its heavy silt and pollution load, the clayey river bed was often carpeted by *L. hoffmeisteri*, *B. sowerbyi* and *B. semperi* in very high densities in spite of the instability of the substrate.

Hirudinea: Leeches were an insignificant part of the fauna at the upper three stations with only occasional specimens of *Glossiphonia weberi* and *Helobdella* ?sp. being collected from beneath stones in the riffle area. A single specimen of the hirudid *Myxobdella* ?*annandalei* was taken at Station I. No collections of *Gastromobdella (Trematobdellidae)* were made although it has been found in hill streams in adjacent areas of Malaya (MOORE 1935) and was reported as the only recorded erpobdellid in Malaya by SHARMA and FERNANDO (1961). At Station IV, three *Placobdella* spp. were taken in large numbers from an aquatic tortoise, *Notochelya platynota*, but these and the large sanguinous hirudid leeches that occur in paddy areas in association with domesticated stock were not really part of the riverine zoome. Also at this station, *G. weberi* and an erpobdellid identified as *Barbronia weberi* formed a small but constant part of the riffle fauna. The latter species, perhaps placed in the wrong genus (see RICHARDSON 1971), became very abundant at Station V, along with a variant (otherwise identical, that had two extra pairs of eyes), *Herpobdelloidea* ?*lateroculata* and two additional rarely found Glossiphonidae, *Batrachobdella reticulata* and *Helobdella* sp. nr *nociva*. The erpobdellid spp. fed mainly on the large Naididae and Tubificidae (hence their primary distribution in the lower river where these attained high numbers), while the glossiphonids preyed on gastropods and other invertebrates as well as oligochaetes (cf. HARDING and MOORE 1927). Density

202

for most species was greater in the more stable riffle areas where many *Barbronia* egg cases and young were collected attached to the stones, but the *Herpobdelloidea* sp. was taken exclusively in bank areas. This species was recorded by Harding and Moore (op. cit) as a predator of Crustacea, but its diet here was worms.

Acarina (Table 64a)

Hydracarina: Six types of definitely aquatic mites were distinguished, but undoubtedly many more were present that were not adequately sampled by the methods employed. The five Torrenticolidae/Atractididae types had variable distributions. Types 2–5 were apparently limited to the upper river and only the type designated 1 was present at the lower stations, reaching a moderate incidence level at Station III. Hydrachnellid H occurred at Stations III, IV and V, but in significant numbers only in the gravels of Station IV. In addition to these, a wide variety of mites from normally terrestrial litter and forest canopy groups occurred, sometimes quite regularly, in bottom samples: Mesostigmata – five different Uropodina, a single gamasid and an Epicriidae type; Sarcoptiformes – one Acaridae/Glyciphagidae type; Oribatei – one Phthiracaridae type and five or six other oribatids. The aquatic status of these is unknown. Many had undoubtedly fallen or been washed into the river, but their occurrence, apparently in good condition, on the riffles of the lower uncanopied river may indicate that some types at least, particularly of the oribatids, were aquatic. Numerically, the total mite catches were a constant and significant faunal component at all stations, but with higher densities recorded at Stations III and IV where the incidence of stone and gravel-dwelling orthoclad prey was highest.

Crustacea (Table 64b)

Few comments can be made on the small forms as no attempt was made to systematically collect or analyse them. Their overall paucity in the river was attributed to the low calcium levels, unstable substrates and scarcity at most stations of suitable cover.

Cladocera: One species, *Moinodaphnia* ?*macleayii* was collected on two isolated occasions only at Stations III and V. These were considered incidental migrants from the standing waters of the paddy fields.

Ostracoda: Three types of Cypridae were infrequently collected: types 1 and 3 from both r and s sediments at the lower three stations, type 2 once only at Station III and type 1 twice at Station II. The numbers involved were small and, except in the erosional areas of Station V, probably insignificant. These again may have originated in pooled or slow-water areas and been washed into the main river.

Copepoda: Cyclopoida were occasionally taken at all stations but the

Table 64

	Station I			Station II			Station III			Station IV			Station V		
	r	s	b	r	s	b	r	s	b	r	s	b	r	s	b
a. ARACHNIDA															
Hydracarina															
Torrenticolidae/Atractididae type 1		×					+++			+				×	
type 2		++			+										
type 3					+		+								
type 4		+			×										
type 5					+										
Hydrachnellid type H		+			+		++			++				++	
Other Hydracarina								++			++			++	
Mesostigmata/Sarcoptiformes/Oribatei[1]		+			+			+			+			+	
Total Acarina	K2.53	F2.66		G2.41	H2.48		K2.95	K2.77		K3.00	J2.74		G2.72	F2.53	

[1] aquatic status of these groups unknown

	Station I			Station II			Station III			Station IV			Station V		
	r	s	b	r	s	b	r	s	b	r	s	b	r	s	b
b. CRUSTACEA															
Cladocera															
Moinodaphnia ? *macleayii*							A1.60						A0.70		
Ostracoda															
Cyprid type 1				×				×			×			×	
type 2					×		×								
type 3							×								
Total Ostracoda				A1.30	A2.20		B1.60	A1.30		×	B2.00		×	A1.30	
Copepoda															
Cyclopoida	B1.60	B2.00					C1.90	A1.30		F2.70	A1.30		B2.20	C2.22	
Harpacticoida					B1.60			B1.60			A1.30				

(× drift only ×)

Amphipoda
  Orchestia anomala — *   A1.30   *   A1.90

Decapoda
  Macrobrachium malayanus — ++   ++   (also Tribs D+E)
  Macrobrachium pilimanus — +   +   (Tribs B, H+J only)
  Macrobrachium geron — (?)   ×
  Atya spinipes — F1.85 G   B1.40 L   ×
  Total Caridea — ×   L   (also Tribs A, B+J)   ×
  Polamon johorense — ×
  Paratelphusa maculata — ×   ×   (also Trib. J)
  Total Potamonidae — B1.00   L   A1.30 D   B   D

numbers were not significant. Several species of Harpacticoida, often considered as psammofauna, were taken in drift samples at Station I.

Amphipoda: No exclusively aquatic types were found. An apparently semi-aquatic talitrid, probably *Orchestia anomala*, was regularly seen at Stations III, IV and V among the marginal grasses and debris accumulations.

Isopoda: No aquatic isopods have been recorded although terrestrial forms were present along the banks and occasionally taken as drift. None ever exhibited amphibian tendencies.

Decapoda: The caridean prawns of the genus *Macrobrachium* (= *Cyrphiops*) constituted a major element in the fauna of the forested reaches to which they were apparently confined. Three species occured. *M. malayanus* and *M. pilimanus* (? = *M. lar* recorded by MIZUNO and MORI (1970) from the Gombak) were common in bankside root-niches, between and beneath boulders and, where current flow was stabilized, on open depositional gravels and sands. *M. geron* was recorded as the only prawn species from the lowland, wooded Tributaries H and J, and with the other two species from the slightly enriched Tributary B by the water-cress farm, into which it may have been introduced with stock plants. Whether a valid segregation of hill and lowland species occurs is not clear. As discussed earlier, benthos sampling methods were ineffective for these animals. Most records were obtained by causal sampling or in association with electrofishing which yielded considerable catches of prawns and provided a good qualitative collection technique. As far as could be determined from stomach examination, the predominant food was detrital although larger leaf particles and invertebrate remains were also found, pointing perhaps to an omnivorous diet. The apparent absence of prawns from the lower river (the electrofishing would surely have caught them if present) probably resulted from loss of bottom and bank cover shortly after the river leaves the forest. The *Saraca*-root habitat disappears and the large boulders that provide dead-water areas and refuges from predators are replaced by smaller-stoned riffles about 2 km above Station III. Also, the increased sand load and substrate instability would make maintenance of position difficult during spates. The presence of *M. geron* in the clean water lower tributaries that in places have rocky substrates, are not subject to violent floods and do provide some cover in the form of *Colocasia* roots tends to substantiate this theory. The effects of increasing mean temperature may also be important but would have to be experimentally assessed. Gravid females of both upper river species were found throughout the year.

Atya spinipes has definitely been collected from the river at Station III and up into the forest but was not recorded during this survey. This species is more cryptic than *Macrobrachium* and may not have been as catchable. Its complete distribution is unknown. NORRIS and CHARLTON

(1962) reported collecting *Caridina weberi sumatrensis*, but without the specimens this record from the lower river cannot be confirmed.

Two potamonid crabs, *Paratelphusa maculata* and *Potamon ?johorense* constituted minor, but significant, faunal elements in the areas where they occurred. There appeared to be a rigid distributional segregation between them. *P. maculata* occurred only in the lowland sections and tributaries of the river from Station III down to the beginning of the polluted area above Station V where it disappeared. This species is the common Malayan rice-field crab, living in mud/sand burrows and feeding omnivorously on plant materials, worms, insects and detritus. *P. johorense*, on the other hand, was found only in the upper forested reaches, more frequently in the tributaries than in the main river. Preference was shown for the *Saraca* habitat, but it was also collected in torrential sections hiding in rock holes in the cascades and beneath stable boulders. The two species occurred together only in Tributary J. As will be seen later in the fish section, this stream, which drains rubber plantations and gardens, had elements of both headstream and lowland faunas present. KARUNA-KARAN (1969) reported both crab species, together with *M. geron*, from a similar stream on granite on Singapore Island. Diets of the African potamonids have been fully described (WILLIAMS et al. 1961, WILLIAMS 1962, 1965) and may be summarized as primarily herbivorous but incorporating up to 30% invertebrate foods, usually Ephemeroptera nymphs, particularly in the juvenile stages. Cursory observations in the Gombak indicated similar feeding habits, with the young, most commonly found actively hunting among roots or in coarse gravel beneath leaf masses, eating some animal foods while the stomachs of the apparently sedentary adults contained masses of detrital material. Young *Potamon* were collected in all months, but nothing is known of the seasonality of the lowland *Paratelphusa* which might perhaps have a life cycle attuned to either floods or rice-field inundation periods.

Collembola (Table 65a)

Collembola were a minor element of the benthos at all stations that probably included many species feeding on detritus or its microflora. At the open stations where agricultural and domestic refuse accumulated at bankside, swarms of neustonic collembolans were sometimes seen feeding on the surface film.

Orthoptera (Table 65b)

Epilampridae: Semi-aquatic cockroaches were conspicuous but numerically unimportant at Stations I–IV where they inhabited the roots, accumulated drift and emergent vegetation at the bank areas. They were not seen at Station V, but no reason for this absence was

Table 65

	Station I			Station II			Station III			Station IV			Station V		
	r	s	b	r	s	b	r	s	b	r	s	b	r	s	b
a. COLLEMBOLA															
Poduroidea	+	×		+	×		+	×		×	×		+	×	
Isotomidae ? type											×			×	
Total Collembola	H2.53	A1.30		D2.45	C2.08		D2.71	C2.45		C1.81	A1.30		E2.64	C2.08	
b. ORTHOPTERA															
Epilampra sp.			×			×			+			+			
Rhicnoda sp.			+			+			+			×			
Other epilamprid type	*														
Tridactylid type			*	*		*									
Total Orthoptera			A			A			D			A			
c. MEGALOPTERA															
Hermes sumatrensis	*		C	*		B									

obvious. *Rhicnoda* sp. was more common in the forest sections, while the *Epilampra* sp. appeared in higher numbers at Stations III and IV. A second swimming epilamprid was collected once at Station II. These species had no apparent morphological adaptations for aquatic living, but they were collected under submerged leaf and detritus accumulations and at Station I often under stones, so respiration was apparently not a problem.

Tridactylidae: Small mole-crickets with distinct natatory lamellae on the tibiae were regularly collected at the forest stations. According to LA RIVERS (1956), this group is fossorial, but in the Gombak, they were more commonly seen on the lichen-covered emergent boulders. When disturbed, they entered the water and burrowed into the bottom gravels. As in the Epilampridae, their amphibious habits made them marginal members of the aquatic fauna.

Megaloptera (Table 65c)

Hermes sumatrensis was a constant member of the Upper Zone riffle fauna to which it was exclusive, but at low population density. The animal diet was apparently catholic, but parts of Plecoptera and Hydropsychidae predominated in the gut. This species was reported from the Gombak River by BANKS (1931) who also recorded *Neochauliodes sundaicus* Weele from lower altitudes in other waters in the Kuala Lumpur area. No other corydalid was seen during this survey. Temperature was probably the factor limiting distribution to the forest zone as found for other Corydalidae by HARRISON and ELSWORTH (1958) in the Great Berg River system.

Ephemeroptera (Table 66)

The mayfly zoome of the Gombak was remarkably diverse for a small river, a function of the wide spectrum of microhabitats available. Of the 44 recorded taxa (there may be additional species in certain families unidentifiable without more adult material) only one, *Prosopistoma wouterae*, has been described previously, and that from a nymph. Dr. PETERS of Florida A. and M. who worked the collections reported that no specific affinity existed between the Gombak fauna and that of the ULMER (1940) records from Java, the only work of real consequence in the Sundanian region on this order.

Siphlonuridae: *Isonychia* sp. A with its large size and distinct form was a conspicuous, constant and characteristic species of the forest river to which it was confined. Single records were made at Stations III and IV, but these were undoubtedly the result of drifting. Its main habitus was *Saraca* roots, although numphs also occurred on stones in the current from where they were able to filter the flow for detritus with their hair-fringed femora and tibiae (cf. JONES 1950). The family is reportedly

Table 66

EPHEMEROPTERA

Taxon	#	I r	I s	I b	II r	II s	II b	III r	III s	III b	IV r	IV s	IV b	V r	V s	V b
Isonychia sp.A	a	B1.48		J	E1.40		L	M3.47	M3.81	A	O4.10	L3.11	A	O4.27	N4.01	H
Baetis (14+ spp.)[1]	a	O3.34	L3.41	E	O3.12	M3.22	H	N3.84	L3.17	E	O3.65	H2.64	F	O3.73	H3.16	D
Pseudocloeon sp.G	a	H2.48	B1.78	*	N2.87	H2.34	A	C2.22	A1.30	B	F2.54		*	J3.37	D2.54	
Pseudocloeon sp.B	a			*	A1.30		A									
Pseudocloeon sp.C				*			*									
Unknown (? Baetopus type)		A2.45	A2.45													
Chromarcys sp.A													A			
Epeorus sp.A	a	(Upper tribs) J2.82	C2.32	I	J2.56	C1.65	L									
Thalerosphyrus sp.A		(Upper tribs)						C1.78	A1.30	F			G			F
Thalerosphyrus sp.B																
Thalerosphyrus sp.C													C			*
Composoneuriella sp.A	a	A0.70	A0.70	B	B1.90		C	B1.85	D2.67	G	I3.04	H1.30	F	G3.05	H2.94	F
Thraulus bishopi	a	A0.70		A	B1.65	C2.44	C	B1.60	B1.00	A	C2.34	B1.78	B	G3.19	A1.60	A
Isca sp.A or spp.?	a	H2.88	D2.31	D	F2.08	B1.60	E	B1.60	A1.30	A	B2.63	A1.30	A			
Habrophlebiodes sp.A	a	F2.38	C2.34	A	B1.60	C1.10		A1.30	A1.60							
Dipterophlebiodes sp.A	a	A1.30	A1.78	A							A1.30	B1.60	A			A
Choroterpes s.s. sp.A		(×)				B2.00	*			*						
Ephemerella (Drunella) sp.A																
Ephemerella (Drunella) sp.B		A1.40				A1.30	A	L3.22	E2.18	F	G2.57	D2.18		D2.75	B2.02	
Ephemerella (Crinitella) sp.A								N3.52	C2.15	D	L2.93	A1.78	B	C1.95		
Ephemerella s.s. type		A1.30	A1.90		B1.60	A1.30	A	A1.65			B1.78	B1.70		B1.08		
Teloganodes sp				*												

Taxon											
Tricorythus sp.A	C1.90	A1.30 A		*	A1.30	A1.90	A0.70	A	A1.30	A	
Tricorythid genus incertus											
Potamanthodes sp.A [a]	B1.78	A	B1.78 D2.15	A B O3.59 M3.37	I M3.77	I2.77 G	I3.56	J3.24 N			
Ephemera (*Dierephemera*) sp.A [a]	F1.95 D2.04	A1.30	* H								
Neoephemeropsis sp.A [a]	B D			(×) C1.40	G		H				
Neoephemeropsis sp.B	+ +	+ +	+	+ +	+	+ +	+ +	+ +			
Caenis sp.A [a]	+ +	+ +	+	+ +	+	*	*	*			
Caenis sp.B	H2.45	H2.45 F2.20	H3.60 H2.42	F F2.26	B	E	E2.92	E2.73 F			
Total *Caenis* spp.	K2.96 C	C A		B							
Caenid genus incertus	A	A1.30	*								
Prosopistoma wouterae	B1.60										

adults collected; [1] See Figure 40

predatory, but there are no fanged mouthparts in this genus as found in some genera of the Ameletopsinae. The adults of this species were the most commonly caught Ephemeroptera at the Station II light-trap throughout the year but were never taken at the lower stations, substantiating the nymphal distribution. The limiting factor was probably substrate instability as food supply did not change appreciably. However, the increase in suspended inorganic material that might hinder feeding, altitude and shade (i.e. temperature) may be important also.

Baetidae: This family included a large assemblage of species that could not be satisfactorily separated. Originally eight categories were set up, but Dr. MÜLLER-LIEBENAU has found at least 14 *Baetis* spp. and two or more *Pseudocloeon* spp. so the genera will be discussed as units.

The actively-swimming nymphs of *Baetis* were facultatively herbivorous, grazing on periphyton and collecting detritus primarily from clean stones in currents (cf. JONES 1950, HYNES 1961, CHAPMAN and DEMORY 1963, COSTA and FERNANDO 1967 and many others). At all stations they were a major component of the insect fauna, becoming a dominant at Stations IV and V where food was very abundant and the number of stenothermal predators reduced. From the numerical data, little real difference in erosional and depositional densities was obvious, but this resulted both from the heterogenous nature of the bottoms which always overlapped in physical properties and the fact that small baetids were commonly found in the interstitial spaces of the gravels, presumably feeding on the microflora growing there as described by BROWN (1961) and MINCKLEY (1963). A definite succession of species was evident (Fig. 40) but without complete associated adult and nymphal records, discussion of this is premature.

Pseudocloeon type sp. G was a tiny 3–4 mm nymph when mature, found with its adult throughout the river in considerable numbers, on both erosional and depositional substrates, and although not recorded in the b numerical data, also on the trailing vegetation in the current. Diet appeared to be similar to that of *Baetis*. *Pseudocloeon* type sp. B was less numerous, but still found at all stations except I. This larger species, 3–6 mm and slightly more dorso-ventrally flattened than sp. G, was almost restricted to epilithic surfaces and at Station II was observed in swarms on the large boulders in the current. Species B and that designated as *Pseudocloeon* type sp. C showed a decided preference for a diatom-Cyanophyta diet. Nymphs of the latter species were large, commonly about 10 mm at maturity, and were found only at the upper forest stations, usually in the splash zone on torrent boulders and at the water line where algal growth and probably lichens were readily accessible with changes in water level. This species was a day-active forager whose distribution was almost certainly limited by shade-temperature and the availability of unsilted rheophilic feeding surfaces.

A very few nymphs of a *genus incertus*, perhaps a *Baetopus*, were found at

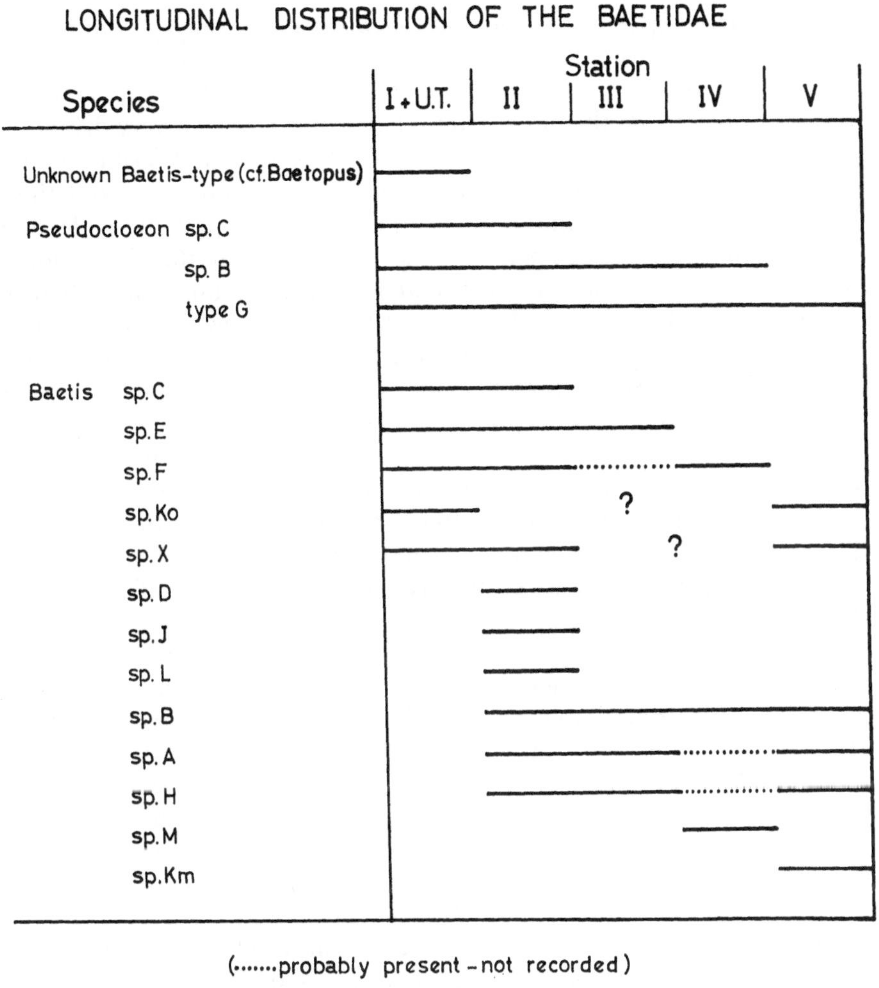

LONGITUDINAL DISTRIBUTION OF THE BAETIDAE

	Station				
Species	I + U.T.	II	III	IV	V

Unknown Baetis-type (cf. Baetopus)

Pseudocloeon sp. C

sp. B

type G

Baetis sp. C

sp. E

sp. F

sp. Ko

sp. X

sp. D

sp. J

sp. L

sp. B

sp. A

sp. H

sp. M

sp. Km

(⋯⋯probably present – not recorded)

Fig. 40. Longitudinal distribution of the Baetidae.

Station I from riffle and sandy pool samples but their status is doubtful.

Oligoneuridae: A single specimen of the rarely recorded genus *Chromarcys* was collected from a rapid-water bank sample at Station IV; no comment can be made on its ecology.

Heptageniidae: *Epeorus s.s.* sp. A nymphs were collected exclusively from torrential rock faces in tributaries high up the valley sides. The species was probably limited to these areas by silting and decreasing frequency of suitable cascades lower down, although competition from the other heptagenids may have been a factor. The few specimens found were in open parts of tributaries where *Spirogyra* grew and this appeared to be its preferred food. A few adults were caught by the Station II light-trap, but these may have come from the tributaries in the area.

Thalerosphyrus sp.A was a dominant element in the mayfly community of the forest river, declining in abundance at Station III and only recorded casually in the lower river, not in the benthos samples. *Thalerosphyrus* sp.B was restricted to the upper tributaries while *T*.sp.C, a very distinctly marked nymph, was confined to Stations IV and V and Tributary H. All were apparently grazers on the algae of stable substrates, but their guts also contained organic detritus and sand. The periodic collection of these normally crevice and rock surface forms in the depositional samples would suggest that leaf material was an important dietary constituent (cf. Jones 1950, Ivanova 1958 for *Heptagenia*), particularly for the smaller nymphs which were the ones found in that biotope. Larger individuals were usually present only on unsilted stones in the current that supported some algal growth.

Compsoneuriella sp. A had the reverse distribution to the common *Thalerosphyrus* sp.A. At Stations I and II, it was collected on isolated occasions, but increased in catch frequency and numbers to a constant status at the lower three stations. It too was collected preferentially on stable substrates and appeared to have similar dietary requirements to the *Thalerosphyrus* species.

Leptophlebiidae: *Isca* sp. A was the most common leptophlebid in the upper river occurring in moderate numbers in about half the collections. Although taken in bottom gravels as reported by Peters and Edmunds (1970) where its detritic food was readily available, its preferred habitat was that of *Saraca* roots and leaf-drifts. This was not apparent from the coarse-net collections as most of the nymphs were able to pass through the mesh, but with the 90 μ net larger numbers were consistently taken. Records for the lower river were sporadic except in the riffles of Station IV where considerable numbers were found on two occasions.

Thraulus bishopi Peters and Tsui *sp. nov.* was only rarely taken at the upper three stations, but was more common in the potamon reaches. In the forest it was restricted to sand and detritus accumulations around the bases of boulders and rooted banks where current effects (0.5–2 m/sec) on the stability of the substrate were minimal. Notably, it was absent from the root habitat. At Stations IV and V, most specimens were not found in the expected fine sand and silt habitats, but in slow current (less than 1 m/sec), often partly shaded stony areas of the river adjacent to the overhanging banks. The few mature nymphs taken were from silted marginal areas, an observation that agrees with Peters and Edmunds (op. cit.). However, these authors described the immature nymphal habitat for a New Guinean species as being torrential rocky bottoms, from which no individuals were recovered here. Guts contained amorphous masses of leaf fragments, detrital material and sand grains. Costa and Fernando (1967) reported an algal, leaf and detrital diet for *Thraulus signatus* from the Maha Oya stream in Ceylon, so it appears that this genus is generally herbivorous.

Habrophlebiodes sp.A had similar distribution to *Isca*, occurring in small numbers in six of 15 riffle samples at Station I, less frequently at Station II and incidentally at Stations III and IV. It apparently had little substrate preference, being found in the interstitial spaces of both gravels and sands. The other two leptophlebid species, a *Dipterophlebiodes* (the first known nymphs of this genus, PETERS 1972) and a *Choroterpes s.s.* were collected on a few occasions only, the first species in the bottom sediments of the Station I tributary and the second at Station II from depositional area and casual bank samples. All three of these species appeared to be detritivores.

Ephemerellidae: The true status of the *Ephemerella* spp. is still unresolved as the few adults collected were not associated with their nymphs. At least five nymphal species in four subgenera were distinguished. *E. (Drunella)* sp.A was collected only rarely in casual samples at Station III; *E. (Drunella)* sp.B and *E. (Crinitella)* sp.A occurred together in low numbers on one or two occasions at Stations I and II, but became constant species of importance at Stations III and IV. Numbers declined at the lowest station. This distribution was dependent on the availability of the preferred epilithic algal food of this genus (JONES 1950, 1951, DITTMAR 1955, HYNES 1961) which became most abundant on the stones in riffles at Stations III and IV. Again, the larger nymphs were always found in stone habitats while the young made up most of the population collected in depositional samples from the sand and gravels. At Stations III, IV and V, another *E. (Ephemerella) s.s.* species (cf. *?Chitonophora*) was infrequently collected in small numbers. This distinctly patterned nymph with a markedly hypognathous head was always taken from the riffle biotope. In addition, a single specimen of *Teloganodes* species unknown (ALLEN, personal communication) was taken from the Station I tributary. No diagnosis of this is possible without adult material.

Tricorythidae: *Tricorythus* sp.A was a significant member of the mayfly community in spite of the low numbers collected by the methods of this survey. It was present at all stations and in all tributaries, inhabiting crevices in the surfaces of boulders and drowned logs or in leaf-drifts from which habitat it was not adequately collected except by hand-picking. Densities in some forest tributaries were empirically quite high, but could not be quantified as distribution was patchy. This species was apparently a facultative herbivore with detritus and algae as common gut constituents. At Station II, a single specimen record of a *genus incertus* belonging to this family was made from a bank sample.

Potamanthidae: Disregarding the *Baetis* spp. complex, the dominant mayfly of the lower river was *Potamanthodes* sp.A. The burrowing/sprawling nymph was collected in all samples at Station III and in the majority at Stations IV and V. In the forest streams it was uncommon and although it was found in four samples in the main river, only one adult was taken in more than a year's light-trap catches at Station II. Fine to coarse sand, with or without silt, whether between riffle stones or in pooled areas, was the substrate of preference as the diet of detritus was readily available in all these locations. Below Station V this species was apparently eliminated either by toxic pollutants or by smothering by very fine tin-mine outwashings which its gill system could not handle. Even at Station V this species was absent from heavily silted pool areas, probably for this reason.

Ephemeridae: A characteristic species of the upper river in the jungle, i.e. a clean unsilted bottom, was *Ephemera (Dierephemera)* sp.A (a new subgenus, EDMUNDS and MCCAFFERTY in MS), that occurred in low numbers in limited and specific substrate areas. The large burrowing nymph was found particularly in the Station I and B tributaries in areas where deep pools with considerable surface velocity were formed (riffle-pools). These had medium-fine sand, some gravel plus occasional large stone bottoms with some accumulation of detritus and leaves. Similar microhabitats were found in the main river at points where tributaries confluenced and it was only at such locations that this species was found. This perhaps indicated that it was really a stable tributary inhabitant and its presence in the main river was incidental. ERIKSEN (1964, 1968) has reported a narrow substrate size preference range of *phi* —3 to *phi* 0 for *Ephemera* and related it to leg morphology and respiratory capacity in interstitial burrows. Here a similar preference probably operates to restrict this species to areas where suitable substrate conditions are provided by particular bottom configurations. Sand with a little detritus comprised the few gut contents examined.

Neoephemeridae: Two species of *Neoephemeropsis*, a genus previously recorded from Java by ULMER (1940), were found in the river quite sharply segregated at the forest edge. *N.* sp.A, a 5–7 mm, dorsally tuberculated and laterally spined nymph, was a constant element of the *Saraca*-root fauna that could be found throughout the year with diligent collecting. A few adults were trapped at Station II in scattered months. The lowland *N.* sp.B, of which only a single specimen was ever found in the forest, was larger, 9–10 mm, without dorsal tubercules or lateral abdominal spines; it occurred at moderate density in leaf and debris piles in quiet bank areas at the three lower stations. No adults were trapped. Most guts contained only sand, but a detritic diet was probable.

Caenidae: Three caenids were recorded from the river. One belonging to an undescribed genus (PETERS personal communication) was found as a single specimen in the Station I tributary. The others were more common, constituting a significant and constant part of the fauna. *Caenis* sp.A was found ubiquitously the length of the river to Station V and in most tributaries, reaching moderate numbers in both depositional and erosional habitats, but favouring areas with aggregations of detritus and leaves. The family is usually categorized as inhabiting silty backwater lenitic areas, but these were not common on the Gombak. Some nymphs were collected from the *Saraca*-root biotope so distribution in the stream bed was probably determined by food availability and habitat stability rather than a particular substrate type. Detritus and sand were the dominant gut contents. *Caenis* sp.B was apparently restricted to Stations IV and V where it was collected in silted bank areas. No discrete numerical data are available as the two were counted together. Sp.A adults were taken at lights only during the December to February and June to August periods, but nymphs were present throughout the year so the life cycle was probably not synchronized with the wet seasons. These species, and also *Neoephemeropsis* sp.B, were not found at the D.I.D. gauging site and were presumed to have been eliminated by adverse chemical or substrate conditions as temperature differences in the lower river were negligible.

Prosopistomatidae: *Prosopistoma wouterae* was a minor but notable member of the ephemeropteran community of the forest river. Very few specimens were collected from riffle gravels or underneath mid-stream boulders, the substrates reported for this species with its original description from Java (LIEFTINCK 1932) and for Philippine, Indian and New Guinean spp. by PETERS (1967). Rather, collections made in the *Saraca*-root habitat with the fine 90 μ mesh net often yielded 3–5 nymphs from a 5 m section of bank, indicating that those in the gravels were probably strays from this habitat. No information on feeding habits was obtained.

Odonata (Table 67)

The larval biotopes of the odonates from the Gombak River have been discussed by FURTADO (1969), so further analysis of these was considered redundant. With the exception of 8 or 10 species, distribution and frequency of occurrence were discontinuous and this, coupled with the problems of larval taxonomy of most genera, makes discussion of longitudinal succession or distribution tentative only. Since FURTADO worked only at the lower edge of the forest, a transitional area between the two main biotopes, his records almost certainly extend the ranges of many species by including what were perhaps marginal distributions. Adult records listed as 'G' in the first column of Table 67 were made by him about 3 km below the B.P. weir site, others in Tributaries D and F as indicated. No attempt was made to record adult distribution in this survey. In some physical aspects, Tributary F was different from the other torrent streams as its course steeply descends a granite peak and in contrast to the usual sandy pool-stony riffle bed, had large areas of bedrock or rounded granite slabs as substrate. Cracks and crevices of these boulders were colonized by the semi-emergent *Piptospatha perakensis* RIDLEY which afforded especially attractive perching and emergent sites for many zygopterans, as evident from the adult records.

Table 67

		Station I			Station II			Station III			Station IV			Station V		
	#	r	s	b	r	s	b	r	s	b	r	s	b	r	s	b
ODONATA																
Drepanosticta 2 spp.		N2.99	I2.28	G	O3.03	F1.90	H	O2.76	A1.30	E						
Prodasineura spp.						(?)				*			*			
P. laidlawii	d															
P. verticalis	d g												A			B
Copera marginipes													A			B
Coeliccia albicauda	f															
Calicnemia chaseni	f															
Ondocnemis orang	f															
Pseudagrion spp.										H	F1.98		O			
P. perfuscatum	g													A1.30		O
?*Ceriagrion* sp.													A			
Devadatta a. argyoides	dfg			E	A0.70		B									
Rhinocypha spp.				E			C									
R. perforata limbata	d g									D	A1.30		F			L
R.fenestrella	df															
Libellago l. lineata	g						A		(?)				*			
Euphaea o. ochracea	dfg	G2.31	B1.40	K	F1.81	A1.30	I			D			A			
Dysphaea dimidiata	d g															
Neurobasis c. chinensis	g						D			G			C			
Vestalis amoena	dfg			G			B			F			K			
Tetracanthagyna sp.				*												
Acrogomphus sp.		E1.70	K2.59	F	E2.23	G2.11	H	E2.10	E2.31	G	I2.46	G2.20	I			J
Burmagomphus divaricatus			A2.00	B										E1.40	I2.47	D
Gomphidia perakensis						A1.48	*			A			B			
G. a. abbotti	g			A												
Heliogomphus kelantanensis																
Ictinogomphus decoratus melaenops								A0.70						A0.70		
Leptogomphus ? *risi*												(?)		A1.30	A0.70	*

Rotated data table (FURTADO, 1966):

Species	d / g								
Macrogomphus sp.									
Megalogomphus sumatranus	d g		(?)	A0.70				A1.30 / A0.70	A
Merogomphus femoralis									
Microgomphus c. chelifer		*	A *						
Microgomphus sp. nov.					*				
Onychogomphus fruhstoferi						E		D	
Unidentified gomphid		G2.43 / A0.70	* F	B1.81 / (?)	A / D	D / A	A0.70	A1.30 / E2.28	H
Paragomphus capricornis	g			A1.90		A	A1.30		A
Phaenandrogomphus asthenes		A0.70		(?)					
Chlorogomphus dyak			B		A				
Idionyx ? *yolanda*	d g	A1.30	A		A	A	A1.30		G
Macromia type	g				A *	B	E		
M. gerstaeckeri					*				
M. arachnomia									
M. callista	g	*							
? *Zygonyx ida*	d g		E	A1.40	G	A0.70	G	A1.90	G · A
Z. iris malayana	d g			A0.70	G		G	C1.40	B1.00 · A
Neurothemis fluctana	d g								
N. fulvia	d g	A1.78	A	A1.60	A	A1.30	B	A0.70	A1.90 · H
Trithemis type	d g						A	A1.90	
T. aurora	d g								
T. festiva	g			(?)		(?)	B		B
Orthetrum type	d g						B		
O. t. testaceum	d								
O. glaucum	d								
O. chrysis	d g			(?)		(?)	K	A0.70	A1.30 · L
Onychothemis type									
O. culminicola	g								
O. coccinea	g								
O. t. testacea	g								

adult records made at Tribs D and F (d, f) and main river at forest edge (g) extracted from FURTADO (1966).

A summary of FURTADO's findings, including modifications based on field observation, indicated the following habitat preferences. Most of the zygopterans, being 'clingers', occupied the root-mat, accumulated debris, emergent or trailing plant habitats in the marginal areas irrespective of water velocity. A few, e.g. *Euphaea* and the two calopterygid spp. were also taken in stony riffle areas, and *Devadatta* was apparently restricted to the *Saraca* habitat. The two *Drepanosticta* spp. (Protostictidae), not recorded by FURTADO, were almost invariably found in sand-gravel riffle areas. The aeschnid, *Tetracanthagyna* sp., had a limited larval habitat at the edge of fast riffles in roots and stable debris-drifts. Among the Anisoptera, only *Zygonyx* and *Onychothemis* (Libellulidae) were found clinging to rock surfaces. The first genus occurred in fast riffles and cascades and clinging to torrential rock surfaces, just below zones of maximum velocity. The species of *Burmagomphus*, *Ictinogomphus*, *Gomphidia*, *Heliogomphus*, *Leptogomphus* and *Microgomphus* were 'hiders' in leaf and stick accumulations and also often the marginal root areas, particularly of slow-flowing reaches, a habitat shared with all the libellulids (*Trithemis* group, *Orthetrum* and the genera discussed above) and with the *Idionyx* and the *Macromia* group (Corduliidae). FURTADO listed *Chlorogomphus* (Corduligasteridae) as a burrower, but the few records made here indicated it too belonged to the above group. The remaining assemblage of gomphids were burrowers, found in sand-gravel substrates, usually in depositional areas, but often extending into regions classified as riffles which, at the microhabitat level where direct current effects are not a factor, were indistinguishable. The bank records for this group indicate burial in the finer sediments among or below the marginal vegetation.

Longitudinal distribution in the river, not considered by FURTADO, is best discussed with reference to occurrences in the three zones and relative changes in population densities. Characteristic of the Upper Zone and more or less restricted to the forest stream were *Devadatta*, ?*Zygonyx ida*, *Chlorogomphus*, *Tetracanthagyna*, *Burmagomphus*, *Heliogomphus*, *Microgomphus*, and the unidentified gomphid (perhaps near *Phaenandrogomphus castor*), the last species being particularly conspicuous. The Middle Zone fauna included *Prodasineura*, *Libellago* and *Neurobasis* but was not markedly distinct.

Taxa found exclusively in the Lower Zone as nymphs were *Copera*, *Orthetrum* and *Onychothemis*. However, FURTADO recorded adults of the last two complexes from below the B.P. weir. A number of species were common to two or more zones, tending to increase in frequency of occurrence and numbers downstream. In this group were the various species of *Onychogomphus*, *Gomphidia*, *Rhinocypha*, the *Macromia* and *Trithemis* complexes and most notably, *Pseudagrion* spp. which appeared at Station III and became the dominant Odonata at Stations IV and V. Antithetically, species decreasing in frequency downstream were *Euphaea*, apparently with a loss of suitably large interstitial spaces as the Md on the riffles decreased below Station III, *Zygonyx malayana* only rarely found at Station V, and the species of *Drepanosticta*. This last genus was not found below Station III and partial segregation of the two species was apparent. Sp.1 occurred only in the tributaries and upper river while sp.2 was collected at Stations II and III, but not from the upper tributaries. *Acrogomphus* and *Vestalis amoena* were found ubiquitously in moderate numbers. The distribution of several other groups, particularly the burrowing gomphids, could not be determined because too few records were available.

The eliminative effects of pollution were not severe in this order. At the D.I.D. gauging site several *Pseudagrion* spp. and a few burrowing gomphids were still present, but as siltation would not interfere with these taxa directly and individuals of select prey species were still available, their presence was not surprising. The coarser-substrate-dwelling Libellulidae were apparently eliminated by the change in bottom type to unstable silt.

Substrate and cover appeared to be the prime factors determining nymphal distribution although temperature, and thus oxygen availability, may have been important in limiting the Upper Zone fauna. Food supply would appear not to be an important factor as the ephemeropteran prey species preferred by most odonates were readily available at all localities as well as accessory trichopteran and plecopteran larvae and nymphs for the 'clingers' and 'hiders' and Chironomidae and/or oligochaetes for the 'burrowers', However, with the complex biotopic demands of the adults, discussion of distributions based solely on the requirements of the nymphal stages is imprudent.

Plecoptera (Table 68)

This order has been collected quite extensively in main range locales, but detailed coverage of the whole peninsula is lacking, and confusion, generated by poor descriptions and therefore unknown synonyms, exists in the taxonomic literature. Previous adult records give a list for Malaya of 22 species dominated by the subfamily Perlinae (see KLAPÀLEK 1909, BANKS 1931, 1938 for references). The 19 species collected in the Gombak, listed in Table 68, added a number of new species and records for Malaya, but still included a considerable proportion of the total known fauna. Species previously reported from the main range hills that might have been expected but were not found included:

Acroneurinae
  Kalidasia kraepelini Klap.   *Neoeuryplax ochrostoma* Klap.
Perlinae (Neoperlinae)
  **Javanita caligata* Burm.   ***Ochthepetina aeripennis* Enderl.
  ***J. fascipennis* Banks   ***O. violaris* Enderl.
  Neoperla fallax Klap.   ***Oodeia dolichocephala* (Klap.)

Table 68

	#	Station I r	Station I s	Station I b	Station II r	Station II s	Station II b	Station III r	Station III s	Station III b	Station IV r	Station IV s	Station IV b	Station V r	Station V s	Station V b
PLECOPTERA																
Neopeltoperla fraterna	a	+		*		(?)										
Peltoperlodes biseata	a	+		+			*									
Total Peltoperlidae		C1.18		D			F									
Amphinemura minuta	a															
Amphinemura nymphs (*A.minuta*?)		+	+	+	+	+	+									
Protonemura jacobsoni	a	+		+		+										
Nemoura sp.																
Rhopalopsole spinata	a	*		*	+		+									
Total Nemouridae	a	F2.13	B2.08	E	C1.60	A1.30	A				(×)					
Perla kelantonica	a															
Perlid nymph K (*P.kelantonica* ?)					×											
Paragnetina sp.	a	×			×	×		×								
Unidentified perlid adult type VS	a		+													
Neoperla luteola	a															
Neoperla nitida	a															
Neoperla sp. nov.	a															
Neoperla nymph L (*N. luteola*?)		+			+	+		+	+		+	+				
Neoperla nymph A					+	+		+	+	+	+	+		+	+	
Neoperla nymph B					+	+		+	+	+	+	+	+	+	(?) +	
Tylopyge helvus	a						+			+			+			
Phanoperla clarissa	a							+	+							
Neoperline nymph J																
Kiotina sp. nymph						+			+					+		
Etrocorema nigrogeniculata	a	+	+			+		+			+	(×)		+		
Etrocorema trapeza	a		+			+					×	×				
Etrocorema sp. (*?ahenobarba*)	a	×			×											
Other neoperline nymphs					+				+			+				
Total Perlidae		O3.24	K2.68	E	O3.69	O3.29	F	O3.76	J3.04	F	O3.82	H2.56	F	M2.69	C2.06	D

adults collected at light-trap near Station II

N. jacobsoni Klap.   **Phanoperla pallipennis* (Banks)
N. sumatrana Klap.   *Etrocorema ahenobarba* Klap.
N. minutissima Enderl.

Nemouridae

Nemoura atrissima Samal.   *Nemoura remota* Banks

The nomenclatural confusion in the Neoperlinae, especially the nominal genus, is obvious from the above. RICKER (1952) considered the genera marked with an asterisk to be subgenera of *Neoperla*, so that with the three identified *Neoperla* spp. and the *Phanoperla* recorded from the Gombak (Table 68), a total of 14 species are now described in one generic group for a relatively limited geographical area. The position is perhaps analogous to that in this same genus in Africa in which a large complex of previously described species was hypothesised by HYNES (1952 et sqq., see 1968a) to be a single polytypic species, *N. spio*, with highly variable, but completely intergrading types. Whether such a controversial situation recurrs in the Oriental *Neoperla* will have to await comprehensive studies.

Peltoperlidae: Two representatives of this family were found exclusively in the forest zone. *Neopeltoperla fraterna* (= *Nogiperla*) and *Peltoperlodes biseata*. The first species was very rare, but the second, although only recorded in a limited number of benthos samples, was a constant and characteristic member of the fauna that was often present in considerable numbers in causal samples from stabilized leaf accumulations between boulders. In these habitats its food of decomposing leaves (cf. WALLACE et al. 1970) and frass was readily available. Its distribution was probably limited by temperature and shade, or more immediately by the availability of its preferred microhabitat which largely disappeared outside the forest where silting of deposited organics took place rapidly.

Nemouridae: Two species were identified from adults, *N. (Amphinemura) minuta* and *N. (Protonemura) jacobsoni*. *Amphinemura* sp., unidentified *Nemoura s.s.* sp. and *P. jacobsoni* nymphs were collected from the larger cobble in riffle and flowing-pool areas in restricted numbers, but the preferred habitats were at the base of the *Saraca* root-mat where it made contact with finer gravels and sand. None of the three taxa was taken outside the forest although their general herbivore diet of rotting leaves and detritus (cf. HYNES 1941, JONES 1950 and others) was available elsewhere. As they formed part of the drifting fauna (see later), some incidental records might have been expected at Station III but none was made. Temperature was probably the factor limiting a wider distribution, exceeding an upper lethal limit in the unshaded river.

Leuctridae: *Rhopalopsole spinata* Kawai (*sp. nov.* in press) was the first record of this family in Malaya. It apparently occupied a similar niche to that of the Nemourinae and was found particularly in the upper tributaries although also present in the main river at Station II.

A single nymph identified as a Capniidae was collected in a small tributary, but no report on it was made by Dr. KAWAI. Capnids have been found on Penang Island (Kʜᴏᴏ personal communication) and a species terrestrial in the nymphal stage has been collected occasionally from cryptic habitats in the jungle in the vicinity of Stations I and II.

Perlidae: *Perla (Kamimuria) kelantonica* and the nymph tentatively ascribed to it were found very rarely in the upper tributaries, including Station I, but were a conspicuous though uncommon find in the upper main river riffles, to which the species was apparently restricted. *Paragnetina* sp., whose adult was never caught, occupied the same habitat as *P. kelantonica*, in among large stones on riffles, but had a different distribution. This species was limited to the main river but apparently only the lower forested section (it was never found in the upper tributaries or in the vicinity of Location A). Its range extended downstream as far as Station III and the torrential sections of the middle tributaries H and J in wooded areas. Three *Neoperla* spp. were identified from adults taken in the light-trap near Station II, *N. luteola* in large numbers and *N. nitida* and *N. sp. nov.* rarely. Three nymphal types based on distinctive frontal markings (which may be totally unreliable) were recognised and these had a wide distribution in the river from Station II down to Station V, occuring predominantly in riffle areas but with small nymphs also being recorded from depositional area samples. Below Station V they were rapidly eliminated by the scarcity of suitable riffle areas and increased silting of their preferred habitat. Only type L nymphs (? = *luteola*) were found in the Station I and other upper tributaries and then only in small numbers. The other two types may have been at the top of their range in the main river in the forest.

Neoperline nymph J also had a river-long distribution, being found most commonly in leaf-drifts at Stations I and II, but also in marginal debris and emergent vegetation at the lower three stations. Two types of this very distinctly marked nymph were found, apparently longitudinally segregated. In the Upper Zone, in the heavy shade of the forest stream, the nymph was a deep reddish-brown, with red cerci and with or without dorsal abdominal dark spots. In the lower river, the frontal pattern was slightly different and the nymph was very much lighter, a dull yellow, always with distinct spots on the abdominal dorsum.

Kiotina sp. nymphs were collected predominantly from sandy habitats in flowing pools and at the lower ends of riffles, although the species was also taken from depositional pockets between stones in stronger current. It was not taken below Station III (except for a single specimen at IV), but was a constant element particularly at Station II where it was abundant.

Only adults of *Tylopyge helvus* Kawai (*sp. nov.* in press), a first record for this genus on the peninsula, and *Phanoperla clarissa* were seen so nothing can be said about their distribution except that they were present in the vicinity of Stations I and II.

At least three *Etrocorema* spp. were present from the adult records,

E. nigrogeniculata, E. trapeza, a new species described by KAWAI, and an unidentified sp. (perhaps *E. ahenobarba*), and possibly a fourth. Nymphs of *E. nigrogeniculata* were the dominant plecopteran at Stations I and II and were present at III and IV in low numbers; none was collected from Station V. Unassociated nymphs of a second type were restricted to the upper two stations and the forest tributaries. Nymphs of this genus were most frequently taken in riffle samples and by manual collection from the undersides of stones in the current.

Overall, the numbers of Perlidae collected were not large, but they were a constant faunal constituent. The considerable catch densities at Station IV were almost exclusively *Neoperla* spp., as were the lesser ones at Station V; the numbers of individuals of the other genera taken at these stations were small. The depositional sample catches at the upper three stations were mainly *Kiotina* sp., but with some young nymphs of the other species.

Food of all these setipalpians included whatever prey was available, mainly *Baetis* nymphs, larvae of Chironomidae and Hydropsychidae, rarely Elmidae, and at the lower stations particularly, Oligochaeta. However, in many stomachs large quantities of sand and detritus in excess of that expected from prey guts were found, particularly in *Neoperla*, *Etrocorema* and *Kiotina*. This partially agrees with the findings of COSTA and FERNANDO (1967) who ascribed a mainly vegetarian diet of algae and detritus to *Neoperla* in the Maha Oya River, Ceylon. *Perla* and *Paragnetina* nymphs were more exclusively carnivorous.

Hemiptera (Table 69a)

Taxonomically the Heteroptera are one of the better worked groups of the Malayan sub-region fauna. The references have been summarized by FERNANDO and CHENG (1963) and CHENG and FERNANDO (1969). In total number of species recorded the diversity in most families is exceeded by no other region of comparable area. However, a large proportion of the species is limited to lenitic biotopes, except in the Gerridae where a number of species are found exclusively in rushing waters. In discussing the aquatic and semi-aquatic Hemiptera, any consideration of population size is pointless as catches made by the sampling techniques used routinely were negligible. Large numbers of the gregarious gerrids and velids could be caught in local areas on occasion by sampling specifically for them, and at other times be completely absent, depending on discharge and weather conditions. In Table 69a, a subjective assessment of overall abundance is given with only presence or absence at a particular location indicated; however, the considerable mobility of most families makes any rigid distribution scheme problematical. No study of diets was made, but the order is exclusively carnivorous (USINGER 1956), with the exception of the Corixidae which may feed on

Table 69

	#	Station I	Station II	Station III	Station IV	Station V
HEMIPTERA						
Hydrometra sp.	×		++			
Rhagovelia femorata	++	+	++	++	+	+
Tetraripis doveri	+		+			
Microvelia sp.	×					
Amemboa horvathi	+	+	+			
Limnogonus fossarum	(×)					(×)
Limnometra anadyomene	++	++	++++			
Metrocoris nigrofasciatus	+	+++	+++			
Pleciobates tuberculatus	+++					
Ptilomera lundbladi	+++		++	+	+	+
Rheumatogonus intermedius	++		++		+	+
Ventidius sp.	+++		+			
Esakia sp.	×					
Cryptostemma sp.	×					
Cercometus pilipes	+	+	+		+	++
Ranatra varipes	×		+			
Enithares malayensis	×		++			+
Enithares mandalayensis	×					
Plea liturata	+	++	++	+	(×)	
Helotrephes corporaali	++					+++
Diplonychus rusticus	×					
Aphelocheirus gularis	++	+++	+++	+++	++	
Laccocoris nervicus	++	+++	++		+	++
Ctenipocoris asiaticus	+					+
Micronecta sp.	×	(×)	(×)	(×)	(×)	(×)

Subjective assessment of relative abundance

b. LEPIDOPTERA

	r	s	b	r	s	b	r	s	b	r	s	b	r	s	b	r	s	b
Cataclysta type 88			(Upper Tribs)															
Cataclysta type 132			(Upper Tribs)															
Cataclysta type 350 ±	+			+			+			+			+					
Paragyractis type	×			+														
Nymphula (*Paraponyx*) type R3									×									
Nymphula (*Paraponyx*) type R5									×			×			×			+
Nymphula (*Paraponyx*) type D5									+			+			+			+
Gill-less Nymphulini type F1						×												
Gill-less Nymphulini type R1																		+

227

detritus, algal protoplasm, microfauna and surface organic ooze in addition to the more usual invertebrate prey (cf. GRIFFITH 1945). The supra-aquatic groups derive much of their food from the surface drift of terrestrial insects, but with the Cryptocerata feed extensively on whatever aquatic invertebrates or vertebrates are available.

Hydrometridae: A single *Hydrometra* sp. was recorded at Station II in a flood-pool area.

Veliidae: *Rhagovelia femorata* was the constant species at all stations in riffles, joined at Station II by *Tetraripis doveri*, which occurred in low numbers. Very occasional records of the latter species were made at Station III but it apparently preferred torrent stream conditions. The most common velids of lowland waters according to FERNANDO and CHENG (1963) are the *Microvelia* spp., but only a few specimens were ever collected, all at Station IV; these were probably visitors from the paddy area.

Gerridae: This was the largest and most frequently seen group with 9 spp. identified and possibly more present. Nymphs and adults are considered together. *Ptilomera lundbladi* was the dominant species present at all stations, but in greater numbers in the torrent sections. It was not seen in the polluted section downstream of Station V. Its preferred feeding station was in fast-flowing smooth water, often at the lower end of a riffle or in the race behind a large boulder, where a group of three or four in the upper river but usually only a solitary individual in open locations would be found collecting drowned and drifting insects. BULLOCK and FURTADO (1968) calculated a population density of 350–500 per 360 m at Station II, but they counted only adults and mature nymphs so that their total was a considerable underestimate. A single record of the lenitic species *Limnogonus fossarum* was made at Station V, probably a transient from adjacent ponded areas. The only other gerrid collected from the lower river was *Rheumatogonus intermedius* which was found in moderate numbers on riffles at Stations IV and V and strangely not found at the higher stations although classified as a slow-moderate flow forest stream species by CHENG and FERNANDO (1969). Taken in the forest river and tributaries in occasional collections were *Pleciobates tuberculatus, Amemboa horvathi* and *Metrocoris nigrofasciatus*, all from fast-water areas adjacent to *Saraca* root-mats and log-drifts. None was common, although a congregation was often caught from one location. *Limnometra anadyomene* was a more frequent catch from slower, often pooled surfaces and below the small backwaters formed by root-masses and gravel bars. Nymphs of the two smaller sized gerrids, *Ventidius* sp. and *Esakia* sp. which could not be specifically identified, were collected only at Station II, perhaps as a result of more intensive sampling effort there. All records were from almost still water behind boulders and roots; the favoured habitat was a large flood-pool that always harboured numerous Collembola on its surface. As far as could be determined, no definite breeding season occurred since for *Ptilomera* at least, small numphs were present at all times.

Dipsocoridae: A rarely caught species (three records all at Station IV) was *?Cryptostemma* sp., found in riffle samples from mid-stream. This genus is nominally semi-aquatic and the individuals may have come from decomposing beds of riparian litter.

Nepidae: *Cercometus pilipes* was collected at Station I in a bankside pooled area, from the flood-pool and *Saraca*-roots at Station II and quite commonly at Station V among emergent grasses and stick accumulations. *Ranatra varipes* was taken in the same habitat at Station V on several occasions. The upper river records were probably incidental.

Notonectidae: *Enithares malayensis* was collected only once from the flood-pool at Station II and *E. mandalayensis* in a few bank samples at Station V. Neither species was part of the true river fauna, but rather was transient from ponds or mining pools.

Pleiidae: *Plea liturata* was found at upper river stations occurring in small numbers in the bank samples at Stations I, II and occasionally III. A single specimen was found at Station IV. The preferred habitat of the genus is standing water (FERNANDO and CHENG 1963) and most species were caught in backwater areas, but its presence in the upper torrent river at all and absence in any number from the slow reaches cannot be explained except by the lack of stable cover in the lower river.

Helotrephidae: *Helotrephes corporaali* was a characteristic species of the reduced current *Saraca* habitat where it attained moderate populations. It also occurred in bottom gravels on slow riffles in the Upper Zone. Further records were made at Station V, but not from the intermediate stations.

Corixidae: An unidentified *Micronecta* sp. was collected, always as single specimens, from all stations. Like the Notonectidae, this was probably a migrant from outside the river system and not part of the lotic fauna although HARRISON (1958a) recorded this genus from sandy substrates in the Great Berg River and CHUTTER (1970) from Vaal catchment areas.

Belostomatidae: *Diplonychus (Sphaerodema) rusticus* was the only species collected. It was recorded at low density in the bankside accumulations of leaves and sticks and in the trailing and emergent vegetation at Station V only, but probably also occurred at Station IV.

Naucoridae: *Aphelocheirus gularis* was a constant species at all stations on the river, primarily inhabiting between-stone habitats in riffle areas although it was also taken in leaf-packets and bank samples. *Laccocoris nervicus* and *Ctenipocoris asiaticus* shared the marginal root-leaf-detritus habitat. Neither species was common in the upper river, but both reached moderate populations at Stations III and IV. *Laccocoris* was not recorded from Station V. It is reportedly a fast-water form, but was taken from pool margins as well as riffle areas.

No Hemiptera were collected in the pollution zone below Station V, but the samples were small and may have missed the cryptic forms. No conclusions can be reached about the effects of the heavy silt load.

Lepidoptera (Table 69b)

Pyralidae: Nine larval heteroceran types were recognized, four living under conspicuous silk-web retreats (Argyractini), and three building tubes and two gill-less forms of unknown habitat (Nymphulini). *Cataclysta* type 88 and *Cataclysta* type 132 (numbers refer to gills; head capsule features also different) were found only on rock faces in torrential upper tributaries. *Paragyractis* type was present in the main river, but was not found below the lower edge of the forest. *Cataclysta* type 350± was omnipresent where submerged rock surfaces were available, out of the main channel. The diet of all these was algae/detritus scraped off the rock surfaces. In the limited number of guts examined, no animal remains were found although FITTKAU (1964, 1967) described this group as *Simulium* feeders in central Amazon streams. The case-retreat builders were restricted to the middle and lower open river where marginal and emergent Graminae were common. *Nymphula (Paraponyx)* type R3 (round grass case, small, tribranched gills) was found only at Station III in several isolated casual bank samples. *Nymphula* types R5 (large five-part gills) and D5 (flat bi-sectioned bamboo leaf larval case, small five-branched gills) were collected in bank samples at the three lower stations feeding on the emergent grasses. Type D5 was still present at the D.I.D. gauging site and was probably indifferent to fine inorganics as it could easily eliminate silt from its large case. The two gill-less species were less common. A single specimen of type F1 was taken in a casual sample from *Saraca* and trailing vegetation in the Station I tributary; no case was collected with it. Type R1 larvae were collected only at Station V on several occasions from marginal vegetation, never with a case. Both these species were presumed to feed on the available grasses, or perhaps on roots of *Colocasia*. The nature of the habitats, large stable rock surfaces or marginal weed beds, made collection of this family with the standard sampling technique unlikely so that the above remarks are based on observations and casual sampling only.

Trichoptera (Table 70)

The difficulty in discussing this order is again the inability to satis-factorily identify, or in the larval stages even to separate, the taxa present. The published work on the group for Malaya is poor, being limited to adult collections made by several workers, most notably BANKS (1931, 1938). The larval/pupal material collected from the Gombak, and probably the whole peninsula, agrees only approximately with that described by ULMER (1951, 1955, 1957) for the Sunda Islands. WIGGINS

230

(personal communication) has indicated that in many of the case-building groups, 'generic characters established for one area are not consistent for the same genera in other areas', so that until considerable associative breeding-out is done, the affinities of the local fauna in the immature stages, even at this level, will remain unknown. Where possible a genus, with subcategories if distinctive features were evident, was assigned to the larval material; many of these are tentative. Others had to be left at the subfamilial or familial level. Because of these problems, only a brief discussion and indication of general distribution of a very rich fauna of at least 79 recognizable taxa or types spread over 15 families is possible.

In Table 70, adult records (a), mostly from Station II, and presence of larvae (+) at a location are indicated. When the numbers collected of each type were small, a family total to indicate relative densities is given.

Philopotamidae: *Chimarra* spp. were not common in the main river, but in some upper tributaries were fairly abundant, found generally in the gravels behind and beneath stable stones in the current. A few specimens only were collected at Station III and the group appeared to be limited to the forest river. Their diet was largely detritic materials captured in oblong nets similar to those described by SATTLER (1962).

Rhyacophilidae: Two distinctive larval types were recognized from the upper tributaries where they were found exclusively. Other types occurred in the riffles at Stations II, III and IV, but the population density of these carnivores (SCOTT 1958 *inter alia*) was low.

Glossosomatidae: *Glossosoma* sp. was collected as individual specimens from all stations except III, attached to the upper surfaces of rocks in fast currents. *Synagapetus* type larvae attained high densities on boulders in partially-exposed sections of Tributary B, below the small watercress farm where algal growth, particularly diatoms, was rich. Elsewhere it was recorded only rarely from the Station I tributary. Both larval types were apparently restricted to surfaces where algal grazing was possible as is characteristic of the family (cf. HYNES 1961, ULFSTRAND 1967).

Hydropsychidae: *Hydropsyche* spp. were present at Stations I–IV in some numbers but were only incidentally recorded at Station V. The routine quantitative methods did not adequately sample the larger stone habitats in which this genus constructed its nets, so the records from casual, coarse-net sampling are a better indication of the distribution. *Diplectrona* was the most obvious genus in the river with at least four species. Larvae were present as dominants at Stations I, II and IV and as a smaller constant component at III and V. Most larvae were collected from riffle areas where they constructed coarse nets between gravel and stone fragments. At the lower three stations, however, where rough substrates were not as common, size segregation occurred with the small individuals being found in the s gravel samples and the larger larvae more common in the r samples. *Cheumatopsyche* type larvae were found at Stations I to IV in low numbers. Macronematinae larvae of three types

231

Table 70

	Station I			Station II			Station III			Station IV			Station V		
	r	s	b	r	s	b	r	s	b	r	s	b	r	s	b
TRICHOPTERA															
Chimarra spp. a, b, c, d, e					a										
Chimarra sp. nr *fulmeki*					a										
Total *Chimarra*	F2.04	H1.90	C	B1.54	A1.30	C	A1.60	C2.16			A1.00				
Rhyacophila spp. a, b		+													
Rhyacophila type		+			+			+			+				
Total *Rhyacophila*	A0.70		A	A1.30			A1.90	B1.48		B1.60		A			
Glossosoma sp.										A0.70		A			
Synagapetus sp.															
Hydropsyche sp. S			*	C1.65									A0.70		A
Hydropsyche sp. a	A1.30	+			+a	*									
Hydropsyche sp. b	A1.30	+			++	*									
Total *Hydropsyche*	E1.86	A1.30	A	E2.04	a	B	A0.70		D	C1.60	A0.70	F	A0.70		A
Hydropsychodes spp. a, b.			C							A1.30		A	A0.70		A
Other Hydropsychinae larvae		*				E				A0.70				+	
Diplectrona sp. a		+			+a			++							
Diplectrona spp. P, B		+			++a			++							
Diplectrona sp. nr *aurivittae*					a										
Total *Diplectrona*	L3.09	B2.41	L	K2.72	C2.00	L	F3.36	E2.30	B	K3.10	F1.85	A	G2.56	G2.80	B
Cheumatopsyche sp.	D2.00	A1.90	C	B1.40		A	C2.28			D1.81	A0.70	E	A0.70		
Cheumatopsyche type larvae					a						a				
Macronemum sp.					a										
Macronematinae larvae ? 3 spp.	C1.81			A0.70	+a	B	H2.32		B	D1.95	a				
?*Amphipsyche* sp.					a										
Psychomyia sp. a					a			a							
Psychomyiella sp. a					a										
Psychomyiella sp. b					a								A1.90		B

Taxon										
Ecnomus sp. a									a	
Ecnomus sp. b								a		
Ecnomus sp. c						a		a		
Ecnomus sp. d						a		a		
Ecnomus sp. e										
Ecnomus sp. f										
Other Psychomyiinae type a				a						
Other Psychomyiinae type b	a									
Other Psychomyiinae type c				a						
Other Psychomyiinae type d				a				a		
Hyalopsychella sp.				a				a		
Pseudoneureclipsis sp.				a				+		
Tinodes spp. a, b, c, d				a						
Nyctiophylax spp. a, b				a						
Polyplectropus sp.				a						
Polycentropus sp.				a		a				
Other Polycentropodinae				a						
Total Psychomyiidae/Polycentropodidae larvae	F2.16	D1.98	H2.40	D1.90	L3.37	I2.87	K3.15	F2.26	I2.91	C2.51
Stenopsyche sp.		B		A		A				
Stenopsyche sp.	*		A0.70	*		C				
Marilia sp.	*	A	A0.70		D1.93	B	B1.40	A		
Goera sp.				a		A				
Helicopsyche sp.	*			*		A1.30				
Helicopsyche sp.	*	A		D		A1.30				
Ganonema sp.	(larval cases only)									
?*Limnocentropus* sp.	A1.30	D	A0.70	A1.90	B2.20	A1.30	A1.30	C		
Tagalopsyche sp.	+			+		+		+		
Oecetis sp. a										
Oecetis sp. b								+		
Oecetis sp. c										C

	Station I			Station II			Station III			Station IV			Station V		
	r	s	b	r	s	b	r	s	b	r	s	b	r	s	b
Total Oecetis		A1.90	F			G	D2.29	B1.93	B	G2.65	A1.00	F	G2.22		B
Setodes sp. a					a			a							
Setodes sp. b					a										
Total Setodes			A		a	A			C	A1.40			A0.70		B2.00
Adicella sp. a, b, c, d					a										
?Adicella sp.														a	
Leptocella sp.															
Triaenodes sp.		+			+										
Mystacidini larvae		+			+									+	
Other Leptocerinae a					+			+			+				
Other Leptocerinae b					+			+			+				
Other Leptocerinae c					+										
Other Leptocerinae d								+			+				
Other Leptocerinae e					+										
Total other Leptoceridae larvae	B1.60					I									
Lepidostomatinae spp.		A1.60	G		*		B1.40	A0.70	D	E2.30	A1.30	C	A1.90		H
Dinarthropsis sp. a, b, c, d					a +										
Neolepidostoma sp.			H	A1.30	A1.30	F	G2.80	A1.78	B	E2.20	A1.30	A			
Plethus sp.			A		*			*			*				
Hydroptila sp. G		*			*										
Hydroptila sp. H		*			*										
Hydroptila sp. X												*			*
Total Hydroptilidae larvae	D2.08	B1.40			D1.90		D1.81	A1.30		C1.90	A1.30	A1.30	A1.90		

a = adult records; Note: Adult records at II may indicate presence at I also as the light-trap was situated between these stations

were found at all stations but reached their highest densities in the sandy riffles of Station III. At this location, the retreats of these larvae were constructed only in the stabilized areas between stones on riffles, never in the shifting depositional zones with similar sized sediments. The chimneys protruded 8–10 mm above the bottom, extending into the faster current zone to facilitate filtering of food and apparently were sufficiently high to keep most of the incessant saltating bed-load from entering the opening. In construction, these funnels, with net-sieve and larval chamber attached below, were identical to those described for this genus by SATTLER and KRACHT (1963). ?*Amphipsyche* sp. was represented by only a single adult and one larva collected at Station II.

The family is omnivorous (see discussion by SCOTT 1958), but the proportions of food types taken here obviously depended on availability. All guts examined contained detritus. *Hydropsyche* from the upper river also contained a few animal remains and several *Diplectrona* from the lower river contained diatoms and cyanophytes. *Macronemum* probably fed exclusively on the drifting organics (see SATTLER 1963), but the other genera were partly predacious, leaving their nets to hunt actively. On several occasions larvae were found at the ends of short silk threads on open rock faces. Increased silting, which seriously affected numbers of all but *Diplectrona* at Station V, eliminated even that genus in the lower reaches of the river, although it reappeared wherever a small zone of riffle was present.

Psychomyiidae/Polycentropodidae: Larvae of this complex could not be differentiated and were treated as a group. They were relatively uncommon at the upper stations, but many adult types were collected at the Station II light-trap. At Stations III and IV, this group became a dominant with numerically important larval populations on the riffle areas and lesser densities in depositional bottoms. Silting at Station V was probably responsible for the reduction in population density there. No diet was determined but both families are carnivorous (DITTMAR 1955, HYNES 1961, ULFSTRAND 1967), feeding opportunistically on all available invertebrates, but particularly Chironomidae.

Stenopsychidae: *Stenopsyche* sp. was the largest caddis found in the river. It was apparently restricted to the main river in the forest (only a single specimen was collected at Station III), where it built large nets beneath stable stones in torrential areas. An omnivorous diet dominated by detritus was indicated for the few larvae examined, but predation may be secondarily important.

Odontoceridae: *Marilia* sp. was only occasionally collected in quiet pool areas at Station II and in bank areas under accumulated leaf material at Station III.

Goeridae: ?*Goera* sp. larvae were found in a few riffle samples at Stations I–IV and apparently occupied a sub-stone habitat.

Helicopsychidae: larvae of this family, perhaps *Helicopsyche* sp., with flatly-spiralled cases in contrast to the more usual conical type, were collected regularly and in moderate numbers from 1° and 2° tributaries in the forest. Empty cases were found at all main river stations, but viable larvae only once at Station III in a sandy back-water area. The larva appeared to have narrow microhabitat requirements as in the small streams it was always collected from fine sand-bottomed pool areas, often under piles of leaves where detritus had accumulated. The detritus constituted its main food and no evidence was found for an algal diet as reported by Costa and Fernando (1967). The limited distribution was perhaps a result of the spate regime in the main water courses that constantly disrupted the interstitial substrates and would have buried or washed out such cased forms. High turbidity in the main river during certain periods might also have been a limiting factor for this closed-case species. The small tributaries in the forest always ran almost clean even after incessant rain.

Calamoceratidae: *Ganonema* sp. was collected regularly in casual samples from *Saraca* roots and stick-drifts at Stations I and II. It was limited to the forest where arboreal cover was adequate and where the petiole it used as a retreat was available. Its diet was not investigated.

?Limnocentropodidae: Distinctive larval cases similar to those described by Wiggins (1969) as belonging to *Limnocentropus* sp. were collected at Stations I and II and the upper torrent streams on several occasions but no larvae were taken.

Leptoceridae: Apart from *Tagalopsyche* sp., *Leptocella* sp. and *Oecetis* spp., the larvae, representing a number of genera, were not separable during the survey. The larval distributions of these were made from records of material sent to Dr. Wiggins. *Tagalopsyche* was a very visible member of the fauna at Stations I and II where it occurred as a constant species. It was most common in the bank biotope clinging to the trailing roots and vegetation. At Station III, a few individuals were collected from marginal grasses. The irregular lower river records may have resulted from drift. A general phytophagous diet was indicated. *Leptocella* sp. was apparently restricted to the *Saraca* habitat where it attained sizeable populations. The three separable *Oecetis* spp. showed some succession; sp.a was found exclusively at the upper three stations, sp.b at Station IV in quite high densities and sp.c at Station V only. *Setodes* spp. and other larvae, some no doubt *Adicella* spp. as indicated by the adult catches, occurred in gravel and coarse sands in low numbers at all stations.

Lepidostomatidae: Four adult types were identified to subfamily only, so the *Neolepidostoma* type larval records probably include several different species. These larvae in their leaf-fragment cases were common among the roots of the bank habitat in the forest where currents were slight. The occasional records from riffle samples were in leaf accumulations. Some larvae were collected mainly from riffles at lower river stations, perhaps as drift victims, but more likely utilizing the leaf fragments and detritus accumulated there. *Dinarthropsis* sp. was collected in low numbers from the *Saraca* roots at Station II only.

Hydroptilidae: Minute, 1–1.5 mm, ?*Plethus* sp. larvae were common in splash zones and on smooth-faced boulders in the current in the forest zone tributaries and at Stations I and II. Where diatom growth occurred, densities of five per sq.cm of surface were sometimes recorded. *Hydroptila* type G was found at the three upper stations, type H in crevices cn the undersides of rocks at Stations I and II, and type X, a fringed-case form, in low numbers on bottom substrates at the lower two stations. In very localized areas, on pilings and smooth logs and sticks in the current where the effects of silting were reduced, this type was often numerous.

Coleoptera (Table 71)

The members of families of this order having aquatic or semi-aquatic stages comprised a conspicuous part of the fauna, but only limited work, none recent, has been done on any of the groups in Malaya and literature on the important lotic families (in the Dryopoidea) is virtually non-existent. FERNANDO and GATHA (1963) have collated specific records and reviewed the known literature for the peninsula. With the present state of the taxonomy, any observations on trophic position and population structure can only be generalizations and must be considered in this light.

Dytiscidae: This is the most commonly found group in lowland waters and has been the best worked with some 38 described Malayan species. However, adults were rarely seen in the river; only one genus of adults, *Hydrovatus*, was identified from Station IV, but larvae of at least three types were occasionally collected in low numbers from Stations I–IV. The majority of specimens were from sandy depositional substrates but riffle habitats among stones and leaf-drifts also yielded a few larvae. The normally carnivorous larvae all had guts packed with sand and some detritus so a general diet is probable, as suggested by FERNANDO and GATHA (op. cit.)

Gyrinidae: Adult *Orectochilus* sp. were collected from all stations. At the lower stations, gregarious accumulations of 5–15 individuals often occurred in smooth-flowing areas among trailing vegetation, while at Stations I and II, only occasional individuals were collected from *Saraca* bank areas. *Dinuetes* type larvae were found in gravel substrates at Stations I–IV on a few sampling dates, but were not common. COSTA and FERNANDO (1967) reported a herbivorous diet of leaves, detritus and algae for *Gyrinus* sp. in Ceylon and most guts examined contained some detritus, but the larvae are normally carnivorous (LEECH and CHANDLER 1956) so that those plant elements may have derived from ingested prey.

Haliplidae: Only a single species has been reported in Malaya,

Table 71

	Station I			Station II			Station III			Station IV			Station V		
	r	s	b	r	s	b	r	s	b	r	s	b	r	s	b
COLEOPTERA															
Hydrovatus sp.															
Dytiscidae larvae (3 types)	A1.30	E2.48		A0.70	G2.45	A	C1.65	B1.60							A
Dinutes type + other larvae		A0.70		F1.90			A0.70	A0.70		C1.18	H1.30				
Orectochilus sp.		a			a			a							
Haliplus sp.							B1.40	a	K	A0.70	a	C		a	
Amphiops sp.															
Enochrus sp.								a			a				
Helochares sp.								a			a			a	
Psalitrus sp.					a										
Hydrophilidae larvae (3 types)	A1.30	A0.70					E2.02	A1.30							
Hydraenidae	J2.62	C2.30	E	L2.85	H2.55	A									
Scirtes type	A0.70			L2.57	E1.85	G			B	A1.30		B			
Ptilodactyla sp.															
Eulichas sp.	H1.88	E1.74	K	I2.04		I	F1.88	B1.40	E	C1.18		B			
Eubrianax type	F2.28	A1.78	*	C1.70		C	E2.31	B1.60	C	C1.95		A	B1.65		A
Psephenus type	G2.35	D2.00	*	E1.98	A1.30 (?)	B	I2.52	B1.60	C	A1.90	A1.30				
Psephenoides sp.	D2.00	B1.60	*				C2.20	B1.60	A	A1.30	A1.30				
Dryopid type 1				A1.00		C	B1.60			A1.30	A1.30	*			
Dryopid type 2				A1.30			E2.31	A1.30	A				A1.30		A
Dryopid type 3				A1.30											
Elmidae larvae	O3.88	O3.66	F	O4.11	O4.60	M	O4.47	M3.72	J	O4.71	O3.92	L	O4.16	N3.70	H
Parnid adults	M2.98	D2.16		N3.41	L2.91	B	O3.63	I2.78	B	O3.80	L2.91	C	N3.35	F2.74	B
Torrindicola sp.	*				*				*						
Lampyrid type 1		*		*											
Lampyrid type 2				*											

a = adult records; (a) = single adult record.

238

Haliplus pulchellus Clark, but the *Haliplus* type collected regularly at Station III and rarely at Station IV did not carry the lateral spinous tubercles figured for that species by FERNANDO and GATHA (op. cit.), so probably represents a second species. The preferred habitat was marginal and the diet detrital.

Hydrophilidae: This family was surprisingly uncommon in the river, even in areas adjacent to paddy fields where a rich fauna of 20 genera and at least 51 spp. are known. Adults of *Amphiops* sp., *Enochrus* sp. and *Helochares* sp. were collected at the lower three stations at lights, and *Psalitrus* sp. from Station II, but all were uncommon. Unidentifiable larvae of at least three types were collected irregularly at the upper three stations, but never from IV or V; no reason for this discontinuity was apparent, unless the lenitic areas present below and adjacent to Station IV were selected preferentially to the river as ovipositional sites.

Helodidae: Larvae of the *Scirtes* type were a dominant in the Stations I and II fauna, but only incidental in the lower unshaded river. Most records were from riffle areas or in bank samples associated with decomposing leaves. Whether these organics or the associated microflora constituted the food is unknown.

Ptilodactylidae: Larvae tentatively assigned to *Ptilodactyla* were collected in riffle samples at Stations I and IV on two occasions. Four species are known in Malaya from adults, but none was collected here. *Eulichas* sp., probably belonging in a family of its own (HINTON personal communication) was the largest, most conspicuous insect of the forest stream. It occurred at low population density in most quantitative samples from coarse gravel and stony areas. Less frequent catches were made in the middle river, declining to three records, probably incidental, at Station IV. The larva was absent from the lowermost station. Gut contents were leaf particles, sand and detritus. Nothing is known of the post-larval stages.

Psephenidae: Larvae of the *Psephenoides* type, *Psephenus* type and *Eubrianax* type were regularly collected in the upper and middle river. Only *Eubrianax* extended down to Station V, in reduced numbers, while the other genera declined downstream to an incidental status by Station IV. Larvae were most common on the underside of clean stones in the current, feeding on fine detritus (cf. COSTA and FERNANDO 1967), but small individuals, less than 2 mm, were found in sandy bottom samples. Distribution was probably dependent on the availability of suitable substrates and increased silting and decreased stone size accounted for the noted decline in the lower river. In the upper tributaries, unidentified larvae much larger than main stream types were collected from cascade and splash areas. These apparently fed on the diatom-detritus layer on the rock faces.

Final disposition of a large material of the parnid families Dryopidae and Elmidae was not available at the time of writing, but at least four spp. of Dryopidae and 20 Elmidae were present (HINTON personal communication). The adults of these families were taken in all r samples and in many from depositional substrates wherever leaves and detritus, the principal food, accumulated. Virtually none were collected from bank or *Saraca* root samples although mid-stream leaf-packets often had considerable populations. Numbers found in the small tributaries were low, but in the main river, particularly at Stations III and IV where the richer algae growth seemed to provide a significant part of their diet (cf. JONES 1951, HYNES 1961), population densities were high. Records for the adults of these families were not separated routinely and are listed as 'parnid adults'.

Dryopidae: Three types of larvae were recognized, none of which was common at any station. Only a single casual record was made in the upper tributaries and the larvae were apparently absent from fast-flowing sections, although their habitus on the underside of stones in riffles would take them out of direct current influences.

Elmidae: Larvae were abundant and a dominant at all stations on the riffles and also occured in the s samples, chiefly in the earlier instars. The most common habitat was in gravel under and behind stones and obstructions where the detritus that comprised the principal food was abundant. The large numbers taken in pool bottom sediments where similar substrate conditions occur were predictable. As with the adults, population density reached a peak in the open, unshaded reaches where algae increased. The high densities recorded at Station IV indicated that the larvae too utilized this food resource. With the demise of riffle areas and occurrence of heavy silting between the stones on the bottom, this fauna disappeared below Station V.

Torrindicolidae: Four minute (1–1.5 mm) larvae of a *Torrindicola* sp. belonging to this myxophagian family were collected from 90 μ mesh casual riffle samples at Stations I and II. No adults of this family, recorded previously only from Brazil and Africa, were found.

Notably absent from an otherwise rich fauna were the Hydraenidae. A single adult was collected from the Station I tributary, but no larvae were seen. FERNANDO and GATHA (1963) also reported only two records for Malaya although they speculated that the fauna might be expected to contain 50 or more species.

Collected from riffle samples at Stations I, II and III on isolated occasions were two related types of larvae tentatively placed in the Lampyridae. Aquatic members of this family have been previously reported from Celebes by BLAIR (1927). Those found had characteristic lampyrid features and in addition, lateral tracheal gills from abdominal segments 1–8. The head capsule was reduced in both types, but in one much more so than in the other.

In addition to the above records, a number of unidentifiable larvae and adults were taken at every station. As indicated at the beginning of this section, the lack of taxonomic information on the aquatic forms precludes any real discussion of factors affecting longitudinal succession in the river with changing conditions.

Diptera (Table 72)

The lotic dipteran fauna of the Malayan sub-region is hardly known except for occasional references to records in the Journal of the Federated Malay States Museum (e.g. EDWARDS 1928) and a few descriptions of Chironomidae summarized by KARUNAKARAN (1969). This work has all been on adults and the immature stages are completely unknown. The fauna is diverse (particularly the Nematocera, with numerous taxa in the Chironomidae and Simuliidae), but not abundant in terms of numbers except in the lowland river where microhabitat uniformity is increased. Because of the insurmountable taxonomic problems in this order, determination to genus or genus group only was attempted, using the keys of JOHANNSEN (1934, 1935, 1937a, b) and THOMSEN (1937) as a basic guide. The Simuliidae have been worked more than the other groups and papers by EDWARDS (1934) and more recently SMART and CLIFFORD (1969) enabled Dr. DAVIES to put tentative specific names on a portion of this fauna. Identifying the Chironomidae, as is the case almost everywhere but Europe, was an impossible proposition. To obtain an idea of the extent of this family, comprehensive general samples were collected with the standard 165 μ net on several occasions. All distinct types were mounted and identified as far as possible, using JOHANNSEN (op. cit.) and MASON (1968). Distinctive types and their distribution were added to this data as they were found in later samples. Undoubtedly many more species and probably genera were present in specialized habitats or were able to pass through the relatively large mesh (cf. JONASSON 1958). To assess the importance of the family in the river community, frequencies in the benthos samples were recorded at the familial level except for three prominent recurrent larval types enumerated separately.

Chironomidae: (Tanypodinae): *Procladius* larvae were found only in the Station I tributary on a single occasion. The two *Pentaneura* spp. constituted a ubiquitous and dominant element, occurring roughly equally in riffle and pool substrates where their prey species, mostly other chironomids, were abundant.

(Chironominae): In the Chironomini the most conspicuous larvae belonged to a complex of *Cryptochironomus* – like spp. (Larva A and variants) which showed succession of the various types down the river. These predatory larvae, which also apparently ingested detrital foods, were abundant at all stations down to V where their numbers decreased, perhaps because of high temperatures, but more likely as a result of heavy silting in the interstitial regions. The *Chironomus s.s.* group of

241

Table 72

	Station I			Station II			Station III			Station IV			Station V		
	r	s	b	r	s	b	r	s	b	r	s	b	r	s	b
DIPTERA															
Procladius sp.		+													
Pentaneura sp. a		+			+									+	
Pentaneura sp. b					+			+			+			+	
Total Tanypodinae	L2.96	M3.20		M2.87	K2.90		L2.70	N3.27		M3.04	J2.73		K3.25	K3.28	
Chironomus s.s. type 1					+						+			+	
Chironomus s.s. type 2															
Chironomus s.s. type 3					+			+						+	
Chironomus s.s. type 4					+			+			+			+	
Chironomus s.s. type 5		+													
Chironomus s.s. type 6		+			+										
Total *Chironomus* s.s. type larvae	C1.78	B2.20		C1.78			B1.65			C2.02	A1.30		J3.39	M3.80	
Larva A															
Larva A type 1		+			+										
Larva A type 2		+			+			+							
Larva A type 3					+			+			+				
Larva A type D		+			+			+							
Larva A type B		+			+									+	
Total Larva A group	K2.54	K2.82		H2.54	L3.03		O3.20	M3.26		M3.18	K3.08		O2.30	C2.31	
Cryptochironomus sp.		+						+			+			+	
Stenochironomus sp.											+				
Xenochironomus sp.											+				
Stictochironomus sp.		+												+	
Glyptotendipes sp.					+			+			+				
Microtendipes sp.		+													
Chironomini type B		+									+				
Chironomini type 527											+			+	

Taxon	O4.91 / G2.26 ‖ O4.83 / A1.90	O4.03 / G3.15 ‖ O3.72 / K3.38	O3.79 / J3.29 ‖ N3.25 / L3.79	O3.72 / N3.25 ‖ O3.44 / O4.12	N3.77 / M3.14 ‖ O3.91 / L3.62
Polypedilum sp. 1	+	+	+	+	
Polypedilum sp. 2	+	+	+	+	+
Paralauterborniella sp.	+	+	+	+	
Tanytarsus sp. 1	+	+	+	+	+
Tanytarsus sp. 2	+	+	+		+
Tanytarsus sp. 3	+	+	+		
Tanytarsus sp. 4					
Tanytarsus sp. 5	+		+	+	
Tanytarsus sp. 6					
Tanytarsus sp. 7					+
Tanytarsus sp. 8					+
Tanytarsus sp. 9					
Micropsectra sp.	+	+	+	+	+
Total other Chironomini/Tanytarsini	O4.91 / G2.26	O4.03 / G3.15	O3.79 / J3.29	O3.72 / N3.25	N3.77 / M3.14
Larva C	O4.83 / A1.90	O3.72 / K3.38	N3.25 / L3.79	O3.44 / O4.12	O3.91 / L3.62
Corynoneura sp. 1	+	+	+	+	+
Corynoneura sp. 2	+	+	+	+	+
Trichocladius sp.	+	+			
Psectrocladius sp. 1	+	+	+		+
Psectrocladius sp. 2	+	+			+
Cardiocladius sp.	+	+	+	–	+
Cricotopus sp. 1	+	+	+		
Cricotopus sp. 2	+	+	+		
Orthoclad type 1			+		
Orthoclad type 2			+		
Orthoclad type 3			+		
Orthoclad type 4			+		
Orthoclad type 5			+		
Orthoclad type 6			+		
Orthoclad type 7	+	+	+		+
Orthoclad type 8		+	+		
Orthoclad type 9			+	+	
Orthoclad type 10		+	+	+	
Orthoclad type 11		+	+	+	+

	Station I			Station II			Station III			Station IV			Station V		
	r	s	b	r	s	b	r	s	b	r	s	b	r	s	b
Orthoclad type 12		+			+										
Orthoclad type 13		+			+										
Total Orthocladiinae	O3.05	L3.34	A	N3.23	N3.23		O3.89	N4.23		L3.66	M3.58		O4.85	O4.37	
?Prodiamesa type		(×)													
Simulium 18 spp. (see Fig. 41)	H3.04	D2.15													
Bezzia group	C1.78	D2.51		E2.42	G2.10	A	C1.78	A1.30		D2.00	C1.90		D2.69	A1.30	
Culicoides type	C1.78			I2.51	F2.43	*	I2.45	C1.78		E2.08	A0.70	*	F2.68	C2.48	
Forcipomyia type						*	G2.42	A1.30		A1.30	A1.30		A1.30		
Dixa type	*			*											
Hexatoma type 1	M2.32	K2.62	K	I2.36	K2.32	E	L2.64	J2.26	G	F2.19	D1.70	A	B1.18	A0.70	
Hexatoma type 2	H2.19	D2.00	B	K2.47	I2.32	A	B1.78	B1.00	A	A1.30	B1.00		A1.30		
Antocha type	B1.40	B1.93		C2.08		B	I2.80		B	B1.40			B2.00	A1.60	
Megistocera type			B	A0.70											
Elliptera type								A0.70							
Limonia type	*														
Other Tipulidae				B1.40	A0.70		A0.70	A0.70							
Apistomyia sp.	*			*						A1.30					
Genus incertus nr Blepharocera	*			*											
Pericoma type								A1.30							
Maruina ? indica	*			*							A1.30		B2.00	B2.00	
?Maruina sp.															
Atherix type	B1.60	B1.00	B	D1.70	A1.31	G	B1.40	A0.70		A1.30					
Empididae sp. 1	C1.78	B1.60	A	J2.49	D1.93	A	D2.65	A1.30	A						
Empididae sp. 2			A			A									
Dolichopodidae			A			A	A1.30		B						
Stratiomys type															
Scyiomyzid type 1															B
Scyiomyzid type 2															A
Eristalia type								A1.30							
Other Cyclorrhapha	B1.78	B2.00		C1.78	C2.08		E2.51	C2.48		C1.90	A1.30	A		A2.20	A

species were present at all stations, but rarely at the four uppermost locations.

Chironomus type 3, a large blood-gilled larva, occurred exclusively at Station V and reached high numbers in silted marginal areas, feeding on the bottom organics. This species assemblage, with the annelids discussed previously, characterized the polluted reaches. At the lowest sampled area of the river, near the D.I.D. gauging site, type 3 was the only benthic insect found. Apart from these types, only the two species of particle-feeding *Polypedilum* among the 10 other genera were numerically significant at all stations, although Chironomini 527 was locally abundant in the riffles at Station IV. Winged hydra-like tubes of the nine *Tanytarsus* spp. were found on all types of surfaces, rock, between gravel particles and particularly silted leaf-beds, throughout the river, often at locally high densities. Some between-taxa differentiation of microhabitat requirements may exist, particularly with respect to current and ability to filter different sestonic foods from turbid water, but delineation of this would require much more investigation. *Micropsectra* was only rarely collected.

Overall numbers for the subfamily were high, reaching a maximum at the lowest station where uniform small Md sediments and rich suspended and benthic organic food supplies provided a near optimal habitat for most of the phytophagous species.

(Orthocladiinae): Larva C, a *Corynoneura* – like sp., was a constant member of the upper and middle river fauna found in both depositional and erosional areas in the interstices of the gravels. Its decline in numbers at Station V was attributed to silting of this habitat. The three types restricted to the upper forest stations, Orthoclad 11 (= ?*Brillia*) and Orthoclads 12 and 13 were uncommon. Orthoclads 1–10 and the other genera listed in the table were most frequently found at Station III and some at Station IV where algal production provided a reliable, if sometimes limited, supply of food. Overall numbers for the Orthocladiinae groups were considerable. Small *Corynoneura* larvae made up most of the population at Stations I and II, while at the other stations, many species contributed to the large populations of this subfamily.

From the number of discontinuities and apparent omissions, it is obvious that the distributions for this family given in Table 72 are incomplete. However, it was thought worthwhile to attempt this separation if only to emphasize the diversity of both fauna and biotopes that characterize the river.

Simuliidae: 18 species were identified from the river and its tributaries and a nineteenth, recorded in the adjacent watershed of the same Forest Reserve, undoubtedly occurs in the Gombak also. Larvae and pupae were collected from emergent vegetation, leaf-packets and mid-current stone habitats by hand as the routine techniques did not efficiently sample the firmly attached fauna of these biotopes. The quantitative collection data given in Table 72 therefore do not indicate the true population densities. In most localities aggregation was normal, with a selected substrate harbouring many individuals and an adjacent, similar area being completely free. Most larvae and pupae were collected from trailing grasses and roots in the more open parts of streams (see also HUGHES 1966a), but in the forested tributaries, leaf accumulations in areas of rapid, but uniform, flow were the preferred substrate. Few immatures occurred on rocks in the current, perhaps because of the molar action during spates in those exposed areas. When they were found, they were confined to the dense dark-coloured granites, almost never being recovered from lighter-coloured quartzitic rocks or flaky schist and sandstone fragments. All attempts at collecting adult material failed. Only three adults were caught by passive Malaise-type traps in the course of the work. Most species appeared to be avian feeding, probably moving directly into the forest canopy on emergence. Identifications by Dr. DAVIES were therefore based on dissected mature pupae. Distributions are illustrated in Fig. 41.

With the notable exception of *Simulium (Gomphostilbia)* sp. 2. which occurred only in the comparatively warm lower river, the Simuliidae were restricted to the jungle-shaded river and tributaries. The main river near Station II, sampled extensively, yielded only four species at overall low densities. Of these, *Simulium s.s. kiuliense* Smart and Clifford (? = *nobile* de Meig.) and *Simulium s.s. striatum* Brun. were found exclusively in larger river habitats. The Station I tributary was particularly rich in species, but these were divided into two groups. The *Gomphostilbia* group of species were collected in the small upper feeder tributaries with emergent vegetation and grasses in more open areas where the highway crossed the watershed, opening up the jungle and permitting entry of light and local warming of the bottom, as described earlier. These species, *Simulium (Gomphostilbia) sundaicum* (= *rayonhense*), *G. pegalanense*, *G.* sp. nr *sundaicum* (cf. *varicorne*) and *G. flavocinctum* also occurred in Tributary B near the watercress farm; two of them recurred in Tributary F which had increased exposure. *G. flavocinctum* was also collected from an open cliff-cascade in a small tributary higher up the valley from Location B. The second group composed of five *Simulium s.s.* spp., *S. kinabaluense*, *S. nitidithorax*, *S. laterale*, *S.* sp. 1a and *S. digitatum*, were recorded in the Station I tributary from leaf accumulations and rocks in the deeply-shaded jungle reaches where diurnal temperature fluctuations were minimal. The collections of these species in the high elevation Tributary B were from similar microhabitats. In the highest sampled tributary, two undefined larvae, *S.* sp. 3 and *S.* sp. 4, were found on one occasion on stones in the current. *S. sabahense* Smart and Clifford has been recorded from the adjacent Templer Park region, but not from the Gombak drainage. In that location it was collected in association with *G. sundaicum!* The three *S. (Eusimulium)* spp., *E. aureohirtum* (= *tuararense*), *E.* sp. A and *E.* sp.3, were found in three different tributaries. Sampling

LONGITUDINAL DISTRIBUTION OF SIMULIUM SPP.

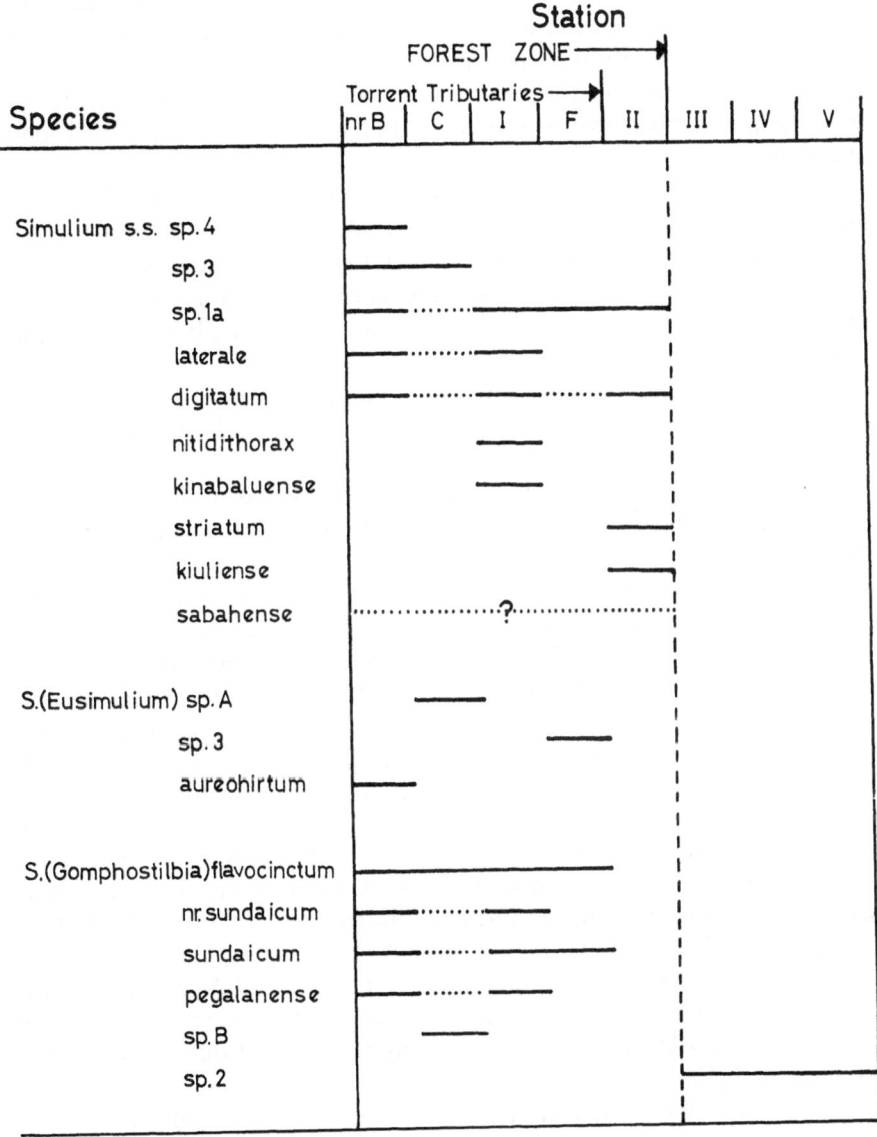

(······ probably present – not recorded)

Fig. 41. Longitudinal distribution of *Simulium* spp.

in the tributaries, other than around Station I, was limited to four or five trips at different times, so that the distributions as discussed above must be considered provisional only, especially as the taxonomic relationships are still tentative. No investigation of food habits was made, but it was assumed that fine particulate detritus (WILLIAMS et al. 1961) or drifting microflora (cf. FREDEEN 1964) were filtered from the current and ingested unselectively. In the watercress beds near Location B where there was supplemental organic enrichment, the population densities of the gomphostilbids became very high with larvae even attached directly to unstable muddy bottoms where the outlet waters carried enriched food resources.

Ceratopogonidae: This family was inadequately sampled by the 165 μ mesh net so the quantitative records are underestimates, but larvae of the *Bezzia* group and *Culicoides* type were found at all stations in bottom samples and from marginal fine sediments taken with the 90 μ mesh net. Only two records of *Forcipomyia* type larvae were made, both from bankside fine silt and leaves at Station II.

Dixidae: *Dixa* type larvae were incidental forms present only on stone surfaces in some upper tributaries and at Stations I and II on rare occasions.

Culicidae: Various mosquito larvae were collected in flood-pools and isolated backwaters among root-mats, but were not considered part of the lotic fauna.

Tipulidae: *Hexatoma* type I was a prominent and dominant faunal element at Stations I–III and *Hexatoma* type 2 at Stations I and II only. Below these locations, their numbers declined to incidental status at Station V. Large larvae tended to occur more frequently in stony riffle areas, but moderate numbers were also collected in coarse depositional sediments. Distribution may have been limited by siltation, but the decreasing abundance of leaf-food was a more likely factor. *Antocha* type larvae were present throughout the river to Station V as a constant, but not abundant, form. *Megistocera* type larvae were very rare, only being collected once at each of Stations I and II. Characteristic of the torrential sections of various tributaries and at Stations I and II were *Limonia* type larvae in gelatinous envelopes on the lee-side of large rocks in the current. These were apparently predatory on chironomids and *Bezzia* spp. larvae occupying the same habitat. A single record of an *Elliptera* type larva was made at Station IV and various other unidentifiable tipulids were found at Stations I–IV.

Blepharoceridae: A constant species, but one with discontinuous distribution even in its usual torrent rock-face habitats, was *Apistomyia* sp., found, with a second rarely collected species belonging to a *genus incertus* (nr *Blepharocera*), exclusively at Stations I and II and the high tributaries. Diatoms scraped from the rock surfaces constituted the main diet although attached detritus was also ingested. Pupation occurred in crevices in the immediate vicinity of the larval habitat. Both species were apparently confined to the hill-rhithron zone by temperature and silt-free substrate requirements.

Psychodidae: The larvae of *Maruina indica* and a rare, unidentified

? *Maruina* sp. were found under a wide range of current conditions as well as in the same fast-flowing shallow water biotope as the blepharocerids. Both were found exclusively in the upper river where temperatures were moderate and constant and silting at the rock-faces was restricted to short periods during heavy spates. In the lower river, a *Pericoma* type psychodid larva was occasionally collected from bottom and more frequently from marginal habitats at Stations III, IV and V.

Rhagionidae: *Atherix* type larvae were a constant part of the benthic fauna at Stations I and II, but were collected only in isolated samples at Stations III and IV and not at all from Station V.

Empididae: Two empid larvae were recognized, both limited to the upper three stations. Type 1 was a regularly encountered species occurring almost exclusively in stony riffles, while the other type was only incidentally recorded. These larvae and those of *Atherix* were at least partially predacious.

Dolichopodidae: A single record of dolichopodid larvae was made at Station III from a riffle sample.

Stratiomyiidae: *Stratiomys* type was collected at Station II from a single bank sample; this was probably an accidental record of a semi-aquatic litter species since no other members of this family were found even in the suitable slow-water habitats of the lower river.

Tetanoceridae: *Sciomyzid* types 1 and 2 occurred only at Station V where their pulmonate snail prey (BERG 1953, 1964) were most readily available.

Syrphidae: *Eristalis* type larvae occupied the silted marginal habitats of the lower river below Station V. (A single record was also made at Station III). The numbers found were never large even in the heavily polluted zone. A number of other unidentified cyclorrhaphan larvae were collected from Stations III, IV and V.

1.2.2. Similarities with the faunal communities of other tropical rivers

The remarkable parallelism in community invertebrate fauna and that reported for similar rain forest rivers in other geographical regions has been intimated. Rivers in the central Congo basin described by HULOT (1951) and MARLIER (1954), on Ceylon by COSTA and FERNANDO (1967) and on Madagascar by RAMANANKASINA (1969), but particularly the small streams of central Amazonia described in general terms by FITTKAU (1964, 1967, 1969) encompass a range of biogenic conditions and habitat types analogous to those in the Gombak, so that this unity and presence of ecologically equivalent types were not unexpected. No detailed exposition of the faunal similarities is necessary, but widespread utilization of allochthonous foods by the herbivores, either directly or after microfloral transformation, and the relatively high proportion of nominally predatory forms in the assemblage are the most striking features. The main biotopes, sandy bottoms, leaf-drifts, root-bank complexes and stones-in-current harbour families and often genera in common. In many cases, the microhabitats to which Fittkau ascribed a particular invertebrate type contained the same or closely related species as those found here in the Malayan biome.

1.2.3. Features of the fauna of various sections

To extrapolate from the mass of data presented in the previous section, the important and distinctive faunal characteristics are summarized for each section.

Upper Torrent Tributaries and Station I

a. All the ephemeropteran families found in the Gombak, except the Oligoneuridae, were present; *Epeorus* and one *Thalerosphyrus* sp. (Heptageniidae) were unique to the torrent-surface biotopes and the Ephemerellidae were rare.

b. Odonata: *Devadatta*, *Tetracanthagyna*, *Euphaea*, unknown gomphid and Protostictidae were prominent; *?Zygonyx ida* was unique.

c. All the Gombak plecopteran genera were represented. *Perla kelantonica* was a notable species.

d. Comparatively few Hemiptera (*Aphelocheirus*, some Gerridae, *Plea* and *Helotrephes*) occurred.

e. Trichoptera: *Synagapetus* and *Helicopsyche* were restricted to these stream types; Philopotamidae and Rhyacophilidae were present along with diverse Hydropsychidae; *Tagalopsyche* and other leptocerids, and particularly *Plethus* sp., were characteristic faunal elements.

f. Elmidae were not a major component. *Scirtes*, Psephenidae and *Eulichas* were prominent and Dytiscidae and Gyrinidae of minor importance.

g. Fifteen species of *Simulium* occurred. Dixidae and *Limonia* were exclusive to this section. Blepharoceridae, Psychodidae, other Tipulidae and Rhagionidae made up a constant dipteran assemblage with a group of small Orthocladiinae, Tanytarsini and some Tanypodinae that were also present.

h. *Hermes* (Corydalidae) was a conspicuous species. *Potamon johorense*, various Palaemonidae, Gordiidae, mites and semi-aquatic Epilampridae and Tridactylidae were constant features. Two species of *Cataclysta* type lepidopteran larvae occurred solely in these streams. *Brotia* and *Ferrissia* spp. were rare, other Mollusca absent. Ranid and rhacophorid larvae were often abundant.

Main River in Forest and Station II

Most features of the above tributaries with some notable omissions recurred in the main river as long as it remained shaded and fast-flowing.

a. *Isonychia* and the Baetidae dominated the ephemeropteran fauna. Leptophlebiidae of all the recorded genera were found here. *Compsoneuriella* (Heptageniidae) was an addition with occasional Ephemerellidae. The conspicuous *Ephemera* was restricted to specialized habitats. Prosopistomatidae, Caenidae and Neoephemeridae were constant groups.

b. An increase in the number of species of gomphids and libellulids was found. *Rhynocypha* and *Vestalis* occurred with occasional Corduliidae and Cordulegasteridae.

c. All plecopteran genera were present; *Paragnetina* was a distinctive but rare species; *Etrocorema* spp. and *Kiotina* sp. and other Neoperlidae were common.

d. Hemiptera: a large assemblage of gerrid genera mostly restricted to corrent niches occurred; the Cryptocerata included species of *Laccocoris*, *Aphelocheirus* and *Helotrephes*.

e. The large and diverse trichopteran fauna contained members of all families, dominated by *Diplectrona* and other hydropsychids and *Adicella* spp., and with *Stenopsyche* as a unique member. Torrential habitats had fauna as in the tributaries. *Saraca* areas contained many cased species of Leptoceridae, *Neolepidostoma*, *Marilia* and *Ganonema*.

f. The Dryopoidea were a constant element of the fauna, with *Psephenus*, *Psephenoides*, various dryopids and *Scirtes* present in small numbers. The Elmidae were diverse, but not abundant. *Eulichas* was a distinctive member of the fauna.

g. The drastic reduction of numbers of *Simulium* spp. was notable; only four were present in the main river, of which two were unique. Some Chironomini appeared in addition to the other midge groups in which *Corynoneura* spp. were dominant. Torrent-specialist Diptera were still present in suitable hygropetric locations. *Atherix* and *Hexatoma* spp. constituted a constant faunal group in riffle habitats.

h. Only a few Annelida were found, Naididae with an occasional *Glossiphonia* sp.. *Macrobrachium* spp. were abundant, but other Crustacea were rare. Molluscs were still represented only by *Brotia* and *Ferrissia*. Lepidoptera added a third *Cataclysta* type and less commonly, a *Paragyractis* sp.. *Amolops* and *Megophrys* tadpoles were a distinctive part of the riffle fauna.

Middle River and Station III

The river in this section was in transition between the fully shaded-swift current-large substrate of the upper reaches and the potamon-like areas of the lower river. The fauna showed considerable reduction in some elements from the upstream assemblage and had a few additions more characteristic of the lower river. The main features of these changes are given below.

a. The *Isonychia, Prosopistoma, Choroterpes, Dipterophlebiodes* and *Ephemera* of the *Saraca* and slow-flowing shaded pool habitats disappeared. Numbers of *Isca* and *Thalerosphyrus* were reduced, the latter species being partially replaced by *Compsoneuriella*. Notable development of the Ephemerellidae, a family which was characteristic of Station III, occurred with an increase also in frequency of the Potamanthidae. *Neoephemeropsis* sp.A was replaced by sp.B.

b. Odonata: Amphipterygidae and Aeschnidae dropped out. Protostictidae and Epallagidae were reduced in frequency. Calopterygidae and some gomphids became common and a few Coenagrionidae appeared.

c. The families Peltoperlidae, Nemouridae and Leuctridae were eliminated and the number of species of Perlidae was reduced. *Perla* and several species of the *Neoperla* complex were eliminated. One *Etrocorema* remained and *Paragnetina* was rare. *Kiotina* sp. was present, but not as commonly as in upstream locations.

d. Most of the torrent-dwelling gerrids, except *Ptilomera*, were absent or occasional. Naucoridae were a constant faunal element.

e. The families Calamoceratidae, Glossosomatidae and Stenopsychidae were eliminated and *Chimarra, Rhyacophila* spp. and *Neolepidostoma* were rarely found. Psychomyiidae/Polycentropodidae became common and the Hydropsychidae, notably the Macronematinae, were characteristic.

f. The large fauna of Elmidae, the absence of Helodidae and the appearance of a few Haliplidae were notable.

g. The dipteran fauna, except for the Chironomidae, was reduced with only one *Simulium* sp. present and the *Apistomyia* of the upper river replaced by a rarely found *Pericoma* sp..

h. Prawns were noticeably absent and *Potamon* was replaced by *Paratelphusa*. Ostracodes were collected rarely. The Corydalidae had dropped out. The nematode *Labronema* was a characteristic species. Oligochaete numbers increased but diversity did not. Several case-building Pyralidae appeared in addition to the omnipresent *Cataclysta* sp..

Lower River (unpolluted) Station IV

Many of the species listed as declining at Station III became even less common in this part of the river, often occurring only in riffle-rejuvenating areas. Among the prominent changes were the following.

a. *Thalerosphyrus* declined and *Thraulus* and *Potamanthodes* occurred more frequently. Ephemerellidae were not as common as at Station III, but another species of this family was introduced here.

b. The Protostictidae disappeared and the Coenagrionidae became prominent along with the libellulid *Onychothemis* and a number of Gomphidae.

c. In the Plecoptera, the *Neoperla* species group became abundant while *Kiotina* and *Paragnetina* dropped out.

d. Hemiptera: *Rheumatogonus* replaced the fast-water species, except *Ptilomera*, as the most common gerrid. Pleiidae were no longer found, but the Naucoridae were common.

e. Hydropsychidae and Psychomyiidae/Polycentropodidae were still abundant and with several species of the Leptoceridae constituted the trichopteran fauna, most of the other families being present only in low numbers.

f. *Eulichas* and the Dytiscidae were no longer found, but the Coleoptera were abundantly and richly represented by the family Elmidae.

g. Empididae and Rhagionidae had disappeared at this station but the single *Simulium* sp. was still present along with a large group of Chironomidae of all types.

h. The richest additions to the fauna were made in the Oligochaeta and Hirudinea, where *Branchiodrilus* and *Allonais* became prominent and several other worm species and their predators were added. The thiarid snails, *Thiara* and *Melanoides*, became common and *Gyraulus*, *Indoplanorbis* and *Siamopaludina* occurred in low numbers.

Lower River (polluted) Station V

This location was characterized by reduced diversity, but increased numbers as expected from an enriched area.

a. The Baetidae with several additional species and Potamanthidae dominated the ephemeropteran fauna with only occasional members of other families present.

b. Libellulids and gomphids were common, with some Corduliidae. The zygopteran fauna was dominated by the *Pseudagrion* spp.. Platycnemididae *(Copera)* occurred occasionally and the Chlorocyphidae and Calopterygidae still had representatives in the marginal area community. Epallagidae were very rare.

c. Only two Neoperlinae, both *Neoperla* spp., were present, *Etrocorema* having dropped out between Stations IV and V.

d. The only notable additions to the Hemiptera were the belastomatid *Diplonychus* and the nepid *Ranatra*. *Laccocoris* dropped out, but *Aphelocheirus* was still present.

e. Net-spinning caddis were few, except for the ubiquitous *Diplectrona*. Psychomyiidae were not common and Leptoceridae, Lepidostomatidae and Hydroptilidae had declined to an unimportant status.

f. The Dryopidae–Psephenidae group and the Haliplidae were absent although Elmidae were still present in reduced numbers.

g. The Chironomidae attained high population densities, particularly the *Chironomus s.s.* spp.; Psychodidae and Tipulidae were minor components. Cyclorrhaphan flies of the families Tetanoceridae and Syrphidae appeared here.

h. Semi-aquatic cockroaches were not found. The oligochaetan

Naididae and Tubificidae became abundant and the Erpobdellidae with them. The Mollusca were further enriched by species of Ampullaridae and Lymnaeidae; the Thiaridae became abundant. The only Crustacea occurring at this station was an amphibious *Orchestia* (Talitridae).

Below this location most families were eliminated. Only a few species in the Gomphidae, Coenagrionidae, Chironomidae, Syrphidae, Naididae, Tubificidae, Thiaridae and Pyralidae were found at the D.I.D. gauging site on the lower river, with occasional slow-water gymnoceratid Hemiptera. No Naucoridae or Belostomatidae were collected, but may have been present.

1.2.4. Factors effecting microdistribution and succession

Horizontal and longitudinal distribution of stream benthic animals are closely related as most taxa select a microhabitat by optimizing the complex of local factors determining niche while being restricted to a physically distinct zone either by inability to tolerate extremes of one or a combination of those factors as they change with location. If transients are disregarded, the community found at a location has reached a compromise with the habitat conditions and in most cases it can be assumed that if a species is not present, it has failed to find a satisfactory niche. With some exceptions, particularly in the non-insects, such absence does not indicate that an organism has failed to reach a biotope (on the contrary, by passive dispersal or adult flight most locations are historically accessible (MACAN 1961b)), but rather that it could not thrive there. However, in some cases, zoogeographical determinants have dictated present distribution through spatial isolation; in a contiguous watercourse this is unlikely. A wide literature exists, much of it analysing single or related factors with respect to a single or group of species, often considered out of context and with disregard for interaction effects. Comprehensive overviews of the more useful work have been made by MACAN (1961a, b, 1962b, 1963), CUMMINS (1966), ULFSTRAND (1967) and HYNES (1970) so discussion here will be limited to general aspects and expansion of the remarks on habitat preferences of the Gombak fauna already made.

The presence, or more usually absence, of many chemical constituents (most commonly calcium, bicarbonate, sodium chloride, pH and toxic materials) has often been correlated with invertebrate distribution, but with respect to local microdistribution these have little influence and may be disregarded. However, overall nutrient poverty, as experienced here, may limit some groups. Where heavy metals and insecticides are found (see e.g. JONES 1958, GOSE 1960, GAUFIN 1965, WARNICK and BELL 1969), elimination of the fauna is often unselectively complete with only a few very tolerant groups remaining below what is usually a man-made polluting source. As far as could be determined, such materials were a minor factor in the river, but as the effects of some are cumulative and a single contamination by others is sufficient to eliminate species, there was no way of evaluating the long-term effects of the low concentrations of copper and zinc occasionally detected in the lower river. No assay for insecticides was possible. Acid extremes of pH are often complicated by the presence of heavy metal leachates (KEMP 1967, PARSONS 1968) so the effects of the acidity alone are masked, but only rarely have studies shown any distribution pattern dependent on pH alone (see HARRISON and AGNEW 1962). Values for pH in the Gombak were very uniform and it is unlikely that this was a factor for any species. However, the normal buffer system, particularly concentration of bicarbonate ion, has been indicated as an important determinant of range for various organisms. MANN (1955) showed that the distribution of leeches in Britain was directly influenced by alkalinity and there may be a similar correlation here as several species were found only at Station V or at IV and V. More likely, however, the presence of this group in any numbers was determined by the availability of snail and worm prey. The Mollusca, because of their medical importance, have been widely studied, especially with respect to calcium concentration and hardness. SHOUP (1943) felt that streams with less than 20 mg/l. bicarbonate were 'unproductive' of snails, a figure also quoted by HYNES (1960) as the cut-off concentration for many groups. HARRISON and SHIFF (1966) reported maximum development of the *Bulinus – Biomphalaria* community in the $CaCO_3$ range 5–40 mg/l., with a decrease in fecundity at either extreme. No absolute low tolerance levels were given although in subsequent papers (HARRISON 1968, HARRISON et al. 1970) an optimal concentration of 35 mg/l. was determined. In very few studies have lower limits been given. MAITLAND (1965) found *Ancylus fluviatilis* absent from waters with less than 2 ppm, but here, as noted, *Ferrissia* sp. was ubiquitous at even lower concentrations. The other pulmonates and some prosobranchs were not found above Station IV, but hardness even at that point was very low. Mollusc distribution may however be secondarily limited by any one or combination of other factors, such as silt, temperature, shade, current velocity, and vegetation (see e.g. BOYCOTT 1936, HUBENDICK 1958, PIMENTEL and WHITE 1959) and according to VAN SOMEREN (1946), CRIDLAND (1958) and BERG and OCKELMANN (1959), by low oxygen tension for some families. The virtual absence of small Crustacea (Amphipoda, Cladocera, Ostracoda) from the river, particularly the lower, more lenitic sections where a zooplankton might be expected to develop, was attributed to the calcium deficiency, but the unstable flow pattern may have been equally decisive.

The factors altitude, temperature, shade and oxygen tension are closely related, the first two not important determinants at the microhabitat level, but often considered the primary factors determining succession in stream faunas. At the biotope level, oxygen concentration can vary over short distances, depending on utilization for decomposition and renewal by diffusion and turbulence so that closely adjacent areas may harbour individuals of different respiratory capacity (MANN 1956, 1961, MADSEN 1968a) in different ontogenic or physiological states (see KNIGHT and GAUFIN 1966). This could have been the factor controlling distribution of plecopteran and ephemeropteran groups that had members with an apparent proclivity for leaf-packets (e.g. *Peltoperlodes*) while others occurred exclusively in gravel riffles, but attempts at micro-oxygen determinations at this level were not successful. In the upper river at least, oxygen conditions probably would not vary between adjacent areas as lotic turbulence would maintain adequate exchange. In the quieter, low reaches it conceivably could be a minor factor controlling distribution, but others, such as the availability of suitable food would probably be more important. The presence or absence of particular chironomid larvae in particular sediments may depend on oxygen availability and the elimination of all but the *Chironomus s.s.* spp. at the D.I.D. site was probably a function of this (see WALSHE 1948, BRUNDIN 1951) as sedimentary oxygen concentrations would have been low at this organically enriched location with little reaerating turbulence. As a factor limiting longitudinal distribution, percentage saturation, not absolute concentration, is the important consideration; as seen, this changed very little between Stations I and III, remaining above 90%, and only lost 5% and 15% at Stations IV and V respectively, although the extremes were greater. Diurnal differences were minor, but dependent oxybionts, some Nemouridae, some Diptera and Leptophlebiidae, may have been limited in their ranges by the downstream decrease, particularly as it occurred in conjunction with elevated mean water temperatures.

Shade as a factor has been little investigated, but as shown particularly by HUGHES (1966a, b, c) who also reviewed the literature, a number of species are restricted to umbriferous sections of streams or dark microhabitats by negative photoaxis. In his Transvaal stream he found *Tricorythus*, several Leptophlebiidae and *Centroptilum* restricted to shaded areas, Elmidae, Simuliidae and Chironomidae shade indifferent, and *Neoperla, Baetis, Euthraulus, Cheumatopsyche* and *Hydropsyche* preferentially in illuminated areas. The equivalent taxa here did not follow this distribution and although the lanceolate-gilled leptophlebids, *Habrophlebiodes, Dipterophlebiodes* and *Choroterpes* were restricted to the forest, along with a number of other taxa, this was probably because of a substrate requirement. *Baetis* would have to be examined species by species. As seen, *Simulium s.s.* spp. appeared to be shade-loving in the Gombak, but ZAHAR (1951) reported the genus to be absent from shaded areas so this factor may prove to be minor and recessive to temperature and detrital availability (ALBRECHT 1968).

Temperature and altitude as factors often, but not always, go together. Differences in temperature requirements have been considered by many authors to account for most longitudinal and successional distribution patterns (see IDE 1935, SPRULES 1947, SCHMITZ 1954, MACAN 1958a, 1961b, 1962c, ILLIES 1961a, MINCKLEY 1963, KAMLER 1967, MINSHALL and KUEHNE 1969, HYNES 1970). In the limited area of a single habitat, temperature changes are usually minimal, so this factor has little bearing on microdistribution. Accounts of altitudinal succession of taxa are common and the dependent temperature relationship is usually held responsible, often with disregard for other factors of substrate. In non-equatorial streams, temperature and the number of available degree days (IDE 1935) determine to a large extent the success of a species at a particular altitude. Diversity generally increases downstream as temperature rises, with little concomitant elimination of species because of lethal temperature conditions until other factors such as oxygen tension and changing substrate have also become limiting. In the Gombak where there is virtually no seasonal alteration, the temperature-time function is non-existent and temperature becomes a significant factor only when it is, in its own right, limiting. Only a few isolated studies have been made on upper heat-tolerance levels in lotic invertebrates and those generally on cold stenotherms at temperatures below the mean minima encountered in the Gombak. Many of the families and genera found restricted to the forest river (*Isonychia*, Nemouridae, Stenopsychidae, *Blepharocera*) are probably warm stenothermal and their distribution is limited by the amplitude of diurnal fluctuations. However, until detailed experimental work is conducted on local insects in the 22–32 °C temperature range and at the near-saturation oxygen concentrations commonly encountered, further speculation is groundless.

For most taxa, the primary factor determining microdistribution is substrate type and for those taxa of the more specialized habitats, longitudinal range is likely limited by changing bottom configuration. The nature of substrate in its widest context, i.e. including both the physical support and the accompanying food resources, is a product of current velocity and channel characteristics; therefore these factors will be considered together. Current velocity has repeatedly been used to explain faunal distribution and extensive experimental work has been done delimiting preferred and optimal ranges and dislodgement velocities (see for example VERRIER 1953, DORIER and VAILLANT 1954, 1955, PHILIPSON 1954, SCOTT 1958, MORETTI and GIANOTTI 1962, BOURNAUD 1963 and review in BISHOP and HYNES 1969b). However, in the natural situation, current is an indefinable factor because of its heterogeneity over small areas and the restriction of most benthic animals by morphological or ethological adaptations to the boundary layers of marginal flow (NEILSON 1950, AMBÜHL 1959, 1961, cf. HYNES 1970) that, except under special circumstances, isolate them from the full mechanical force of the current. For most taxa, current acts directly only to maintain a zone of 'physiological enrichment' (*sensu* RUTTNER

1963) of gases and nutrients in the stratified current shell immediately surrounding the individual; thus, as found by Wu (1931) and FELDMETH (1970) for *Simulium* and various caddis larvae respectively, a respiration requirement for current *per se* exists, but velocity is relatively unimportant. The only groups for which the direct effects of current are important in determining distribution are the net-spinning Trichoptera and Simuliidae which utilize sometimes specific currents and dynamic pressures to maintain efficiency in their seston sieving devices (ZAHAR 1951, PHILIPSON 1954, 1969, EDINGTON 1965, 1968, HARROD 1965, CARLSSON 1967). Both groups may have been restricted by current in the Gombak as neither was common in the reduced-flow, depositional areas. However, other factors such as degree of siltation of the microbiotope and temperature which increased downstream or comparative stability of the substrate, which MAITLAND and PENNEY (1967) found essential for *Simulium*, were probably more important in determining succession. For the surface-dwelling Hemiptera, particularly the gerrids, current velocity and the formation of suitable feeding niches directly control distribution as noted in the descriptive section. Current velocity has been described, e.g. by PETR (1970), as determining the distribution of fauna along a river whereas other workers (KAMLER 1967, CHUTTER 1969a) have refuted any dependent longitudinal relationship, although at any one location such a regime might exist. The indirect effects of current that to a degree determine the size structure, available pore space and stability of the bottom substrate as well as its capacity to hold detrital food and develop a periphyton are much more important (LINDUSKA 1942 and see MACAN 1963, THORUP 1966, ULFSTRAND 1967, 1968a for reviews). These factors are of immediate relevance to any individual organism and primarily determine its presence or absence and selected microhabitat at a location and ultimately the longitudinal dispersion of the species in other suitable biotopes wherever they occur. The substrate preferences for most of the Gombak fauna as far as could be determined were given earlier, but these were only empirical and must be regarded circumspectly. Knowledge of specific tolerance and optimal ranges for each species that determine its distribution can only result from pertinent single or multi-variable experimentation of the type carried out by CUMMINS (1964), ERIKSEN (1964, 1966), CUMMINS and LAUFF (1969), McLACHLAN (1969) or from attempting to analyse factor combinations from field data (ULFSTRAND 1967).

In the Gombak, the alteration of the bank biotope from *Saraca* root-mat to emergent grasses and trailing vegetation accounted for many differences between the communities at the respective stations, but other substrate modifications as a result of increasing sand and silt fractions and the availability of different foods also resulted in changes in the fauna of the s biotope. As silt loads increased, the characteristic silt-mud bottom and its attendant fauna developed and supplemented the clean-water sand associations. Other progressive changes and replacements in the fauna attributed to this factor were already noted.

The direct effects of large erosion loads on invertebrates are little known, but abrasion and interference with respiration, particularly in gilled taxa, are likely. HARRISON and FARINA (1965) demonstrated that planorbid eggs were adversely affected by highly turbid water and some bivalve molluscs were limited by excessive silting (ELLIS 1936). Indirect effects (see HYNES 1960 for examples) alter microhabitats by filling-in interstitial areas and burying the fine detritic materials that make up a dominant diet item of many taxa. Effects on the seston feeders depend on the size selectivity of their nets, most of the fine-net spinners being restricted to moderately clean water. Epiphytic food resources are affected by decreased light and abrasion and the consequences of increased load were obvious at Stations IV and V where algal mean biomass declined from its maximum value at III as silt load increased, in spite of nutrient enrichment. In South African rivers where high erosion loads are common, CHUTTER (1969b) indicated that changes, particularly of the stones-in-current fauna may ensue, not from smothering effects, but from the decreased stability in the biotope. In a later paper (1970), he considered the amount of sand and silt present in the river bed as the major factor determining faunal distribution. Although the divisions of the Gombak fauna are complicated by some temperature, shade and chemical effects, intolerance to increasing silt load may be the restricting factor for a number of the Upper Zone species.

1.2.5. Biotopic associations

When the sampling was initiated it was hoped that differentiation of the faunal associations of the various biotopes and correlation of these with the physical parameters governing the structure and diversity of such communities would be possible. However, as was evident from Tables 62 to 72, the majority of taxa were rarely limited to only one of the defined biotopes. The exceptions were noted in preceeding sections. This difficulty in assigning a taxon to a 'typical' substrate or habitat was a function of three related factors. First: the sampling methods used necessitated a random visual selection of the general substrate type of r or s dependent on sediment size and current velocity. This sometimes resulted in marginal areas or ecotones being selected for sampling. Second: the substrate distributional pattern was so variable, particularly in the forest river, that over very short distances the microhabitat type altered, depending on bottom configuration. This problem was particularly acute in sampling erosional substrates, as at the invertebrate level, such a bottom is merely a succession of depositional zones, between and beneath the larger sediment particles. Hence, the fauna may locate in and be sampled from an s microhabitat within a much larger r area. Third: the bank samples always contained many elements of the eroding stone-in-current type of fauna as well as species more characteristic of the root-leaf assemblage; current and feeding niches were almost identical in some situations, e.g. at the exposed margins of a root-mat where conditions were similar to those on riffles. These were inevitably sampled with the un-specific net method used. Moreover, the deeper zones of the mat and lower areas of leaf accumulations often became heavily silted i.e. analogous to depositional areas. These too would be sampled indiscriminately as b with the methods used. Overall, the sampling techniques were inadequate for such microhabitat differentiation and, to enable satisfactory conclusions to be drawn, environmental parameters would have to be determined at the intimate invertebrate level. With the techniques presently available, this is not possible in the field. A further factor, illustrated particularly by the stone and gravel fauna, was that of different habitat requirements and selection during ontogeny. The problem is again one of scale, necessitating consideration at the invertebrate-substrate level and differentiation between 'optimal' and 'acceptable' ranges of biogenic conditions. Md of the substrate, interstital pore space and volume of available food in a microhabitat are the limiting factors of primary importance to any invertebrate. For example, in the early instars of some of the Hydropsychidae, nets were constructed in apparently unsuitable areas of reduced flow between small gravel substrates classified as depositional. At the insect level, water movement and food availability were undoubtedly adequate for the small caddis. Such a situation clearly operated at Stations IV and V, where particulate suspended food was abundant, since the nymphs moved into the larger substrates of the riffle areas only as they and their spatial requirements changed. Analogous distribution patterns occurred in the Baetidae at all stations where early instars were collected in fine gravel-sand, while the more mature nymphs lived in the 'characteristic' stone-top and vegetation biotopes. DÉCAMPS (1967) reported a similar preference for different biotopes by different sized conspecific larvae. What was evident, partly from the data but more from field observations, was preference by individual taxa or larger group for a certain gross substrate type, light or current conditions. Because of these factors and the acute taxonomic problems, statistical assessment of distribution and diversity at the biotope level was not justified.

1.2.6. Factors effecting segregation between taxa

In the benthic community of the Gombak, notably in the insect Orders Ephemeroptera, Plecoptera and Diptera, but also in the Oligochaeta, a number of potentially competitive taxa coexist. If the generally accepted premise, that coexisting species are separated ecologically (cf. DE BACH 1966), holds for lotic communities (see IDE 1935, BOTOSA-

NEANU 1960, ILLIES 1961a, MACAN 1962c, BERTHÉLEMY 1967, ULF-STRAND 1968a, 1969, GRANT and MACKAY 1969, *inter alia*), segregating mechanisms operate. HARRISON and AGNEW (1962) pointed out that the ecogeography of stream-breeding insects is the study of two different animals, the larvae and the adult. This reality was emphasized by ULF-STRAND (1969) who felt that in the amphibiotic insects competition may, and interspecific segregation must occur in each of the life phases which take place in completely separate communities.

At the nymphal level, segregation is effected by several means: differences in habitat preferences and spatial distribution; in the timing of the life cycle and activity patterns; in feeding habits and ability to exploit the various food resources. Spatial separation by station, i.e. at a gross level, has been dealt with earlier and for some taxa may be the end result of competitive displacement rather than preference for particular conditions. At the medium level of biotope, a number of abiotic environmental factors such as light, temperature, current, but more important substrate and the type of food available operate to separate taxa, but at this level, the ecological requirements of related or congeneric species are usually, but not always, similar. GRANT and MACKAY (op. cit.), using a coefficient of ecological separation, concluded that separation of most homogeneric species was weak at this habitat level. At the microspatial level of the niche, which cannot as yet be adequately examined, small-scale interspecific differences in requirements of the above physical factors probably act to prevent mutual interference between taxa in the exploitation of food, which in this nymphal stage is the primary *raison d'être*. In the Gombak, several prominent groups of species or similar taxa occurred in, as far as could be determined, similar habitats, e.g. the *Baetis-Pseudocloeon* complex, the many species of *Elmidae*, the *Ephemerella* spp. at Station III, *Nemoura* and *Amphinemura*, the *Saraca*-root assemblage of Leptophlebiidae, *Diplectrona* spp. and *Hydropsychodes* spp. and the chironomid genera. Whether there is competitive interference between members of these assemblages is unknown, although in the Hydropsychidae at least it is likely, but they must be segregated by fine microhabitat or nutritional preferences to enable them to coexist. For example, the filter-feeding hydropsychids build nets with different sized mesh, coarse in most genera *(Cheumatopsyche* and *Diplectrona)*, but some fine enough (2-4 × 14-33 μ in some Macronematinae (SATTLER 1963, 1968)) to filter bacteria and spore-sized particles from the water column. In the Heptageniidae, where *Thalerosphyrus* spp. and *Compsoneuriella* sp. occurred together, as at Station II, a behavioural microhabitat difference may exist, similar to that described by MADSEN (1968b) for two co-habiting Danish *Heptagenia* spp..

BRINKHURST (1969) reported that cohabiting oligochaetes of different genera had specific bacterial flora and hence selective enzymes in their guts that enabled exploitation of different parts of the available organic food. Further, in an elegant series of experiments on three sympatric tubificids, he and CHUA (1969) showed that there was no difference in the food ingested, but that the exploited bacterial species were different for each worm species and that one of the worms could even assimilate amino acids directly from the muds. No doubt similar capabilities exist in the rich oliochaete fauna of the lower river where several species in different families apparently occupied a single habitat. Complementary feeding niches may be common but determination of these for the herbivorous (particularly detritivorous) and euryphagous forms poses considerable difficulties, and has not yet been tackled successfully. For carnivores, size-selective predation at both intraspecific and related species levels has been demonstrated by SHELDON (1969) and DODSON (1970). The members of the species-rich setipalpian plecopteran fauna of the Gombak, among which most of the *Neoperla* spp. are found cohabiting, may exploit different foods, but no data could be obtained on this. The *Simulium* spp. within the shade and open stream groupings already described, appeared to be homogeneous in their ecology, with larvae of several species and sizes found side by side on the same leaf or stone. The filter-feeding mechanisms within this family are almost identical, and as suggested by ULFSTRAND (1968a), this group needs further study.

The above discussion has ignored temporal segregation which in temperate streams is usually the paramount factor (see ULFSTRAND 1967, 1968a, b, MINSHALL 1968, GRANT and MACKAY 1969, HYNES 1970 from an extensive literature). However, as will be shown by the quantitative data, there was little cyclicity to life histories of most taxa in the Gombak. Segregation of similar taxa at either inter- or intrageneric levels to maximize the exploitation of seasonally available food resources was unnecessary as both algal and detrital food were in continuous, albeit often precarious because of floods, supply. Moreover, as there were only marginal differences seasonally in the terrestrial environment, temporal spacing in the immature stages to synchronize emergence with favourable flight and oviposition periods was superfluous and the necessary controls for this, temperature or photoperiod changes (see HYNES 1970), were not available anyway.

In the adult community, interspecific interaction for resources is minimal in most groups as the imaginal stage is ephemeral, often not feeding at all. The Plecoptera and Odonata with longer-lived adults that do actively feed are the exceptions. Spatial isolation of winged adults cannot seriously be considered and ULFSTRAND (1969) concluded that differences in the timing of flight-periods afforded the easiest mechanism of isolating congeners at the adult stage. The rationale for segregation is to avoid interspecific matings and there may be selective pressure to achieve this (MAYR 1963). In a few groups, the Odonata particularly but also in the setipalpian stoneflies as shown by RUPPRECHT (1969), behavioural mechanisms prevent hybridization. Such species-specific pre-copulatory displays may exist in other aquatic insects and largely eliminate the need for any other isolating mechanism.

In the context of the Gombak fauna, much of the above discussion may be academic, as to a degree, these mechanisms are all density-dependent. As will be seen later, populations of most of the benthic fauna (Oligochaeta and Chironomidae perhaps excepted) probably rarely attain stability for long enough to build densities up to a level where contact interactions become common. Many species are facultatively polyphagous and occur throughout the year in all instars (see later section) so that there may be ecological overlap, but insufficient pressure from demand for limiting microhabitats, restricted food supply or temporal conditions to necessitate either homo- or heterogeneric segregation in the aquatic stages. If environmental resources are not limiting, as is likely, the need for adult temporal segregation which is probably an adaptation to ensure maximum exploitation of biotic resources over the extended, long-term immature stage, is no longer present. DITTMAR (1955) and ULFSTRAND (1968a) have commented that in disturbed biotopes where communities are not in a steady state, considerable interspecific ecological overlap may be expected. With the flow regime and non-seasonality of conditions in the river, this probably applies to the rhithron community of the Gombak.

1.2.7. Diversity

Table 73 was drawn up to give a conspectus of the distribution of numbers of taxa of the main families and groups at the various locations. All taxa classified as significant at a station were included irrespective of the biotope from which they were recorded. Some families were incompletely separated (Elmidae) or the distribution of individual species in groups of taxa was not known (Trichoptera, Odonata). Fig. 42 summarizes these relationships and shows the general trend for increased non-insect diversity and decreased number of insect species passing downstream. If the 79 separated Trichoptera and 53 Odonata taxa and the 20 species of Elmidae are excluded, the remaining number of taxa distinguished from the river to date was 291, of which 162, 158, 132, 127, 122 occurred at Stations I–V respectively. With the differentiation of the unassignable taxa, the numbers at the upper stations would be likely to increase disproportionately with respect to the lower stations as the Trichoptera particularly, from adult records, were very diverse at Station II. The Elmidae, with fewer species, probably reached their apogee of development as a zoome at Stations III and IV. Thus for the river as a whole even without the effects of effluents at Station V, there was a pronounced loss of species with distance from the forest margin. In many rivers, niche diversity and concomitantly species number increase with width and depth and elaboration of the bank area (MACAN 1961b),

Table 73. Distribution of numbers of taxa of the main groups at the various stations (p indicates that the group was present but could not be sorted; *indicates other unclassified specimens present; records for headwater torrent streams included under I)

	I	II	III	IV	V	Total
Coelenterata	1			1	1	1
Turbellaria	2	2	2	2	2	2
Nematoda	1	1	2*	2*	2*	2*
Gordiida	3	3	3			3
Sphaeriidae		Trib. H only				1
Unionidae				1		1
Viviparidae				1	1	1
Ampullariidae					1	1
Thiaridae	2	2	3	5	5	5*
Lymnaeidae				1	1	1
Planorbidae			1	2	2	2
Ancylidae	1	1	1	1	1	1
Naididae	3*	4*	6*	7*	7*	8*
Haplotaxidae			1	1	1	1
Phreodrilidae				1		1
Tubificidae				1	4	4
Enchytraeidae					1	1
Megascolecidae	1*	1*	1	2	2	3*
Lumbricidae				1		1
Glossiphonidae	1	2	2	2	4	4
Erpobdellidae				1	3	3
Hirudidae	1					1
Hydrachnellae	4*	4*	3*	2*	2*	6*
Cladocera			1		1	1
Ostracoda		1	3	2	2	3
Copepoda	p	p	p	p	p	p
Amphipoda			1	1	1	1
Decapoda	3	4	2	1		6
Collembola	p	p	p	p	p	p
Epilampridae	2	3	2	2		3
Tridactylidae	1	1				1
Megaloptera	1	1				1
Siphloneuridae	1	1				1
Baetidae	9	13	7	7	7	17*
Oligoneuridae				1		1
Heptageniidae	4	2	2	3	3	5
Leptophlebiidae	5	4	3	3	1	5
Ephemerellidae	3	2	4	3	3	5
Tricorythidae	1	2	1	1	1	2
Potamanthidae	1	1	1	1	1	1
Ephemeridae	1	1				1
Neoephemeridae	1	1	1	1	1	2
Caenidae	2	1	1	2	2	3
Prosopistomatidae	1	1				1
Protostictidae	1	2	1			2
Protoneuridae		?p	p	p		2
Platycnemididae			p	p	p	4
Coenagrionidae		?p	p	p	p	2*

Table 73. continued

	I	II	III	IV	V	Total
Amphipterygidae	1	1				1
Chlorocyphidae	p	p	p	p	p	3
Epallagidae	2	2	1	1		2
Calopterigidae	1	1	2	2	1	2
Aeschnidae	1	1				1
Gomphidae	8*	8*	7*	6*	6*	16
Cordulegasteridae	1	1				1
Corduliidae	p	p	p	p	p	4
Libellulidae	p	p	p	p	p	12
Peltoperlidae	2	2				2
Nemouridae	3	3				3
Leuctridae	1	1				1
Perlidae	10	13	7	5	4	13
Hydrometridae		1				1
Veliidae	1	2	2	2	1	3
Gerridae	5	7	1	2	2	8
Dipsocoridae				1		1
Nepidae	1	1			2	2
Notonectidae		1			1	2
Pleiidae	1	1	1			1
Helotrephidae	1	1			1	1
Belostomatidae					1	1
Naucoridae	3	3	3	3	2	3
Corixidae	1	1	1	1	1	1
Pyralidae	5	2	4	3	4	9
Philopotamidae	p	p	p			6*
Rhyacophilidae	2	3	1	1	1	3*
Glossosomatidae	2	1		1	1	2
Hydropsychidae	8*	12*	6*	5*	5*	14*
Psychomyiidae	p	p	p	p	p	p
Polycentropodidae	p	p	p	p	p	p
Stenopsychidae		1				1
Odontoceridae		1	1			1
Goeridae	1	1	1	1		1
Helicopsychidae	1	1				1
Calamoceratidae	1	1				1
Leptoceridae	p	p	p	p	p	13*
Lepidostomatidae	1*	2*	1*	1*		4*
Hydroptilidae	3*	3*	1*	1*	1*	4*
Dytiscidae	p	p	p	p	p	p
Gyrinidae	2	2	2	2	1	2
Haliplidae			1	1		1
Hydrophilidae	3*	3*	3*	2	1	4*
Helodidae	1	1	1	1		1
Ptilodactylidae	2	1	1	2		2
Psephenidae	3*	3*	3*	2*	1*	3*
Dryopidae	1*	3*	2*	2*	1*	4*
Elmidae	p	p	p	p	p	20
Torrindicolidae	1	1				1
Hydraenidae	1					1

Table 73. continued

	I	II	III	IV	V	Total
Lampyridae	1	2	1			2
Chironomidae	25*	22*	32*	30*	28*	59*
Simuliidae	15	4	1	1	1	18*
Ceratopogonidae	2*	3*	2*	1*	1*	3*
Dixidae	1*	1*				1*
Tipulidae	5*	4*	3*	4*	2*	6*
Blepharoceridae	2	2				2
Psychodidae	1	2	1	1	1	3
Rhagionidae	1	1	1			1
Empididae	2	2	2		1	2*
Dolichopodidae			1			1
Tetanoceridae					2	2
Syrphidae					1	1
Totals	184	183	156	153	141	344

Note: These totals do not include any Elmidae (20 + spp.) or Trichoptera (79 + taxa).

but in the Gombak, with the loss of slope (the many-faceted riffle-boulder biotope) and shade (the complex of *Saraca*-root niches), the number of microhabitats apparently decreased. In addition, the considerable bed and suspended loads dictated and established a conservative bottom type devoid of emergent phanerogams and dominated by interstitial sands even on the riffles. In these areas the fauna remained abundant, as allochthonous foods were plentiful, but tended to be relatively species poor as microhabitat diversity was low. A parallel situation was found in streams draining heavily eroded watersheds in South Africa (CHUTTER 1969b) and by MACKAY and KALFF (1969) in Canada.

PATRICK (1964) enumerated the species of both algae and animals in some Peruvian headwater streams of the Amazon. Several of these, sampled at 600 m altitude in rain forest, were probably comparable to the Gombak in structure, being fast, shallow-water, circum-neutral, and by her classification, mesotrophic. The 'lower invertebrate' fauna (Platyhelminthes, Annelida, Crustacea) was poor with 5–15 spp. and no Mollusca present, while the insect community comprised only 57–102 species. The latter total seems particularly low in view of the apparently much greater diversity reported in the Chironomidae (100+ spp. from small streams by FITTKAU (1964) and up to 1000 spp. in the Amazon Basin (ILLIES 1964)) and in the Trichoptera by MARLIER (quoted by ILLIES). In PATRICK's results, both groups were relatively poor in species with a maximum of 33 Diptera spp. in any one river (admittedly inadequately collected) and only 17 spp. of Trichoptera! The Gombak,

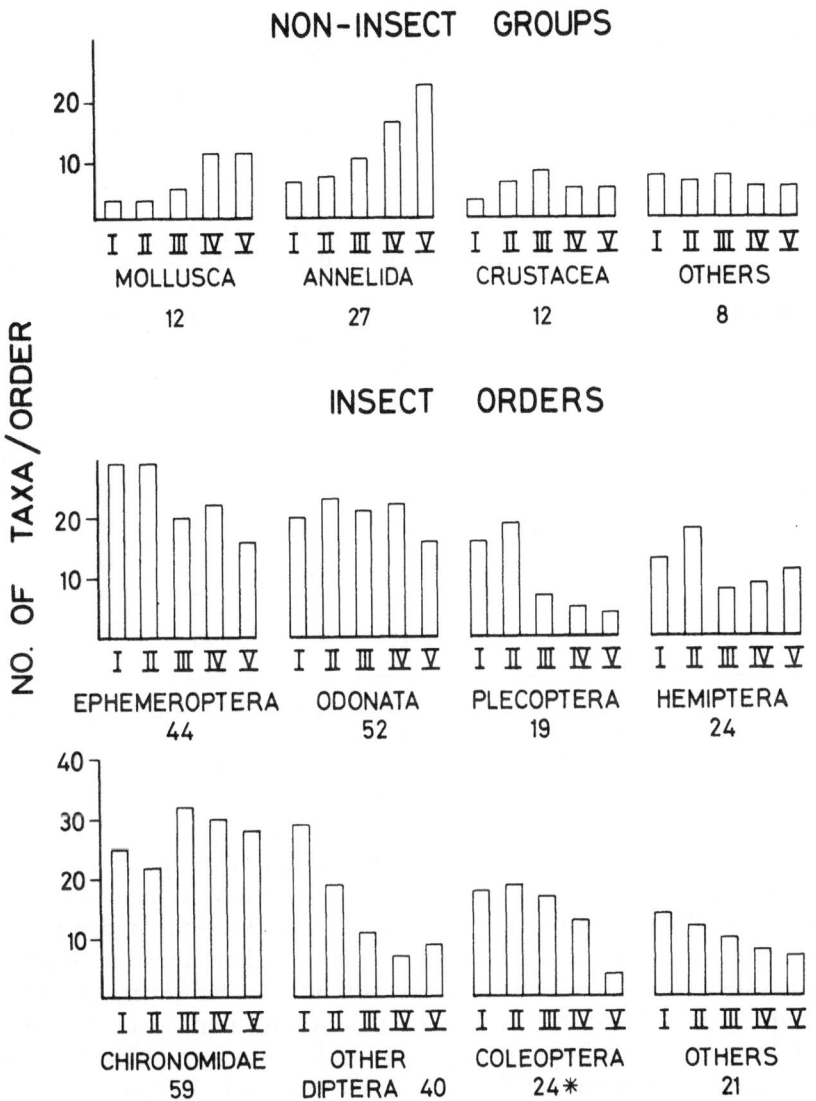

Fig. 42. Number of recognized taxa per Order per station (number under Order is total number of taxa separated); * Elmidae (20 spp.) omitted.

as seen, has a considerably richer fauna and the totals given in Table 73 are provisional, particularly in the above two groups. Considering only the forest-stream totals, upward of 60 trichopteran taxa and some 54 Diptera were recorded, and the number of chironomid types, often containing more than one species, would undoubtedly have exceeded this if more extensive collecting and breeding-out had been done. The general conclusion reached by PATRICK, that the ranges of the numbers of species found in tropical rivers (albeit just the ones she studied) were similar to

Table 74. Association tables and coefficients of association of the numbers of significant taxa found at the sampling stations in all substrate types*

NON-INSECTS		I**	II	III	IV	V
	I	19	17	13	11	10
	II	0.85	21	14	11	10
	III	0.55	0.57	28	22	21
	IV	0.42	0.41	0.72	33	29
	V	0.34	0.33	0.62	0.79	40
INSECTS		I	II	III	IV	V
	I	143	104	62	51	40
	II	0.74	137	73	54	43
	III	0.50	0.61	104	71	56
	IV	0.43	0.47	0.71	94	57
	V	0.36	0.39	0.60	0.65	82
TOTAL		I	II	III	IV	V
	I	162	121	75	62	50
	II	0.76	158	87	65	53
	III	0.51	0.60	132	93	77
	IV	0.43	0.46	0.72	127	86
	V	0.35	0.38	0.61	0.69	122

* Trichoptera (79 + taxa), Odonata (53 taxa), Elmidae (20 + taxa) not included because of unknown taxonomy or uncertain distribution of the components of large species complexes.
** Station I includes records for headwater torrent streams.

those in rivers of temperate areas, would seem not to apply in the Malayan context. Here the numbers of species of Odonata, Coleoptera, Lepidoptera and Ephemeroptera are considerably greater than in comparable Holarctic rivers and almost certainly greater than those in southern temperate areas for which data are still scarce. However, the Order Plecoptera is reduced in diversity, particularly if the neoperline complex is, in fact, a single polytypic conglomerate. ULFSTRAND (1968a) collated the records of important rhithron biotope studies in Europe and found the maximum number of Ephemeroptera recorded from any one river to be 22 (DITTMAR 1955) and for Plecoptera 40 (SOWA 1965) with average records for most rivers between 15–20 and 15–25 species in these Orders respectively. HARPER and MAGNIN (1969) reported 24 Plecoptera and BISHOP (1968) found 45+ spp. of Ephemeroptera from small streams in the species-rich Great Lakes region of North America, where the Northern and Eastern faunal distribution zones meet. Unfortunately, comprehensive faunal lists are not yet available from other rain forest localities, so that comparison of diversity between equatorial lotic biocoenoses is not possible.

In Table 74, the association tables and coefficients of association of the numbers of significant taxa in all biotopes at the sampling stations are given. The coefficients were calculated as in the earlier section on algae. As expected, adjacent stations showed the greatest affinity, with Station III usually being more closely associated with Station IV than with Station II. The association between the faunas of Stations I and III was marginal, but greater affinity between the lower Stations III and V was evident.

Objective evaluation of changes in community structure with longitudinal distance by quantification of the commonness of species, perhaps best carried out by using the truncated log-normal distribution of PRESTON (1948) and WILLIAMS (1953) or a modification of it with other geometric size classes, would have been attractive, particularly so as such analyses at each location also indicate indirectly the degree of modification of the abiotic environment. This technique has been applied with some success to the analysis of community structure in South African rivers by ALLANSON (1961) and CHUTTER (1963) and in the assessment of pollution-induced changes in algal communities (see references in HOHN 1966). However, to be able to correlate changes in the number of species in the modal interval, slope and degree of truncation of the generated curves with modifications in physical conditions between stations or biotopes, the areas sampled must be comparable and efficiency of sampling reproducible. Unfortunately, these preconditions, particularly in the *Saraca*-root and large boulder habitats, could not be met. Moreover, satisfactory division of some species complexes was not possible. Recurrent group analysis (FAGER 1957) using the data of the alphabetical indices was not feasible. The main drawbacks to this were the non-uniform sampling and the indeterminate and periodically variable size and composition of the often large sub-familial groups. Any evaluation of the data in Tables 62 to 72 using either method would have been biased and a distortion of the actual situation. The descriptive assessment given will therefore have to suffice.

At this level the riffle community of the Upper Zone may tentatively be considered as a climax (*sensu* ENGELMANN 1966) in that it is composed of numerous relatively small populations. Out of the forest zone this community, with a decrease in species and large increases in the numbers of a few taxa, becomes increasingly disclimactic as a result of the abiotic factors discussed in the previous sections. This introduces the anomalous situation, discussed by MINCKLEY (1963) and found in many rivers, in which a stable or mature community with specialist adapted and stenotopic forms is present in the youthful eroding parts of a watercourse, while a faunal disclimax of eurytopic forms is present in the mature, base-levelled reaches!

1.2.8. Life history data

Although all immature specimens collected were sized and a limited number of adult flight records were available for some amphibiotic groups, little useful life cycle information was extractable. For many taxa the systematic position was too uncertain to make any assessment worthwhile and for most of the species that could be defined too few records were available from the benthos sampling for analysis. No intensive collecting was carried out especially for life cycle studies and often fewer than 10 specimens, even of recurrent taxa, were found in a sample. For most groups, all size classes of immature stages and, in the hemimetabolous insects, mature nymphs were present throughout the year. Leech egg cases were always in evidence; gravid prawns and juvenile crabs were collected in every month; prepupae and pupae of several recognizable types of Trichoptera were apparent throughout the year; pupae of the common Diptera (Simuliidae and Blepharoceridae) were collected on almost every sampling date. The general conclusion was that, for the greater part of the fauna, hatching was continuous and development asynchronous and independent of any cyclical pattern, i.e. that at any particular time multiple cohorts of any taxon, all in individual growth phases, comprised the population of that species. A similar conclusion was reached by PETR (1970) for the riffle fauna of the Black Volta. Growth rates even at the almost uniform temperatures of the river may however be governed to a degree by other extrinsic conditions, so that the time needed to complete development could be variable. Many tropical lenitic aquatics require about four months for a complete cycle (CORBET 1956), but in a changeable river system, the rate of growth may depend more on the availability of food than on time. Populations of immatures extant during stable periods when both detritic and algal food are available undoubtedly develop at a more rapid pace than those hatched out into or subjected to denuded or impoverished substrates following a period of spates. This is analogous to the situation in temperate rivers where diet preferences and optimal growth periods are adjusted to the seasonal availability of food, i.e. algal grazers develop most rapidly in phototrophic periods of summer and detritivores in winter and early spring when leaf-fall materials are most abundant (cf. HYNES 1961, ULFSTRAND 1968a, b). Here where light and leaf-fall are not seasonal, both categories of food vary in abundance together, in response to discharge conditions. If development is rapid (3–5 weeks under optimal conditions for some Caenidae and Baetidae (HYNES and WILLIAMS 1962)), maximum use of resources would be possible between adverse periods, with several generations completing their cycles during a prolonged stable period. Nymphulae or nymphs present during the sub-optimal periods of food availability would develop much more slowly, but would still provide all stages to the population. For predators, regulation of growth rate might also be influenced indirectly by the availability of prey animals; however, as most have longer development periods, the short-term effects on population density might not be important. Some comparable

Table 75. Percentage frequency distribution of size classes of a) *Chironomus* s.s. sp. ♂ larvae at Station V (presence of pupae indicated by P) and b) *Neoperla* sp. nymphs at Station IV and V

Size class mm	19 X	14 XI	12 XII	9 I	6 II	9 III	3 IV	1 V	29 V	26 VI	24 VII	21 VIII	18 IX	16 X	13 XI
		1968							1969						
a. 0–2						0.5									
2–4	16.7			17.4	12.3	13.0	10.0	5.6	60.0	6.3		12.5	4.1		
4–6	41.7		50.0	43.5	21.9	34.0	30.0	21.1	20.0	10.9		12.5	18.9		12.5
6–8	41.7	100.0	14.3	34.8	30.6	15.8	50.0	59.0		32.8	75.0	25.0	21.3	100.0	43.7
8–10			35.7	4.4	22.8	28.0	10.0	8.7	20.0	50.0	25.0	37.5	50.8		43.7
10–12					10.5	8.7		5.6				12.5	4.9		
>12					0.9										
Pupae	P	P	P	P	P	P	P	P	P	P	P	P	P	P	P
Total larvae	12	2	14	23	114	577	10	161	5	64	4	8	122	1	16
b. 0–1	6.3	5.0	13.5	20.0	7.1	2.4	8.6	16.7	9.4	13.8	8.2	7.7	4.2		16.7
1–2	31.2	46.0	24.4	45.0	29.6		24.1	12.5	25.8	17.2	14.3	27.0	20.8		8.3
2–3	37.4	22.0	24.4	15.0	21.2	21.4	41.4	8.3	9.7	6.9	24.5	30.8	12.5	25.0	33.3
3–4	6.3	10.0	8.1	15.0	16.9	30.9	12.1	16.7	19.4	24.2	18.4	3.8	29.2	25.0	8.3
4–5	12.5	8.0	2.7		8.5	23.8	8.6	20.8	9.7	3.4	6.1	11.5	4.2		8.3
5–6	6.3	8.0	5.4		4.2	4.8	1.7		6.4	10.3	8.2	11.5	8.3		8.3
6–7		20.0	5.4		4.2	4.8	3.4	8.3		20.7	10.2		4.2	25.0	8.3
7–8			2.7		4.2	4.8		4.2			4.1	3.8	4.2		8.3
8–9			2.7		2.8	4.8		12.5	3.2			3.8	8.3		
9–10			2.7		1.4	2.4			6.4	3.4	2.0		4.2	25.0	8.3
10–11			5.4												
11–12				5.0											
12–13			2.7								4.1				
13–14															
Total nymphs	16	50	37	20	71	42	58	24	31	29	49	26	24	4	12

271

flexibility in life history as a result of exogenous factors has been documented even in seasonal rivers, e.g. DÉCAMPS (1967) for *Rhyacophila* and MACKERETH (1960) for *Glossosoma*, and in those localities it represents a great advantage over other species whose life cycles are rigidly limited by temperature or photoperiod. Such a system would account for both the fluctuations in population size, discussed in a subsequent section on abundance, and the omnipresence of all instars. However, it totally invalidates the normally employed methods of analysing life cycles. By way of example, the records for *Chironomus s.s.* sp.3 from all types of samples at Station V and for *Neoperla* from Stations IV and V are presented in Table 75. In the latter genus, the situation was complicated by the differences in size between sexes with the female nymph generally 1–2 mm larger than the male at the same point of development (except in the early instars). Since sexing was not possible until the pre-emergent stage (male 6–8 mm and female 10+ mm size classes), two groups of nymphs of the same age but different size were present from any brood. This confounded any attempt to follow a generation through development to maturity. Recruitment of young instars into the population was apparently continuous and emergence took place from variable nymphal size classes depending on sex. Adults of *N. luteola* and pupae of the *Chironomus* sp. were collected over most of the period, so no phenological life cycle was indicated for either one.

To determine natural growth patterns for the aquatic insects, it is imperative to find a time reference (very difficult in the non-seasonal tropics), or to impose one artificially on the population under study. A natural flood of catastrophic effect would provide this, as would a poison treatment that annihilated the standing population (cf. HYNES and WILLIAMS 1962). After such an event, provided that recolonization by drift was prevented and stability of the substrate maintained, the initial group of larvae/nymphs hatching from diapausing, quiescent or newly-deposited eggs could be followed in their development by the conventional techniques, taking into consideration the effects of substrate factors on the intensity of production.

Adult flight records might appear to offer potentially more useful information for elucidating insect life histories. However, emergence may be modified in some species by a lunar periodicity (CORBET 1964), adult longevity is an unknown factor (see HYNES 1970) and flight-periods and catchability with light-traps are subject to various ethological and physical limitations. The records obtained are presented in Tables 76 and 77. For many species only single or scattered catches were made and these cannot be discussed. As the results were from only the single location near Station II, they may not be representative of the whole river as harsher temperature conditions (higher maxima and greater fluctuations) in the lower reaches may have imposed more cyclical effects on the fauna.

Table 76. Light-trap catch records for adults of a. Plecoptera and b. Ephemeroptera per two week period

| | 1968 | | | | | | | | | 1969 | | | | | | | | | 1970 |
	A	M	J	J	A	S	O	N	D	J	F	M	A	M	J	J	A	S	May
a. PLECOPTERA																			
Neopelloperla fraterna								× ×											
Pelloperlodes biseata															× ×	×			
Protonemura jacobsoni						× ×	× ×						× ×	×					
Amphinemura minuta					× ×	× ×	× ×	× ×	×	×			×		× ×	× ×	× ×	× ×	
Rhopalopsole spinata							× ×	× ×			× ×								
Neoperla luteola	×	× ×	× ×	× ×	× ×	× ×	× ×	× ×	× ×	× ×	× ×	× ×	× ×	× ×	× ×	× ×	× ×	× ×	
Neoperla nitida								× ×											
Neoperla sp. nov.														×					
Etrocorema nigrogeniculata	×	×					× ×	× ×		× ×	× ×	× ×	× ×	× ×	× ×	× ×			
Etrocorema trapeza														×			×		
Etrocorema ? sp.							× ×					× ×							
Phanoperla clarissa							× ×						× ×	×					
Tylopyge helius	×	×			× ×			× ×		× ×			× ×	× ×			×		
Perla kelantonica										×	× ×	×					×		
Unknown Perlidae																			May

Table 76. (continued)

b. EPHEMEROPTERA

	1968									1969									1970
	A	M	J	J	A	S	O	N	D	J	F	M	A	M	J	J	A	S	
Neophemeropsis sp. A						×						×							May, June
Potamanthodes sp. A		×															×		
Isonychia sp. A			×	×	×	×	×	×		×	×	×	×	×	×	×		×	
Caenis sp. A		×			×	×			×		×	×		×	×		×		
Thalerosphyrus sp. A						×	×				×								
Epeorus sp. A								×		×			×			×			
Compsoneuriella sp.											×				×	×			
Ephemera (Dierephemera) sp. A																×			Feb.
Teloganodes sp.																			Apr., May, J
Thraulus sp. A	×	×															×		
Choroterpes s.s. sp. A									×										Apr., May
Habrophlebiodes sp. A				×	×														
Dipterophlebiodes sp. A																		×	Jan., Apr.

N.B. *Baetis* and *Pseudocloeon* were not separable to species. Adults of both genera were taken in all months.

Plecoptera (Table 76a)

Nemourids were caught only during the June–September period, but mature nymphs were collected throughout the year. *Neoperla*, as indicated previously, had continuous emergence and flight as reported for this genus in Africa (Tjönneland 1961). *Etrocorema nigrogeniculata*, the only other common perlid, also had an extended flight-period.

Ephemeroptera (Table 76b)

Records for this family were fragmentary. *Isonychia* sp.A adults were captured in all months except December and January. *Caenis* sp.A had distinct flight-periods, being caught in December–February and June–August only, i.e. during the drier seasons. The limited nymphal records of this species indicated no growth pattern as a heterogeneously-sized population was present in all months. Synchronous emergence may have occurred, although no accumulation of mature nymphs was found, not even in the *Saraca* areas, prior to the flight periods. Records in the Baetidae could not be assigned to the many species types later recognized, but adults of both *Baetis* and *Pseudocloeon* were taken throughout the year with peak numbers emergent in March and April.

Trichoptera (Table 77)

Numerous specimens of this Order were collected, but a large part was from the two genera *Diplectrona* and *Adicella*. Many other species were caught only on single occasions and cannot be considered, but are listed here to complete the record. The flight-periods of the several *Diplectrona* spp. were concentrated in the August–May period, but un-identified adults were taken in the intervening months. Pupae of all the Hydropsychidae were more obvious during June–July, but this was a result chiefly of lowered mean water levels rather than of abundance, as they were collected in all months. *Stenopsyche* sp. with a May–November flight period may have a seasonal cycle. Larvae were not found during the July–September period, but were collected throughout the rest of the year. The numbers found were too small to permit presentation, but empirically the large larvae (about 20 mm) were present only at the end of the rains. Pupae were not found. The *Adicella* group of species were apparently present throughout the year. *Hydropsychodes* sp.a had a flight-period in August–September of both 1968 and 1969, but only a single larvae was identified. Solely as a generalization, many of the Psycho-myiidae and Polycentropodidae species were captured only during the July–September 'dry' season.

The inconclusiveness of most of the above section illustrates the amount of work that will remain even after the taxonomy is worked out. Before

Table 77. Light-trap catch records for adults of Trichoptera per two-week period

	1968								1969								
	M	J	J	A	S	O	N	D	J	F	M	A	M	J	J	A	S
Setodes sp. b	X																X
Setodes sp. a		X											X			X	
Psychomyiella sp.	X												X			X	
Tinodes sp. a	X															X	X
Tinodes sp. c																X	X
Tinodes sp. d														X			
Tinodes sp. b							X									X	
Diplectrona sp. b	X				X					X	X				X		
Diplectrona sp. a					X	X	X	X	X						X		X
Diplectrona ?						X	X	X		X							
Diplectrona sp. nr aurivittata							X	X	X	X	X	X	X		X	X	X
Diplectrona sp. c												X					
Stenopsyche sp.	X	X	X	X	X	X	X	X	X		X			X			X
Adicella sp. a	X	X	X	X	X	X	X	X	X		X						
Adicella ?				X	X												
Adicella sp. b							X	X	X	X	X		X	X	X	X	
Adicella sp. d																X	
Adicella sp. c							X			X							
Cheumatopsyche sp.	X					X	X	X									
Ecnomus sp. a	X	X															
Ecnomus sp. c	X												X			X	
Ecnomus sp. e													X				
Ecnomus ?										X							
Ecnomus sp. b																X	
Ecnomus sp. d																X	X
Ecnomus sp. f																X	X
Chimarra sp. a		X						X									

Chimarra sp. nr
fulmecki
Chimarra sp. e
Chimarra sp. b
Chimarra sp. d
Chimarra sp. c
Hydropsyche ?
Hydropsyche sp. a
Hydropsyche sp. b
Hydropsychodes sp. b
Hydropsychodes sp. a
Pseudoneureclipsis sp.
Psychomyia sp. a
Leptocerinae a
Leptocerinae b
Leptocerinae d
Leptocerinae c
Leptocerinae e
Ganonema sp.
Lepidostomatinae b
Lepidostomatinae a
Lepidostomatinae c
Lepidostomatinae d
Glossosoma sp. a
Helicopsyche sp.
Amphipsyche sp.
Psychomyiinae a
Psychomyiinae d
Psychomyiinae c
Psychomyiinae b
Polyplectropus sp.
Nyctiophylax sp.
Philopotamidae a
Macronemum sp.
Hyalopsychella sp.

277

any realistic assessment of secondary productivity can be made, the extent of polyvoltinism and the factors controlling it will have to be elucidated.

1.3. Quantitative assessment of the fauna

1.3.1. The depositional (s) and erosional (r) biotopes

As previously discussed, quantitative collecting from the bank and boulder biotopes was not possible. Any discussion of standing stock and biomass must therefore be limited to the depositional and erosional areas that could be sampled by a reproducible and at least semi-random technique. Fig. 43 to 47 show the percentage composition by Order of the benthic fauna based on 15 four-weekly series of samples collected at the five main stations. The 'Others' category includes Mollusca, Crustacea, Vertebrata (where present) and also the other minor Orders recorded as present in section B. Detailed discussion of these on a date-by-date basis is not warranted, particularly of the biomass data, as the actual relationships were often distorted by the presence of a single large *Eulichas* or *Brotia*. Seasonal differences in abundance of the individual significant faunal components will be dealt with in the next subsection. The figures do however indicate the proportional composition of the fauna at the different stations, the relationship between numbers and weights for the various Order categories and the variability of community structure at a gross level over the short periods between sampling dates. Some of this variability was a function of sampling aggregations or recently ruptured and denuded bottom area and was inevitable with the small areas that could be worked on each date. The variation is therefore an indication of the resistance of the fauna to erosion or its recolonization capability either through drift or migration. Because of changeable discharge pattern, with a spate in almost all weeks or certainly between the sets of samples, assessment of this is difficult. Not all spates of comparable size had comparable effects on the invertebrates. For example, after the flood of 8 May 1969, Trichoptera at Station III were practically wiped out and subsequently recovered slowly, while the other Orders, although adversely affected, were not decimated by the breaking of substrate stability. The extent of denudation depended upon the antecedent flow conditions, i.e. the state of stability of the substrate before the change in level occurred, and the duration of the flood crest; short periods of high flow, as often happened at the upper locations, caused less disruption than extended spates during which much of the original bottom structure became modified.

The differences and similarities between the s and r biotopes were largely ones of degree of occupancy for a faunal element rather than presence or absence, as discussed. Table 78 summarizes these relationships for all samples. At Station I the Diptera, mostly Chironomidae, were numerically dominant in both s and r, but more so in the fine sediments where they constituted more than half the total fauna. Coleoptera formed the subdominant group and had the reverse abundance, with proportionally more Elmidae occurring in the riffle areas. In terms of biomass, even including the numbers of large Tipulidae, the Dipteran contribution was less important, being dwarfed by the large contributions of the Coleoptera, especially the occasional *Eulichas* specimens. Trichoptera, mostly filter-feeding Hydropsychidae, and Plecoptera were more abundant on the riffles, but were only minor fractions of the total weights. Ephemeroptera were evenly distributed, both numerically and in biomass. The Naididae made up an important percentage of the total catch on the riffles, but were of insignificant weight.

Much the same pattern applied at Stations II, III and IV, with the following important deviations. At Station II, the Ephemeroptera were less abundant. The Leptophlebiidae spp. were root-dwellers and the Heptageniidae, Tricorythidae and *Pseudocloeon* occurred in the boulder habitats and thus did not get sampled. Most of the mayfly nymphs in both s and r were *Baetis* spp. with a few of the bottom-dwelling Leptophlebiidae *(Habrophlebiodes* and *Thraulus)*. In the IIs samples, almost 25% of the biomass was *Macrobrachium*. In the IIr samples, elmid beetles were more abundant than the Diptera and with the other Coleoptera made up almost 80% of the standing biomass. The Ephemeroptera became a major faunal element in both s and r samples at Station III, with *Pseudo-*

Fig. 43–47. Percentage benthos composition by Order of depositional (s) and erosional (r) samples at four-weekly intervals; a. numbers per sq.m, b. biomass mg per sq.m, for Stations I, II, III, IV and V.

	DIPTERA		ODONATA
	TRICHOPTERA		COLEOPTERA
	EPHEMEROPTERA		ANNELIDA
	PLECOPTERA		OTHERS

Note: * at Station II (Diptera + Ephemeroptera + Odonata + Annelida) = 2.4% of biomass in r on 3-IV-69. * at Station III (Trichoptera + Ephemeroptera + Coleoptera + Annelida + Others) = 3.1% of numbers in s on 14-XI-68.

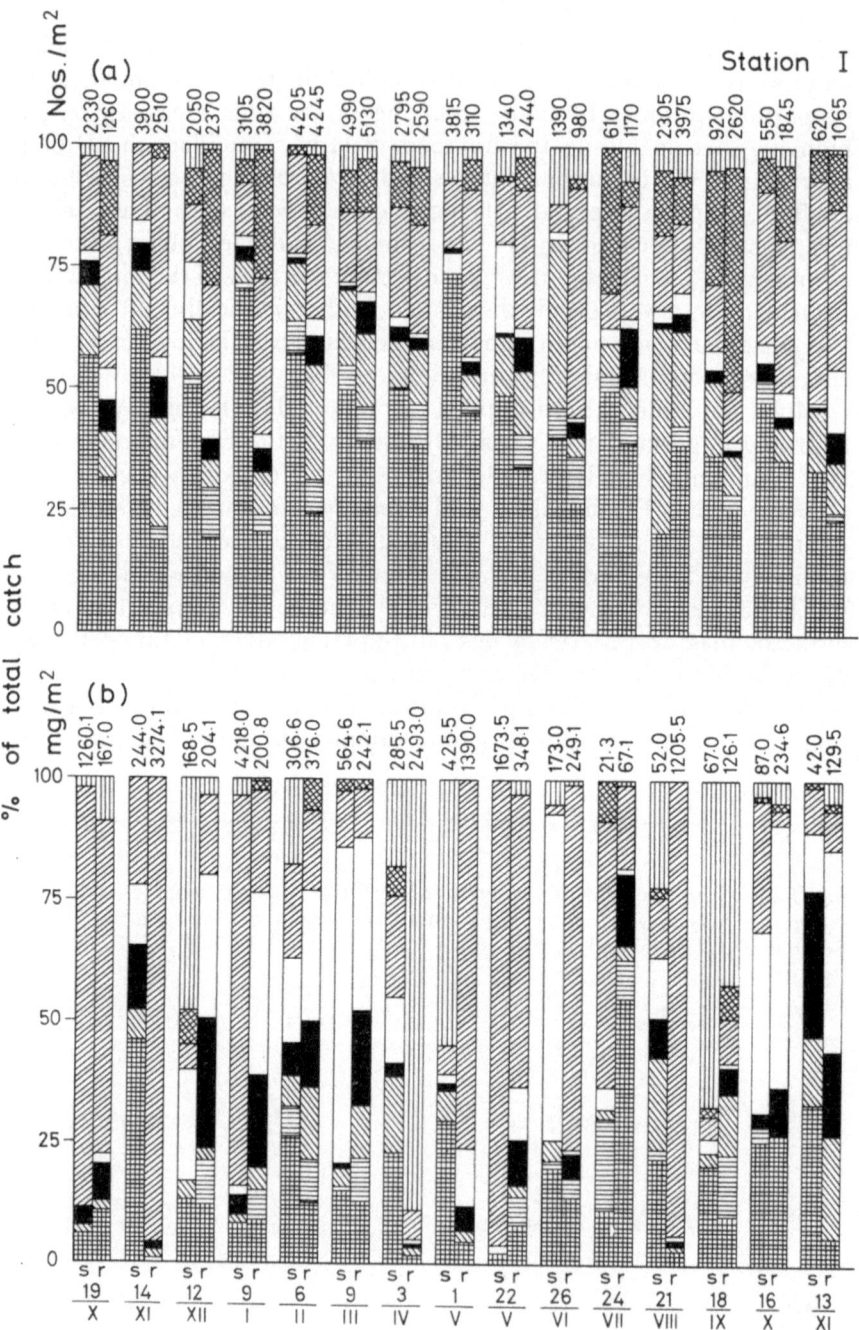

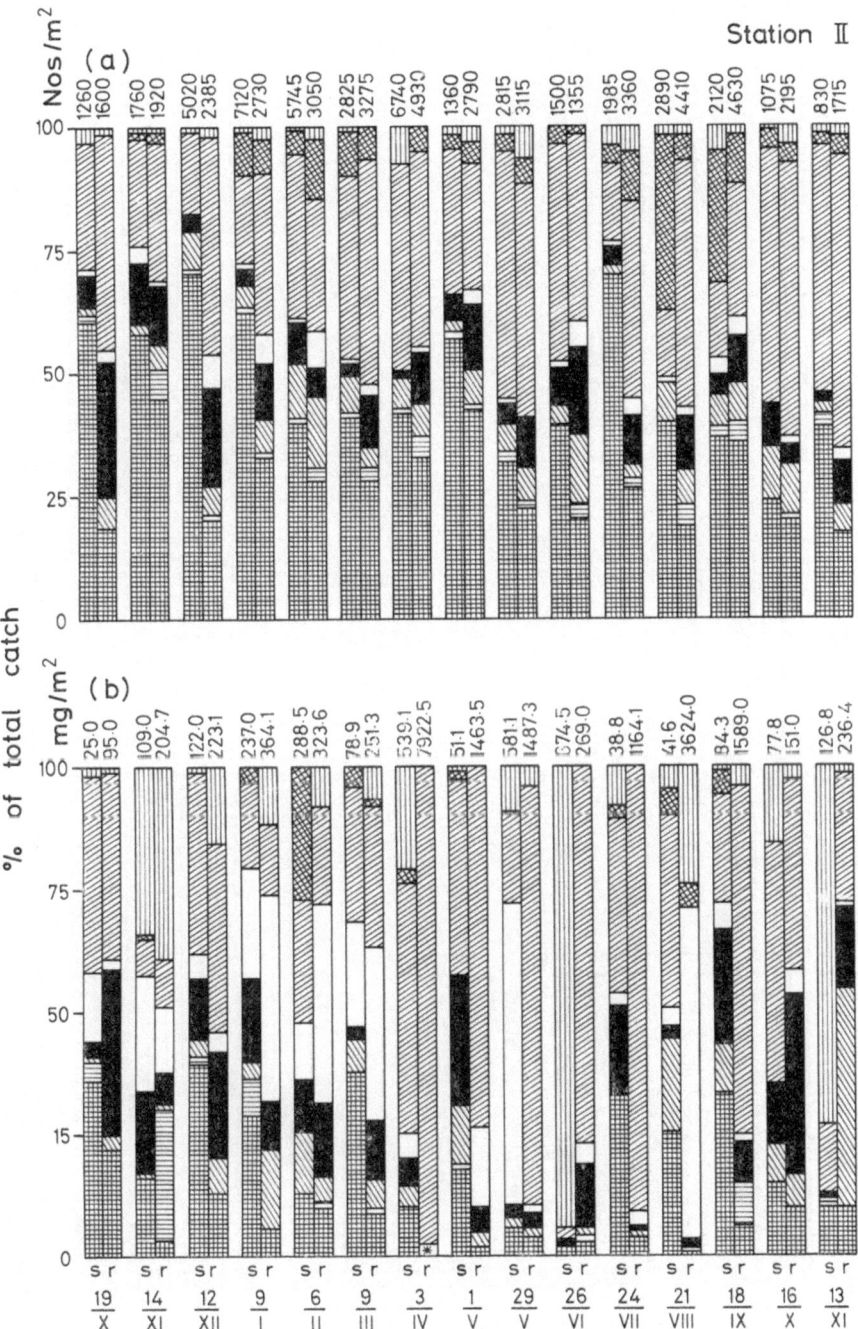

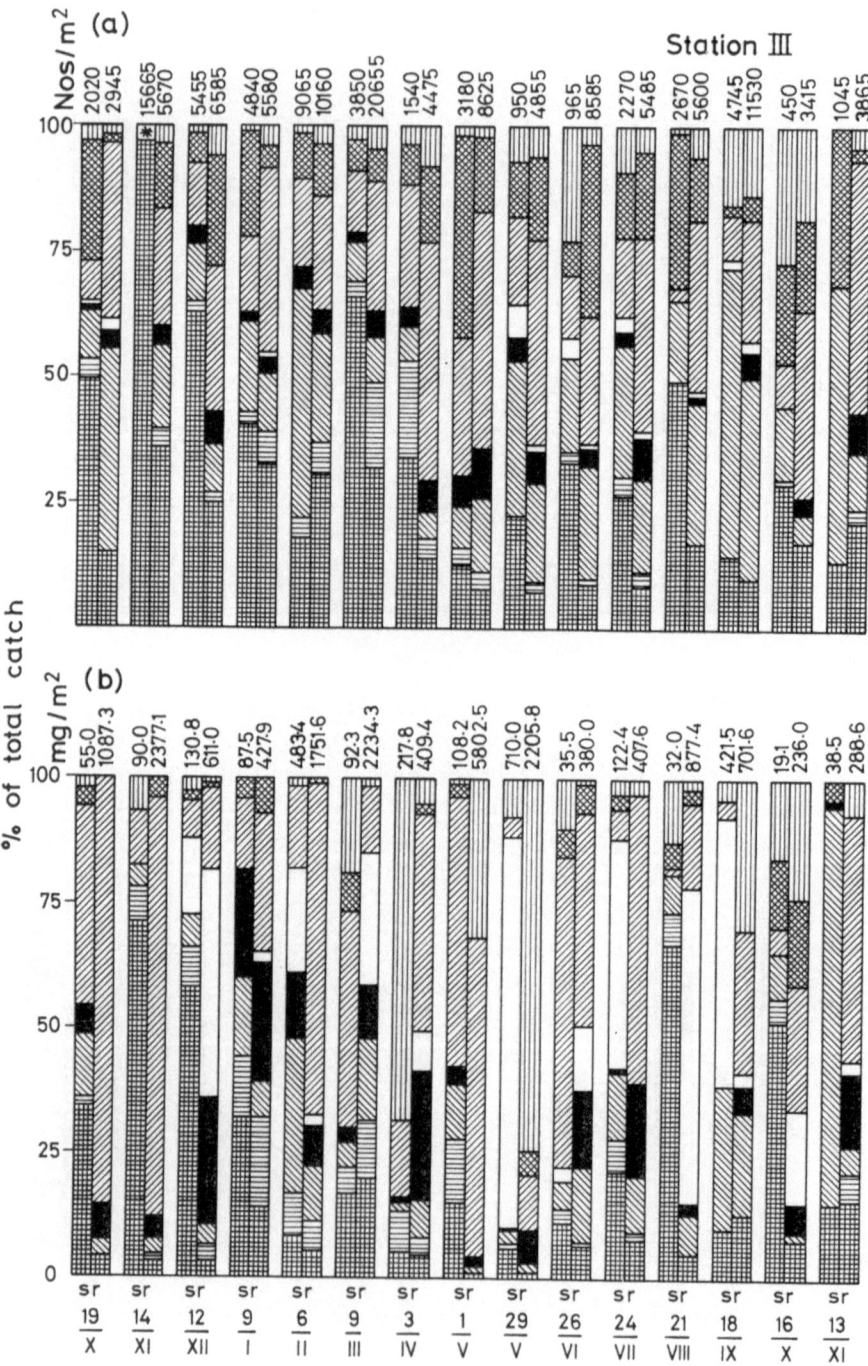

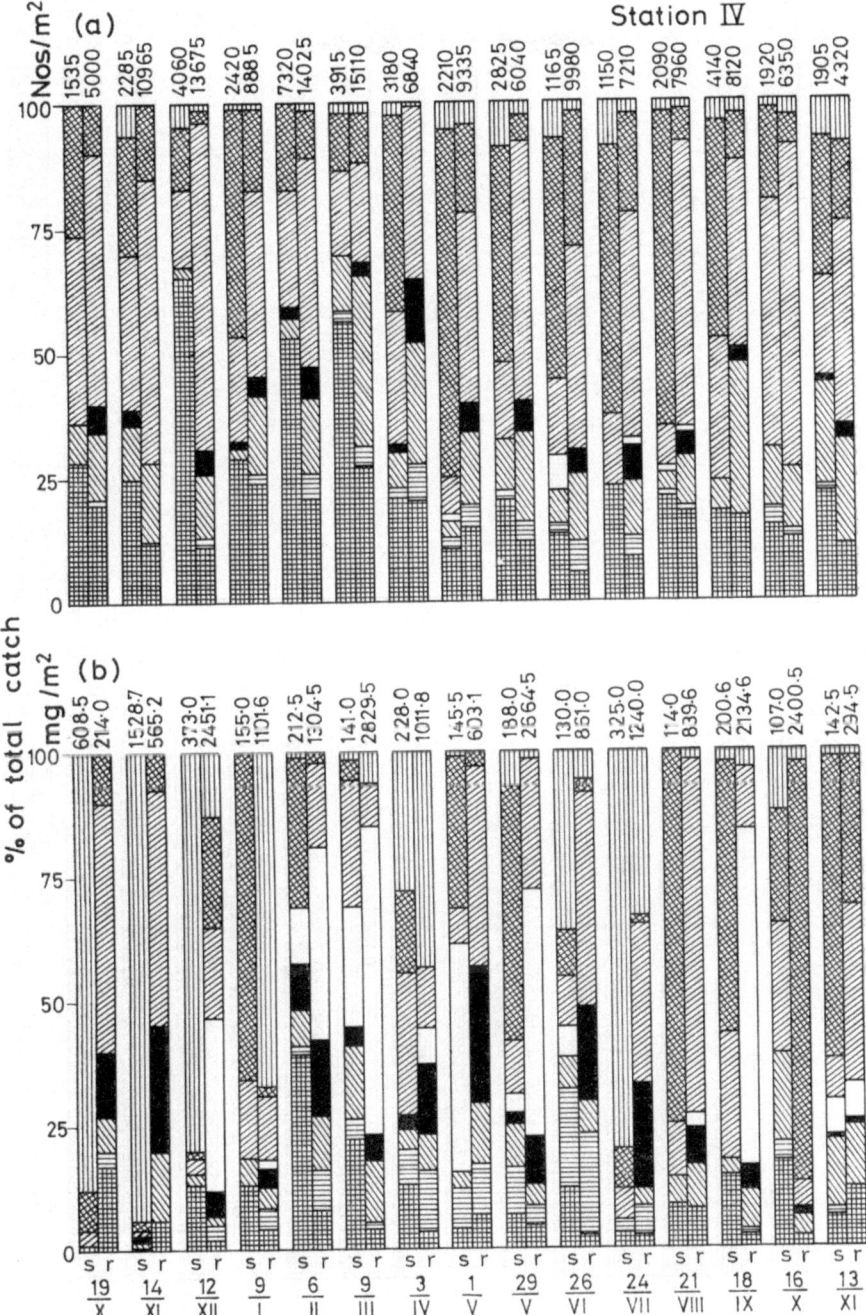

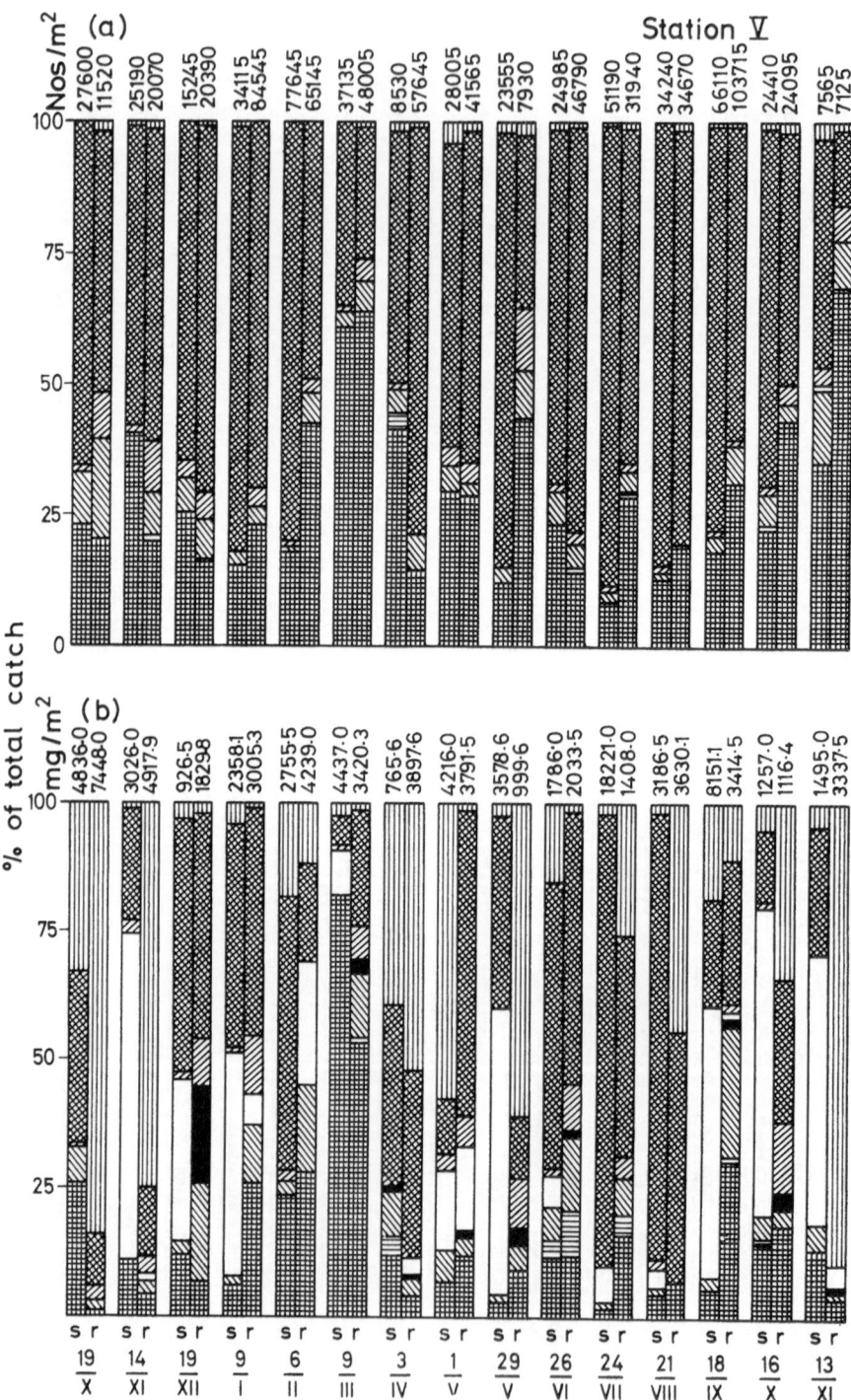

Table 78. Summary of benthic catches a. Orders as percentage of numbers; b. Orders as percentage of biomass. (Mean values calculated from 15 sets of four-weekly samples)

	Station I		Station II		Station III		Station IV		Station V	
	s	r	s	r	s	r	s	r	s	r
a.										
Diptera	54.66	31.63	49.45	28.41	51.33	20.47	33.97	16.52	23.07	29.84
Trichoptera	2.03	4.95	0.89	2.52	2.33	5.54	0.88	3.01	0.30	0.28
Ephemeroptera	13.47	13.00	6.30	7.34	19.57	18.12	6.66	18.07	3.18	5.47
Plecoptera	1.70	4.82	4.35	11.39	1.89	5.35	0.87	4.98	0.02	0.08
Odonata	3.32	3.27	0.91	3.19	0.55	0.72	0.38	0.33	0.11	0.03
Coleoptera	15.92	24.52	28.98	39.32	10.45	32.59	21.94	43.42	1.15	2.94
Decapoda*	0.20	0.03	0.10				31.89	11.88	71.32	60.55
Annelida	5.86	15.13	7.03	5.69	10.44	11.52	0.80	0.04	0.50	0.12
Mollusca					0.14	0.00		0.08	0.00	0.03
Vertebrata*	0.05		0.21	0.23	0.01	0.01			0.34	0.66
Others	2.83	2.62	1.78	1.91	3.29	5.67	2.62	1.73		
Total Numbers	34925	39130	45045	43460	58710	108030	42120	133815	485520	605190
b.										
Diptera	11.07	4.38	13.19	2.66	16.08	5.58	7.52	4.31	13.57	14.47
Trichoptera	0.49	1.61	0.84	1.54	4.78	2.88	2.09	4.14	0.24	0.75
Ephemeroptera	2.73	2.81	3.77	1.79	14.91	6.00	3.39	6.83	3.10	7.49
Plecoptera	3.00	4.22	6.89	3.55	4.49	7.32	1.17	8.92	0.02	1.66
Odonata	7.98	7.21	17.28	4.00	35.73	8.42	3.34	29.52	21.50	4.46
Coleoptera	67.35	57.55	23.58	79.43	12.79	48.43	8.12	21.09	0.86	4.20
Decapoda*	4.89	21.13	24.79				16.17	13.97	49.51	27.92
Annelida	0.71	0.58	3.77	1.07	1.49	1.87	56.92	4.25	12.11	35.20
Mollusca					5.99	9.28		6.59	0.05	2.11
Vertebrata*	0.06		2.19	5.87	1.80	0.43			0.13	1.75
Others	1.78	0.46	3.69	0.10	1.94	9.78	1.27	0.38		
Total Biomass (mg)	9588.6	10707.1	3075.5	19368.5	2644.0	19798.1	4599.3	20515.6**	60985.9	48489.2**

* Not sampled adequately by box; ** Contribution of a few large *Brotia* not included (21525 mg/m² @ IV, 31300 mg/m² @ V)

Cloeon and *Ephemerella* abundant in the riffles and *Potamanthodes* and the young of the other two genera present in the s habitat in numbers. More than 10% of the total population were Annelida, mostly small naidids, whose biomass contribution was insignificant. At Station IV, the Coleoptera dominated the riffle catches and Diptera were proportionately reduced. Ephemeroptera, mostly *Potamanthodes* and baetids, again made up a major part of the riffle fauna, but were reduced in the pooled sand habitats in which Oligochaeta were numerous (sub-dominant). The presence of large naidids made the biomass of these important. Mollusca were numerically insignificant, but a few large individuals of *Brotia* made up more than 50% of the mean biomass.

At Station V, the distribution was altered by the large worm population that relegated all Orders, except Diptera and Ephemeroptera, to minor numerical status. In terms of biomass, however, the Mollusca made a major contribution. At both Stations IV and V, the weight contributions of the very few large *Brotia costula* specimens were not included in the percentage calculations because they grossly distorted the results.

In the river as an entity, Diptera declined in percentage importance in both s and r habitats down the river except in Vr samples, where the abundant *Chironomus s.s.* restored the status of the Order. Trichoptera were a relatively minor faunal component at all stations, particularly in the s areas. However, at Station III, the Macronematinae were moderately numerous. The Hydropsychidae were a constant riffle element, but nowhere approached the importance found in Japanese rivers (TSUDA 1961, KAWAI 1966), in the Upper Volta (PETR 1970) or in areas where seston of lenitic origin was available (CHUTTER 1963, and many others). Plecoptera were nowhere common in depositional samples and declined downriver in the r areas (except at Station IV) to almost disappear by Station V. Numerically, the Odonata were insignificant except at Stations I and II where the *Drepanosticta* spp. were a constant component, but in terms of biomass, this Order made relatively large contributions. The Coleptera (Elmidae) increased in frequency from Stations I to IV, then rapidly declined. Most of the beetles were restricted to riffle areas, but the overlap in substrate made them a regular catch from s areas. Annelida in the upper river were equally abundant in both biotopes, but, at the lower two stations where they dominated, most of the worms were found in the softer, fine sediment areas where organic materials were abundant. Vertebrate larvae were present in appreciable numbers only at Station II.

Table 79 presents the total numbers/sq.m for the s and r biotopes over the study period. Overall, the riffle areas supported a larger standing stock than did the pool areas, except at Station II where there was little difference in mean biotope densities. The seasonal fluctuations were mainly a result of sustained periods of wash-out and substrate instability during the inter-monsoonal 'wet' periods, i.e. numbers were at a maximum

Table 79. Total benthos numbers per sq.m × 10 cm deep from depositional (s) and erosional (r) biotopes for the 15 sets of samples

	Is	IIs	IIIs	IVs	Vs	Ir	IIr	IIIr	IVr	Vr
19–X–68	2330	1260	2020	1535	27600	1260	1600	2945	5000	11520
14–XI–68	3900	1760	15665	2285	25190	2510	1920	5670	10965	20070
12–XII–68	2050	5020	5455	4060	15245*	2370	2385	6585	13675	20390*
9–I–69	3105	7120	4840	2420	34115	3820	2730	5580	8885	84585
6–II–69	4205	5745	9065	7320	77645	4245	3050	10160	14025	65145
9–III–69	4990	2825	3850	3915	37135	5130	3275	20665	15110	48005
3–IV–69	2795	6740	1540	3180	8530	2590	4930	4475	6840	57645
1–V–69	3815	1360	3180	2210	28005	3110	2790	8625	9335	41565
29–V–69	1340	2815	950	2825	23555	2440	3115	4855	6040	7930
26–VI–69	1390	1500	965	1165	24985	980	1355	8585	9980	46790
24–VII–69	610	1985	2270	1150	51190	1170	3360	5485	7210	31940
21–VIII–69	2305	2890	2670	2090	34240	3975	4410	5600	7960	34670
18–IX–69	920	2120	4745	4140	66110	2620	4630	11530	8120	103715
16–X–69	550	1075	450	1920	24410	1845	2195	3415	6350	24095
13–XI–69	620	830	1045	1905	7565	1065	1715	3865	4320	7125
Total	34925	45045	58710	42120	485520	39130	43460	108030	133815	605190
Mean	2328	3003	3914	2808	32368	2609	2897	7202	8921	40346

* Taken on 19–XII–68 because of high water at Station V on 12–XII–68

Table 80. Standing biomass in mg dry weight per sq.m × 10 cm deep from depositional (s) and erosional (r) biotopes for the 15 sets of samples

	Is	IIs	IIIs	IVs	Vs	Ir	IIr	IIIr	IVr	Vr
19–X–68	1260.1	25.0	55.0	608.5	4836.0	167.0	95.0	1087.3	214.0	7448.0
14–XI–68	244.0	109.0	90.0	1528.7	3026.0	3274.1	204.7	2377.1	565.2	4917.9
12–XII–68	168.5	122.0	130.8	373.0	926.5*	204.1	223.1	611.0	2451.2	1829.8*
9–I–69	4218.0	237.0	87.5	155.0	2358.1	200.8	364.1	427.9	1101.6@	3005.3
6–II–69	306.6	288.5	483.4	212.5	2755.5	376.0	323.6	1751.6	1304.5	4239.0
9–III–69	564.6	78.9	92.3	141.0	4437.0	242.1	251.3	2234.3	2829.5	3420.3
3–IV–69	285.5	539.1	217.8	228.0	765.6	2493.0	7922.5	409.4	1011.8	3897.8
1–V–69	425.5	51.1	108.2	145.5	4216.0	1390.0	1463.5	5802.5	603.1	3791.5
29–V–69	1673.5	581.1	710.0	188.0	3578.6	348.1	1487.3	2205.8	2664.5	999.6
26–VI–69	173.0	674.5	35.5	130.0	1786.0	249.1	269.0	380.0	861.0	2033.5
24–VII–69	21.3	38.8	122.4	325.0	18221.0	67.1	1164.1	407.6	1240.0	1408.0
21–VIII–69	52.0	41.6	32.0	114.0	3186.5	1205.5	3624.0	877.4	839.6	3630.1
18–IX–69	67.0	84.3	421.5	200.6	8151.1	126.1	1589.0	701.6	2134.6	3414.5
16–X–69	87.0	77.8	19.1	107.0	1257.0	234.6	151.0	236.0	2400.5	1116.5
13–XI–69	42.0	126.8	38.5	142.5	1495.0	129.5	236.4	288.6	294.5	3337.5@
Total	9588.6	3075.5	2644.0	4599.3	60985.9	10707.1	19368.6	19798.1	20515.6	48489.2
Mean	639.2	205.0	176.3	306.6	4065.7	713.8	1291.2	1319.9	1367.7	3232.6

* Taken on 19–XII–68 because of high water at Station V on 12–XII–68; @ Contributions of a few large *Brotia* not included (see Table 78)

in January, February, March and July, August, September 1969, during periods when a degree of substrate stability was established and maintained for relatively long periods. This was most evident in the results from riffle samples where spate disrupture effects were greatest. But it must again be emphasized that effects in the short-term period (2 or 3 days) antecedent to sampling were more critical than weekly or monthly patterns as recovery and recolonization were rapid (see also next section). Severe wash-out of benthic communities by floods is well documented and the references numerous (e.g. ALLEN 1951, HYNES 1961 and see references in BISHOP and HYNES 1969a). PETR (1970) has recently discussed the effects of monsoon flood disruption of the riffle biotope in a tropical watercourse and showed that maximum population densities were attained at the end of the dry season in the Black Volta.

Comparison of numerical population densities between rivers has little relevance unless the sampling conditions (mesh size, area sampled and efficiency) and the faunal composition (size of animals, activity patterns and catchability) are comparable or at least similar, as between watercourses in a narrow geographical area. In an extensive literature on rivers and streams of greatly differing physical and chemical characteristics, large variations have been reported, from the low hundreds to $50+$ thousand per sq. m, so that the only germane comment to be made here is that the unpolluted stations were in the lower end of this range, comparable to most oligo- and meso-trophic rivers and the mean densities at Station V were at the other extreme, as found in many enriched streams. Further elaboration will have to wait for more studies on the particular fauna of Southeast Asian rivers of comparable structure and biota.

In terms of standing biomass (Table 80), the relationships were similar. Considerably more was present per unit area in the erosional than depositional substrates, except at the enriched Station V where the annelid and mollusc faunas of the fine sediments gave a higher mean value. However, if the gross record for Vs on 24 July 1969, on which date there was an abundance of Tubificidae is ignored, the mean for the other dates is about 3055 mg/sq.m, slightly less than that of the riffle. The anomalously high mean at Is resulted from large contributions by *Eulichas* sp. on 19 October 1968, 9 January and 29 May 1969. Seasonal variations, not surprisingly, followed the numerical patterns, with the highest standing stocks recorded during the less flood-prone monsoon periods.

These mean standing biomasses of 0.2–1.4 g dry weight/sq.m for the unpolluted river and even the 4.1 g/sq.m at Station V are very modest when compared with the values up to 200 g/sq.m reported from some other lotic systems. Again, no realistic comparison between rivers can be made until mesh sizes and weighing methods are standardized. For reasons outlined in the next section and because of the limited nature of the areas represented by the s and r samples in the forest river, these data are not indicative of the true trophic status of the Gombak as much of the population and hence production was not included in the areas sampled. The only comparable data from tropical rivers appears to be that of PETR (1970) who found 4.5–43 g wet weight/sq.m in Black Volta River rapids of varying current velocity, of which weight the largest part was net-spinning Hydropsychidae. However, the Volta is chemically much richer (pH 7.7–8.6, $CaCO_3$ 76–90 mg/l.) flowing through savanna forest. FITTKAU (1964, 1967) gave densities of 0.5–1 g/sq.m for pooled, sandy-bottomed areas, with and without leaves, 1–10 g/sq.m for leaf accumulations in sand and 15–20 g/sq.m for the root-bank biotope of the streams he studied in Amazonia. However, he gave no quantitative data, no indication as to how these estimates were arrived at or whether they were wet or dry weights. The *Saraca*-bank biotope, unquantified in this survey, supported at least 10 g/sq.m standing stock, particularly with the contributions of the relatively abundant Decapoda and *Isonychia*. The s area samples covered an area similar to FITTKAU's slow-flow biotope and yielded similar low or lower standing stocks.

1.3.2. Seasonal variation in the abundance of principal benthic invertebrates

The abundance of a particular taxon at a station on a sampling date was not so much a result of life history, as little cyclicity was evident, but rather of changes in habitat stability in the period immediately prior to that date. The build-up of a population to a high density over several months occurred if continuous emergence, oviposition and immature recruitment took place (as was apparently the case judging from the distributions) with no intervening catastrophic wash-out of the habitat. After such a substrate disruption, recolonization was probably rapid and continuous, but specialized food supplies of algae or benthic detritus in a particular state of decomposition or microfloral colonization required a certain period for re-establishment to optimal levels. Severe spates affected the fauna of both riffle and pool biotopes, but in different ways. The molar action of unstable substrates and the erosive action of suspended sediments drastically reduced the number of riffle-dwelling invertebrates and the shifting and smothering effects made life in the depositional areas untenable. The bank-root biotope, however, was little altered, unless as occasionally happened, it was physically disrupted or buried by sand. The populations of those areas were relatively stable as far as could be determined by the available methods.

In this regard, a further uninvestigated factor determining abundance in the areas that could be sampled was lateral migration, which may occur as a response to seasonal changes in substrate conditions or food availability (see BISHOP and HYNES 1969a). In temperate streams, the generalization can be made that the gravel-riffle habitat often supports a widely tolerant community that acts as a source of migrants for the colonization of more specific habitats that offer seasonally optimal but vicissitudinous conditions for particular species. In a river subject to violent scouring of the bottom, such a stock reservoir is obviously important and the riffles in the lower reaches, and undoubtedly also the root-bank biotope in the forest, fulfil this function. Variations in apparent overall density in a given area as a result of movement of this type would not, of course, represent real population fluctuations.

In Fig. 48 to 52, the variations in population density of the principal taxa or groups, in numbers/sq.m, calculated as the mean of the s and r catches at each station, are presented. If a taxon appeared at a station only intermittently or in insignificant numbers, it was not plotted. The orthoclad maxima (Fig. 48) coincided with the end of the January–March dry season during which the diatom flora at Stations III to V and interstitial detritus at all stations built up and no severe wash-out was experienced. With the onset of the short inter-monsoon period of April–May, these population densities were decimated and gradually recovered in the June–September period. Rupturing of the bottom and destruction of food supplies followed the onset of the major wet season of October–November, reducing the fauna to very low densities. At Stations I and II where substrate size and reduced primary production provided only sub-optimal conditions for this group, the erosive effects of prolonged high water had less impact on population densities. The Chironominae, a very mixed group including the Chironomini common at the lower stations and the many tanytarsid species at all locations, appeared to be less affected by variations in discharge. In the stable depositing area of Station I, this group reached considerable densities during the January–May period, apparently unaffected by the spates of the short rains. Massive populations (more than 13000/sq.m) of this assemblage developed at Station V with organic enrichment, and the larvae in their retreats were little affected by erosion. The numbers in December and late May coincided with maximum numbers of leeches (see Fig. 49) which were major predators on this group.

Larva C (Fig. 49), a common *Corynoneura* type larva at all stations except V, had a similar pattern to the rest of the Orthocladiinae. The predatory Tanypodinae and Larva A (near *Cryptochironomus*) group of taxa reached their greatest densities when other chironomid prey species were abundant, the latter group particularly in March and September. *Chironomus s.s.* sp.3, a large conspicuous red larva of the lowest polluted reaches, occurred in the s and r samples in reasonable numbers only intermittently, when stable conditions existed, but it was always present in marginal muddy areas as collected in the bank samples. The phytophagous *Hexatoma* spp. were present throughout the year at the forest locations, where they attained moderate maxima in March and September, but were found less regularly in the lower river.

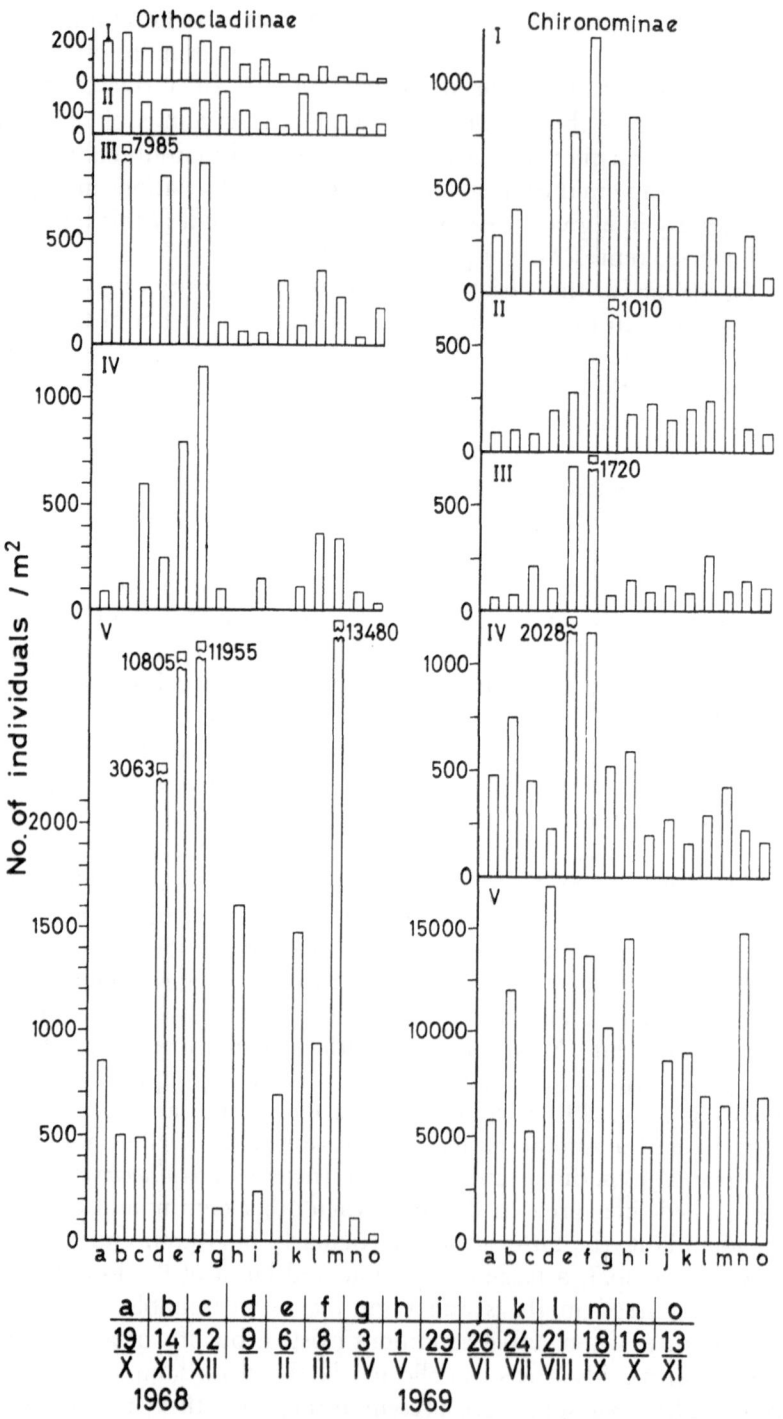

Fig. 48. Seasonal abundance of Orthocladiinae and Chironominae (number of individuals per sq.m — mean of s and r samples).

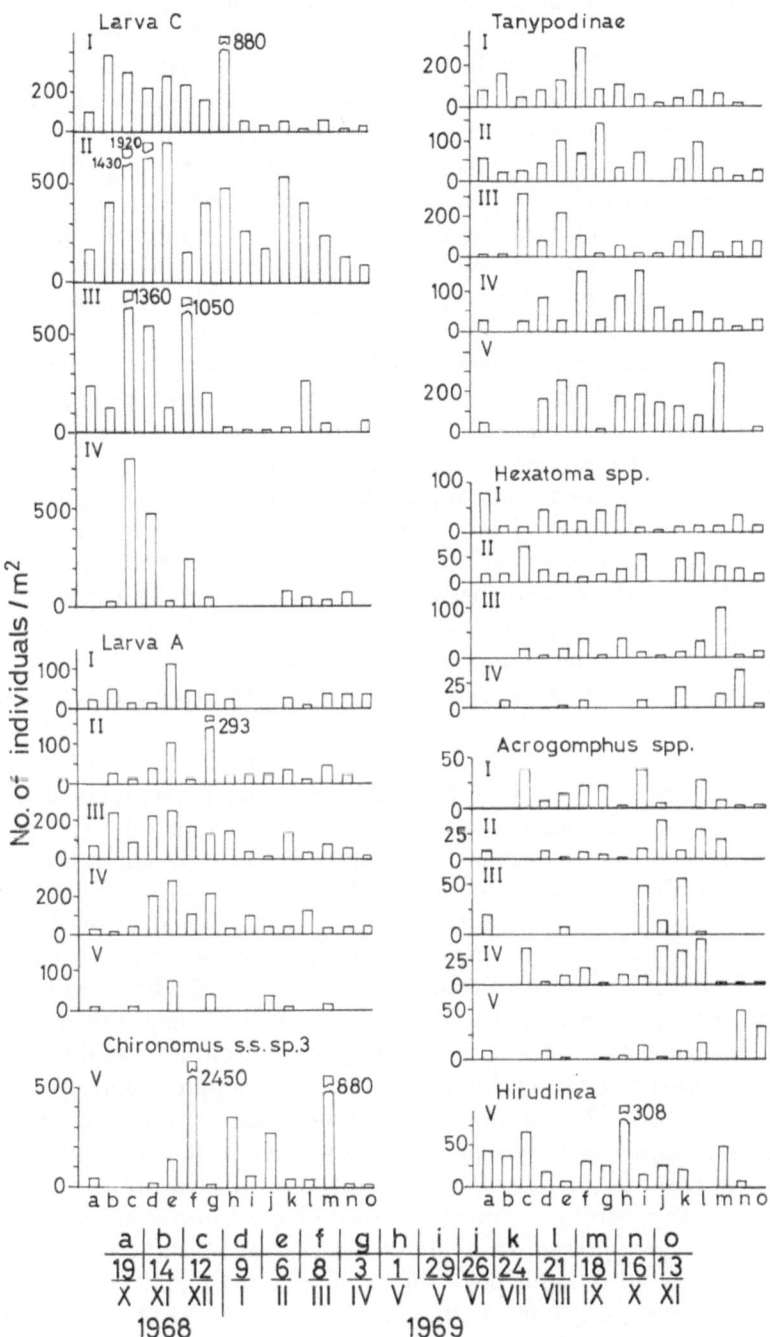

Fig. 49. Seasonal abundance of Larva C, Larva A, *Chironomus s.s.* sp.3, Tanypodinae, *Hexatoma* spp., *Acrogomphus* spp. and Hirudinea (number of individuals per sq.m–mean of s and r samples).

The fluctuations in the abundance of *Caenis* at Station I (Fig. 50) may support the speculated life cycle. Numbers were low through the wet season and increased sharply in January after adults had appeared; they subsequently declined to the July–August adult emergence period. However, all sizes of nymphs were present on all occasions so that there may have been two overlapping, out of phase populations. No pattern was evident at Stations II, III and IV. At Station V where spp. A and B occurred together, the gap in occurrence was unexplained unless the November spates eliminated the population entirely and there was no recolonization until after the second flight period in July–August.

Thalerosphyrus spp. (Fig. 50) were most abundant in the monsoon season when detritus and algae supplies were more available in the stone-face habitats. The *Baetis* constellation of species attained high population numbers in the stable periods, particularly at the three lower unshaded stations. This apparent correlation with algal food availability was even more evident for *Pseudocloeon* spp., whose population was reduced in November and May (Fig. 50), although migration to sheltered bank areas at that time might also account for the low numbers. An inverse relationship may be evident between the sizes of *Baetis* and *Pseudocloeon* populations, particularly at Stations III and IV. Both appeared to occupy the same spatial and food niche. To determine if such a relationship existed, very detailed work would be necessary to separate the many species involved.

The *Ephemerella* spp. of both subgenera (Fig. 51) reached sizable populations only at Stations III and IV; seasonal fluctuations of these roughly followed the spate frequency regime. The apparent cyclicity in abundance is an artefact as all instars were present throughout the year. Snails were abundant only at Station V; the changes in the sampled population (Fig. 51) were more a function of location than of absolute numbers. *Thiara* and to some extent *Melanoides* were present mostly in the marginal emergent grass and trailing vegetation areas when small; after the frequent-spate period of October–November, they subsequently moved out into the silt and gravel substrate, from where they were collected as mean water levels declined and food supplies built up in the bottom sediments. The reason for the decline in the period after July was not readily apparent unless predation by leeches and sciomyzid flies became intense. Numerous empty shells were always present so that detection of any increase in mortality would have required special study. Oligochaetes were present throughout the year at all stations (Fig. 51). The predominantly naidid populations at Stations I and II fluctuated with the rainy season, but whether this was a function of reproductive activity or of physical effects on the unstable substrates could not be determined. At Station V, peak numbers of both the Naididae and Tubificidae occurred during the stable flow periods, but as it did for the *Chironomus*, this may indicate migration out of drying bank areas into

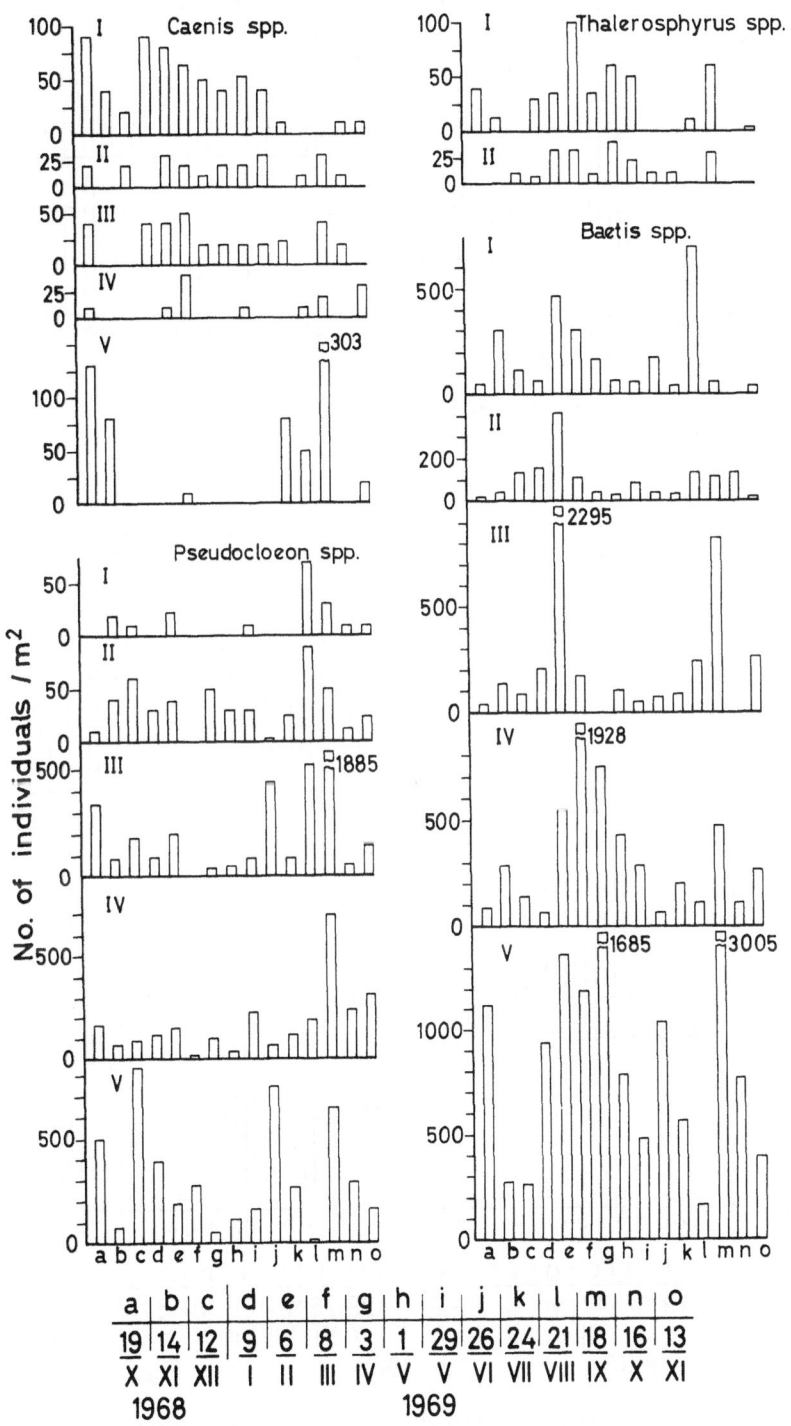

Fig. 50. Seasonal abundance of *Caenis* spp., *Pseudocloeon* spp., *Thalerosphyrus* spp., and *Baetis* spp. (number of individuals per sq.m–mean of s and r samples).

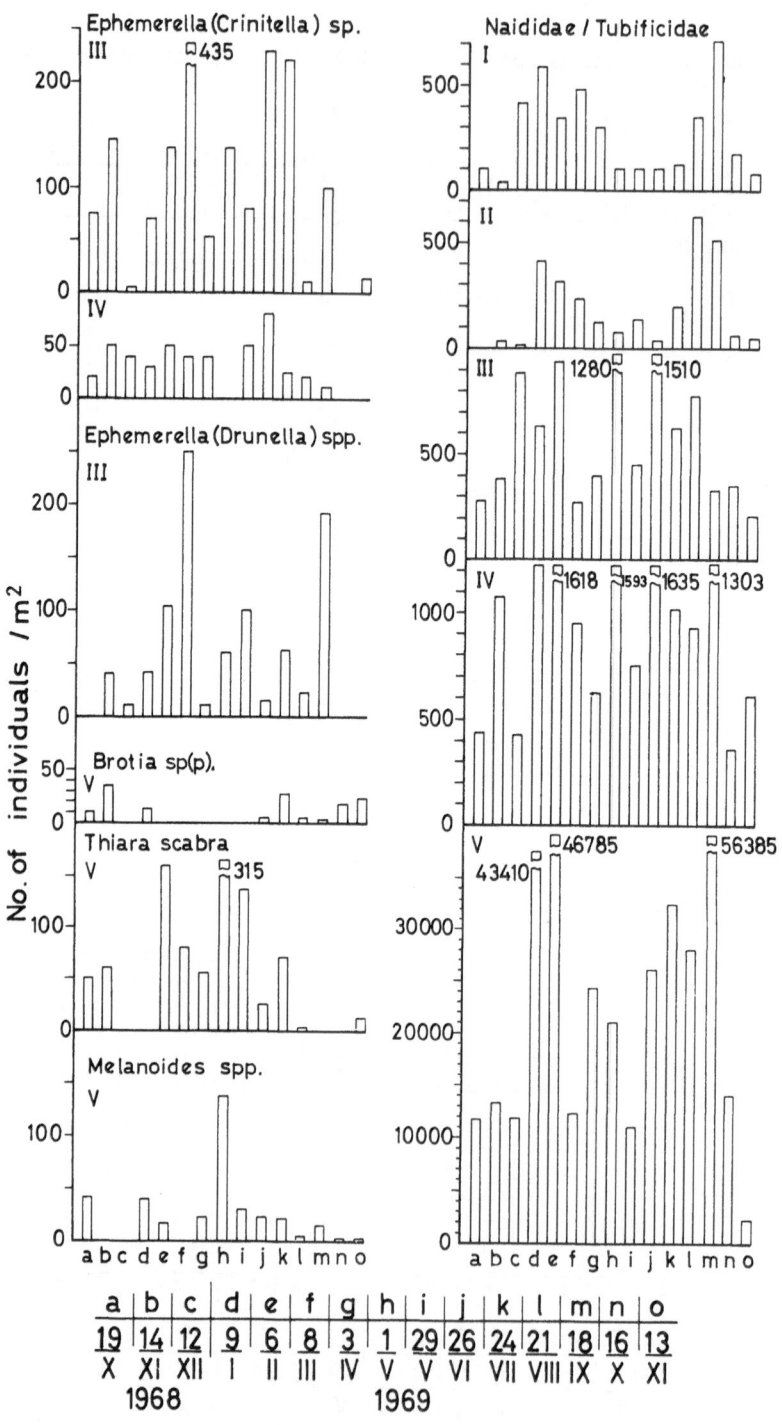

Fig. 51. Seasonal abundance of *Ephemerella (Crinitella)* sp., *Ephemerella (Drunella)* spp., *Brotia* sp(p)., *Thiara scabra*, *Melanoides* spp. and Naididae/Tubificidae (number of individuals per sq.m–mean of s and r samples).

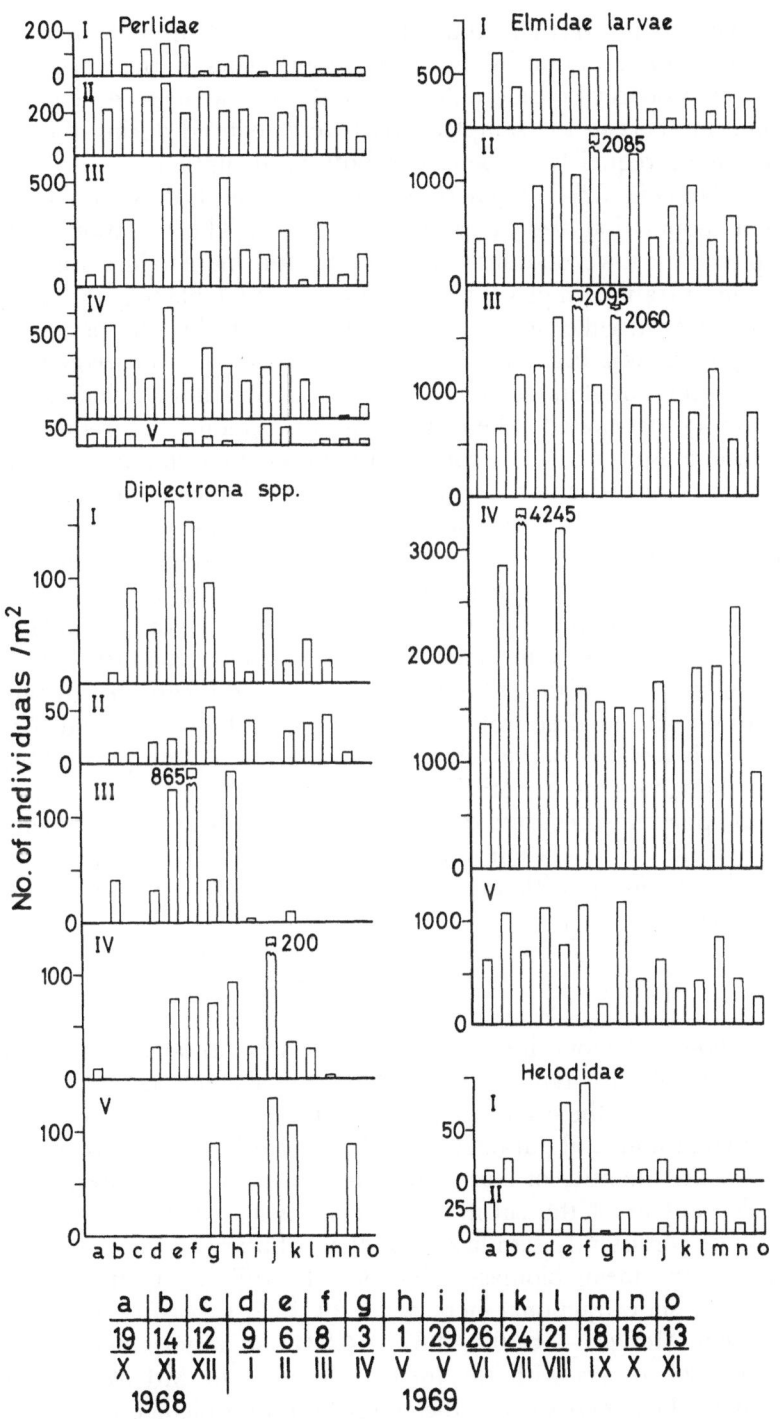

Fig. 52. Seasonal abundance of Perlidae, *Diplectrona* spp., Elmidae larvae and Helodidae (number of individuals per sq.m–mean of s and r samples).

the main stream. Sexual tubificids were collected throughout the year. The extent of predation by the erpobdellid leeches on the worm population was probably considerable, but could not be assessed. Total oligochaete numbers in the sampled areas were reduced during the periods with a high frequency of spates and bottom disruption.

Perlidae (Fig. 52) (all genera together) showed little seasonal fluctuation in population size at Stations I and II. The variations between samples at the lower stations were probably due more to local distribution and concentrations in places where prey were available (e.g. the orthoclad chironomids in riffles at Stations III and IV during the periods when algal growth was high) than to any real differences in frequency. *Diplectrona* spp. populations built up over the stable flow periods and were apparently decimated when substrate disruption was frequent. The discontinuous occurrence records at the lower three stations, with larvae appearing then declining sequentially, may indicate the presence of several species with different life cycles; however, fluctuations were more likely to have resulted from subtle progressive changes in bottom conditions in the areas sampled that first rendered conditions suitable and then less than optimal for this genus. The other Hydropsychidae were taken too infrequently to indicate any population dynamics. The larval stages of the riffle beetles (Elmidae) were only slightly affected by physical conditions with modest declines in numbers at Stations III and V in the October–November periods, but unaccountably a slight peak at Station IV at the same time in both years. At Stations II and III, optimal conditions apparently occurred during the February-May period when algal and detrital accumulation was high. The *Scirtes* type helodid larvae showed a build-up with the elmids at Station I, but no variation at Station II.

1.3.3. Synthesis

The host of factors discussed in the previous sections that affect the magnitude of the standing stock and biomass and prevent accurate measurements of these make any attempted correlation with the determined environmental parameters of the river difficult. The relationships between the quantity of fauna and Md *phi*, interstitial pore space and organic content of the substrate were investigated, but none was explicitly or consistently significant. If the tributary at Station I is disregarded, the mean biomass values for the riffle areas at Stations II, III and IV show fractional increases (1.29, 1.32, 1.37 g dry wt/sq.m with decrease in Md (*phi* — 6.30, —4.62, —4.28) and a sharp increment from 1.37 to 3.23 g/sq.m with the further decrease to Md *phi* — 2.63 at Station V. This increase was the result both of a change in population density and alteration of the specific composition and structure of the community, a transition from an insect dominated between-stones fauna to one of worms and molluscs living in the fine sediments of the enriched biotope. On a qualitative basis, many taxa were common in stabilized areas rich in detritus and leaf frass, but there was no quantitative correlation between overall populations and the sedimentary organics measured. MINSHALL (1968) made similar observations, in apposition to those of EGGLISHAW previously discussed. There was no correlation at all between the parameters and the mean biomass values of the s areas. However, these factors cannot really be eliminated from consideration. They were probably not adequately defined by the gross measuring techniques used and the observed faunal changes may have been the result of interaction between several of them as described by ULFSTRAND (1967). Perhaps at the level of the individual or even species group more satisfactory results could be found if the effects could be assessed with refined techniques under controlled field or laboratory conditions, as advocated by BUSCEMI (1966). In the river where additional factors, notably spates, constantly disrupted or modified whatever interactions took place, the faunal state, which could only be measured periodically, probably bore little relation to the biotopic conditions in force at the time of sampling; rather, it was the dynamic result of previous variable and unmonitored conditions.

2. Vertical distribution of the Benthos

A recurring problem in stream benthic studies, particularly those oriented towards assessment of standing stock and production relationships is that of obtaining unbiased quantitative samples. Most of the sampling devices used in lotic situations, reviewed by MACAN (1958b), ALBRECHT (1959), CUMMINS (1962) and SCHWOERBEL (1966) have inadequacies of one kind or another, but are chiefly deficient in that they sample only the uppermost substrate layers. Exceptions to this are coring apparatuses (see HOPKINS 1964, MAITLAND 1969) which, however, generally enclose small volumes, are difficult to use in coarser sediments and are subject to error (BRINKHURST 1967) and the freezing corers of EFFORD (1960), OHLMACHER & GLEASON (1964) and STOCKER & WILLIAMS (1972) which have manipulative shortcomings. ANGELIER (1953, 1962), SCHWOERBEL (1961, 1964, 1967), BERTHÉLEMY (1968) and HYNES and COLEMAN (1968) have reported that stream insects regularly occur in considerable numbers in the deeper gravel layers beneath and beside stream channels and COLEMAN and HYNES (1970) demonstrated that only about 20% of the total fauna of the first 30 cm occurs in the top 7.5 cm, with approximately 26% in each of the next three 7.5 cm layers. Any quantitative assessment of the benthos should, therefore, include sampling of this hyporheal biotope. COLEMAN and HYNES (op. cit.) used long-term, deep colonization pot-samplers in which supplied substrate could be freely colonized at all levels (down to 30 cm).

However, their system suffered from a gap between sampler and stream substrate that may have served as a vertical migration pathway and was subject to siltation. In addition, they found that surface samples obtained with their pot-sampler contained significantly more benthos than was recovered by conventional, size-selective, net-sampling techniques used in parallel. RADFORD and HARTLAND-ROWE (1971) presented data showing that a surface Surber sample collected only 10% by numbers and 53% by weight of the fauna recovered by their 15 cm deep 'juice-can' sampling device. A mesh factor was certainly involved in the difference, and their sampler, although simple to use, probably suffered considerable draining losses of invertebrates on recovery as it had no enclosure mechanism.

In an attempt to assess the vertical distribution of the biomass in a region of stable sediments, in the forest stream, a simple apparatus, utilizing the colonization of a known volume of supplied substrate, was designed. Sampling devices containing a variety of materials and substrate to simulate natural benthic conditions have been widely used, particularly in connection with pollution studies where the chief objectives have been comparative assessment of the fauna, the rates and extent of population movement and recolonization of denuded areas. These apparatuses have ranged from 'trays' (MOON 1935, LINDUSKA 1942, WATERS 1964, ULFSTRAND 1968a) to more elaborate devices (WENE and WICKLIFF 1940, CIANFICCONI and RIATTA 1957, HENSON 1965, MASON et al. 1967, ARTHUR and HORNING 1969, HILSENHOFF 1969) that collect adequate comparative samples, but ones often representative of only a select portion of the fauna, usually the amphibiotic insects of coarser, surface, riffle-type substrates. However, none of these techniques involved vertical sampling within the bottom layers and almost all constituted artefacts in the habitat subject to colonization by specialized faunal elements.

2.1. THE SAMPLER

The apparatus (Fig. 53) consisted of a rectangular box $20 \times 10 \times 10$ cm (i.e. 1/500 cu.m), made of 23 gauge (about 1 mm) galvanized iron sheet. Top, bottom and ends were solid, but the front and back were free to move vertically in slides incorporated in the ends. These slides extended below the bottom of the chamber to support the plates of the front and back when in the open position. At the top of the sampler, flanges received the plates, ensuring a tight seal, and at the bottom, a strip of compressible foam rubber minimized leakage. To close the plates, stainless steel wires were passed through holes in the top of the box and attached to the upper edge of the plates.

The method of operation was as follows: a hole of the required depth was excavated in the stream bed, the sampler was inserted facing upstream-downstream with the plates in the open position, filled with sub-

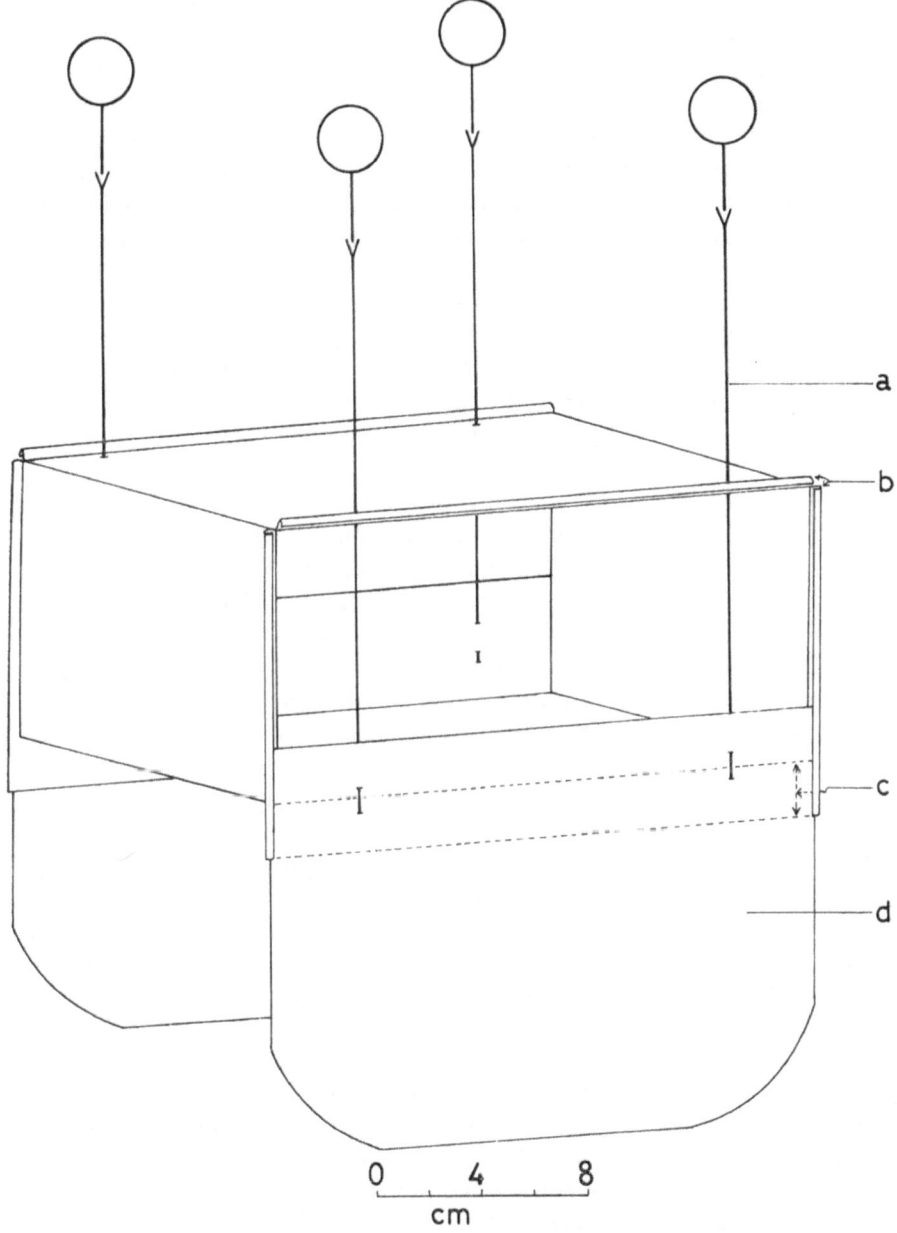

Fig. 53. The vertical depth sampler; a. wire handle for closing plate, b. flanges to receive plate, c. anti-leak compressible strip, d. plate in open position.

strate materials from that depth and then burried with the looped wires protruding to the surface. Coincident series of four samplers were dug in so that the bottoms of the boxes were at 10, 20, 30 and 40 cm. The wire handles had to be fully buried and the location marked by survey only as human interference, even in the forest, was a problem. After a stabilization period during which equilibrium levels of fauna, detritus, silt, etc., were theoretically attained with the surrounding bottom, the sampler was closed *in situ* by pulling on the loops until the plates reached the top flanges. The entire box was then either pulled or dug out, depending on the weight and depth of overlying substrates, and a polyethylene bag was slipped around the sampler just prior to removal from the water to collect any leakage. Jamming of the plates in the slides was minimized by ensuring, during the setting of the sampler, that large stones would not hinder closure and by filling the grooves in the end slides with wax to prevent sand from interfering with free plate movement. The samples were preserved and processed as before using calcium chloride flotation and elutriation and a 90 μ sieve to separate the fauna. No fractionation of the samples was necessary and all animals were enumerated and weighed. Chief limitations of the method were the arduous physical work involved in setting the apparatus and the relatively long period before a sample could be recovered. COLEMAN and HYNES found a period of 30–40 days sufficient to establish equilibrium (personal communication), but in the Gombak, minimum times of two months were used. Two series of four samplers each were lifted after eight weeks on 17 April and 23 May 1969, two series after four months on 7 August and 4 December 1969 and a single series after 24 weeks on 4 December 1969. The selection of site is critical to any overall estimation of distribution and the usual difficulties inherent in any benthic sampling program with respect to aggregation and non-uniform substrate apply equally here. To reduce these effects, a visually and texturally uniform, 25–30 m long, flowing riffle area below Station II (Plate 3) was selected where the sediments were at least 2 m deep and subject to almost no severe transport erosion. This habitat was comparable to the transition area between depositional and erosional habitats, having some fine sediments mixed with coarser sands and gravels, but lacking most of the larger sizes (cobble or greater). Locations for insertion were selected randomly within the site limits and the depths arbitrarily assigned to these locations.

On two occasions, the surface sampler (0–10 cm) was vandalized or, less likely, washed out of position; to complete the series on those occasions, samples were taken in adjacent areas on the date of removal using the box-net sampler (1000 sq.cm × 10 cm deep), equipped with a 90 μ mesh to obtain a comparable sample.

2.2. RESULTS AND COMMENTS

When the total number and biomass of the fauna found in the samples,

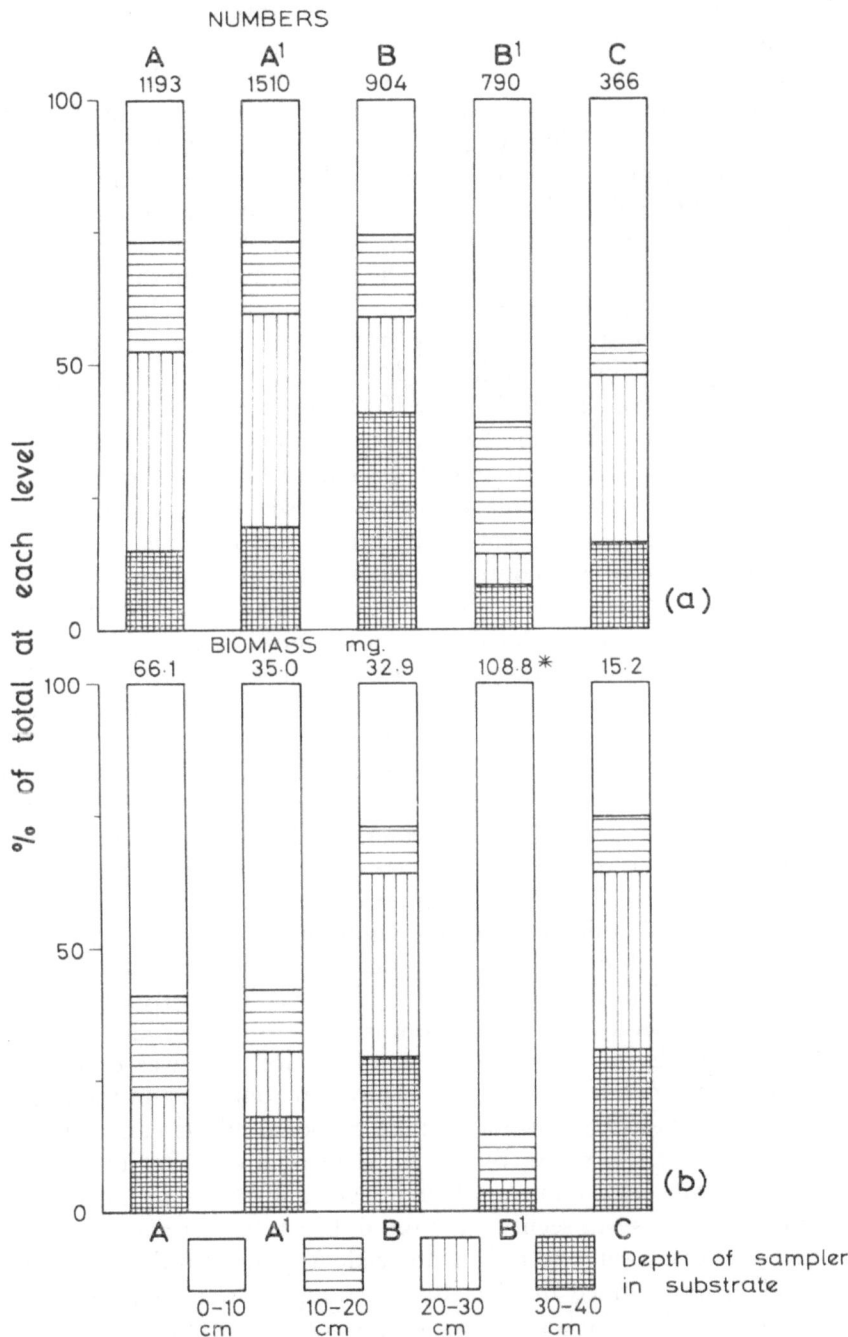

Fig. 54. Vertical distribution of benthos; a. numbers, b. biomass. (Percentage of total catch at each level.); * includes 80.1 mg contribution from a single large *Eulichas* in the 0–10 cm layer; A + A¹ after 8 weeks; B + B¹ after 16 weeks; C after 24 weeks.

expressed as numbers or mg/(0.1 sq.m × 10 cm deep), were considered, significant portions were found in the lower three levels. Fig. 54 shows the percentage distribution of numbers and biomass for the five series. The data at the top of the histogram columns represent the total numbers and weights of organisms recovered in the four samples of that series. The comparatively high weight in series B^1 was due to a single large *Eulichas* larva in the surface sample which contributed more than 80 of the 108.8 mg. This individual will be omitted from further consideration as the medium-sized gravel was not its usual habitat and its biomass obscures the general relationships. The high numbers in this same sample were the result of a mass eclosion of Chironominae. The upper 10 cm contained a mean percentage by numbers, calculated from the percentages present in each series, of 37.4% and by corrected weight (disregarding the *Eulichas*) of 42%; the 10–20 cm stratum 15.9% and 14.2%; 20–30 cm layer 26.7% and 22.1%; 30–40 cm depth 20.0% and 20.7%. The apparent discontinuity of distribution at the second level may have resulted from physical influences on the substrate. Under normal flows, the top layer of about 5 cm of the gravel was kept relatively clear of fine silt and detritus, maintaining open interstitial spaces and a fauna of filter-feeders and algivores that preferred this habitat. At lower levels, silt and detritus accumulated to form a more compact substrate, with decreased oxygen conditions (very local areas of hydrogen sulfide were evident) but with more available detrital food. The 10–20 cm stratum was probably transitional between these, changing character with stage. With rising surface currents and pressure, the open silt-free zone moved down as the gravels were vibrated and the fine particles picked up by the current; after a spate this was obvious. During low flows, the fine sediment-filled layer moved upward toward the surface. Under the flow regime of the Gombak, this zone must have been unstable and probably did not establish a definite faunal community.

From the totals data in Fig. 54, a decrease in both catch numbers and biomass is evident from series A to B to C exposed for two, four and six months respectively. This decrease was general in all groups (except for the peak in chironomid numbers in B^1 surface) and was probably the result of accumulated differences in physical conditions in the samplers over the extended colonization times. Seasonal effects cannot be ruled out as the last two series, B^1 and C, were removed after the October–November wet season during which in 1968 and 1969 some fluctuation in benthic numbers was seen. In the top 0–10 cm, a decrease in fauna was attributed to current effects from antecedent spates, but the deeper layers would have remained unaffected. The samplers were removed only at modal flow when the water was clear, so a time lag of two to three days since the last spate was usual, and recolonization of the surface layer by drifting fauna likely.

Distribution of the invertebrates by group (Table 81) was spread

Table 81. Total numbers and percentage vertical distribution of invertebrate groups.

Group	Total Catch*	%	% of each group at each level (cm)			
			0–10	10–20	20–30	30–40
Chironomidae	2667	56.0	30.8	13.9	36.9	18.4
Other Diptera	234	4.9	50.9	25.6	17.1	6.4
Ephemeroptera	292	6.1	29.8	30.8	3.4	36.0
Plecoptera	153	3.2	31.4	13.1	22.9	32.7
Trichoptera	18	0.4	16.7	27.8	27.8	27.8
Coleoptera	788	16.5	39.7	17.1	25.4	17.8
Odonata	41	0.9	63.4	12.2	24.4	0.0
Others**	570	12.0	33.3	20.2	17.5	28.9
Total	4763		33.8	16.8	29.1	20.4

* Sum of numbers/0.1 sq m for each 10 cm level; ** Includes Hemiptera, Naididae, Hydracarina and lesser groups.

irregularly over the four depths with the 0–10 and 20–30 cm levels generally containing higher proportions than the other strata. Except in the Odonata, with too few specimens from which to draw valid conclusions, and Diptera, the top layer, sampled by most workers using conventional methods, always contained less than half the fauna.

Among the Chironominae, the orthoclad larvae (including Diamesinae) comprised 58.4% of the total immatures, Chironominae 26.6% and Tanypodinae 12.6% with only 2.4% pupae. Only 8.5% of the orthoclad group and 13.4% of the tanypodinids were present in the catch from the 10–20 cm level. The Chironominae, mainly filter and surface-feeding Tanytarsini and a few *Cryptochironomus*, were more common in the upper two layers. However, FORD (1962), in a study of the vertical distribution of chironomids in finely divided stream muds similar to those at the lower stations, found the larvae confined to the top 5 cm where available food and suitable respiratory conditions and substrate occurred. The other Diptera, Empididae and infrequently Tipulidae and Ceratopogonidae, showed a slight preference for the surface stratum where leaf food was available. No Simuliidae occurred in this habitat.

Baetidae and Caenidae were confined to the upper level; *Thraulus* showed a slight preference for the second layer, but the Leptophlebiidae as a group occurred with highest frequency at the 30–40 cm depth. The single ephemerid nymph taken was from the surface. Ephemerellidae and Heptageniidae were not found in this habitat. COLEMAN and HYNES (1970) found the Caenidae restricted to the surface and the Leptophlebiidae most common in the lower levels. However, they reported increasing frequency of Baetidae nymphs with depth. This seems strange in a free-swimming, delicately-gilled group.

Table 82. Size distributions of fauna at the sampling depths

Level (cm)	Total Catch*	Percentage of each size (mm)					
		0–1	1–2	2–3	3–4	4–5	> 5
0–10	1608	22.3	53.0	14.1	4.7	2.7	3.2
10–20	800	18.1	41.9	24.4	8.8	1.9	5.0
20–30	1385	30.7	50.9	8.7	4.3	2.5	2.9
30–40	970	28.4	47.9	15.5	6.7	0.5	1.0
Total 0–40	4763	25.3	49.4	14.5	5.7	2.1	3.0

The Perlidae were found at all levels in low numbers, and the single nemourid nymph taken was in the lowest stratum. Only a few Trichoptera were taken at each level. The Odonata (all but two specimens Zygoptera) were predictably caught preferentially from the 0–10 cm layer. The Coleoptera, mostly larval Elmidae, occurred in considerable numbers at all levels, but showed a preference for the surface. The dytiscid and helodid larvae, present only in low numbers, occurred at all depths. Naididae dominated the 'Others' category, and were found almost uniformly at all levels.

The distribution calculated on total catch at each level, irrespective of series, gave slightly altered percentages from those stated previously, because of the bias caused by large numbers of one group on one date (e.g. Chironomidae in B¹ surface sample), but indicated essentially the same vertical trend.

Size distribution of the fauna at each depth (all groups considered together), given in Table 82, showed few differences between depths. No general utilization of the stable bottom layers as a 'nursery' nor aggregation of a particular size group at any given depth was apparent. The anomaly of a larger total number of 1–2 mm than 0–1 mm animals was probably a result of using a 90 μ sieve which might have allowed the smallest nymphs and larvae to pass through during the flotation-elutriation separation.

2.3. Discussion

The results obtained confirmed the observations of Coleman and Hynes that stream animals can occur in significant numbers deep in the bottom substrates and emphasized the need for further evaluation of this part of the fauna. Many of the studies to date, quantifying invertebrate availability to fish and estimating production of stream benthos, have probably produced considerable underestimates, except in areas where bedrock is close to the surface, as the mean biomass recovered would have been too small (see HYNES 1970), perhaps by a factor of three or more from the evidence now available. Their results must therefore be regarded with circumspection. This would account for the paradoxical finding of many fishery investigators (ALLEN 1951, HORTON 1961) that fish consume many times the available insect standing stock. From COLEMAN and HYNES' results, there did not appear to be a decrease in numbers with depth (to 30 cm) and from the data in this study, only a slight falling off in numbers and biomass at the deepest level was evident. Therefore, under normal conditions of oxygen availability in the sediments (POLLARD 1955, ERIKSEN 1963, 1966, VAUX 1968), i.e. unless strong reduction and anaerobiosis occur, stream insects probably inhabit a large part of the sub-surface gravel beds characteristic of many stream valleys. This fauna would use the stream bed itself only for emergence and oviposition, but much of its production would occur in areas adjacent to the immediate stream bottom substrates. Since most particulate food probably enters through the open stream surface and percolates downward into the gravel through which it is further distributed by interstitial ground-water movements, a decrease in population with vertical and lateral distance from the stream bed would be expected and with adequate sampling techniques may be found. The existence of such a large potential zone of colonization and growth for stream insects, hinted at by the discovery of nymphs of *Habroleptoides modesta* in the hyporheal (SCHWOERBEL 1967), *Leuctra major* deep in the gravels (BERTHÉLEMY 1968) and various insect immature stages in the pots of COLEMAN and HYNES and in this study, would explain the enormous numbers of univoltine insects that are often reported to emerge semi-synchronously from limited stream areas in temperate latitudes, numbers far in excess of the immature standing stock found in the vicinity just prior to the mass emergence. In addition, the apparent rapid recolonization of desiccated stream beds by some faunal elements (HYNES 1958, HARRISON 1966) could be easily rationalized by upward and lateral movements of such sub-surface inhabitants in streams with deep gravel beds that are unlikely to dry out completely. During peak discharge periods, much of the larger fauna and some of the smaller is lost to the drift through physical disruption of the substrate surface. However, the extent of denudation is reduced by the migration of some faunal elements downward into the expanded substrate zone where the interstitial spaces have been cleared of fine particles (CORDONE and KELLEY 1961, BUSCEMI 1966). Subsequently, as detritus and silt are replenished, these invertebrates recolonize the surface with a fauna whose size and age distribution is similar to that prior to the spate, i.e. one not derived from a new generation cycle based on quiescent eggs. A recolonization of just this type has been described by HYNES (1968b) in the Afon Hirnant, Wales.

The sampling system as it stands has many imperfections, the chief being the length of time required to obtain a sample. As a routine procedure in comparative surveys this system is not very feasible, but in a long-term study, especially involving life-history definition and production assessment, one or two series of such samplers withdrawn at appropriate seasonal intervals would indicate the extent and composition of the deep-living population.

A second deficiency results from the disruption of natural bottom conditions in installing the samplers. The length of time needed to restore 'normal' silt and detritus conditions in and around the sampler is unknown, but it was assumed that in an agitated lotic substrate the stabilization period would not exceed 4–8 weeks. As seen, this assumption may have been erroneous as the fauna decreased with time after the initial two months, perhaps as a result of continued filling-in of the interstices and living spaces with silts or consumption of the non-renewable organic particles trapped in the sediments as the sampler was buried. However, under the near-optimal conditions for decomposition, these trapped organic materials were more likely to have disappeared after a few weeks rather than persisting for months, thus pointing to a continual renewal of food resources to sustain the fauna and leaving spatial effects as the most likely explanation for the decline. When the sampler was inserted, usually against an undisturbed vertical face of sediments at the upstream side, a considerable area of silt-free substrate was replaced laterally and downstream of the box. This empty habitat, probably with improved oxygen and food supply, may have proved particularly attractive in the immediate short-term, with lateral movement into it by most insect taxa as described by BISHOP and HYNES (1969a) and COLEMAN and HYNES (1970). Subsequent slow filling of the interstices would have reduced the available space and resulted in a decrease in the fauna present. Since COLEMAN and HYNES found that the number of colonizers was still increasing at 28 days, it seems likely that this decrease in habitat area does not begin to limit the fauna for some time, perhaps until some critical pore size or oxygen concentration is passed. Microhabitat conditions were obviously never unfavourable, or the considerable fauna present at all depths (nearly all young growing stages) would not have moved in, but were probably just spatially limited after the longer four and six month periods. COLEMAN and HYNES in long-term colonization experiments found no decrease in faunal numbers comparable to that seen here. With their pot-sampler, the problem of a large disturbed zone was reduced by leaving the outer case installed semi-permanently in the river and inserting and removing an inner sampling cylinder. However, the gap between the two parts may have allowed vertical migration and acted as a silt trap; this was the most serious shortcoming of their technique which the present sampler was designed to overcome.

Comparison of the data obtained for the surface 0–10 cm layer (on a per sq.m basis) with the routine benthic collections made at Station 11 with the box-net sampler showed few differences between the standing stocks sampled. Detailed comparison was not possible because sampling dates did not correspond and substrate conditions were not parallel. However, the mean number of individuals collected from the riffle habitat at Station II was 2900/sq.m and for the pool samples 2955, ranging from 1715–4930 and 830–6740 respectively for the months when the vertical sampler collected 1710–4800 individuals with a mean of 3216/sq.m. The taxa taken in the deep sampler were not those characteristic of either of the other two habitats, but contained some species found in each of them. The above comparison simply gives an indication of the efficiency of the sampler.

In some aspects this is only a field experimental technique measuring colonization, rather than a sampling method (ULFSTRAND 1968a), but until an adequate assessment of the hyporheic habitat is possible by other means, this at least gives an indication of the extent and composition of this obviously quite significant faunal component.

3. Secondary production

The standing stock data obtained in this survey do not permit any estimation of secondary production, even for the limited riffle or pool areas from which they were obtained. For reasons already given, the per sq. m values are unreliable and incomplete and with the added complication of inadequately known vertical distribution of the benthos, the errors cannot be assessed. Moreover, even if these data were acceptable, neither the turnover ratio (TR) nor instantaneous growth rate (G) (cf. MANN 1966) could be calculated for the whole or individual components of the fauna because of the lack of knowledge about life histories. Demographic analysis using growth-survivorship methods based on the models of RICKER (1946) and ALLEN (1949) and now widely used in fish production work (CHAPMAN 1967) has been applied to invertebrate populations by NEESS and DUGDALE (1959), COOPER (1965), NEGUS (1965) and GILLESPIE (1969). However, in these studies, the life histories and/or age structure of the chironomid, amphipod and mollusc populations were well-defined, cohorts discrete and recruitment limited in time, all conditions that cannot be met by most elements of the Gombak fauna. For estimations of production based on instantaneous growth rate, as used by WATERS (1966) on *Baetis*, at least the length of the life cycle must be known and preferably also the type of growth pattern. These parameters are not known for any Gombak invertebrates and, as discussed, may be variable depending on extrinsic factors. Intrinsic effects cannot of course be estimated. WATERS (1969a) in a comprehensive review of these two techniques concluded that the use of an annual TR estimated for a complete fauna would be the most useful proposition and, from a review of temperate studies, that this TR was in the range 2.5–5 (mode 3.5) for mixed invertebrate populations and 3–4 for aquatic insects alone on a univoltine basis. If this were to hold for tropical faunas, and there seems little reason to expect otherwise, some first order estimation of production would be possible knowing standing stock ($\bar{B}$) and the number of generations per period. For univoltine populations, annual TR is approximately equal to life cycle TR, but for the multivoltine insects, mainly Ephemeroptera and Diptera, and non-cyclically breeding decapods, the annual ratio is directly dependent on the number of generations. Conversely, the longer-lived groups, Odonata, some Coleoptera and Plecoptera, Megaloptera etc., would have correspondingly lower TR. Thus, to obtain a mean annual TR for a stream community, either the component parts of production from groups of different longevity would have to be estimated separately, or a compound ratio weighted to take into account the relative contribution of each group would have to be applied to a mean biomass figure.

An alternative method of determining TR directly, based on the average size frequency distribution of a population, i.e. avoiding the problem of cohort discrimination by estimating an 'average cohort' for the population as being equal to a size category, has been proposed by HYNES and COLEMAN (1968) and modified by HAMILTON (1969). However, this technique that uses the biomass losses between successive size classes during development as an estimate of mortality equal to production (sensu IVLEV 1966) assumes that equal populations, i.e. classes, with the same growth and mortality rates, are sampled quantitatively. Besides the observed differences and variability in growth rates between taxa and the problem of sampling, the 'real loss' to a size class could not be measured in the Gombak fauna as considerable loss not equal to mortality often resulted from spate action between sampling dates. This completely altered the structure of the population since some stages were more susceptible to damage than others. The inability to estimate such losses, even by measuring drift (see following section), the continuous recruitment to all size classes and the variable length of life cycle invalidate this model and the equations suggested by MATHEWS (1970) which use the difference between maximum and minimum $\bar{B}$ for a first order estimation of production. In the non-insect invertebrates (particularly the Oligochaeta and Gastropoda, but to a degree also the prawns and crabs), which make up a large part of the total biomass, production estimation in apparently continuously breeding but variable-sized populations is another problem that at present defies solution because of the inability to define $\bar{B}$ for these groups.

Little or no work has been done on secondary lotic production in the tropics (RZÓSKA 1967) and probably cannot be initiated until life cycles of the major faunal elements and the effects of environmental changes on them are known (PETR 1970). Speculation on possible production rates based on WATERS' modal TR of 3.5 and 2–3 generations per year for the riffle fauna of the Gombak are intriguing but indefensible. The baetine mayflies and some chironomids probably complete more than three cycles and some of the Plecoptera, Coleoptera and Trichoptera certainly fewer than one in a 12 month period; the contribution of the Decapoda and Odonata is unknown. The above model would give riffle production rates in the range 6–12 g dry wt/sq.m/yr for the unaltered river using the $\bar{B}$ values in Table 80 and $2.5 \times$ this for the enriched lower section. Production in the depositional areas would range downward from 6 g/sq.m/yr at Station I, where riffle and pool were similar, to 1.75 g/sq.m/yr at Station III. The enriched fine sediments with their large worm population might have a rate as high as 35 g/sq.m/yr if the TR for the Oligochaeta is in the 2–3 range. If the findings of the vertical distribution section are accepted, the values for the upper stations might be doubled, but they still indicate only very modest production rates.

4. Drift and Vectored Imaginal Flight

In the last decade a large number of investigations on the drift of aquatic invertebrates have been carried out, mainly in Europe and North America, but also in Japan and New Zealand. The realization that drift plays a significant role in population dynamics and in the life histories of a number of species has led to measurement of the extent, composition and variation of drift as a routine adjunct to other forms of sampling in many lotic ecological studies. Most of the pre-1968 studies on quantitative aspects and control mechanisms were discussed in previous papers (BISHOP 1969, BISHOP and HYNES 1969b). Subsequent reviews and assessments of various aspects of drift have been made by ELLIOTT (1968a, 1969a, 1970a, b), KUBÍČEK (1968), CHASTON (1969a, b) SCHMIDT (1969), McLAY (1970) and SCHWARZ (1970), and of the overall significance of the phenomenon by ULFSTRAND (1968a), WATERS (1969b) and HYNES (1970).

Compensating mechanisms, upstream migration by aquatic forms and/or flight by the terrestrial stages of amphibiotic insects, have also received considerable attention (see review in BISHOP and HYNES 1969a). More recently, SCHUHMACHER (1969), HULTIN et al. (1969), HUGHES (1970), SCHWARZ (op. cit) and ELLIOTT (1971) have investigated upstream movements of the nymphal forms and ELLIOTT (1969a), MACKAY (1969), THOMAS (1969a), LEHMANN (1970) and SCHWARZ have recorded upriver flights of adults of various insect species.

The sampling in the Gombak had several objectives. First: drift monitoring was considered a supplementary qualitative method of assessing the composition of the fauna in the upper river where the nature of the substrates and bank-root habitats made sampling by the conventional techniques difficult. In most previous drift studies, almost all elements of the fauna have been taken, over an extended period, and the usefulness of the technique in this regard has been established. Second: drift has not been assessed to date in a non-seasonal river in which temperature and day-length variations are practically non-existent and the usual annual or even synchronous multivoltine generations of the fauna are absent. It was hoped that, in the Gombak with a seasonally homogeneous fauna comprising all developmental stages of most insects simultaneously, the effects of density-dependent factors hypothesized to control the amount of drift would be more easily differentiated. It was also hoped that if selectivity for drift existed it would be more obvious than in the temperate situation where in any particular season much of the population is in one stage and its real susceptibility or propensity to drift is masked by physical changes in temperature, flow and substrate. Third: drift can indicate the amount of food potentially available to the fish, both the surface drift of lotic insect emergents, supra-aquatic arthropods and terrestrials from the forest and the sub-surface drift. Fourth: with the obvious difficulties attendant on productivity assessment in a fauna in which recruitment is continuous and cohorts inseparable, some measure of annual production and turnover ratio estimation might be possible if a measurable stable drift rate were a reflection of excess of population over carrying-capacity of the bottom as suggested by WATERS (1966).

The preliminary studies on the movements of imagines of selected amphibiotic groups were an attempt to clarify the postulated 'colonization cycle' relationship in what is almost an ideal location. There is little or no horizontal wind (the complicating factor in most studies) under the forest canopy, so that the only permanent air-flow along the river is a downstream micro-wind, generated by the flowing water. The tunnel-like nature of the stream bed through the forest limits wide dispersal of the imagines, so that flights up or down the river, if they occur, must be carried out in the immediate area where assessment should, theoretically, be simplified.

4.1. METHODS

4.1.1. Drift

Numerous techniques and apparatuses have been described for sampling the drift as almost all workers have developed their own systems. These have been critically reviewed by WATERS (1969b). The methods and nets used in the Gombak were similar to those used by BISHOP and HYNES (1969b) based on WATERS (1962a). The much smaller mid-water sampling nets used by ELLIOTT in his various studies or by McLAY (1968, 1970) were judged unsuitable for local conditions where large leaves and debris would easily block the small orifice (2 × 10 cm) and where the water depth on the riffles was usually less than 25 cm, eliminating the need to sample only part of the column. The nets were made of terylene mesh of pore size 165 × 560 μ, 30 × 30 cm at the mouth, tapering to 3 × 3 cm at 2 m length. These were attached by light-weight canvas to aluminum rod frames that were held in position in the river by wires looped over steel rods driven into the substrate. These nets sampled a column of water from bottom to surface except at peak flows when the depth exceeded the net height. The nets were sufficiently long that the standing wave caused by resistance to flow was always more than 50 cm down the net during modal flows, even at the end of a 3 h sampling period. On rapidly rising stage, with accompanying loads of leaves and sand, the nets occasionally became blocked and had to be cleared. Nets were set in mid-channel in a section of uniform rapid flow at both Stations I (see Plate 2) and II for a 24 h period once every four weeks, weather permitting. On a number of occasions, a set of readings had to be abandoned when severe thunderstorms caused excessive rise in stage and rupture of the nets or loss of one or more subsamples through erosion. Complete substitute sets of data were obtained as soon as the floods subsided. The long period of 36 days between sets eight and nine was unavoidable, caused by political disturbances! The following period was shortened to 20 days to return to the 4-week schedule. Because of the constant bed and saltating sand movement at the three lower stations, no drift sampling was possible. As described earlier, 3.5 kg of sand accumulated in a half-hour net setting and human activity in the river during daylight hours and interference with the equipment precluded any useful collecting. Samples were collected every three hours (03:00, 06:00, 09:00...) except for the period around sunset which was divided into two 90 min periods (18:00–19:30, 19:30–21:00) to monitor the effects of changing light intensity (see BISHOP 1969). Since sunset at the latitude of the river varies only about 20 min over the year, the collecting periods were not adjusted seasonally. Under the heavy forest and in the shadow of the steep hills, darkness at the river level developed rapidly with only a very short, 10–15 min twilight period between 18:15 and 18:45 h. The tributary, under a complete canopy, was always sampled first and Station II, with slightly prolonged daylight, about 15 min later, the time taken to walk the 0.8 km between the stations. The volume of water passing through each net and the total flow in the channel were measured at the beginning and end of each sampling period. The contents of the net were rinsed into heavy gauge polyethylene bags and preserved in the field with 5% formalin. In the laboratory, the sample was washed on a 165 × 165 μ or finer sieve and the large pieces of plant materials separated under a jet of water. Flotation techniques were useless as the drifting organic debris was of similar density to the organisms; sub-sampling of the considerable organic material, using techniques described earlier for the benthos, was usually necessary. Identification and sizing were made at ×20 magnification over a mm grid and counts multiplied by the subsampling factor. Organisms in convenient taxonomic groups were weighed to 0.1 mg after drying over silica-gel desiccant for 24 h. No allowance was made for changes in weight caused by the preservatives. For each station on each sampling date, a drift factor was calculated that took into account the filtered volume and the total discharge of the river. This enabled estimates of total 24 h drift past each sampling point to be calculated.

4.1.2. Adult flight movements

Because of the short-term variations in stage and discharge, a trapping device at water level was not practical. An insect trap working on a similar principal to those used by Roos (1957) and Elliott (1967a) was designed to hang just above the usual high water level in the centre of the river. Fig. 55 shows the main details of construction: entrance 2 × 1.7 m, upward-sloping sides and top, killing chamber 1m above the top of the trap. Two such nets made of nylon netting (mesh size approx. 0.3 × 0.3 mm) were used in tandem, suspended side by side (one facing up-, the other downstream) from a footbridge crossing the river at Station II with the bottom of the entrances 1 m above the modal flow level. To ensure that all insects inside the trap at the time of collection were captured, a rope was attached to the centre of the lower spreading pole and, prior to emptying, the whole floor of the trap was moved up and down several times to induce the insects sitting in the net to enter the killing chamber. Air movements were recorded as total feet of flow by a Negretti and Zambra Air Flow Meter located immediately above the traps.

In use, the effectiveness of the traps was not entirely satisfactory. As will be seen from the results, very few insects were caught, perhaps because the nets presented an obstruction in the normal flightway which the insects may have avoided by flying over or around. No visual sightings of such evasion were made, but the total number of imagines seen in flight was also very small. Any such effects probably applied equally to both nets so that upstream and downstream results were comparable.

Only the Orders Trichoptera, Ephemeroptera and Plecoptera were sorted and categorized as male (♂), female with soft abdomen, fat body obvious (♀), ovigerous female with ripe eggs in hard, distended abdomen, fat body absorbed (♀) and spent female with most or all eggs laid (○); sub-imagines of the Ephemeroptera were counted as adults. Large collections of Lepidoptera, Diptera and some Odonata and Coleoptera were taken but could not be easily worked. Many collections contained large ants that entered the traps either by walking down the ropes or by cutting through the walls. These became trapped in the killing chamber, but if insufficient killing solution was used, these scavengers removed the catch in short order; such disturbed catches were discounted.

314

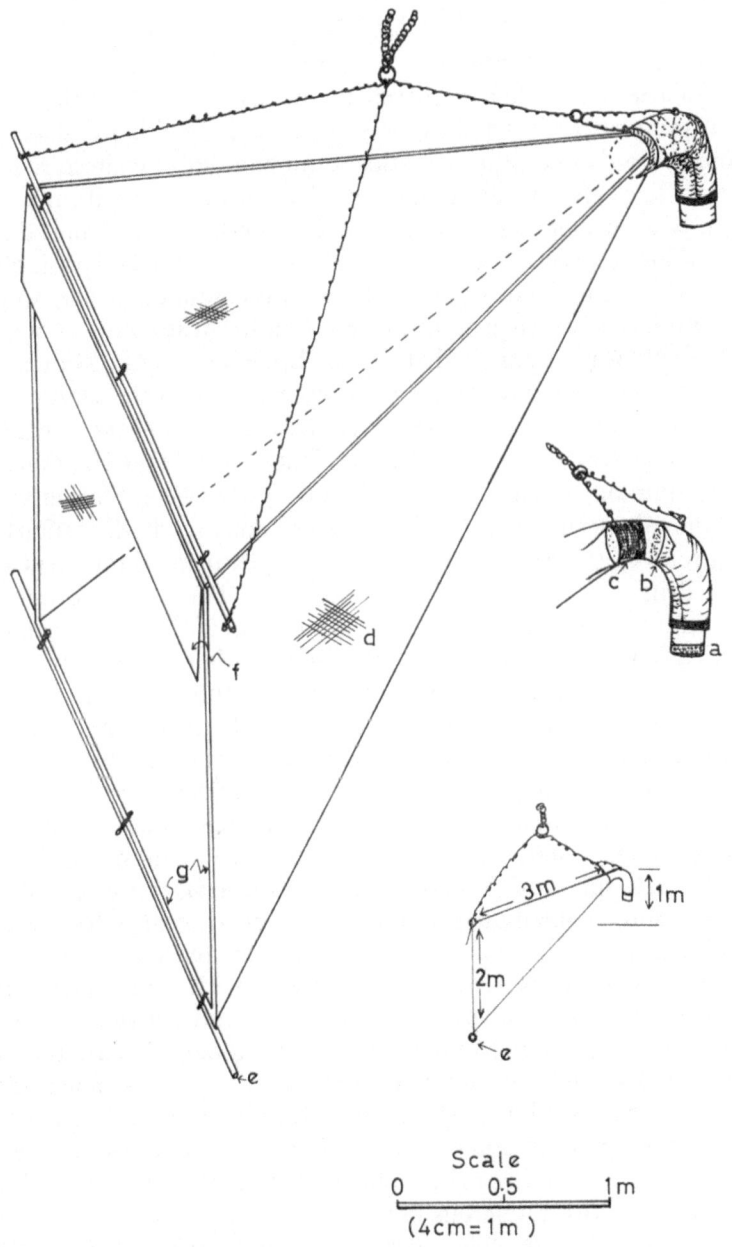

Scale

0   0.5   1m

(4cm=1m)

Fig. 55. Construction of the adult insect trap used for upstream-downstream movement studies; a. killing jar attached to galvanized collector (10 cm diameter), b. non-return funnel, c. sailcloth collar, d. terylene mesh (approx. 0.3 mm) with sailcloth edges, e. extension poles, f. 0.5 m flap, g. reinforcing tape.

4.2. RESULTS AND OBSERVATIONS

4.2.1. Drift

Most of the taxa which occurred in the river at the stations where drift was sampled were found in the benthoplankton either casually as an occasional record or constantly in most or all samples, often in considerable numbers. The terrestrial component was usually significant. Ants, Isoptera, various lepidopteran and coleopteran larvae from the leaf canopy, a few spiders, grasshoppers, Collembola and, in surprisingly low numbers, the spent or emerging imagines of the aquatic Orders were the more common components of the exogenous drift. The endogenous aquatic drift was dominated by the Ephemeroptera, Diptera and Coleoptera. *Baetis* spp. and *Pseudocloeon* spp. were by far the most common forms, with a persistent but small contribution from such genera as *Caenis*, *Isca* and *Habrophlebiodes*. *Thraulus*, *Tricorythus* and the Heptageniidae were less common, along with *Isonychia* and *Ephemerella* at Station II only; the occasional specimen of *Prosopistoma* was taken at both stations. The many species of the Simuliidae were the pre-eminent component of the dipteran drift, followed by the Chironomidae. There was no dominant group within this latter family; the proportions of the three major sub-families varied from month to month. Orthoclads were usually, but not always, more common at Station I; there was no apparent pattern at Station II. Psychodid, ceratopogonid and tipulid larvae were occasionally caught, but these were uncommon. The coleopteran element was dominated by the Elmidae, mostly larvae, but with a significant number of adults as well. Hydrophilids, dryopids, gyrinids, dytiscids and helodids were frequently taken, usually as larvae, but only in low numbers. Perlidae, mostly *Neoperla* spp. and Nemouridae (*Amphinemura*, *Nemoura* and a few *Protonemura* spp.) contributed a small but recurring drift fraction. The various Hydropsychidae, particularly *Diplectrona*, and the Psychomyiidae were the most frequent drifting Trichoptera, with some Leptoceridae, Philopotamidae (*Chimarra* spp.), *Neolepidostoma* and occasional members of other families present. Among the other aquatic Orders, the gerrid *Ptilomera lundbladi* was a common catch, *Plea* spp. and naucorids less common. Mites, Gordiidae, Decapoda (*Macrobrachium* spp.), Naididae, a few Odonata (mostly Gomphidae) and the occasional Copepoda were recurring members of the drift along with tadpoles of various families and in some months, numerous fish fry.

Drift Rate and Density: The amounts of total drift past a transect at the sampling locations at Stations I and II, the 'drift rate' of WATERS (1969b) are listed in Table 83 for the fifteen sampling periods. Included in these totals are the terrestrial contributions to the drift. The variation at Station I, from 7,000 to 41,200 individuals/24 h, was a function both of existing discharge rate and, more important, antecedent flow con-

Table 83. Total drift past a transect at the sampling locations-Stations I and II

| | Station I | | Station II | |
	Number	Weight	Number	Weight
5–X–68	20065	19677	347711	625482
30–X–68	7754	4656	254200	97976
28–XI–68	12018	10625	203660	54907
1–I–69	20480	8369	348659	419534
24–I–69	18947	7499	191745	28229
23–II–69	14163	5624	246500	72558
19–III–69	5955	1045	237138	38834
17–IV–69	41193	13319	170533	34236
23–V–69	31816	7059	351765	91863
12–VI–69	11350	1546	133679	13498
10–VII–69	20726	2992	141034	16344
7–VII–69	6927	3045	166942	40990
3–IX–69	16448	1801	282395	74067
2–X–69	9043	3987	131720	46980
30–X–69	16986	16012	124605	24224
Total	233391	107256	3342286	1679227
Mean	15559	7150	222819	111981
Estimated total/year	5.68×10^6	2.6 kg	81.33×10^6	40.9 kg.

ditions as discussed later. The proportional variation in drift at Station II was less, from 125,000 to 350,000 inds/24 h, reflecting the greater stability of the physical habitats in the main river. Comparison of biomass has little meaning as a few large Coleoptera or Lepidoptera larvae or a single tadpole obscured any relationship between sampling dates.

The 'drift density', number of individuals per unit volume, is perhaps a more valid indication of drift activity and provides the only rational means of comparison between different watercourses (WATERS 1969b). These data are summarized in Table 84 expressed as numbers/100 cu. m for the major Orders of the drifting fauna, calculated from the actual catch and filtered volume. There were obvious differences between drifting frequency of various components, a function of activity patterns, feeding habits and propensity to resist wash-off and entrainment by physical forces (see DORIER and VAILLANT 1955). Overall, 1.79 and 1.56 individuals/cu.m were found for Stations I and II, respectively, of the same order as the slightly < 1 reported by ELLIOTT (1967a) and ELLIOTT and MINSHALL (1968) and the 1.9–4.9 (aquatics only) found by ULF-STRAND (1968a), but lower than the 4–7 (extremes 1.4–10) recorded by McLAY (1968, 1970) and 5.5/cu.m by WATERS (1966). They were much less than the mean of 22/cu.m (maximum 45) calculated from the data of BISHOP and HYNES (1969b) and the high of 46 reported for the

Table 84. Drift density of the major groups in numbers per 100 cubic metres

Station I	Cumec River discharge	Volume filtered	Ephem.	Diptera	Coleop.	Plecop.	Trich.	Other aquatics	Terrest.	Total
5-X-68	0.160	0.0262	31.8	26.9	8.5	3.5	2.8	10.1	61.5	145.1
30-X-68	0.101	0.0284	26.7	21.5	1.3	2.6	3.3	5.3	28.1	88.8
28-XI-68	0.120	0.0299	25.4	18.0	6.2	7.4	10.5	8.2	40.3	116.0
1-I-69	0.118	0.0252	55.8	19.1	8.8	8.2	43.4	16.9	48.8	201.0
24-I-69	0.113	0.0182	71.4	35.6	7.5	7.1	34.3	13.8	24.3	194.0
23-II-69	0.086	0.0292	64.1	52.1	3.9	9.6	22.3	14.6	23.7	190.3
19-III-69	0.042	0.0220	22.2	43.4	10.2	0.8	0.9	22.8	63.7	164.0
17-IV-69	0.067	0.0126	502.4	111.1	6.1	0.2	31.0	50.2	10.4	711.4
23-V-69	0.264	0.0268	36.0	18.0	29.1	2.1	12.4	13.5	28.4	139.5
12-VI-69	0.075	0.0298	55.4	25.0	22.5	2.7	8.8	21.9	38.6	174.9
10-VII-69	0.105	0.0207	87.9	53.8	19.8	10.1	11.8	11.3	33.8	228.5
7-VIII-69	0.078	0.0290	14.2	20.0	12.2	4.2	7.8	15.2	29.3	102.9
3-IX-69	0.189	0.0350	45.0	16.4	13.8	6.4	3.7	4.7	10.7	100.7
2-X-69	0.148	0.0420	16.9	14.9	5.3	3.6	4.9	4.6	20.6	70.8
30-X-69	0.322	0.0400	15.8	6.5	2.8	2.8	3.8	8.5	20.8	61.0
Mean	0.133	0.0277	71.4	32.2	10.5	4.8	13.4	14.8	32.2	179.3
Station II										
5-X-68	2.05	0.0319	48.1	26.1	14.5	9.9	6.4	10.5	80.7	196.2
30-X-68	2.34	0.0566	71.7	12.4	8.8	4.3	5.2	3.4	20.0	125.8
28-XI-68	2.15	0.0400	25.0	24.5	9.7	4.7	3.3	8.5	34.7	110.4
1-I-69	2.38	0.0431	53.0	28.8	15.5	5.3	15.0	13.6	38.3	169.5
24-I-69	2.00	0.0530	32.6	20.0	7.0	4.1	8.8	8.0	30.5	111.0
23-II-69	1.37	0.0320	65.4	29.6	16.3	10.7	15.6	12.4	58.3	208.3
19-III-69	0.79	0.0180	151.6	53.8	33.7	4.8	42.3	6.4	54.9	347.5
17-IV-69	1.02	0.0260	78.3	25.1	20.0	8.3	13.2	8.1	40.5	193.5
23-V-69	3.16	0.0540	47.0	12.8	22.7	3.6	19.3	7.5	17.1	130.0
12-VI-69	1.33	0.0276	56.5	8.7	14.9	3.4	6.8	5.6	20.3	116.2
10-VII-69	1.58	0.0438	39.9	14.4	12.0	2.4	3.6	9.0	22.1	103.4
7-VIII-69	1.25	0.0320	67.5	22.7	18.7	9.6	12.3	6.2	17.7	154.7
3-IX-69	1.79	0.0350	75.4	26.1	24.0	2.4	12.9	15.0	26.9	182.7
2-X-69	1.28	0.0320	40.9	19.7	9.3	6.2	10.7	3.0	29.2	119.0
30-X-69	2.20	0.0600	21.0	4.4	10.0	3.4	6.2	2.9	17.6	65.5
Mean	1.78	0.0390	58.3	21.9	15.8	5.5	12.1	8.0	33.9	155.5

Table 85. Biomass (mg) of drift caught per 24 h sampling period (Factor = total discharge/total volume filtered)

	Factor	Ephem.	Diptera	Coleop.	Plecop.	Trich.	Other aquatics	Terrest.	Total
Station I									
5–X–68	6.11	70.4	35.2	16.8	2.4	3.2	2955.2	300.8	3384.0
30–X–68	3.56	24.8	19.2	4.8	3.2	0.8	475.8	779.8	1308.4
28–XI–68	4.01	30.4	11.2	20.0	80.8	21.6	2192.8	292.8	2649.6
1–I–69	4.68	80.0	27.2	31.2	29.8	88.8	788.0	743.4	1788.4
24–I–69	6.21	116.0	27.2	19.2	43.3	106.4	654.8	240.8	1207.7
23–II–69	2.95	69.0	39.5	16.2	16.2	26.3	1545.2	193.4	1905.8
19–III–69	1.91	9.0	10.6	23.8	36.9	1.3	336.6	129.9	548.1
17–IV–69	5.32	280.3	24.1	2.6	0.1	17.5	2210.3	11.5	2546.4
23–V–69	9.85	66.5	22.4	78.5	19.2	48.0	364.6	117.5	716.7
12–VI–69	2.52	38.5	13.6	160.9	12.6	6.8	216.5	164.6	613.5
10–VII–69	5.07	114.5	56.2	54.6	14.2	21.3	136.4	192.8	590.0
7–VIII–69	2.69	12.5	22.7	43.6	13.6	37.5	342.0	659.8	1131.7
3–IX–69	5.40	46.8	32.0	92.8	13.5	4.9	14.7	128.9	33.63
2–X–69	3.52	42.0	36.2	21.7	19.8	10.0	61.3	941.7	1132.7
30–X–69	8.05	13.1	9.9	35.6	5.0	7.4	936.7	981.4	1989.1
Station II									
5–X–68	64.26	81.6	33.6	30.4	56.8	19.2	1541.3	7970.7	9733.6
30–X–68	41.34	249.6	28.8	53.6	16.8	35.2	318.8	1667.2	2370.0
28–XI–68	53.37	42.4	79.2	23.2	82.7	28.6	127.1	645.6	1028.8
1–I–69	55.22	158.2	77.9	60.4	177.7	37.6	5541.0	1544.7	7597.5
24–I–69	37.76	103.1	73.9	52.8	40.4	42.2	235.0	200.0	747.7
23–II–69	42.81	142.1	45.4	100.5	159.9	126.9	293.4	826.6	1694.9
19–III–69	43.89	201.5	32.0	72.2	64.0	57.6	367.0	90.5	884.8
17–IV–69	39.23	144.3	22.5	48.8	84.2	43.1	301.7	228.1	872.7
23–V–69	57.98	215.4	30.1	154.0	19.6	131.5	896.1	137.7	1584.4
12–VI–69	48.19	83.7	8.0	24.6	17.6	17.1	59.2	69.9	280.1
10–VII–69	36.07	71.8	25.8	23.1	20.5	16.8	116.9	178.2	453.1
7–VIII–69	39.06	106.5	38.7	41.0	68.5	26.9	691.6	76.2	1049.4
3–IX–69	51.14	133.0	50.0	43.7	20.9	38.1	855.0	307.6	1448.3
2–X–69	40.00	128.1	20.5	49.0	41.9	36.5	375.6	522.9	1174.5
30–X–69	36.67	101.9	11.7	42.5	18.7	26.4	232.7	226.7	660.6

Kakanui River, New Zealand, after a flood (McLay 1968). The rich waters of Valley Creek, Minnesota, had a mean drift density of about 40 inds/cu.m of *Baetis* sp. alone, calculated from the data given in Waters (1962a); with the addition of the other drifting components, this total would have been even higher. The overall rate for a river must be dependent on richness and productivity, the types of organisms present and temporal effects of life stage and physical parameters. For example, on 17 April 1969 at Station I, the drift density of Ephemeroptera was exceptionally high (> 5 inds/cu.m) as a result of low discharge and a flock of active *Baetis* and *Pseudocloeon* nymphs.

Composition of the Drift: Some indication of the relative composition of the drifting fauna was seen in the numerical data of Table 84. In Table 85 the weights taken per sampling period for the dominant groups are given. Overall, the total invertebrate drift past a point for all sampling dates had the following percentage composition:

	Ephem.	Coleop.	Dipt.	Plec.	Trich.	Other Aquatics.	Terrest.
Numbers							
I	34.6	6.8	18.1	3.2	7.8	8.5	21.0
II	37.1	10.3	13.9	3.6	7.6	5.5	22.0
Weights							
I	4.9	2.8	1.7	1.4	2.0	62.1	25.1
II	5.4	2.3	1.6	2.5	1.9	37.4	49.0

The 'Other Aquatics' and 'Terrestrial' components comprised less than a third of the numbers but more than 85% of the biomass at both stations. The aquatic invertebrates, particularly the Ephemeroptera, were only a minor part of the drifting biomass in spite of their high numbers. Most of the large weight of Other Aquatics at Station I was *Macrobrachium malayanus*, tadpoles and gerrids; these were proportionately less frequent at Station II. In the main river, the terrestrial component, mostly lepidopteran caterpillars and large ants, comprised almost half the total weight of the drift. It must be emphasized here that these drift catches were made in the undisturbed river, i.e. with the natural predator associations intact, so that the measured drift was that which remained after partial cropping and was not an indication of the gross drift of either terrestrials or aquatics. As will be seen in the fish section, no surface-feeding insectivore was found at the main river forest station, as against two species present in the tributary. The surfeit of terrestrial drift at Station II may be a result of this, although there were mid-water insectivore-omnivore fish present to take the food as soon as it drowned. The large population of *Ptilomera* would deplete the surface drift to a degree, but as they feed in the smoother water areas at the head of pools, predation from the area immediately upstream of the drift nets, located

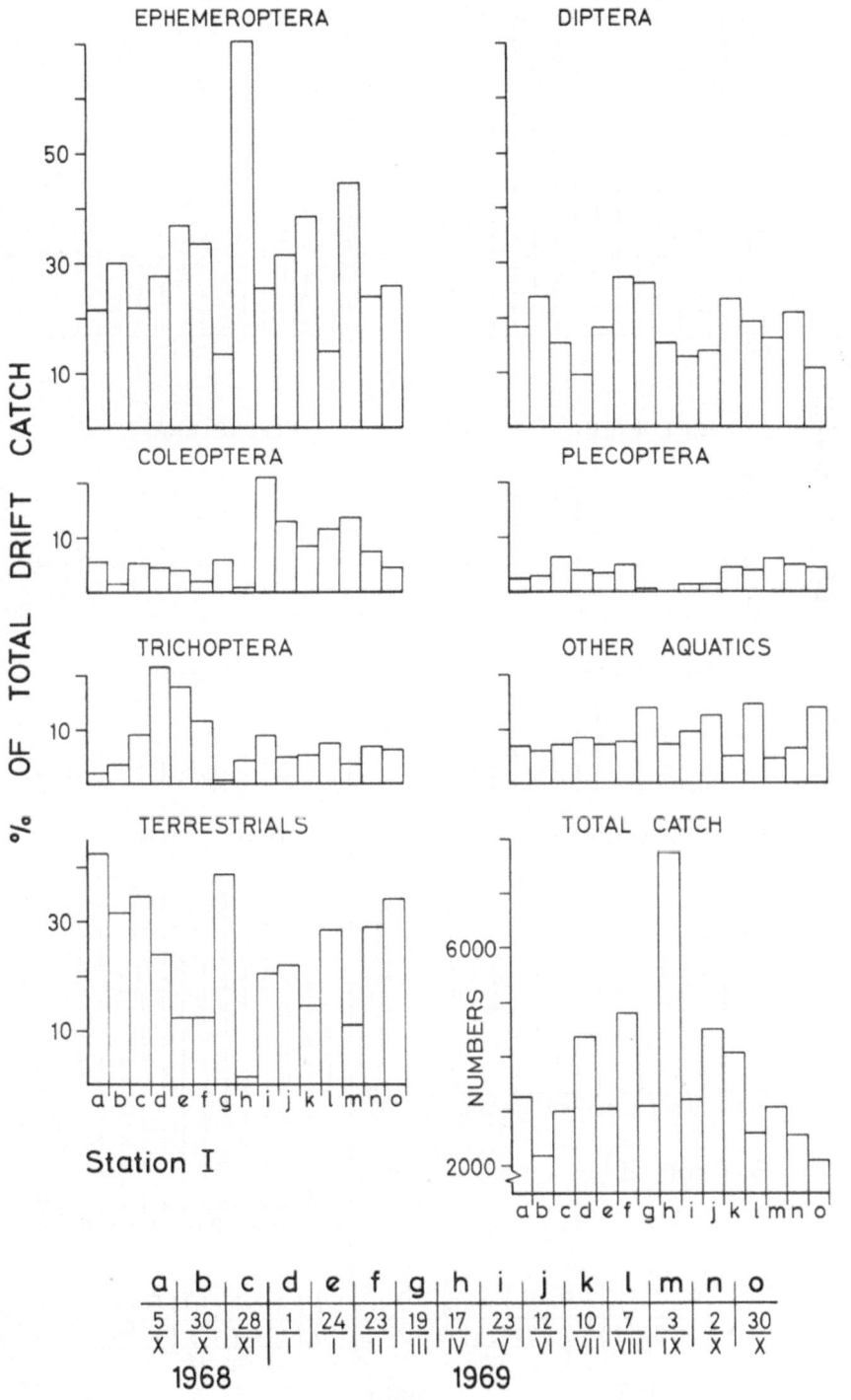

Fig. 56. Percentage composition of the drift catch by Orders per sampling date at Station I.

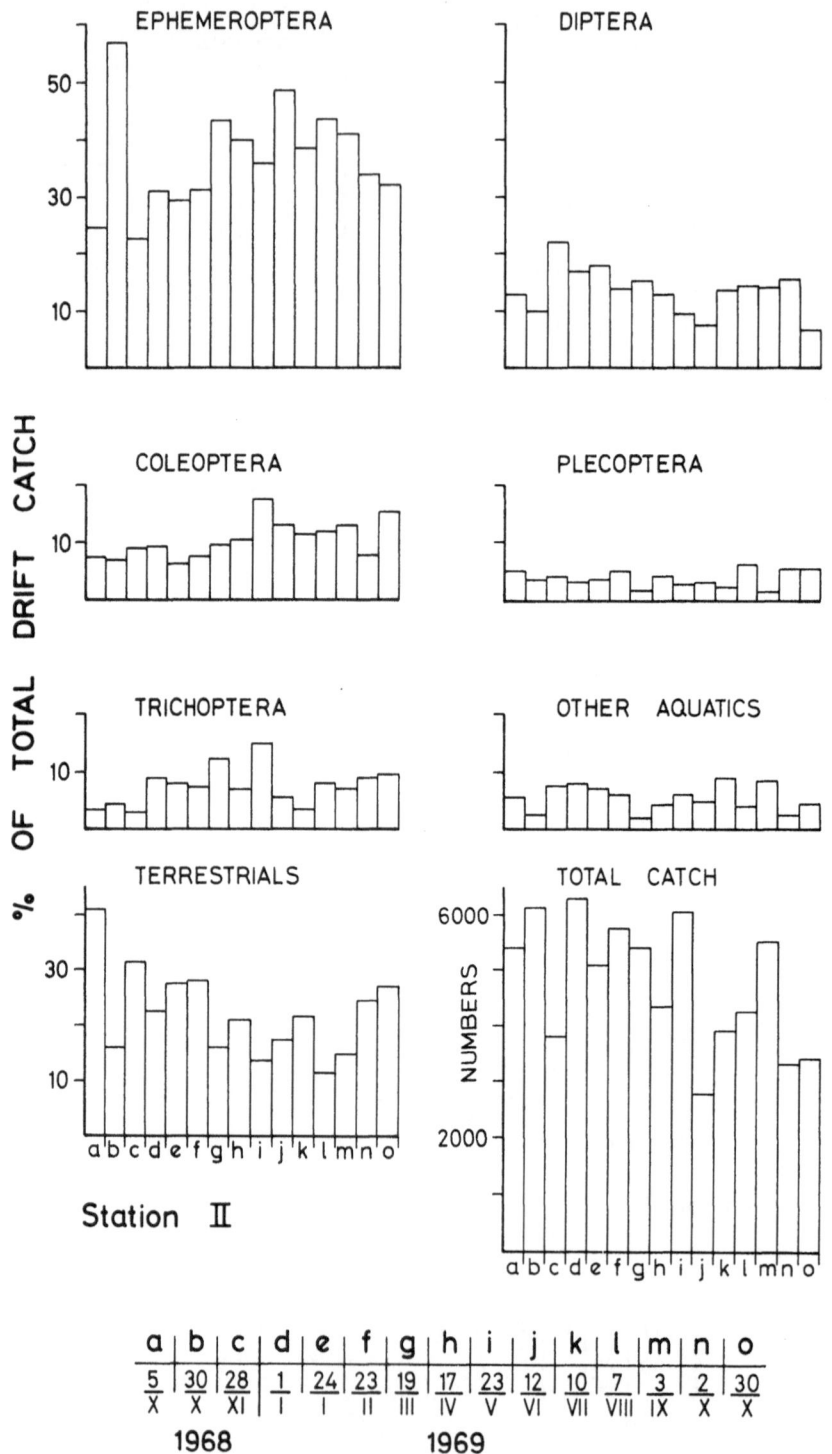

Fig. 57. Percentage composition of the drift catch by Orders per sampling date at Station II.

on riffles, was not severe. The composition of the drift on each sampling date is summarized in Fig. 56 and 57 in which the percentage contributions of the major groups are illustrated. Considerable variation from month to month was apparent with little ready explanation for these fluctuations. The high drift of Ephemeroptera at Station I on 17 April 1969 was not matched at Station II, and the outburst of *Pseudocloeon* drift at Station II on 30 October 1968 did not occur concomitantly at Station I. On 1 January 1969 at Station I, a considerable drift of hydropsychid larvae, normally bottom-dwelling net-spinners, occurred for no apparent reason. Some of the percentage differences were due to fluctuations in the size of terrestrial drift which was dependent on wind action in the canopy and precipitation effects. Generally, but not always, a decrease in the exogenous component was seen on overcast, wet days indicating perhaps that travelling activities were curtailed in the wet by groups, notably ants and spiders, normally active in the marginal and overhead foliage. The very low terrestrial drift at Station I on 17 April 1969 cannot be satisfactorily explained.

In Appendix D, the composition at generic or higher level and relative frequency of all drifting organisms caught at both stations are given. Breakdown into lower taxonomic groups of doubtful status was avoided.

Drift Rate and Benthic Standing Stock: Comparison of the total drift (Table 83) with the mean benthic density on the closest antecedent date (Table 76) showed no correlation. The wide fluctuations in the bottom fauna resulting from frequent spates made it impossible to construct a function of the type found by DIMOND (1967) that would relate these to indicate a density-dependent relationship.

Diel Periodicity: On all but one occasion (Station II, 28 November 1968) the numbers of drifting aquatic organisms taken at night were greater than during the day; no particular pattern was evident for the terrestrial drift (Table 86). In Fig. 58, the total drift per period is divided into exogenous and endogenous components and the decided nocturnal increase in stream animal drift is illustrated. When the drift was divided into components as percentage caught per 3 h period (Fig. 59 and 60), the commonly reported pattern of an increase in activity immediately following sunset was found. The nocturnal increase in drift was generally unorganized, exhibiting no consistent peaking with elapsed time after onset of darkness and only occasionally showed the often reported 'bigeminus' pattern (ASCHOFF 1966), for example on 28 November 1968, 1 January 1969 and 3 September 1969 at both stations. This overall pattern was predictable as light is now considered the primary factor controlling drift activity in many taxa (see reviews by MÜLLER 1966, BISHOP 1969, BISHOP and HYNES 1969a, b, CHASTON 1969b, WATERS 1969b and others). On several dates there was an unexpected post-sunrise peak in drift; this was particularly noticeable at Station II on 5 October 1968, 23 February, 12 June, 10 July and 7 August 1969 and

Table 86. Diel periodicity of drifting aquatic and terrestrial animals (numbers per collection).

| | STATION I | | | | STATION II | | | |
| | Aquatic | | Terrestrial | | Aquatic | | Terrestrial | |
	Day	Night	Day	Night	Day	Night	Day	Night
5-X-68	594	1298	544	848	957	2229	1185	1040
30-X-68	528	961	208	481	1344	3928	400	576
28-XI-68	624	1333	384	656	1344	1272	560	640
1-I-69	880	2434	576	486	1760	3126	608	820
24-I-69	850	1819	128	254	1297	2384	624	733
23-II-69	1312	2892	256	341	1297	2848	1072	541
19-III-69	770	1137	784	427	968	3581	449	405
17-IV-69	1393	6237	65	48	1033	2405	226	683
23-V-69	832	1741	401	256	1619	3650	468	330
12-II-69	832	2677	481	514	1186	1103	225	260
10-VII-69	738	2746	261	343	1024	2048	674	164
7-VIII-69	577	1263	402	333	870	2915	320	169
3-IX-69	705	2016	65	260	1538	3172	385	427
2-X-69	576	1245	288	460	739	1746	160	648
30-X-69	418	972	386	344	898	1486	354	560
Total	11629	30771	5229	6041	17874	37894	7710	8036
Mean/24 h	775.3	2051.4	348.6	402.7	1191.6	2526.3	514.0	535.7
Mean % of discharge filtered		25.2				2.3		

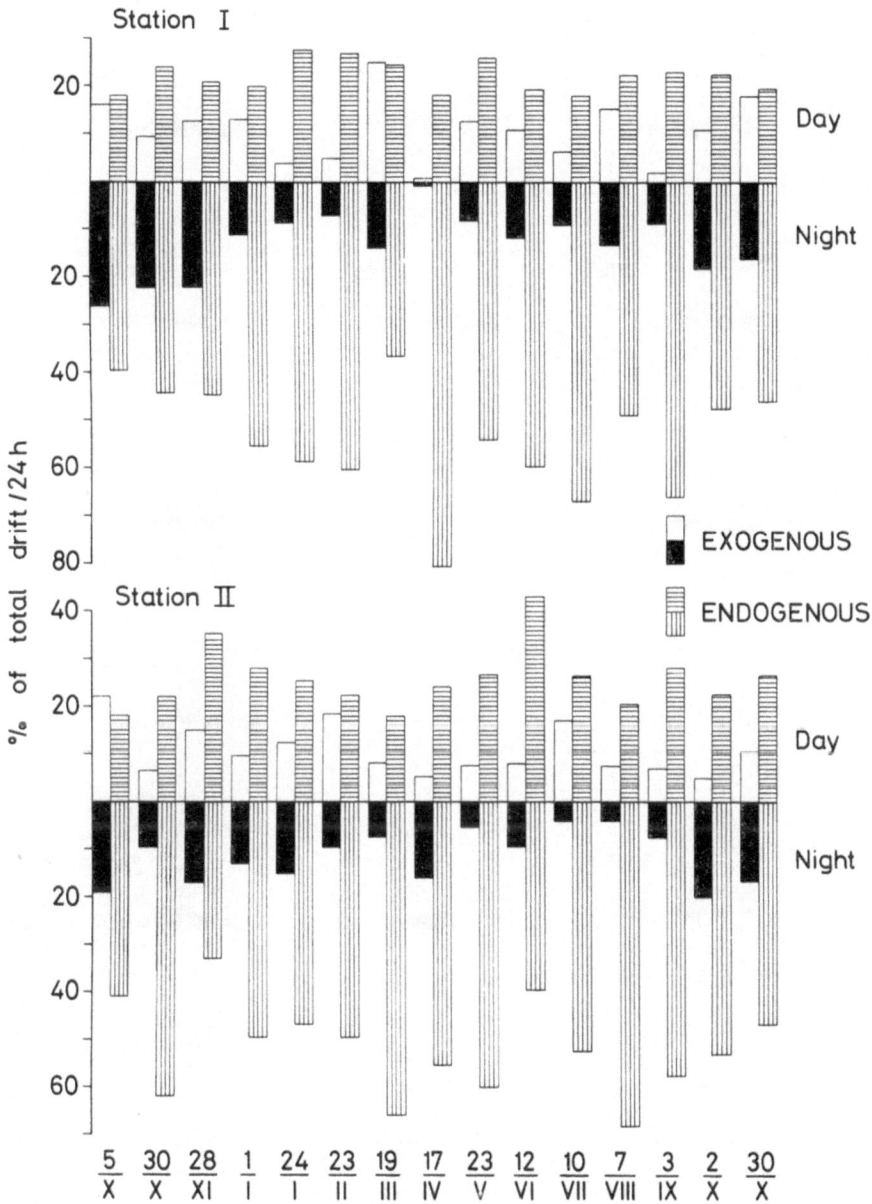

Fig. 58. Percentages by numbers of total day and night drift of exogenous and endogenous components at Stations I and II.

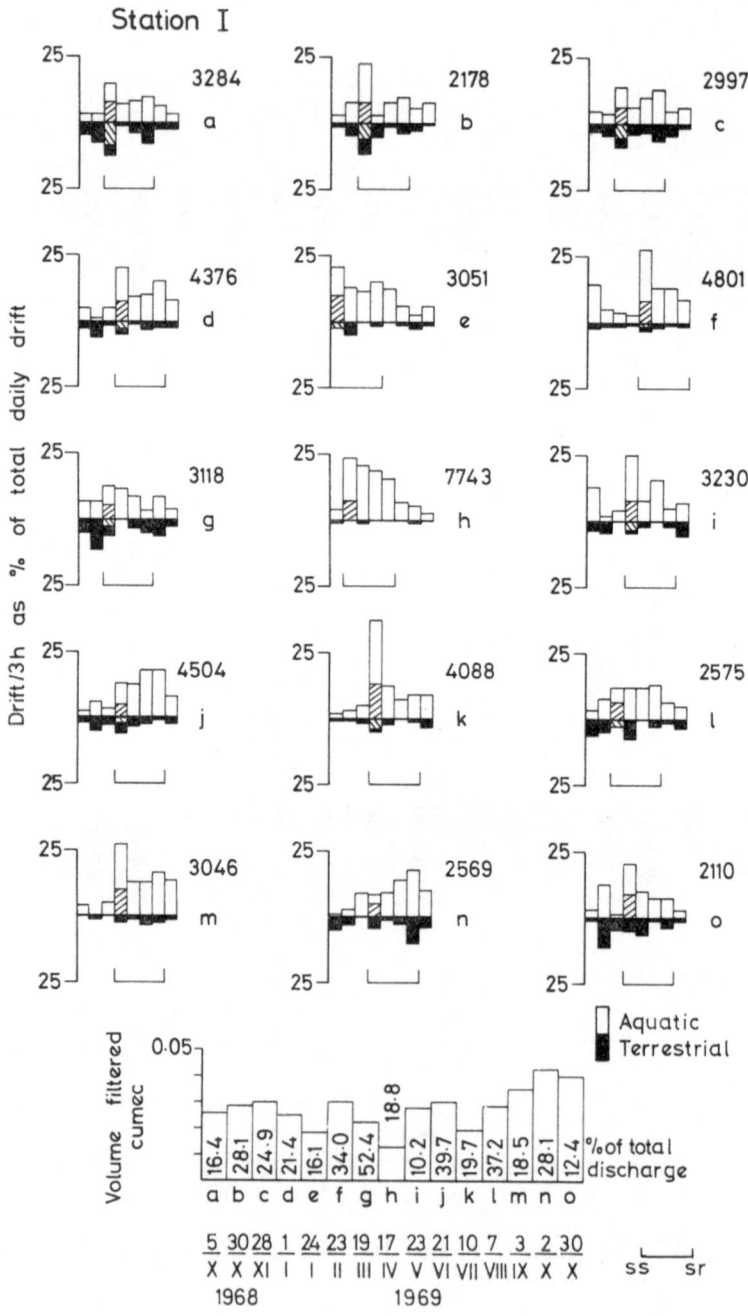

Fig. 59. Diel periodicity of drift at Station I (drift/3h as percentage of total daily drift as shown). Note: First period post-sunset is divided into two halves (18:00–19:30 h : striped; 19:30–21:00 h : solid).

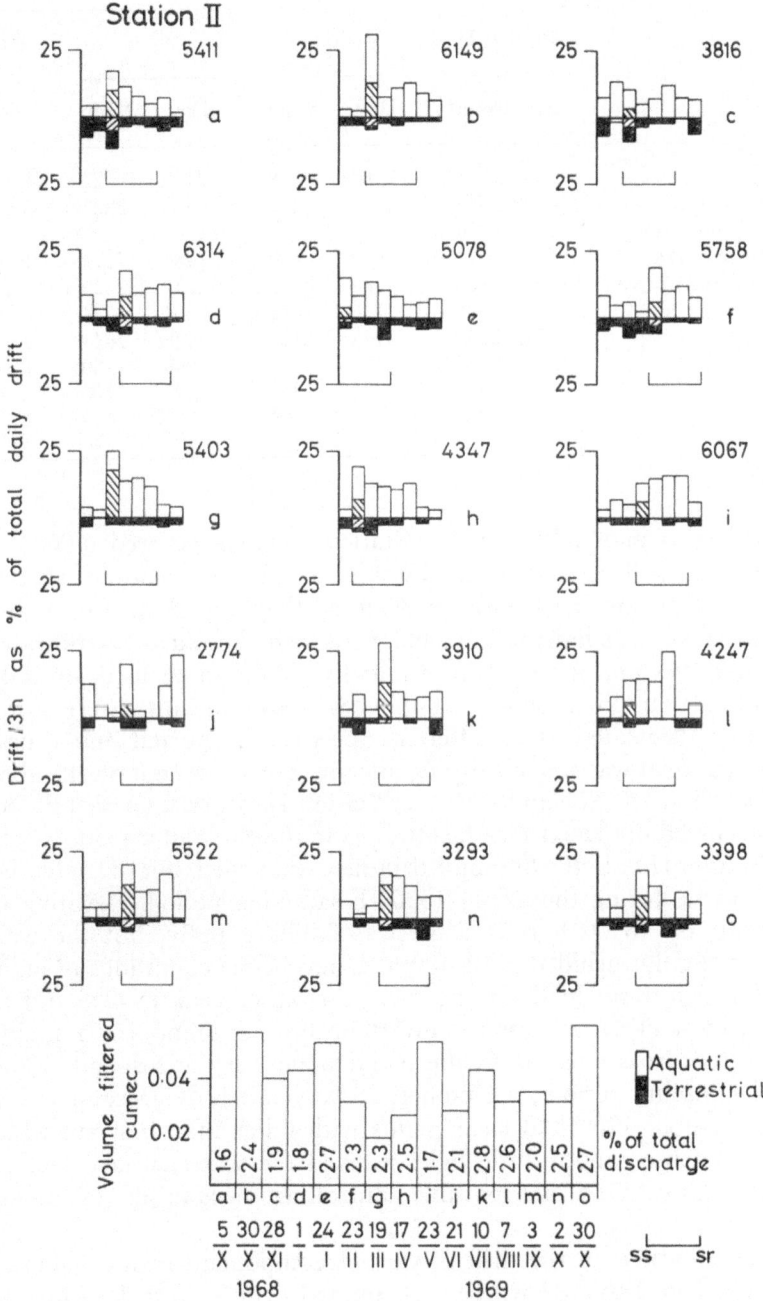

Fig. 60. Diel periodicity of drift at Station II (drift/3 h as percentage of total daily drift as shown). Note: First period post-sunset is divided into two halves (18:00–19:30 h : striped; 19:30–21:00 h : solid).

Table 87. Day and night drift (total numbers for 15 sampling dates) of selected invertebrate types

	Station I		Station II		Total		Ratio
	Day	Night	Day	Night	Day	Night	D:N
Baetis spp.	2753	7844	2385	4379	5138	12223	1 : 2.4
Pseudocloeon spp.	1056	2762	5290	10170	6364	12932	1 : 2.1
Caenis sp. and other							
Ephemeroptera	593	3578	688	3596	1281	7174	1 : 5.6
Total Ephemeroptera	4402	8363	14184	18145	12765	32329	1 : 2.5
Simuliidae	624	2099	1440	2615	2064	4714	1 : 2.3
Elmidae	704	2504	1744	5137	2448	7641	1 : 3.1
Mites	384	745	512	544	896	1289	1 : 1.4
Perlidae	145	342	352	1378	497	1720	1 : 3.5
Nemouridae	112	964	145	695	257	1659	1 : 6.5

on 23 February and 23 May 1969 at Station I. Dawn occurred at 06:15 ± 15 min, but sufficient light to distinguish substrate features was not available at stream level until 06:30 h, so that any change of activity with the advent of light, e.g. orthokinesis and skototactic withdrawal, would have been recorded in the nominally 'day' sample 06:00–09:00 h. The two half-sampling periods (90 min each) around sunset were plotted together as indicated. The difference between nocturnal and diurnal endogenous drift was statistically significant (χ^2 test) except on the above noted occasion (28 November 1968 at Station II) when a severe but local storm occurred in the extreme headwaters in the early afternoon. No rain fell at Station II or in the Station I tributary watershed, but the main river began to rise during the 12:00–15:00 h sampling period, elevating the total drift. By 18:00 h river stage was falling rapidly, but the water remained highly turbid until 24:00 h. Under these conditions of falling stage and heavy silt load, the normal post-sunset peak in drift did not develop, most likely because the invertebrates had found and remained in protected areas away from the molar action of the suspended load. Several sampling periods, 5 October, 28 November, 26 December 1968 and 30 October 1969 had clear nights with fuller than half moon, but under the dense canopy little light reached the water surface and this factor was discounted as a possible controlling mechanism (cf. BISHOP and HYNES 1969b).

Diel fluctuations in the drift of various components of the fauna are summarized in Table 87 where the summed drift at both locations are given. The principal aquatic members of the drift are listed except for the Chironomidae which appeared to have no definite pattern and the Trichoptera, whose large number of species, many inseparable in the

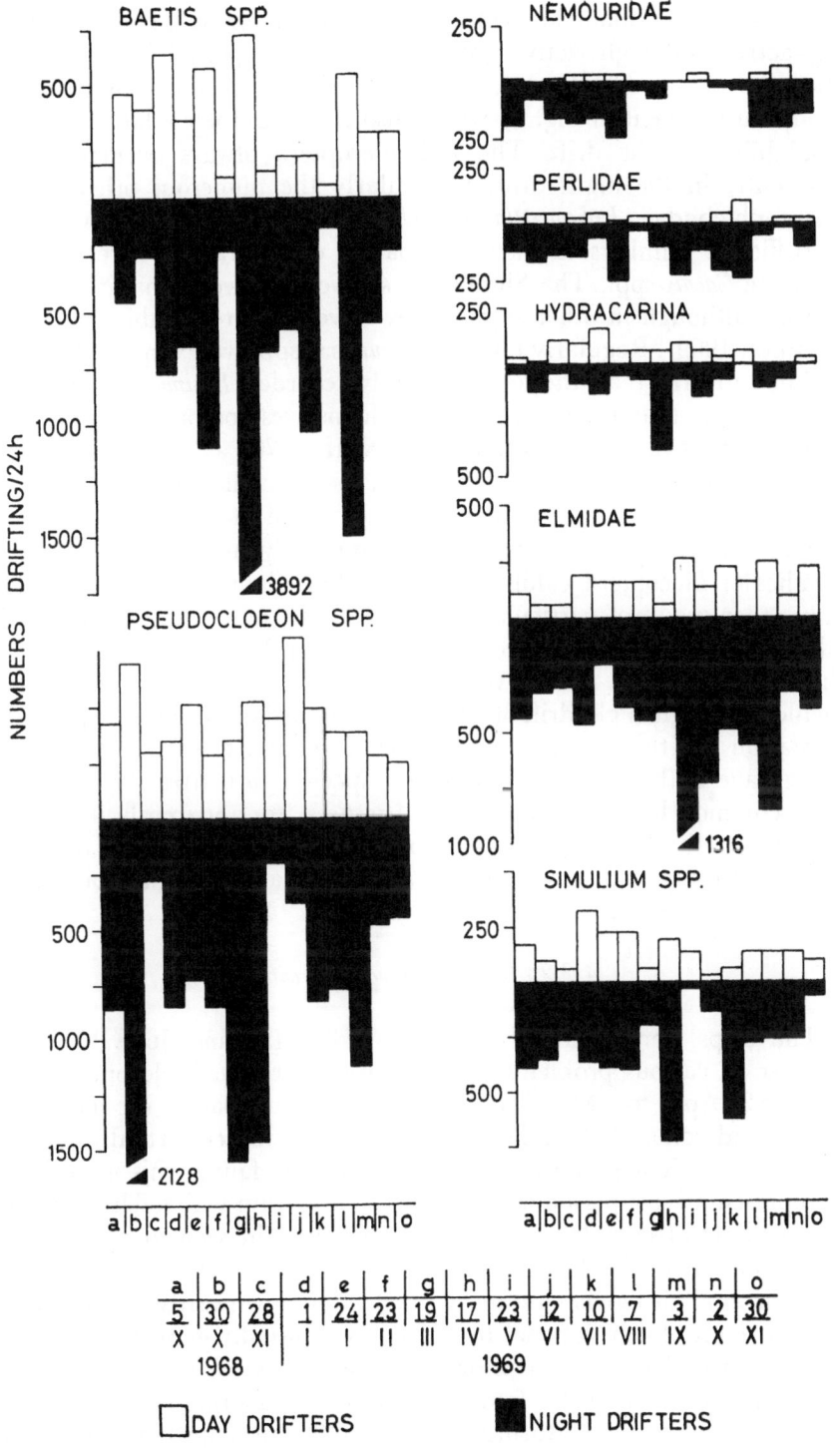

Fig. 61. Day:night drift of selected taxa.

immature stages, made any comparison meaningless as members of both day-active and night-active groups have been reported in this Order. Variations from date to date in the diel drift ratio for these principal groups are evident in Fig. 61 which also indicates their relative seasonal availability to the drift. The Ephemeroptera always occurred more frequently in the night drift, particularly the numerous other species (Leptophlebiidae, *Ephemerella*, *Isonychia*, Heptageniidae, etc.) not taken in sufficient numbers to warrant separate consideration, here included with the *Caenis* spp.. The Simuliidae showed an overall habit of night-drifting although in other studies they have exhibited ambivalence (see KURECK 1969). Regularly taken *Simulium* s.s. spp. were apparently more inclined to drift than the less commonly recorded *Eusimulium* and *Gomphostilbia* spp. that appeared not to let themselves float at the end of long threads (ZAHAR 1961, TARSHIS and NEIL 1970) as much as the first genus. This observation will need much more detailed work for verification. The Elmidae covered a large assemblage of species all living in similar habitats and apparently predominantly active at night. Both families of Plecoptera exhibited the familiar night-active pattern. The only other group of particular interest was the mites taken in the drift. About half were Parasitengona, mostly Hygrobatidae/Torrenticolidae, the others Oribatei, etc.. Overall a ratio of 1:1.4 was found, but this included a single night drift of 376 newly-hatched larvae on 17 April 1969 at Station I. If this count is omitted, the ratio drops to 1:1.1. The aquatic mites have both day- and night-active species (SCHMIDT 1969) but the most common lotic genera, such as *Atractides*, are apparently day-active (see BISHOP and HYNES 1969b for references). The activity patterns of terrestrial mites that expose them to the river are unknown, but would appear to occur at night from the data here.

4.2.2. Bilateral insect trap catches

The traps were operated through April, May and June 1970, but because of various problems, reliable results are available only for the periods 25 April to 3 May and 6–20 June, covering part of the short 'wet' season and some of the following 'dry' period. Since virtually nothing is known of flight patterns of Malayan aquatic fauna, this is probably as representative a period as any other. Heavy rain which fell on 27 and 30 April resulted in low catches, but gentle rain seemed to stimulate flight, notably of the Trichoptera.

Horizontal air movement, mostly deflection downwards by the canopy or as a result of convectional turbulence was usually very slight, 10–300 m/10–14 h period, but just prior to rainstorms, winds of short duration, but considerable velocity, frequently occurred. During one such storm in early afternoon on 30 April, more than 25,000 m of upstream movement was registered in about 70 minutes. Almost invariably air movements that were recordable were upstream during the day and downstream at night. The constant downstream wind at water level was too light to measure.

The total catch (Table 88) was small, only 931 specimens of the three Orders in 24 days of trapping, reflecting in part the low benthic density and the probable year-round emergence of most species. This would severly limit the adult population abroad at any particular time and the mass emergence and swarming so common in seasonal rivers was a rare occurrence on the Gombak. Only on a single occasion was a mass flight experienced. On 20 August 1969 during a torrential storm between 21:00 and 21:30 h, thousands of adults of *Caenis* sp. A, about 70% ovigerous females, were attracted to a light trap set up adjacent to a long riffle at Station IV. There was no concerted direction to the flight and no wind at the time of collection, although prior to the storm the normal strong convectional winds were experienced. The very low total catch for Plecoptera suggests either that the traps were not in a suitable location for trapping these adults or more likely, that flight by them was very limited. The only adults collected by hand-netting (one *Neoperla* and a number of *Protonemura*) were from the root-masses right at the water's edge. If dispersal is normally by walking (cf. THOMAS 1966, 1969a, HYNES 1970) or by short flights close to the banks, as is likely, the traps in mid-stream would under-represent the movements of this Order.

Night catches of Ephemeroptera and Trichoptera were slightly larger than day ones, perhaps because of cooler air temperatures, but more probably because the diurnal fliers were decimated by the abundant insectivorous bird fauna that constantly patrolled the river. Bats exploited this food resource at night. Six times as many Plecoptera were taken at night but the total numbers were too small to be reliable. The overall night/day catch ratio was 1.8. Downstream flight dominated both day and night by a factor of more than 6.5 for the total catch. At night, down-valley movement with the normal air flow was nearly $10 \times$ the upstream movement (see BROWN 1970), but in the daytime, against the slight prevailing up-valley wind, this was reduced to about four times. The velocity of these winds down at water level was unmeasurable, even when air currents above the tree canopy were quite strong.

The predominant direction of flight was definitely downstream in the Gombak in contrast to a number of earlier studies (cited in BISHOP and HYNES 1969b) and conflicting with ELLIOTT (1967a) who found that flight patterns were generally with the prevailing wind. Here, there was flight against the wind, particularly by the Trichoptera as reported by ROOS (1957), but in a downstream direction! The finding by ROOS that the up-valley fliers were dominated by mature, ovigerous females was not corroborated by the data here. The various sexual categories exhibited no deviating pattern of flight from that of the whole collection.

Table 88. Summary of catches of adult Ephemeroptera, Trichoptera and Plecoptera in upstream (U) and downstream (D) traps. (♂ male, ♀ immature female, ●♀ ovigerous female, ○ spent female)

	EPHEMEROPTERA				TRICHOPTERA				PLECOPTERA				TOTAL			
	Day		Night		Day		Night		Day		Night		Day		Night	
	U	D	U	D	U	D	U	D	U	D	U	D	U	D	U	D
25-IV – 3-V-70																
♂♂	1	1	1	4	1	25	5	74	—	—	4	—	2	26	10	78
♀♀	—	1	1	4	—	3	1	12	—	—	3	13	—	4	5	29
●♀	—	—	—	2	1	15	8	94	—	—	—	4	1	15	8	100
○○	2	8	1	15	1	11	1	14	—	—	1	6	3	19	3	35
6-VI – 20-VI-70																
♂♂	—	1	3	8	22	75	10	89	1	2	1	2	23	78	14	99
♀♀	1	2	—	3	3	5	2	20	1	3	1	5	5	10	3	28
●♀	—	4	2	1	22	57	5	121	—	—	—	1	22	61	7	123
○○	4	11	4	9	6	37	3	45	—	—	—	1	10	48	7	55
Total	8	28	12	46	56	228	35	469	2	5	10	32	66	261	57	547
Total catch	94				788				49				931			

4.3. Discussion

The characteristics of drift and its significance both to the ecology of the invertebrates involved and to the stream community as a whole have been described at a preliminary level and first hypotheses made, but a comprehensive assessment of the phenomenon must be delayed until further definitive data are available on compensating mechanisms and on the numerous exceptional cases that defy the various conjectured schemes. The subject in its several aspects was comprehensibly reviewed in earlier work and by several other authors so that repetition here is unnecessary except where a postulated mechanism is applicable to the present results.

Two limitations must be recognized for the data as presented. First, the mesh size of the nets used, 165 μ, puts a restriction on the fauna caught, undoubtedly of little importance in biomass measurement, but perhaps significant in respect to the numbers of early instars in the drift and for such groups as the Ceratopogonidae and Naididae. Second, as in all studies to date, the proportional loss of drift to fish or other predators prior to collection was not assessable, but cannot be ignored in considering the composition and size distribution of the drift in view of the known selective feeding habits of both species and age classes of some fish.

As far as could be determined by empirical observation, the populations in the benthos and those in the drift were the same. A number of studies (MÜLLER 1966, ANDERSON 1967, ELLIOTT 1967a, b, LEHMANN 1967, ULFSTRAND 1968a, SCHWARZ 1970) have shown that propensity to drift is greater in the later instars in a number of species, i.e. that the larger nymphs are over-represented in the drift with respect to their proportion in the bottom. However, ANDERSON (op. cit.) and BISHOP and HYNES (1969b) have reported drift exclusively in the earliest instars of some taxa, so the above generalization does not hold. For the fauna of the Gombak, a quantitative size-group comparison between benthic and drifting populations was not practical. The problems of specific identification of nymphs, necessary for such as exercise, were insurmountable and the fauna routinely sampled may have been deficient in numbers of the late pre-emergent instars if these migrated into the bank-root habitats as is reported for some species (LILLEHAMMER 1966, ULFSTRAND 1967, BISHOP and HYNES 1969a). Annual periodicities in drift composition commonly observed in temperate rivers probably coincide with particular stages in life history of the species involved (MÜLLER 1966); ELLIOTT (1967a, 1968a) has postulated that drifting is a function of increased activity during maximum growth periods which often occur during the later instars. Such a cyclical phenomenon would be difficult to detect in a non-seasonal fauna in which there is no large concentration of any size-group at any time. Increased behavioural drift as a response to both inter- and intra-specific competition for physical resources of food or emergence/pupation sites, a density-dependent factor to be discussed in more detail below, would not be as important in the tropical context where the density of any particular cohort is low within a larger heterogeneously-sized population of the species.

The nocturnally periodic pattern exhibited by 'non-catastrophic' drift (WATERS 1965) in the Gombak was similar to that described from a wide selection of temperate lotic waters (see summary in BISHOP and HYNES 1969b). This periodicity was governed principally by light as a phase-setting mechanism (see ELLIOTT 1968b, BISHOP 1969, CHASTON 1969a, b for reviews) but endogenous activity rhythms in the fauna undoubtedly played a role not directly apparent from the results. Magnitude of the drift at any time was governed by discharge-velocity conditions with minimal effects from temperature variations.

Most of the taxa that recurred in the drift exhibited negative photo-taxis which with positive thigmo- and skototaxis maintains them quiescent in low density areas, often attached to the undersides of stones or leaves during the hours when light intensity reaching the substrate exceeds the threshold level. This intensity is of the order of one lux for many amphi-biotic species, but the depressant energy level is probably unique for each species and perhaps varies depending on location or life-stage. As indicated earlier, there were no apparent depressant effects of moonlight on the drift rate as reported in some open-surfaced streams by WATERS (1962a), ANDERSON (1966) and BISHOP and HYNES (1969b), but not in others (ELLIOTT and MINSHALL 1968). In the canopied Gombak, the intensity reaching the stream surface was probably less than 1% of the already very marginal radiation energy available in moonlight, so this observation was predictable.

Under the forest canopy a conspicuous member of the fauna, but notably absent from the drift, *Pseudocloeon*? sp.C, appeared to be either light-indifferent or to have a very high inactivation threshold. Individuals of this species could always be found on the wet, subaerial surfaces of rocks during the daytime, although not in the densities seen at night. A number of invertebrate species are known to be phototactically positive (BISHOP 1969, SCHMIDT 1969, WATERS 1969b, ELLIOTT 1970a, b) and most day-active drifters are probably indifferent to light effects. In these species temperature has been shown to control activity and drift (cf. WATERS 1968). When exogenous light control is removed with rapidly decreasing intensities at sunset, the endogenous activity rhythm found in many stream insects, but in only some Crustacea (THOMAS 1969b, HUGHES 1970), and almost invariably suppressed by light (CHASTON 1968), becomes the controlling factor in drift. A number of activity peaks of various frequencies have been reported by ELLIOTT (1965a, 1969a, b), ASCHOFF (1966), MÜLLER (1966) and others; these outbursts are responsible for the initiation of high drift rates in excess of the moderate, continuous, passive drift that results from 'accidents' and mechanical disruption of the substrate. The post-sunset maxima follow locomotor and feeding activities that take the invertebrates out of their sheltered niches into areas where food is more available, but current forces and propensity to drift are greater. Entrainment in the drift is normally passive although under certain conditions behavioural action may initiate take-off into the water column, particularly as a response to contact between individuals (McLay 1968). For example, *Hydropsyche* larvae may exhibit territoriality around their net retreats, enforced by stridulation; with growth and enlargement of the catching area of these nets, the frequency of contact with other individuals increases (Elliott 1968a). Such interaction, clearly density-dependent (Glass and Bovbjerg 1969), very likely leads first to dispersal into marginal habitats and then to drift of the displaced invertebrates. Similar interactions occur periodically in particular micro-habitats where locally crowded conditions exist, but the contentious issue of whether high drift rate *per se* is a density-dependent phenomenon as maintained by Waters (1964) and Dimond (1967) probably cannot be resolved, as drift rate must be a reflection of immediate local conditions.

Elliott (1969b), by sampling at 30 min intervals, obtained a hierarchy of short, periodic activity peaks which would be obscured by the longer filtering periods used in this study. The commonly reported bi- and sometimes trigeminus patterns during long nights, with two or three drift maxima subsequent to the initial peak, may reflect endogenous activity rhythms or, as suggested by Bishop (1969) but so far unproven, may represent renewed feeding activity after partial or complete digestion of earlier 'meals'.

The chief factor governing the magnitude of the drift, apart from the numerical and spatial availability of the fauna, is discharge in the river. Under severe spate conditions, when the bottom structure was ruptured, large-scale drift occurred as described by Maitland (1966), Anderson and Lehmkuhl (1968) and others; this could not be reliably assessed with the small nets available because of the forest debris carried by the river in flood. Numerous reports of total drift increasing with a rise in stage and current velocity have been made (e.g. Elliott 1967a, Ulfstrand 1968a); in the present study some linear correlation between drift rate and total flow was found (Fig. 62). No relationship could be expected for the terrestrial component which is independent of river influences, so only the aquatics were considered. The derived equation for rate versus discharge volume at Station I, $y = 0.89 x + 0.21$ with a significant $(P < 0.01)$ r value of 0.70, did not include the two aberrant readings obtained after exceptional antecedent conditions. The abnormally high attrition rate at low discharge was recorded on a rising stage on 17 April 1969 following an extended period of low water during which the fauna became dense and exposed to the currents in vulnerable niches. The other point, registering a low drift rate at high discharge came on 30 October 1969 after a prolonged high water period when the

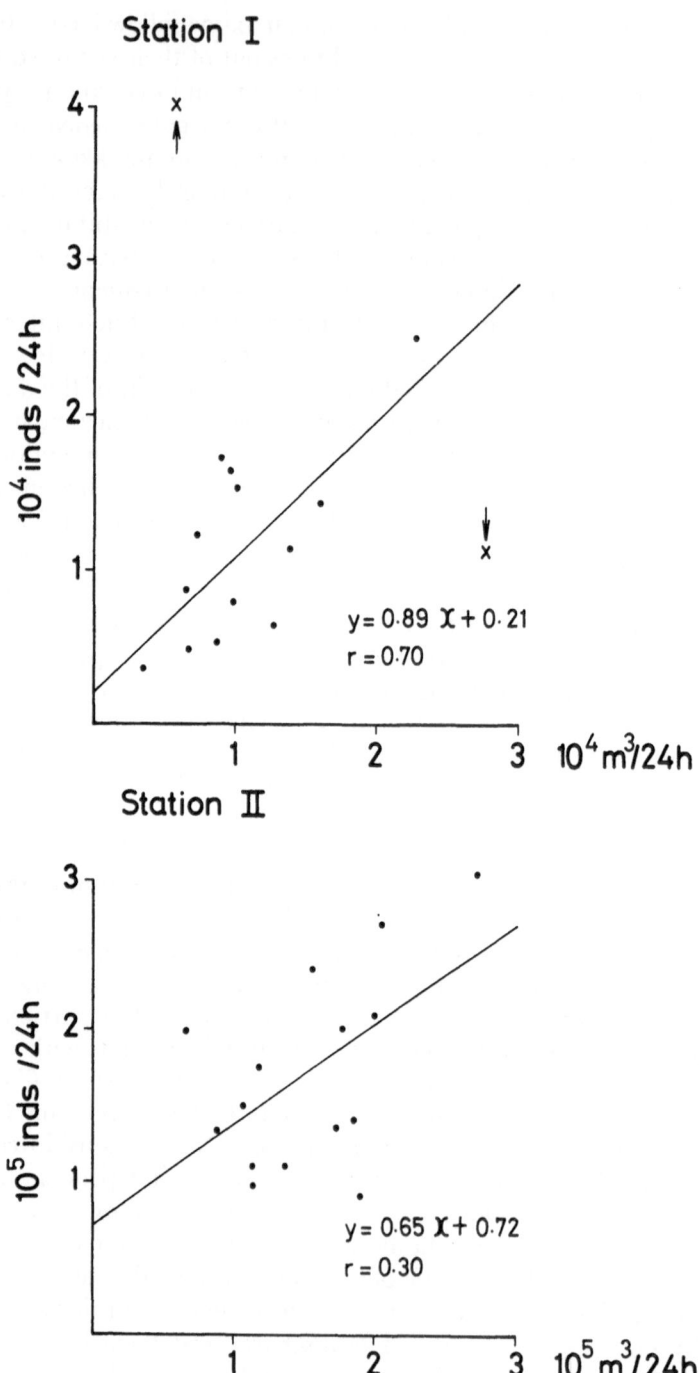

Fig. 62. Relationship between drift rate (number of individuals past a point in 24 h) and discharge (m³/24 h); x–point not included in equation (see text).

population had become numerically depleted or had achieved current-resistant niches. At Station II with more variable microhabitats and a much wider area for drift recruitment, the correlation coefficient was not significant (r = 0.30), but some direct relationship between the quantity of drift and the amount of water passing the transect was still apparent (y = 0.65x + 0.72). ELLIOTT (1967a, b) found that drift density did not increase parallel to increments in current velocity, not even at abnormally high discharges. A similar result was obtained for the Gombak drift and was predictable if, as discussed previously and in BISHOP and HYNES (1969b), direct current effects at the individual invertebrate microniche level are minimal. Plots of density versus total discharge at both stations showed an almost constant number of aquatic drifters per cu. m (Fig. 63), irrespective of stage conditions in the river. The two exceptional values were both at low discharge when behavioural changes or severe spatial or feeding interactions between individuals may have resulted in unusually high drift. HUGHES (1966b, c), ELLIOTT (1967a) and MINSHALL and WINGER (1968) have described, under reduced flow conditions, breakdown of the normal photo- and thigmotaxis and active movements of insects into the water column, perhaps to seek respiratory relief or to escape continual contact with other individuals.

The rapid changes in velocity that accompanied the transient freshets in the river had different effects on drift rate depending on their time of occurrence. Rain most commonly fell in the mid- to late afternoon, but unless this was after 18 :00 h, coinciding with the release of phototactic control, or was very heavy, little enhancement of the drift was seen, for example in the afternoon of 23 May 1969. At peak flows or on falling stage, the drift rate decreased considerably as on 30 October 1969. This agrees with observations of McLAY (1968) and WENINGER (1968), who hypothesized that the fauna sought shelter under such conditions, and with the conclusions reached earlier with respect to deep colonization of the substrates, i.e. that as the bottom becomes unclogged, some of the fauna moves into the interstitial spaces and becomes inaccessible to erosional forces. Moreover, the amount of detritus carried by the current, a reflection of wash-out of bottom and root habitats and earlier shown to vary with flow, showed no consistent correlation with drift rate. This indicated that erosion of the substrates may not necessarily result in loss of rheophilous fauna to the drift.

The other factor that could conceivably affect the magnitude of the nocturnal drift is temperature as reported by MÜLLER (1963a, b, c, 1966), PEARSON and FRANKLIN (1968) and WATERS (1968). However, in most investigations (see ELLIOTT 1967a, BISHOP 1969, BISHOP and HYNES 1969b, SCHWARZ 1970, *inter alia*), temperature has been shown to have little or no effect on drift rate, except of day-active species where activity is temperature rather than light controlled. Under normal conditions, diel temperature variations in the forested river were negligible,

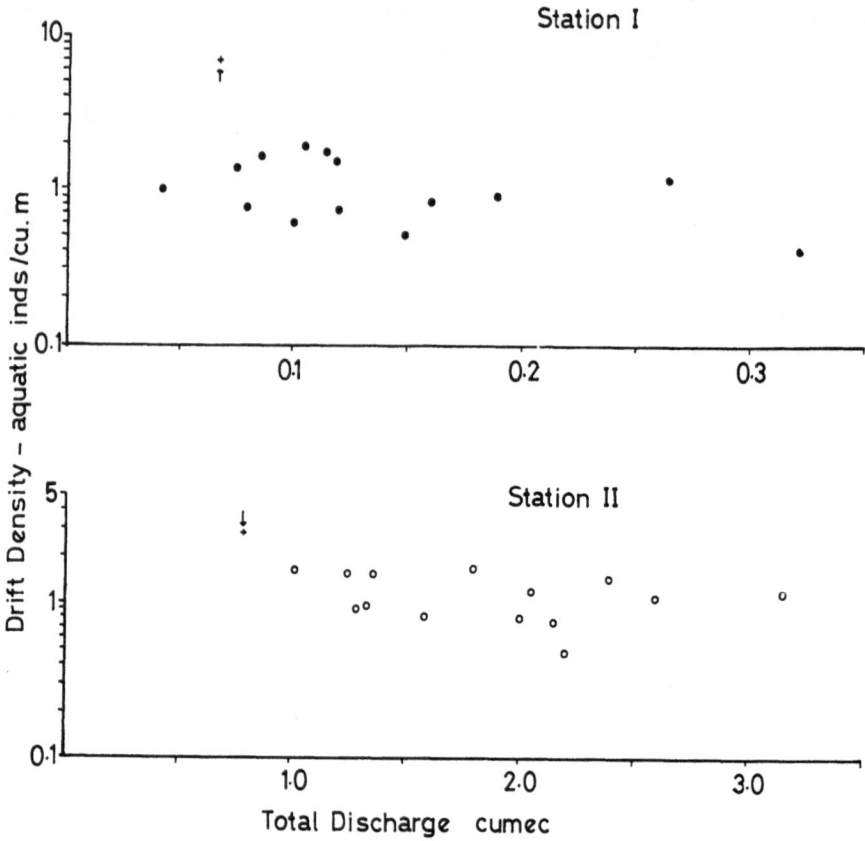

Fig. 63. Relationship between drift density (aquatic individuals/m³) and discharge (cumec).

but with heavy rain, water temperature changes of about 1.5C° were recorded and may have had a small effect on drift rate although this is doubtful.

The distance of daily drift and the proportion of the population participating in the downstream movement must be known before any plausible comparison between drift and upstream movement can be made or net benthic population losses to the drift assessed. The distance travelled before reattachment is a function both of the physical conditions of the river (current velocity, presence of vegetation and roughness of the channel) and of innate qualities of the drifters themselves (physical configuration and swimming ability, capacity to reattach to the substrate on contact and most important, the behavioural reaction to being free in the current). The values obtained by different authors indicate the variability that results from different combinations of these factors. CARLSSON (1967) indicated that *Simulium* tagged with ³²P moved hundreds of metres; WATERS (1965) concluded that *Gammarus* and *Baetis* drift was generated from about 50 m of upstream river; Elliott (1967a) found that the area 10 m upstream of his nets probably supplied most of the drift, a figure close to that deduced by McLay (1970) from a sophisticated model of the problem. However, Madsen (1966) found that *Baetis* sp. and *Ephemerella ignita* attached after 2–3 and 5–6 sec in 30 cm/sec current and Schwarz (1970) reported that for the active stonefly *Diura bicaudata* the mean drifting distance was only 50–100 cm with an extreme distance of 5 m in exceptional current conditions. It is obvious, therefore, that the drift captured at any location is a function of a complex series of upstream interactions between the drifting fauna, particularly the rheophilous species which may react to the current stimulus, and the physical environment, about which no generalizations can be made. For example, the two species of *Hexagenia* investigated by Madsen (1968b) reacted in different ways to suspension in the water column; one, *H. fuscogrisea* swam and searched for dead-water, the other *H. sulphurea*, drifted passively and reattached to the bottom at the first thigmotactic stimulus. In the Gombak, with an alternating bottom pattern of short pool-riffle sections, turbulent water and rough substrate, reattachment would appear relatively easy in spite of considerable current so that the mean distance drifted was probably short. On long stable riffle sections supporting a uniform fauna, the ability to reattach from the drift might become density-dependent if all suitable feeding and perching niches were occupied. Drift distance under such conditions would be much longer than in a riffle-pool system where some of the drift is removed or settles across slow-water areas (cf. Waters 1962b, Bailey 1966).

These difficulties point up the futility of estimating the loss to the standing stock by drift on a percentage areal basis. Headwater areas of rivers with high drift rate have not been shown to be significantly depopulated even by real loss/day rates estimated at about 0.4% by Elliott (1967a) and 1–2% by Ulfstrand (1968a). These rates, if constantly applied in a simplistic model, would deplete a population by roughly 20 and 40–60% over an average generation time of 50 days (assuming no recruitment during the interval) as calculated by Ulfstrand (1968a). These figures are unrealistic without an assessment of compensating upstream movements. In the Gombak where continuous regeneration occurs, measurement of net loss would be extremely difficult.

The proportion of the benthic population taking part in the drift at any instant was small. Values calculated on a per sq.m basis using the bottom collections nearest the drift date and the relationship of Elliott (1965b)

$$P = \frac{xD.100}{X - xD}$$

where

$$X = \text{benthos no./sq.m of surface}$$

339

Table 89. Percentage of the bottom fauna in the drift at any instant

	Station I	Station II
5–X–68	0.0093	0.0323
30–X–68	0.0038	0.0230
28–XI–68	0.0069	0.0082
1–I–69	0.0088	0.0107
24–I–69	0.0080	0.0073
23–II–69	0.0066	0.0197
19–III–69	0.0075	0.0201
17–IV–69	0.0405	0.0295
23–V–69	0.0118	0.0152
12–VI–69	0.0230	0.0269
10–VII–69	0.0438	0.0122
7–VIII–69	0.0047	0.0150
3–IX–69	0.0102	0.0185
2–X–69	0.0084	0.0220
30–X–69	0.0095	0.0151
Mean	0.0135	0.0184

$$x = \text{no./cu.m drift}$$
$$D = \text{mean depth}$$

were as listed in Table 89. These are comparable to those found by
ELLIOTT (1965b) and ULFSTRAND (1968a) for Scandinavian streams
and about $\times 50$ higher than those found by ELLIOTT (1967a) and
BISHOP and HYNES (1969b).

However, the P values obtained here were calculated on the mean
benthos from riffle and pool areas only and could not take into account
the faunal density in the root and boulder habitats. Many of the drifters,
particularly the Leptophlebiidae and *Baetis* spp., were found with higher
frequency in these latter areas. If a true standing stock could be estimated,
including these elements, the proportion in the drift would be even lower.
These negligible percentages and the lack of correlation between drift
rate and benthic standing stock, also shown by the data from WATERS
(1964, 1966), MÜLLER (1966), ELLIOTT (1967a) and BISHOP and
HYNES (1969b) (PEARSON and FRANKLIN 1968, however, were able to
demonstrate a dependence), contraindicate the use of drift as an indirect
measure of production rate as proposed by WATERS (1961b, 1962b, 1966).
Behavioural drift, particularly from localized habitats, may be a function
of excess production rate over carrying-capacity, but in a river in which
niche stability is ephemeral, saturation probably occurs only rarely.
DIMOND's (1967) work, showing drift to increase only after full restora-
tion of the bottom fauna to control levels after severe insecticide treatment,
supports this theory, but his physically stable streams were a special case.
A final important point is that in the Gombak, because oviposition tends

to be semi-continuous, asynchronous and almost independent of season, cohorts are numerically small and overlapping, so that interactive effects are temporally minimized. This is in contrast to temperate locations in which the greater part of a generation often develops almost synchronously, placing great demands on the spatial and nutritional capacity of the habitat. Extended emergence and delayed hatching may partially alleviate this, but competition and density effects (including drift losses) are severe, more so because in these seasonal rivers many diverse faunal elements have parallel life histories increasing interspecific interactions as well.

The adult flight data does not support the 'colonization cycle' concept suggested by MÜLLER (1954a, b) and reinforced by the results of a number of workers reviewed by BISHOP and HYNES (1969a). This concept postulates that adults fly upstream to oviposit, some proportion of the eggs and immatures drift downstream, the latter in response to population pressure and as a dispersal mechanism to ensure full utilization of habitats and adequate *lebensraum* (cf. BOVBJERG 1959, 1964) and the adults on emergence again move upstream. Such a system obviously cannot exist for Crustacea, mites or other wholly aquatic groups and probably operates only for a limited number of amphibiotic species in certain localities. The extent of upstream migration of immature forms is insufficiently known, except for the gammarid amphipods, to be able to compare rates with drift losses but in some cases (e.g. SCHWARZ 1970) in small streams, such movements may fully or even over-compensate for drift.

There was no rationale for the observed downstream flight pattern. In the limited area where trapping was carried out, perhaps only a small part of what is a scarce adult population at any time disperses, most individuals completing their life cycle in the immediate area of emergence. With the minor proportion of the benthos found to be involved in the drift, for reasons given above, there may be no need for a cyclical re-colonization mechanism.

Behavioural drift, as opposed to 'accidental', is obviously an integral part of the population ecology of the lotic species to which it is peculiar. Besides providing a ready dispersal mechanism to colonize transient habitats with their briefly available resources and to repopulate areas denuded by pollution, flood or desiccation, it may ensure optimal production rates if the excess population from an area drifts and is cropped by the fish. Drift increases the availability of the benthos to fish predation by transporting the invertebrates from their restricted microhabitats, often on shallow riffles, into pool areas where a wide spectrum of fish species can utilize them and where visual feeding is easier. The cropping of drifting foods by fish is an obvious but little quantified relationship as not much is known of the diel feeding patterns of stream and river fish (see JENKINS 1969, TUSA 1969, JENKINS et al. 1970, TANAKA 1970).

ELLIOTT (1967c, 1970b), CHASTON (1968b), BISHOP and HYNES (1969b) and WATERS (1969b) have reviewed the literature on drift feeding, which generally records a considerable utilization, but the degree of exploitation varies greatly both between species and between age groups. As will be seen in the section on fish, there are many omnivorous, opportunist feeders as well as several specialized insectivores in the Gombak which undoubtedly utilize the drift fully. However, it is unlikely that any feedback mechanism as proposed by STRAŠKRABA (1965) and CHAPMAN (1966) would indirectly regulate aquatic rates by controlling the density of the benthos by extensive foraging, as drift in the Gombak is apparently independent of benthic abundance.

V. THE VERTEBRATE FAUNA

1. Pisces

The ichthyofauna of the Indo-Malayan region has been worked extensively taxonomically and, although many nomenclatural problems exist, particularly in the Cyprinidae, and undoubtedly a number of species are as yet undescribed, most fish can be adequately defined. HERRE (1940) gave a comprehensive bibliography to that date, ALFRED (1964, 1966) reviewed most of the recent literature in his enumeration of the species found on Penang and Singapore Islands, and MIZUNO and MORI (1970) compiled a reference list for the Southeast Asian region. The chorological patterns of the freshwater fish of the Malayan region are less well known because of irregular coverage by collectors and incompleteness of records even in the well-worked areas. The more common species with wide distribution have been neglected and often unrecorded in the pursuit of the rarer, systematically more interesting species or the faunal associations of special, limited habitats. The comprehensive surveys of freshwaters of Thailand (SMITH 1945) and North Borneo (INGER and CHIN 1962) have no counterpart for Malaya whose fauna, although sharing numerous elements and habitats, is in many respects different from those areas peripheral to the central Sundanian region (JOHNSON 1960, 1964). JOHNSON (1967b) summarized most of the records for specific habitats and proposed two large distribution areas: a southern crescent-shaped zone extending in a thin strip up both east and west coasts, characterized by low altitude, nutrient-poor, often acid or blackwater streams, and a northern area of harder, richer waters, occupying the upper third of the West Malaysia political region. This scheme leaves a large part of central Malaya unclassified. North-south distribution, rather than an east-west separation by the main mountain range, was justified by Johnson on meteorological as well as chemical criteria, as the northern part of the peninsula is subject to definite, and regularly severe, wet and dry seasons. The Gombak watershed is out of, but adjacent to, the coastal strip designated as southern, and as will be seen has a number of affinities with both northern and southern faunas.

In this survey, an attempt was made to assess the longitudinal succession of species in the Gombak and its tributaries and to relate this to available habitats. The only published record of sequential sampling in Southeast Asian rivers or streams is data given by INGER and CHIN (1962) for an unnamed stream at Deramakot, North Borneo, which differed physically and faunistically from the Gombak and cannot really be compared with it. JOHNSON (1967b) discussed generally defined habitat types and their species associations, but treated these in isolation and did not give comprehensive records for any individual stream. The Gombak is particularly suited for such a study as it encompasses a mozaic of water types in a short distance. First estimates of the standing stock and yield to fishing by the riparian population were also made. This kind of data has not been available for any small rivers in this area to date.

Identifications were checked by Mr. E. ALFRED, National Museum, Singapore, whose assistance is gratefully acknowledged.

1.1. METHODS OF ASSESSING THE ICHTHYOFAUNA

1.1.1. Electrofishing

Fish were collected, mainly in June and December 1969, at the main river sampling stations and from tributaries considered representative of the zones. Where feasible, depending largely on accessibility, such collections from the tributaries were made away from their confluence with the Gombak to minimize occurrence of stray records. Quantitative sampling was attempted wherever possible using a 500 volt a.c. electrofisher built by the Department of Fisheries, Penang, or a Cybertronic Mk. XII pulsed d.c. shocker (Marine Electrics, Killybegs, Eire). Where stream width was less than 10 m, in 2°, 3° and some 4° streams, stop nets (5 mm seine mesh) were set above and below a selected reach. Sections 30–60 m long with both riffle and pool areas were fished moving upstream; hand nets and scoops were used to collect the stunned fish. Three separate passes, 30 minutes apart, were usually made. In wider reaches, only the deepest channels were blocked and two or three large hand nets were used in conjunction with the electrode to minimize losses. Fish were killed immediately in 25% formalin and subsequently transferred to 5% formalin for storage. The use of respiratory or other poisons (rotenone, 2–4–D or cresol) was undesirable in the upper Forest Reserve reaches because of its severe effects on the insect communities (see COOK and MOORE 1969) and was impermissible in the lower inhabited zone. These are undoubtedly the most efficient, and perhaps the only quantitative, methods of obtaining representative catches in these waters and rotenone was used to effect by INGER and CHIN in the North Borneo surveys. The nature of the bottom, large boulders and snags, and extensive areas of root and bank habitat made active netting impossible as a general collecting technique; this was effective only in the sandy-bottomed pool areas.

Limitations of the technique: The electrofishing methods had severe limitations but were the best available. The efficiency of the fishers was reduced by the low ionic composition of the water. ALABASTER and HARTLEY (1962) set the lower limit of effectiveness at a specific conductance of 70 μmho and the manufacturers of the Cybertronic apparatus gave its useful range as 50–1000 μmho, outside that found for the Gombak stations. The effective shocking distance was short, 0.5–1.5 m only, so that stealth and judicious control of the current pulse was necessary to catch the active species. With this set of conditions, considerable selectivity for both species and size was inevitable. A current with characteristics effective against the small cyprinids, for example, often had little effect on the siluroids (see LAGLER in RICKER 1968). To minimize this problem, pulse frequency was varied between successive sweeps. MEYER-WAARDEN et al. (1960), BOCCARDY and COOPER (1963) and LIBOSVÁRSKY (1966, 1967), among others, have fully discussed the utility of the technique in population assessment and its general limitations.

In the Gombak, other factors also affected the efficiency of catch. In the upper zones, rapid current velocities and the roughness of the channel made recovery of stunned fish difficult, while turbidity had similar effects in the lowland reaches. The characteristics of the fish, drab coloration and reduced swim-bladder in the torrent Bagaridae, for example, and habits of living in dense leaf and root masses (e.g. *Betta*) or in holes behind rocks added to the problem. LARIMORE (1961), in discussing electrofishing success found an overall recovery of only 51% of numbers (54% of biomass). This average figure applied directly to Cyprinidae, but for catfish (Ictaluridae), the rate was only 10% while for other bottom habitat families (e.g. Catastomatidae) approximately 70% were captured. In this study, in the small upper tributaries where stop nets could be applied effectively, the efficiency after three runs was probably between 90 and 100%. Some losses at the main river stations in the forest resulted from fast currents and frequent rupture of the downstream nets by the large volumes of leaves and debris kicked up by the fishing crew. In lowland areas where complete blocking of the river was impractical, some losses over shallows occurred as the shocker approached, although the fish were generally reluctant to enter these areas. At the lower stations with deeper water and a wide channel, evasion of the electric field was not difficult.

Irrespective of these technical problems, the assessment of standing stock, unless carried out over long stretches of river for extended periods, will remain only a rough estimate in these waters as distribution of the fish in any river section, no matter how uniform the habitat, is discontinuous, particularly in time. For the bottom-dwelling forms with definite home areas, estimates are probably realistic, but for the Cyprinidae which constitute the bulk of the fauna, schooling, nomadic wandering in and out of a section in response to stage and chemical conditions and non-territoriality make standing stock/area data of questionable value. However, the samples obtained probably gave a reliable qualitative picture of species composition and perhaps also of diversity, if size and species selectivity were not too severe.

1.1.2. Methods of surveying the fishery

To obtain estimates of the degree of exploitation of the fish by the rural population, a census of fishermen, frequency of effort, methods of extraction, composition and extent of catch was made (see standard questionnaire, Appendix E). Almost all of the riverside kampongs upstream of Station V were contacted and surveyed and where possible complete data collected. To facilitate identification and to avoid the confusion caused by a multiplicity of vernacular names for some species, a sample specimen of each of the principal fish types, displayed in liquid, was carried and catch records discussed with respect to these types. Because of the problems inherent in interviewing fishermen, compounded by the need for an interpreter, and because continuous census could not be made, certain criteria were applied to the data obtained from the once-only interviews: 1. Most fishermen gave their fishing frequency as number of times per week; for extrapolation purposes, a year was considered to have 40 weeks to allow for generalities and for harvest and festival periods when fishing was unlikely. 2. Almost all fishing was carried out by parties of two or more so that all catch per fisherman records supplied were considered to be the total catch for such a group, i.e. two men fishing together probably both reported the same total catch, so it was counted only once. 3. Fishing efforts of boys under 12 years, sometimes considerable, were counted as half, i.e. 4 boys = 1 party. 4. The interviewees were asked to report numbers and sizes of each fish type caught during an 'average' fishing session. Where such claims, even after repeated questioning, were clearly the maximum sizes and numbers ever caught or even imaginary catches, these were disregarded and a 'probable' catch based on other individuals substituted. This is the area of maximum error, but unless detailed records of catches can be obtained in the field, there is no way of correcting the data. Observations of actual catches were made infrequently as most fishing was carried out either between 17:30 and 19:30 h or after rain when turbidity was increased, at these times access to villages was difficult. The catches that were seen usually had size and species composition similar to that reported and since cooperation from the villagers was generally good, the data may have some relevance.

1.2. THE FISH

1.2.1. Composition of the fauna

The freshwater ichthyofauna of Malaya comprises more than 250 known species at present, of which a large number, 100–150 species, have limited distribution and are considered uncommon (JOHNSON 1967b). When more extensive work is carried out, particularly on the lenitic habitats, the total will probably exceed 300. This fauna is numerically poorer than that of Brazilian rivers (1400+ spp.), the rivers of the Congo Basin (408+ spp.) or even adjacent Thailand (546 spp.) (LOWE-MCCONNELL 1969), although this last total includes a number of taxa found almost exclusively in lakes and impoundments, habitats that occur infrequently in the other areas. However, on a comparable areal basis, the Malayan diversity is remarkable. The problems of speciation in tropical American and African lotic waters have been summarized by LOWE-MCCONNELL (op. cit.) and much of the wide diversity found can be attributed to the vast extent and variation of habitat available. Geographical speciation in physically or chemically isolated sections of rivers has occurred with different frequency in various subfamilies (rather than at the generic level), giving rise to stenotypic species that exploit each microhabitat within the wide range presented by the environment. In the seasonally flooded tropical rivers, temporal isolation and ethological differences have also developed enabling species to evolve with one or more narrow specializations in spite of facultatively sharing the same general food niche in many cases. In the Malayan context, potential spatial isolation has been and is much reduced by the limitations in size of the drainage basins. Extrinsic temporal isolation is also reduced except in the localized areas where an annual inundation lasting weeks or months, comparable to that of Amazonia, provides diverse, seasonally available feeding and breeding niches. Within the relatively stable hill-stream habitats, as opposed to the lowland swamp rivers, a restricted association of species with little overlap in habitat requirements, except perhaps for food type, might be expected to have developed in isolation. In the lowlands with some seasonal variation in habitat, a larger total number of species, many only marginally competitive, could build up as in the MACARTHUR (1969) model. However, in any one locality, diversity will not be great because of strong interactive segregation (*sensu* NILSSON 1967) between species. Overall, a large number of species will be irregularly distributed depending on local succession; this was evident in JOHNSON's (1967b) data. However, too little is known of endemism in the ancient drainage basin systems to comment further.

Diversity in most Malayan lotic habitats is likely to be lower than that found in South America (e.g. 60± species from a single seining in a small brook (EIGENMANN 1912, FITTKAU 1967)) for another reason also. There is not and has not been continuing inter-drainage basin exchange of species through either stream piracy or extensive flooding, as, except in the southern crescent which does have a more diverse fauna (JOHNSON op. cit.), the major drainages have been topographically isolated since the Pleistocene. Stream capture has probably been infrequent so that species exchange and sympatric associations rarely occurred, unlike the Congo-Zambezi and Amazon-Orinoco drainages with common swampy headwaters and continuous species exchange that has led to the development of high diversity. Recent introduction (in historical times) of species from the more northern fauna of continental Asia, particularly lowland species such as *Betta splendens* and various carps, along with other exotics such as *Poecilia* and *Gambusia*, has probably constituted the main increment in genetic stock.

Published accounts of the fish fauna of specific stream systems or drainage areas are few and when given are usually acknowledged to be incomplete. In the Bornean streams surveyed by INGER and CHIN (1962), the largest species association in any one tributary system was 28 spp. and in an entire drainage basin, 58 spp.. ALFRED reported only 25 and 42 extant species from all catchments on the islands of Penang and Singapore respectively, which totals included a considerable number of secondary and vicarious freshwater forms. In the tree-country streams of the southern peninsula, considerably more diversity (109 spp.) was found by JOHNSON, but the number of species occurring in any one stream was never more than 40. From the limited detailed records available, it appears that congeners rarely occur together in the same water body, or when they do, one species is present at very low frequency. In INGER and CHIN's records, the two *Mastacembelus* spp. which were found together in low numbers and the genus *Rasbora*, which on occasion was represented by several species, always had one numerically dominant. JOHNSON's observations showed almost rigid mutual exclusion between the two most common species in each of the genera *Rasbora* and *Puntius* in the waters studied by him. Unfortunately, he gave no further details of associations in particular streams, all his records being for water types only. In the Gombak, these restrictions in species diversity were obvious. Appendix F gives a checklist of the 27 (28?) spp. from 21 genera and 12 families recorded from the Gombak River system. The Cyprinidae dominate with 30% of the species, the two families Channidae and Anabantidae of the Labyrinthici comprise a further 26% and the remainder is distributed among the other nine families. Only four genera have more than one species: there is altitudinal and/or stream size segregation between the multiple *Channa*, *Clarias* and *Mastacembelus* spp. and although the two *Glyptothorax* spp. are found in the same location, one is rare and limited to a definite zone. In comparison with other hill-streams, the Gombak fauna may be impoverished, probably as a result of human activities. Several taxa that might be expected to be present are conspicuously absent: the cyprinodontid *Aplocheilus panchax* is often found in lowland streams and rice-field habitats; loaches of either or both genera *Homaloptera* and *Noemacheilus* have been reported from several adjacent drainages; the bagrid catfish genus *Leiocassis* is commonly found in hill and lowland streams. *Cyclocheilichthys apogon*, almost omnipresent in other freshwaters in Malaya, was not seen or caught during the survey, although it has been recorded from the lower river (FURTADO 1966). It is unlikely that the whole population was eliminated by poachers using

348

tuba, a *Derris* derivative, even though this species is particularly sensitive to poison (ALFRED personal communication) and, being a large channel form, may have been destroyed in the main river with no peripheral population in the tributaries to repopulate. Another possibility is destruction of breeding areas by increased silt load, but it is not likely that all suitable sites would be lost in a watershed that has been disturbed only in its lower courses. The disappearance of this species cannot be readily explained unless the original records were erroneous. However, the specimens are no longer available and this cannot be verified.

The fauna of the Gombak has few affinities with what JOHNSON designated as the species group confined to his southern crescent: *Silurichthys hasseltii* is the only common species, and no representative of the assemblage centered on *Danio* spp., *Labiobarbus* spp. and *Rasbora borapatensis*, with a number of other species, which he considered the core of the restricted northern area fauna is present. Most of the species are part of the ubiquitous aggregation found in hill-streams and rice-lands down the west coast of Malaya from Penang to Singapore, where intensive collecting has been done. JOHNSON felt that chemical composition of the water, relatively rich in the north, poor and acid in the south, was the main difference between the zones. The Gombak, situated between the two, with chemistry of intermediate character, appears to have many of the eurytopic species and few of the specialists from either area.

1.2.2. Longitudinal succession and feeding segregation

The longitudinal distribution of species, summarized in Fig. 64, illustrates the spatial isolation of the few congeners, the influences of channel size (hence current, substrates and type of food available) and the differences in feeding habits that result in succession of species down the river.

Channa gachua and *Clarias teijsmanni* were restricted to the Upper Zone while their congeners were lowland river species. *Channa lucius* and *Channa melasoma* were found cohabiting in a 4° lowland tributary, but only one specimen of *Channa melasoma* was taken, so the relationship may have been ephemeral. *Glyptothorax major* was found ubiquitously while *G. platypogonoides*, in limited numbers, was taken from the upper torrent zone of the main river only. The other pair of species, *Mastacembelus armatus* and *M. maculatus*, appeared to be segregated by a current factor. The former, which occurred only at low frequency in the lower river, was generally found in slow-water, sandy pool-bank areas, while *M. maculatus*, a constant species throughout the river, usually favoured faster stone-riffle habitats.

Geologic factors were considered paramount in the development of longitudinal succession of stream fish by SHELFORD (1911). HUET (1959) ascribed the fish zonation found in European rivers, described by many authors and fully reviewed by ILLIES and BOTOSANEANU (1963), to such

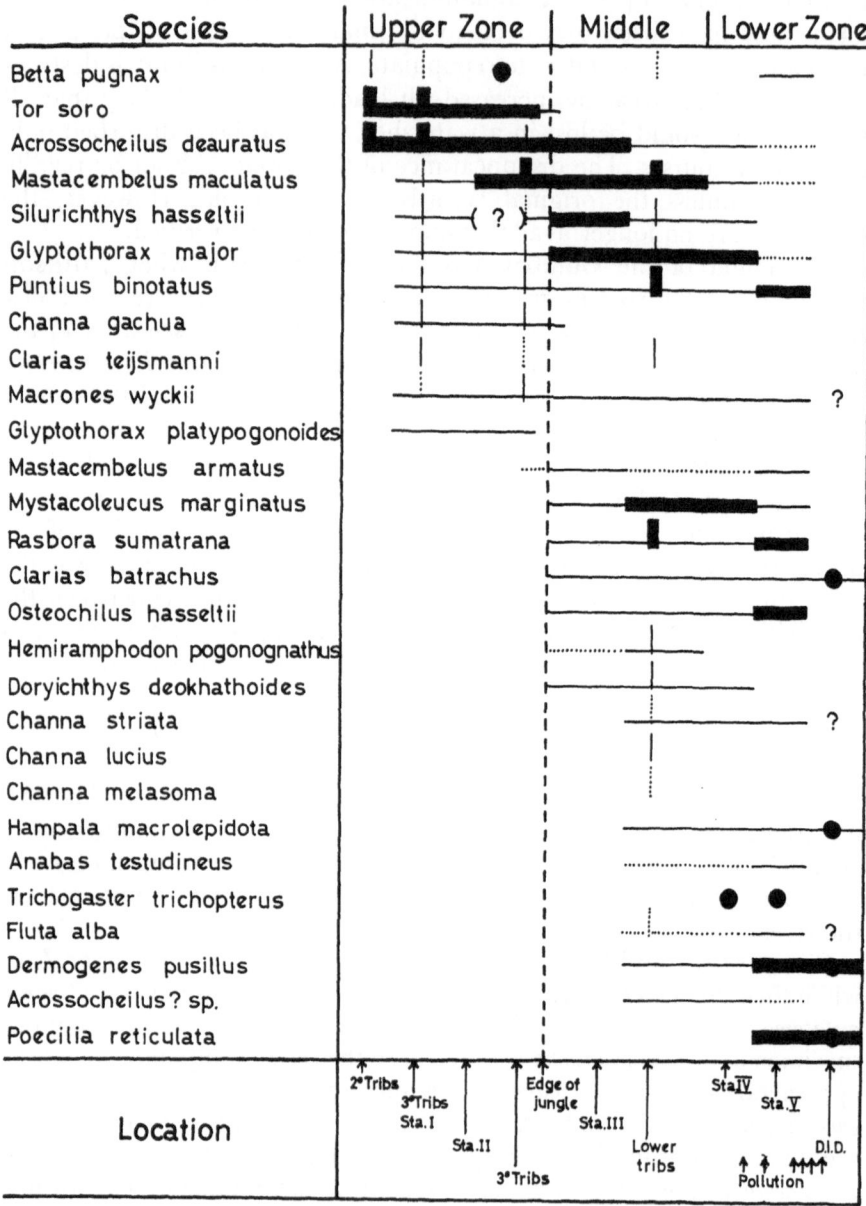

Fig. 64. Longitudinal distribution of Gombak River fish. Horizontal dotted lines denote rare occurrence; narrow line, constant species; wide bar ≥ 10% of catch; vertical symbols indicate occurrences in tributaries; solid circles, species observed but not caught, frequency unknown.

physical factors as substrate type, gradient, temperature and to the biotic conditions resulting from them. In North America, gradient and substrate, the degree of pooling and temperature have been considered most important in determining the distribution of stream fishes (see MINCKLEY 1963 and review on altitudinal distribution by VINCENT and MILLER 1969). SHELDON (1968) discussed the critical parameter of water depth which he correlated with increasing diversity downstream.

In the Gombak, succession was primarily characterized by addition of species downstream, although in several genera or feeding and position categories, replacement of one species by another was notable. Many of the relationships seen can be explained by feeding habits. Analysis of stomachs of the fish taken in this survey was of limited value as a large proportion were empty and many of the remainder were from small cyprinids which reduce ingested food to a mass of fine particles unidentifiable except as animal or vegetable. Fishing was only carried out between 10:00 and 17;00 h; thus the large number of empty stomachs was ascribed to a general lack of feeding during the day by most species. Only *Puntius*, *Mastacembelus* and *Tor* were seen actively feeding in the shaded upper river. Fishing at night would be necessary to determine diets with any accuracy, but this was impractical with the techniques available. Information on the diets has, therefore, been extracted from the papers of INGER and CHIN (1962), COSTA and FERNANDO (1967) and JOHNSON (1967b) and checked where possible by examination of available stomachs. The vertical distributions, closely related to the type of food commonly taken, are from observations made in the field and from data of the above authors. These records and the relative length of the digestive tract give a good indication of the trophic position of each species, so the following observations are probably pertinent. Trophic differences between adults and juveniles could not be considered, so the comments apply to adults only. INGER and CHIN felt that the herbivores and insect predators probably utilized one food throughout life. The situation in the omnivores and larger predators is unknown. Much more work, far beyond the scope of this preliminary study, must be carried out to confirm and extend the feeding relationships of fish in Malayan streams, especially as so many of the Cyprinidae are facultative omnivores.

In the small headwater tributaries, only three species were caught. *Betta pugnax*, which was generally found in leaf-packets, root-masses, and log drifts, feeds on terrestrial insects (mostly ants) spiders and aquatic invertebrates. *Tor soro* and *Acrossocheilus deauratus* both appeared to be general omnivores (gut: body length ratios of approximately 1.3 and 1.5 respectively) feeding on whatever is available, both in the water column and on the bottom. *Tor* generally held a mid-water position in pools, but in these shallow habitats, vertical stratification is irrelevent. (cf. KNÖPPEL 1970 for Amazonian fishes)

In the larger forest tributaries, Station I, B, D and E, two more predators were found alongside *Betta pugnax*, one, *Mastacembelus maculatus*, primarily a bottom feeder on endogenous aquatic invertebrates, and the other, *Silurichthys hasseltii*, whose stomach and gut often contained only exogenous food, mainly Hymenoptera which it collects from the surface, in competition with *Betta*. Also commonly added were two species of omnivores, *Glyptothorax major* and, in the pooled areas of some tributaries, *Puntius binotatus*. *Glyptothorax* has a feeding habitus either on riffle boulders, using its ventral adhesive disc to maintain position, or in the bottom between boulders and gravel, taking in invertebrates preferentially, but also detritus and 'aufwuchs'. INGER and CHIN classified *P. binotatus* as a secondary herbivore ingesting exogenous fruit, leaves, etc. from the water surface, but in the Gombak, this species feeds both in mid-water and off the bottom detritus. *Channa gachua* and *Clarias teijsmanni* are also omnivores, but feed primarily on endogenous insects and Decapoda. Both these species and *Macrones wyckii*, a facultative predator utilizing insects, Crustacea, frogs and small fish, are bottom dwellers.

The main river in the forest, including collections made from the B.P. weir to Location A, had a similar fauna to that of the tributaries. *Clarias teijsmanni*, *Betta pugnax* and *Silurichthys hasseltii* however, were not collected from the main stream although the last species at least is probably not restricted to the tributaries. The only specimens of *Glyptothorax platypogonoides* (2) were caught near Location A and a single *Mastacembelus armatus* was collected at the lower edge of the forest section; both were considered rare.

Station III in the centre of the foothill zone had a transition fauna. *Channa gachua* was no longer found and apparently not replaced, leaving the *Macrones* and *Clarias batrachus* (replacing the *C. teijsmanni* of the upper river) as the dominant bottom predators. The diet of the clarids has received considerable attention, particularly by African workers. THOMAS (1966) considered *C. senegalensis* euryphagous, but primarily a piscivore, feeding in mid-water. *C. gariepinus*, the common East and South African clarid, was reported as a general carnivore feeding on zooplankters, insect larvae and various exogenous foods, frogs, adult insects, etc. by GROENEWALD (1964) and MUNRO (1967). The whole genus probably feeds opportunistically. Those in the Gombak were found with plant materials, Crustacea, insects and fish in their stomachs. *Tor* was not present in this zone, apparently being limited to the forest river, but was replaced by *Mystacoleucus marginatus* with similar euryphagous feeding habits and gut: body length ratio of 1.5. *Rasbora sumatrana* was recorded in low numbers. This species was reported by JOHNSON to feed primarily on exogenous insects from a mid-water position, but some stomachs also contained seeds and leaves in agreement with INGER and CHIN's characterization of it as an omnivore. The exclusive detritivore, *Osteochilus hasseltii* with digestive tract: length ratio about 6.8 was added in this section where bottom and suspended organics became finely fragmented. *Mastacembelus armatus* was fairly common, occupying a similar food niche to *M. maculatus*, but probably not interacting with it for spatial reasons already described. *Betta* was not present, its bankside insect and spider niche, exposed and variable because of agricultural activities, being occupied by *Hemiramphodon pogonognathus* and *Doryichthys deokhathoides*, both found infrequently.

The lowland tributaries, H and J, draining mostly plantation rubber and orchards, contained a species assemblage that included elements from both the forest and lower zones. *Betta* and *Clarias teijsmanni* were again present, *Tor* and *Mystacoleucus* were absent, but *Rasbora sumatrana* was abundant and shared the mid-water stratum with *Puntius* and *Acrossocheilus*. *Fluta alba*, a bottom and bank-living mud and benthos feeder, was an addition from the lowland fauna with *Doryichthys* and *Hemiramphodon*. In Tributary H, two additional wide-spectrum predators, *Channa lucius* and *Channa melasoma*, were present; only one specimen of the latter was found however. In Tributary J, only *Channa striata* occurred, occupying the same community position. The other bottom predator, *Macrones*, was missing from both these tributaries. The absence of the Channidae, and perhaps also the *Betta pugnax*, from other areas in the Middle Zone was attributed to the lack of cover and bankside debris. In both these tributaries there were deep pools with bank overhangs and some emergent *Colocasia* and *Cyperus*.

The Lower Zone, with Station IV at the top end and Station V below various pollution sources, had the greatest diversity of 19 species. *Channa striata* became the largest bottom predator and only channid present; the two smaller species found in the tributaries were not caught in the main channel. The two anabantids, *Anabas testudineus* and *Trichogaster trichopterus*, both primarily exogenous invertebrate feeders characteristic of slower-flowing rice cultivation areas (COCHE 1967), were occasionally found in the submerged grass and reed beds at both localities. *Betta* was recorded twice from causal samples at Station V. An *Acrossocheilus* species, perhaps different from *A. deauratus* but presumed to occupy a similar niche, was added in this zone, as was the predatory *Hampala macrolepidota*. This last species fed on the bottom and bank-feeding clarid fry, on *Dermogenys pusillus* found commonly at both stations, and on *Poecilia reticulata* which was abundant only at Station V in bank and mud-flat areas feeding on the Oligochaeta and other invertebrate fauna of these enriched habitats. Several species had been eliminated at one or both stations. *Silurichthys hasseltii* and *Doryichthys deokhathoides* were not present below Station IV and *Hemiramphodon* was not found at either station, their insect feeding niche being occupied by the anabantids and guppies mentioned above.

Below Station V pollution conditions became serious enough to deplete the fish community. At the D.I.D. gauging site, only four species have been definitely recorded, *Poecilia reticulata*, *Dermogenys pusillus*,

Table 90. Probable trophic positions of Gombak fish

Dominant feeding category	Species	Feeding stratum
HERBIVORES		
1. Detritivore	*Osteochilus hasseltii*	Bottom + mid-water
OMNIVORES		
1. Herbivore dominant	*Puntius binotatus*	
	Mystacoleucus marginatus	
	Acrossocheilus deauratus	Mid-water
	Acrossocheilus sp.?	
	*Tor soro	
2. Predator dominant	**Fluta alba*	Bottom + banks
	Rasbora sumatrana	Upper + surface
	Glyptothorax major	Bottom
	G. platypogonoides	
	Anabas testudineus	All
	Trichogaster trichopterus	
	Clarias teijsmanni	Bottom
	C. batrachus	
CARNIVORES		
1. Exogenous arthropods	*Betta pugnax*	
	Silurichthys hasseltii	
	Dermogenys pusillus	Surface + upper
	Hemiramphodon pogonognathus	
	Doryichthys deokhathoides	
2. Endogenous invertebrates	*Mastacembelus maculatus*	Bottom
	M. armatus	
	Poecilia reticulata	All
3. Crustacea, fish, etc.,	*Macrones wyckii*	
	Channa gachua	
	Ch. lucius	Bottom
	Ch. melasoma	
	Ch. striata	
	Hampala macrolepidota	Mid-water

* The position given to these spp. is tentative

Clarias batrachus and *Hampala macrolepidota*, the last by Norris and Charlton (1962) but not in this survey. *Fluta alba* probably occurs in this habitat and *Channa striata* might be expected, but neither was seen. With the exception of *Hampala*, these are either surface swimmers or have accessory respiratory apparatus enabling them to live in the polluted water.

Morphological adaptations, e.g. adhesive organs (SAXENA and CHANDY 1966) or body shape and swimming proficiency (NIKOLSKI 1933), may be minor factors in determining the occurrence and distribution of species in any water type. The presence of an adhesive disc in *Glyptothorax* is useful in the upper torrent rapids, but the common *G. major* was a ubiquitous form even in the lower river. The cyprinids of the upper river, *Tor* and *Acrossocheilus*, have the torpedo shape characteristic of swift water species and the *Mystacoleucus* and especially the *Osteochilus*, found only in the lower river, are deeper-bodied, laterally compressed fish. In the alternating pool-riffle habitats characteristic of the Gombak and hill-streams in general, reduced profile and greater swimming ability are probably more effective as local niche-segregating agents than as distribution factors as in low-stage periods current pressures are much diminished. On several occasions, high profile *Puntius* that had been swept downstream by floods were seen ascending immediately after the stage dropped. The triangular shape of bottom feeders, largely predators, the ribbon-like form of the interstitial and root feeders *(Mastacembelus* and *Silurichthys)* and the obvious assets of the half-beak system *(Dermogenys* and *Hemiramphodon)* provide local advantage in a particular habitat, but are probably of little consequence in longitudinal distribution.

In Table 90, a general summary of feeding relationships is given with the apparent dominant feeding category listed for each species. Only a few of the species designated as carnivores are strictly so, many feed euryphagously but apparently prefer an invertebrate diet. The organization of the community in all localities appeared to be based on these trophic relationships, but the presence of particular species at any specific site was primarily dependent on the physical characteristics of the location. In all but the smallest channels, each category of food community found was used. *Mastacembelus* and/or *Poecilia* exploited the aquatic invertebrates; *Silurichthys* and/or *Betta* were semi-constant terrestrial insect feeders at all stations except II; *Acrossocheilus* and *Puntius* with either *Tor* (upper) or *Rasbora* (open area) were the mid-water euryphagous species, with *Glyptothorax* an omnipresent bottom-feeding member of the assemblage. *Macrones* was present in all streams of sufficient size to support a mixed predator, often with a *Channa* or *Clarias* sp. also exploiting various components of the available prey. The only dependent herbivore, *Osteochilus*, became common when adequate decomposing detritus became available. The absence of a specialized exogenous invertebrate predator in the Upper Zone river, where terrestrial insects, particularly ants, were, from the drift results, an important available food, was perhaps caused by the fast currents and disturbed surface uncharacteristic of the smooth, generally slow-flowing areas favoured by such species. The omnivores with carnivorous partiality must have utilized this resource which would have drowned and reached them rapidly in the turbulent riffles. Adding into these basic groups as the niches became large enough or variable enough to support further species were the other members of each category. This enrichment of the fauna generally occurred when channel order changed, and with it, various parameters of size, gradient, substrate and sometimes temperature. Also at these points, some established niches were lost or reduced in extent. This partially accounted for the demise of such species as *Tor*, *Channa gachua* and *Clarias teijsmanni* at the edge of the forest and the apparent downstream reduction in numbers of *Acrossocheilus*, *Mastacembelus maculatus* and *Glyptothorax* (see Table 91) as fast water and riffle areas were reduced and succeeded by slower-flowing, sand and mud-bottomed reaches.

1.2.3. Diversity

In Table 91, the frequency and biomass of the fishes caught in streams of various orders are given. The limitations and selectivity for size and/or species inherent in the fishing method must be kept in mind when looking at these data.

Several authors (KUEHNE 1962, 1966, HARREL et al. 1967, WHITESIDE & McNATT 1972) have been able to correlate fish diversity, which may be considered a valid parameter of community succession (PATTEN 1962), and stream order. The catches from all tributaries and main river section of the same order were combined and the species diversity determined using the $d = s\text{-}l/\ln N$ relationship of MARGALEF (1968). The index increased linearly ($y = 0.545 x - 0.258$, $r = 0.93$, significant at 1% level) with stream order (Fig. 65), with index values of 0.61, 1.61, 1.95, 2.60, 2.84 for 2° to 6° channels. The 4° lower tributaries were omitted from the regression analysis because at both Locations H and J the overhanging banks and roots of emergent *Colocasia* prevented capture of many of the small cyprinids that were shocked, thus resulting in a falsely elevated index of diversity. HARREL et al. found a similar increase in species diversity with stream order ($r = 0.96$) in Otter Creek, Oklahoma, and attributed it to an increase in available habitat and decrease in environmental fluctuations. In the Gombak, the lower river was physically and chemically less stable, but certainly had increased complexity of spatial and feeding niches. In their studies on Bornean streams, INGER and CHIN (1962) felt that a vertical compression of the mid-water habitat occurred towards the headwaters with loss of species from the associations found in this zone, but with only minor change in the communities of the physically little-altered bottom and surface habitats. This was partly true in the Gombak, but with increasing channel size the bottom feeders, particularly in the predator group, as well as the mid-water omnivores added species. NIKOLSKI (1937) postulated that headwater species had longer average gut lengths than the lower course and delta fish (235% : 170% of body length), corresponding to a diversification of feeding. Such a comparison has

a.

Species	UPPER ZONE						MIDDLE ZONE				LOWER ZONE	
	Trib. nr A		I + Upper 3° Tribs		II + Upper main river		III		Tribs H + J		IV + V	
	No.	%	No.	%	No.	%	No.	%	No.	%	No.	%
Rasbora sumatrana							1	2.1	9*	14.1	57	21.3
Mystacoleucus marginatus							2	4.3			47	17.6
Hampala macrolepidota											20	7.5
Puntius binotatus			3	2.4	1	1.7	1	2.1	13*	20.3	35	13.1
Osteochilus hasseltii							2	4.3			42	15.7
Tor soro	13	50.0	30	23.6	24	40.0	13	27.7	3	4.7	6	2.2
Acrossocheilus deauratus + sp.	10	38.5	51	40.2	25	41.7	5	10.6	2	3.1	2	0.7
Silurichthys hasseltii			3	2.4			3	6.4	3	4.7	5	1.9
Clarias batrachus									2	3.1		
Clarias teijsmanni			5	3.9								
Macrones wyckii			5	3.9	1	1.7	1	2.1			4	1.5
Glyptothorax platypogonoides			7	5.5	2	3.3						
Glyptothorax major			6	4.7	3	5.0	12	25.5	4	6.3	14	5.2
Channa gachua					1	1.7			1	1.6		
Channa melasoma									1	1.6		
Channa striata									3	4.7	5	1.9
Channa lucius											3	1.1
Anabas testudineus	3	11.5										
Betta pugnax			1	0.8					2	3.1	**	
Trichogaster trichopterus					1	1.7					4	1.5
Mastacembelus armatus			16	12.6	2	3.3	6	12.8	13	20.3	3	1.1
Mastacembelus maculatus									3	4.7	10	3.7
Fluta alba											10 (80)	3.7
Dermogenys pusillus							1	2.1	2	3.1		
Hemiramphodon pogonognathus									3	4.7	(15)	
Dorichthys deokhathoides											(200)	
Poecilia reticulata												

* Small Cyprinidae difficult to recover at H because of vegetation; ** Present in side channels and paddy fields–not seen in river; ( )
Estimated numbers not caught–not used in percentage calculation.

Table 91. continued, (b) biomass and (c) diversity of fishes caught in each zone.

b.

Species	Trib. nr A g	%	I + Upper 3° Tribs g	%	II + Upper main river g	%	III g	%	Tribs H + J g	%	IV + V g	%
R. sumatrana	61.2	46.2					4.8	0.6	18.9	1.5	312.9	5.2
M. marginatus	58.5	44.1					40.5	4.7			1005.0	16.7
H. macrolepidota											352.1	5.8
P. binotatus			75.0	3.1	18.6	2.1	30.8	3.6	173.3	13.4	397.5	6.6
O. hasseltii					385.6	42.9	66.3	3.7			2267.5	37.6
T. soro			530.4	22.1	262.7	29.2						
A. deauratus + sp.			868.9	36.3			224.8	26.1	75.4	5.8	256.2	4.3
S. hasseltii			60.0	2.5			61.9	7.2	16.6	1.3	22.4	0.4
C. batrachus							234.0	27.2	118.1	9.1	165.1	2.7
C. teijsmanni			217.0	9.1					98.7	7.6		
M. vyckii			246.5	10.3	14.9	1.7	37.5	4.4			88.0	1.5
G. platypogonoides					36.8	4.1						
G. major			30.8	1.3	27.0	3.0	15.4	1.8	4.3	0.3	22.0	0.4
C. gachua			118.4	4.9	46.8	5.2						
C. melasoma									28.1	2.2		
C. striata									29.0	2.2		
C. lucius									533.7	41.2	649.8	10.8
A. testudineus	12.9	9.7									91.2	1.5
B. pugnax			5.0	0.2					2.2	0.2		
M. armatus					88.4	9.8			158.0	12.2	68.3	1.1
M. maculatus			244.8	10.2	18.6	2.1	144.5	16.8			29.6	0.5
F. alba									34.5	2.7	293.5	4.9
D. pusillus											2.9	0.0
H. pogonognathus							0.4	0.0	0.5	0.0		
D. deokhathoides									2.8	0.2		

c.

	Trib. nr A	I + Upper 3° Tribs	II + Upper main river	III	Tribs H + J	IV + V
Stream order	2	3	4	5	4	6
No. of individuals (N)	26	127	60	47	64	562
No. of species (s)	3	10	9	11	15	19
Diversity $\left(d = \dfrac{s-1}{\ln N}\right)$	0.61	1.61	1.95	2.60	3.37	2.84

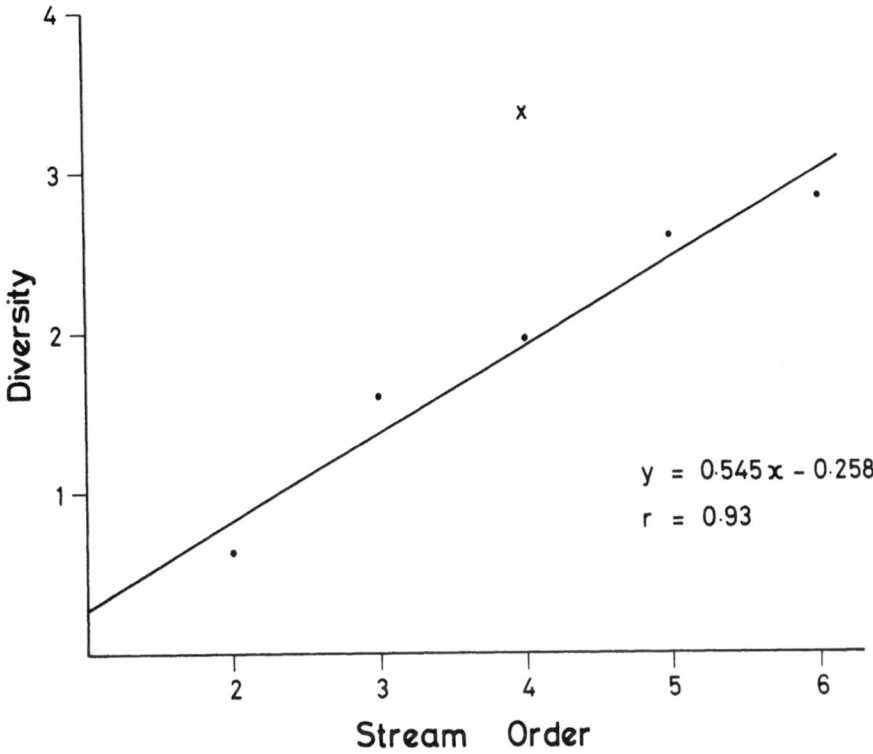

$$y = 0.545\,x - 0.258$$
$$r = 0.93$$

Fig. 65. Relationship between fish species diversity and stream order; 4° tribs indicated by × omitted from regression expression (see text).

little relevance in a river dominated in all zones by euryphagous species (cf. KNÖPPEL 1970).

1.2.4. Trophic relationships

The role played by allochthonous materials in the Gombak is manifested in the summary of the percentages of the population in each broad trophic group, given in Table 92. The detritivore, *Osteochilus*, although only 16% of the population, constituted the largest part (38%) of the biomass at Station V. The preferences of the omnivorous fish could not be determined with any accuracy as they all undoubtedly fed facultatively, as is common in most stream fishes (ALLEN 1969), and the proportion of each food in the diet would have depended almost entirely on availability and competition between the fish present. The categorization given is the best possible at present. Almost all the available food was of exogenous origin. Drifting plant materials (the surface load of seeds, fruits and leaves and the detritus fraction, both of which were considerable, as estimated earlier) and the smaller drift load of animal origin, of which

Table 92. Percentage of fish population by (a) numbers and (b) weight utilizing each main food type

| Feeding type | UPPER ZONE | | | MIDDLE ZONE | | LOWER ZONE |
	Trib. nr A	I + Upper Tribs	II + Upper main river	III	Lower Tribs H + J	IV + V
a. Detritivore	—	—	—	4.3	—	15.7
Herbi-omnivore	88.5	66.2	83.4	34.1	25.0	32.9
Carni-omnivore	—	9.4	8.3	34.0	32.9	33.2
Exog-carnivore	11.5	3.2	—	12.7	14.0	4.4
Endog-carnivore	—	12.6	5.0	12.8	20.3	2.6
Macro-carnivore	—	8.6	3.4	2.1	7.9	10.9
b. Detritivore	—	—	—	7.7	—	37.6
Herbi-omnivore	90.3	61.5	74.2	34.4	19.2	27.6
Carni-omnivore	—	10.4	7.1	29.6	21.2	14.7
Exog-carnivore	9.7	2.7	—	7.2	1.7	0.4
Endog-carnivore	—	10.2	11.9	16.8	12.2	1.6
Macro-carnivore	—	15.2	6.9	4.4	45.6	18.1

a part was endogenous, were equally accessible to the omnivores. The utilization of invertebrate drift food by stream fish has been reviewed by BISHOP and HYNES (1969b), WATERS (1969) and ELLIOTT (1970b). The fish classified as preferential herbivores were a major part of the fauna at all stations, both numerically and by biomass. The preferential predators were less important in the upper sections, but gained equal prominence at the lower stations. Of the carnivorous types, the exogenous arthropod feeders (taking ants, spiders and emergent aquatics) were mostly little fish in low numbers which made up a small part of the standing stock. This group particularly may have been underestimated by the electrofishing as indicated by the visually estimated numbers present at Stations IV and V that could not be caught for the population assessment (Table 91). The same was true for the endogenous invertebrate predators, as the large population of *Poecilia* at Station V was not included in the percentage calculation. The other two species in the last category, both *Mastacembelus*, apparently preferred the less powerful riffles of the tributaries over the main stream, as from the limited samples obtained, their frequency in the upper and lower tributaries was higher than at adjacent main river sites. In the main river, much of the invertebrate fauna living on stones or between cobbles and large gravel was relatively inaccessible to such a bottom swimmer whereas the animals in the leaf-packets and finer gravels found in the smaller channels were more easily taken by the digging and rooting-out feeding habits of this genus with its highly sensitive prehensile upper lip. The species predatory on large invertebrates, particularly Decapoda and other fish, formed a numerically minor but significant part of the population in all except the smallest tributaries. The high biomass percentage in the lower tributaries, and in consequence distortion of the contributions of the other categories, was a result of the collection of a few very large *Channa* in an otherwise normal population.

The foundations of this inferred food web were terrestrial plant materials and exogenous insects. The contributions of autochthonous organic production to the detritus was negligible and the part of the fish population dependent on the aquatic invertebrate fauna small. Many of the omnivores exploited this endogenous animal resource, but were not dependent on it. This scheme is similar to that found in Amazonian streams by FITTKAU (1967, 1969) and KNÖPPEL (1970) where the small fish are primarily surface insectivores or drift feeders preyed upon intensively by larger species. The almost complete dependence of the fish fauna on allochthonous energy sources (cf. HULOT 1951, for tropical African streams and LOWE-MCCONNELL 1967), with the correspondingly small role played by benthic invertebrates, and absence of algal feeders contrasts the normal situation reported in temperate rivers or high altitude tropical streams (VAN SOMEREN 1945), although there is increasing awareness of the importance of plant detritus to the non-

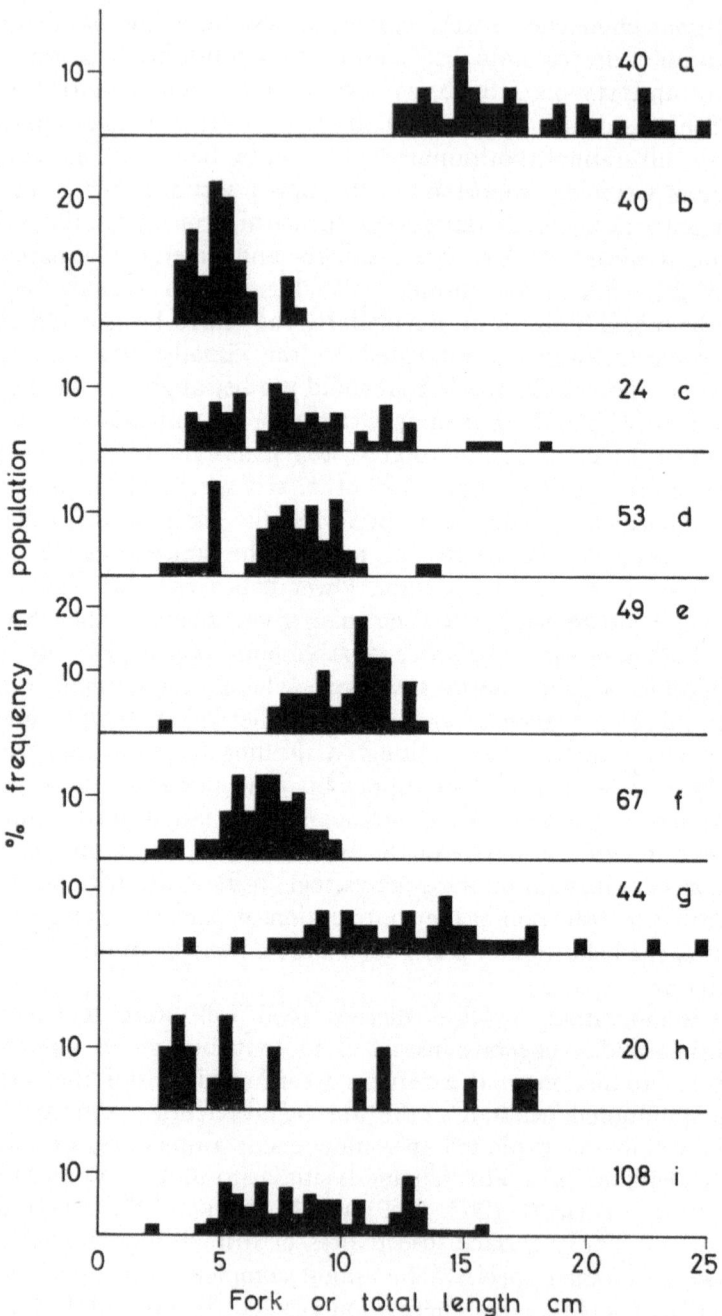

Fig. 66. Length frequency distribution of fish in the Gombak River and tributaries; (Distribution for all species of which 20 or more fish were caught as indicated), a. *Mastacembelus maculatus*, b. *Glyptothorax major*, c. *Tor soro*, d. *Puntius binotatus*, *e. *Mystacoleucus marginatus*, f. *Rasbora sumatrana*, *g. *Osteochilus hasseltii*, *h. *Hampala macrolepidota*, i. *Acrossocheilus deauratus*, * caught in main river only.

salmonid fish of those areas also (DARNELL 1964, MANN 1965, 1967, McCONNELL 1968).

It was initially hoped to be able to determine the age structure and timing, extent and success of breeding activity in the Gombak, but the difficulties in obtaining representative samples prevented this. The length frequency distribution of fish caught in the main river and tributaries (Fig. 66) illustrates the selectivity of the electrofishing for larger fish. Comparatively few young fish were sampled from a population in which most cyprinid species at least appeared to have fish in breeding condition most of the year (see also SOONG 1968, TAN 1968) and whose structure from direct observation was weighted tremendously to the young classes. Consideration of growth rates and migratory patterns was beyond the scope of this initial survey.

1.3. APPRAISAL OF FISHERY

1.3.1. Yield

Table 93 summarizes the numbers and biomass of fish reportedly caught by the Upper, Middle and Lower Zone inhabitants on an annual basis. All size groups have been summed. Biomass was estimated per 5 cm size category based on weight/length relationships obtained from fish caught for Part B above, in the appropriate zone. These results are subject to the limitations mentioned under methods, and data for the Upper Zone particularly may represent fish caught outside the zone of residence. The aborigines who formed the largest proportion of fishermen there often worked long sections of the river, taking fish from the Middle Zone as well. The large number of *Rasbora* taken by them is a function of this as this genus was found only toward the bottom of that zone. Of the 277,500 of this genus caught, more than 87% were in the smallest 5–10 cm size group. At least two groups of fishermen (2–3 men each) made their livelihood from fishing in the river. They used cast nets and spears, fishing both Lower and Middle Zones and often working 4–5 km upstream in a day. Their considerable annual catch was apportioned equally between the two zones. Throughout the river, all fish caught were retained, irrespective of size. *Glyptothorax*, the only possible trash fish, was reportedly sometimes eaten, but small specimens were often discarded or given to children as toys, so mortality after capture was undoubtedly high.

The upper river catches of *Hampala*, in low numbers in the top section above Station III but a considerable yield from the Middle Zone, may have indicated a periodic (because of adverse water conditions) or reproductive migration (DAGET 1957, WELCOMME 1969) out of the turbid and polluted lower section. A few *Osteochilus* were reported caught in the Upper Zone, also above the range found by the electrofishing, but again

Table 93. Numbers and weights of fish reportedly caught per year in the Upper, Middle and Lower Zones of the Gombak

Genus	Upper Zone		Middle Zone		Lower Zone		Total	
	No.	Kg	No.	Kg	No.	Kg	No.	Kg
Rasbora	277500	1225.7	59320	427.9	89410	1091.8	426230	2745.4
Hampala	860	11.1	37220	742.5	93180	3580.7	131260	4334.3
Puntius	1760	15.1	25200	474.4	25400	1616.6	52360	2106.1
Osteochilus	200	4.8	6820	239.5	55600	2715.1	62620	2959.4
Acrossocheilus	16600	132.3	2620	44.1	6600	242.9	25820	419.3
Tor	60422	420.4	—	—	—	—	60422	420.4
Glyptothorax	2450	59.0	9180	74.2	43360	439.9	54990	573.1
Clarias	9711	349.5	53240	1436.6	71080	1941.6	134031	3727.7
Fluta	—	—	6400	53.5	10020	75.0	16420	128.5
Macrones	3608	290.0	36380	2109.9	94540	5704.1	134528	8104.0
Mastacembelus	2530	24.4	19380	160.3	56860	557.1	78770	741.8
Channa	5569	139.0	4760	262.3	60240	984.0	70569	1385.3
Silurichthys	8920	45.8	10080	42.8	19920	174.7	38920	263.3
Anabas	—	—	5880	161.1	40360	971.3	46240	1132.4
Dermogenys	2000	1.3	6600	3.7	36900	17.8	45500	22.8
Total	392130	2718.4	283080	6232.8	703470	20112.6	1378680	29036.8
No. of fishing hours	11235		25100		78640		114975	
Catch/unit effort	34.9	242 g	11.3	248 g	8.9	256 g	12.0	253 g

these may have been strays or fish on a spawning run. Catches of anabantids, various catfish and *Fluta* from the paddy fields could not be
assessed because surprisingly few farmers bothered to fish their drainage
sumps, and those who did reported catches in total volume and weight
and were uncertain as to species composition.

Fishing methods in use in all the sections were cast net (particularly
efficient for the schooling Cyprinidae), rod, spear and occasionally seine
netting. The aborigines in the top zone used long, unbaited, split-cane
traps which often caught 50–100 fish per set. This method was confined
to the forest streams. Fishing with poison and explosives, both illegal as
discussed by ALFRED (1969) were admitted to in all zones. *Derris* deriva-
tives and synthetic insecticides were used, but apparently not very
frequently. Casual fishing with rod and line, particularly for the large
siluroids and *Channa* was carried out intensively during periods of turbid
flow. The riverine people here claimed, as did the Sabahians (INGER
and CHIN 1962), that *Macrones* and *Clarias* are not likely to take bait
except under these conditions.

The annual catch figures of 30 metric tons from the total 0.333 sq. km
water surface are considerable for a small river, but divided up on a *per
caput* basis for an estimated immediate riverine population approaching
8,000 (1,000 in the aboriginal settlement, 2,000 in the Middle Zone,
5,000 in Lower Zone) represents only about 3.6 kg/year.

The catch per unit effort was high. The figure of 35 fish/h for the
Upper Zone reflects the use of traps as well as cast net fishing but may
be too high as catches of small fish *(Rasbora, Acrossocheilus)* were reported
in katis (about 600 g) and converted to equivalent numbers later. The
mean catches of 11 and 9 fish/h for Middle and Lower Zones tallied with
observed creels of cast net fishermen. The weight of catch per unit effort
was uniform for the three sections, about 250 g/h, giving mean weights
per fish of approximately 7,22 and 28 g and emphasizing that all caught
fish were retained, irrespective of size. This large catch represents a yield
of more than 87 g/sq.m of bottom/yr over the whole river.

1.3.2. Synthesis

To enable a rough comparison to be drawn between standing stock
and yield, the data from the electrofishing have been summarized in
Table 94. The total estimated numbers and biomass based solely on these
catches are obviously underestimates, being only about one twenty-fifth
of the supposed yield. If, as found by LARIMORE (1961), efficiencies of
only 10–50% were obtainable for a mixed fish population in turbid
waters with reasonably high specific conductance, no better figure can be
expected in the Gombak with its marginal conductivity as far as electrofishing is concerned. The selectivity of this technique against minnows
(clear from the mean weight per fish caught in the upper section of more

Table 94. Synopsis of electrofishing catches and estimated standing stock

Zone	Electrofishing Catches						Estimated Standing Stock		
	Area fished sq.m	No.	Wt.g	Mean wt./fish	No./sq.m	g/sq.m	Total area sq.km	Total no.	Total wt. kg
Upper	776	213	3428.8	16.1	0.27	4.4	0.130	35100	572.0
Middle	700	111	2155.0	19.4	0.16	3.1	0.075	12000	232.5
Lower	1710	267	6024.0	22.6	0.16	3.5	0.082	13120	287.0
Total	3186	591	11607.8	19.6	0.19	3.6	0.287	60220	1091.5
				Corrected totals*			0.333	401100	5957.5

* adjusted for electrofishing efficiency and unfished 2° stream contributions

366

than 16 g compared to the weight of about 7 g taken by the natives trapping) results in an indeterminable but considerable underestimation of stock numbers and a smaller error in the biomass. In addition, the areal estimates per section included streams only down to $3°$. In the permanent $2°$ streams electrofished, a density of approximately 2 fish/ sq.m (biomass about 10 g/sq.m) was found. The total length of these channels for the whole watershed was about 66 km, mean width 0.75 m, i.e. containing about 100,000 fish weighing approx. 500 kg. Assuming mean electrofishing efficiency of 20% in the higher order streams, and adding the standing stock of the $2°$ streams would give a corrected standing stock of 1.2 fish/sq.m and about 18 g/sq.m which would reduce the discrepancy between fished yield (about 87 g/sq.m/yr) and standing stock to a factor of four to five. If exploitation is not permanently depleting the fish population (no information is available on past catches but the riverine population and presumably fishing pressure have increased only slightly in the last 20 years), a minimum production rate of 90 g/sq.m/yr would be necessary to sustain the fishery. Such a rate is about twice that found for a mixed population of coarse fish in the Thames by MANN (1965); his population density (3–6 fish/sq.m) and therefore interspecific interaction were higher, resulting in a $P/\overline{B} < 1$, but less than that reported by MATHEWS (1971) also for the Thames (39 g/m²/yr for 0 yr catchables and 83.3 g/m²/yr for > 1 yr fish). If the reported catch here is an over-estimate, as is quite likely, the required production rate could be reduced accordingly. Generally production is far in excess of cropped yield as only a proportion of the population is vulnerable to fishing gear. Productivity for temperate fish populations in seasonal oligo-mesotrophic waters usually lies between 1–2.5 times the standing stock (HUET 1964, MANN 1967 and many references cited by CHAPMAN 1967), but much higher turnover ratios have been reported, e.g. 8–15 for trout in Walla Brook from HORTON's (1961) data. Comparable data for tropical streams are not available. In a warm eurythermal river with continuous recruitment of young fish to both total and exploitable populations, and with a constant exogenous food supply, a population turnover ratio of about five would not seem impossible. HICKLING (1962) reported *Pangasius* culture in flow-through cages in Cambodian streams with average growth from 100 g to more than 1 kg in 8–10 months with large amounts of food provided, and enormous productivity (0.7–1.5 kg/sq.m/season) in flowing pond culture of carp in Japan with semi-continuous supplementary feeding. In cage culture in the Tjibunut River, West Java, with no accessory nutrients, dependent only on the drift food primarily of Oligochaeta and Chironomidae and a heavy sewage and garbage load, VAAS and SACHLAN (1956) and also HUET (1956) reported growth of carp from 20–30 g to 180–200 g in four months and up to 3.5 kg in less than two years. Nowhere on the Gombak, except perhaps below the D.I.D. gauging site, did pollution conditions approach those reported in the Tjibunut, and most of the available food in the other sections was of plant origin. These examples are given to show the potential production in situations where continuous supplies of easily accessible food, adequate oxygen conditions and sustained high temperatures are available. The considerable cropping pressure in the Gombak would tend to maintain production at a high level by keeping the mean age of the fish population within the more highly productive young age groups (HAYNE and BALL 1956, fully discussed by BACKIEL and LE CREN 1967 and GULLAND 1967).

As a speculation, a theoretical production rate for the Gombak was calculated using the coefficient of HUET (1964) which MANN (1965) found to give a reasonable approximation. $K = B \times L \times k$ where
K = annual production kg/km of river,
L = stream width in metres,
B = biogenic capacity on a scale 1–10 for oligo-eutrophic waters,
k = $(k_1 \times k_2 \times k_3 \times k_4)$

    k_1 (temperature) $10\,°C = 1$, $16\,°C = 2$, $22\,°C = 3$, $28\,°C = \underline{4}$
    k_2 (pH) $\leq 7 = \underline{1}$, $> 7 = 1.5$
    k_3 (depends on fish type) Salmonidae $= 1$, Other $= \underline{2 \text{ or more}}$
    k_4 (age of fish) 1 yr $= \underline{1.5}$, 2+ yr $= 1$

Assuming a mean stream width of 8 m, $B = 3$ for trophic conditions based on available nutrient salt supply, and the k values underlined, an annual production of 288 kg/km or 36 g/sq.m would be indicated. This is only about twice the corrected standing stock. However, if the k_3 factor (roughly equivalent to the turnover ratio) were five, as discussed above, a production theoretically sufficient to sustain the fishery would be possible (90 g/sq.m). The biogenic capacity B should probably be higher, four or five, as the exogenous food supply, a factor not considered by HUET in the definition of B and certainly of equal importance to water chemistry, is large; then the required turnover ratio would be about four.

To summarize this conjectural section: it appears possible that the reported high per unit area fish yield could be maintained by the fish fauna, assuming a high production rate and low exploitation by natural predators. LOWE-McCONNELL (1967) reported a comparable situation in which a small river fishery based on traps was taking out more biomass than the obvious production, but she attributed this to recruitment from a large area of tributaries.

2. Other vertebrates

A synopsis of the fauna of the forest zone has been drawn up by MEDWAY (1965/66), but only those species observed to have interaction with the river are discussed below. Observations were made incidental to the other studies and information on frequency of occurrence was obtained from MEDWAY (personal communication).

2.1. Mammalia

The only mammals of consequence were otters, *Amblonyx cinera* (Illiger), *Lutra sumatrana* (Gray) and *Lutra lutra* (L.). The first two species were fairly abundant, the last very rare. The extent of exploitation of the fish resources was indeterminable.

2.2. Aves

Heron, *Butorides striatus* (L.), have been recorded on isolated occasions, but there was apparently no resident population. Other piscivorous birds included a fish hawk, *Spizaetus alboniger* (Blyth) and the White Breasted Water Hen, *Amaurornis phoenicurus* (Pennant). Various kingfishers were common along the river in the forest and much rarer in the lower reaches. *Alcedo euryzona* (Temminck), *Halcyon concreta* (Temminck) and *Lacedo pulchella* (Horsfield) were frequently recorded residents. *Ceyx erithacus* (L.) is a forest species with marginal contact with the stream and *Halcyon coromander* (Latham) and *H. pileata* (Boddaert) are winter migrants; the last species is probably chiefly insectivorous. A variety of domesticated *Anas* spp. were occasional invertebrate predators at Stations III, IV and V.

2.3. Reptilia

The only reptiles likely to have any influence on the aquatic fauna were species of *Varanus*. All four monitors known from Malaya have been observed in the Station II area, but only a single sighting of *V. salvator* (Laur), reported by Harrison and Lim (1957) to be piscivorous, was made in the water. The common Speckled-bellied Keelback, *Natrix chrysarga* (Boie) was the only other reptile seen in the water. Inger and Chin (1962) reported this as a fish-feeder, but its main diet was more likely the various frogs common along the river banks. A single observation of the aquatic tortoise, *Notochelys platynota* (Gray) was made at Station IV; this species was herbivorous.

2.4. Amphibia

A number of Anura were recorded, but the contribution of all but a few species to the river ecology was obscured because of identification problems with the larvae. The most abundant tadpole in the forest streams, confined to torrential zones, was *Amolops laruensis* Blgr, often found in groups of 2–10 in restricted areas on both the lower and exposed surfaces of boulders with some algal growth. The pelobatids *Megophrys monticola nasuta* (Schlegel), *M. longipes* Blgr and *Leptobrachium hasseltii* Tschudi were common to abundant in some small forest tributaries with moderate velocity, particularly in open spaces where light promoted local diatom and blue-green algal growth. The larvae of bufonids were found throughout the river, but were especially abundant at Station IV. *Bufo asper* Gravenhorst, *B. parvus* Blgr and *Cacophryne borbonica* (Tschudi) have variously reported distribution, but *B. melanostictus* Schneider is a restricted lowland cultivated-area species. No specific identification of any of the tadpoles collected could be made. The tadpoles of the other frogs entered the river only by accident as most developed in forest pools or slow backwaters and were occasionally washed in with floods. In the forest areas, the following species regularly occurred:

Microhylidae   – *Kalophrynus pleurostigma* Tschudi
                  Microhyla berdmorei (Blyth) and spp.
Rhacophoridae – *Racophorus colletti* Blgr and spp.
Ranidae   – *Rana blythii* Blgr   R. *chalconota* (Schlegel)
                  R. *hosii* Blgr   R. *kuhli* D. and B.
                  R. *laticeps* Blgr   R. *tweediei* (Smith) (rare)
                  R. *luctosa* (Peters) (rare)

Ubiquists, generally rare, but more frequently found in lowland and cultivated areas were *Rana erythrae* (Schlegel) common in ricefields, *R. limnocharis* Boie, and *Kaloula pulchra* (Gray), a temporary-puddle and overflow-pool breeder.

A single Gymnophionan, the caecilian *Ichthyophis* ?*glutinosus* (L.), was occasionally caught in leaf and detritus drifts in pooled areas of the forest tributaries.

VI. ZONATION

Classification of a watercourse as an entity may be useful in a general descriptive survey, but the categories are of necessity wide and subjective. On the other hand, a zonation scheme delimits areas within the system and permits a second order categorization of faunal groups and a degree of objective comparison between rivers in different geographical regions. The third order refinement, involving definition of the biogenic relationships within the habitat mosaic of a zone, requires an intimate knowledge of the ecological characteristics of the biota which, even for the well-studied European region, is not yet fully available. The desirability of classifying rivers or sections of watercourses has been challenged by a number of authors (e.g. BADCOCK 1954, ARMITAGE 1961, THORUP 1966) on the grounds that all divisions are arbitrary and not representative of any true discontinuity in the biotope. Others (HYNES 1960, 1970, MACAN 1961a) have indicated that in their opinion just not enough is known of the fauna to enable differentiation of zonal biocoenoses in the complex and intergrading mosaic of microhabitats that constitute most river channels. Both points are valid, but it is also relevant to note that when discernible changes in faunal associations occur in structurally similar biotopes (as far as can be determined), there may be a case for designation of a zonal boundary. In many cases, an alteration in the relative abundance of the species present will be more important than actual faunistic changes. Any zonal designation necessitates a considerable degree of refinement in faunal identification, but as pointed out by ILLIES and BOTOSANEANU (1963), an open nomenclatural system as had to be used here will provide this. In the present study, taxonomic inadequacy was serious, but to allow comparison of the Gombak with other rivers, at the community level, some preliminary attempt at zonation combining physiographic and faunistic criteria was thought to be desirable.

The three main zones, Upper, Middle and Lower, arbitrarily established at the beginning of the survey on the basis of stream and watershed morphometry, current velocity, gross substrate type and temperature, divided the river into sections that were apparently different if not discrete. Chemically, there were no sharp break-points until the inflow of waste products in the Lower Zone in the vicinity of Station V. In terms of the biota, the selection of sites for the principal stations may have influenced the species found there, but from field work in peripheral areas, they appeared to be representative. However, Station I might

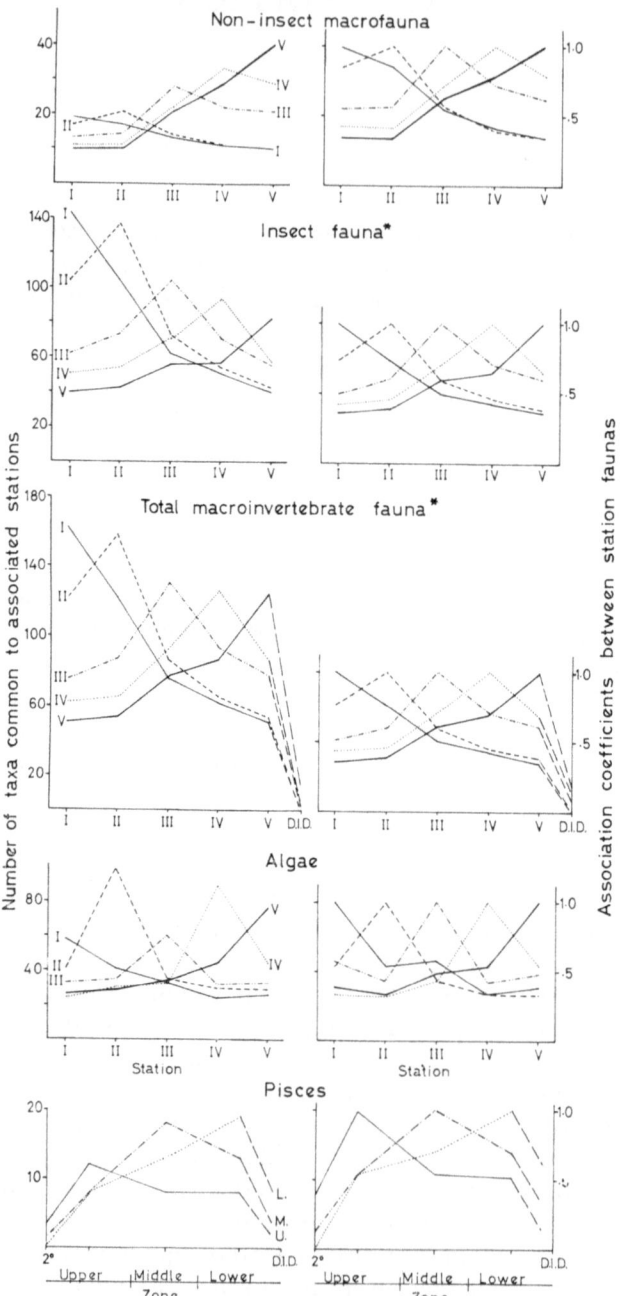

Fig. 67. Summary of the zonation of the biota. The number of taxa found at each station and the number of the same taxa present at the other stations plotted in diagrams on the left. Coefficients of association between each station and the others plotted in diagrams on the right; * does not include Trichoptera (79 taxa), Coleoptera (Elmidae– 20 spp.) or Odonata (53 taxa).

profitably have been located on a smaller, higher altitude forest stream to differentiate better the communities within the Upper Zone.

The invertebrate fauna, as noted previously, changed radically in composition down the river with numerous additions, eliminations and replacements. These succession relationships were collated for all biotopes in Table 74. Treating these data after the method proposed by ILLIES (1961a) and further expanded by ILLIES and BOTOSANEANU (1963), produced the graphs on the left of Fig. 67. In these, the curves indicate the number of species found at each station and how many of the same species are present at the other stations. Because the insects and non-insects apparently responded in different ways to the physical changes along the profile, they were treated separately. The progressive increase in numbers of non-insect species and decline in insect diversity downstream are obvious.

Among the non-insects, two distinct assemblages are clear, Stations I with II and IV with V, whose affinities with the other stations are low, generating similar curves. Station III is intermediate in total number of species. The slope of its curve through Stations IV and V is negligible, indicating that its fauna is close to that of the lower section.

The curves for the insect fauna give separate peaks at each station and as the vertical distances between curves at contiguous stations are large, considerable differences between them are indicated. Stations I and II still form a loose entity, as their joint affinities with the lower stations are parallel; the lower three stations apparently have discrete specific assemblages in addition to a core of ubiquitous taxa. The addition of the Trichoptera and Odonata not included in the analysis because of uncertain taxonomy and/or distribution would probably enhance the separation of the two station groupings, as the first Order showed greatest diversity in the forest river and the second in the lower unshaded reaches.

The curves for the total fauna indicate, with distinct and steeply-sloped peaks, the extent of association among the stations. The Stations I and II curves, and to a lesser degree those of IV and V, each share a common but longitudinally diminishing group of species with the other stations and again segregate as pairs. The Station III curve is almost a composite of the other groups, reflecting the transitional nature of the community there. If the curves are extended to include the D.I.D. gauging site in the zone of pollution and heavy silt deposition, based on the casual and undoubtedly incomplete samples collected there, they all show a precipitous decline. Fewer than 15 invertebrate taxa were collected out of what might have been a total faunal list of 20–25 species if more intensive collecting had been carried out. All types collected were also present at Station V, but none was common with the Station I fauna.

In his discussion of the ILLIES system, THORUP's (1966) main criticism was that biocoenoses were differentiated by the height of the peaks and that this criterion, the number of species at a location, may be controlled by exogenous factors such as pollution, rather than by a natural zonation resulting from physical factors. Also, he felt correctly that if the number of species present at each station decreases only gradually at the other stations, i.e. neither rapidly nor completely, the concept of biocoenosis as a separate and unique association was invalid. However, this in no way detracts from the usefulness of the technique in demonstrating zonation. To overcome the problems inherent in comparing numerically different faunas (species number alone not being an adequate criterion for separation), the association coefficients from Table 74 were plotted to give the right-hand figures. This presentation has the advantage that all peaks are the same height and the association relationships between stations are numerically comparable. Hence, the distance between and slope of curves connecting stations are valid differentiating criteria. These curves emphasize the separate grouping of Stations I with II and IV and V (with an association coefficient between groups > 0.5) and Station III as a transitional assemblage. Within the two groups, considerable differences between the faunas of the pairs are evident, particularly IV and V where a definite break in the curve, attributed directly to the pollutional effects, is seen. The sharp declination of all curves with respect to the D.I.D. site was apparent.

The algal flora showed surprisingly little affinity between stations as summarized in Table 48, for a number of reasons already discussed. Division of the river into fish zones on the basis of the data available would be presumptous, especially as there are apparently no real physical barriers of temperature or gradient to limit distribution. A preliminary, but not very conclusive designation of characteristic species was already made, but its wider applicability without data from other watercourses is of course unknown. In any case, fish, because of their mobility and considerable plasticity of response to changes in environmental conditions, are not really part of any sectional assemblage. Rather, as discussed by MINCKLEY (1963), they are superimposed on the other communities and thereby utilize a greater proportion of the available resources than either the invertebrate fauna or the algae.

The data for algae and fish were plotted the same way as those for invertebrates, and emphasize the differences (= zonation) in the biota of the river sections. The fish were compared only on a zonal basis with the relationships to the 2° tributaries (3 species) and D.I.D. population (6 species) at the extremities of the studied part of the river, indicated by the sharply declining dashed lines.

The zonation scheme as proposed here therefore consists of the Upper Zone with a possible subdivision to form an Upper (Torrent) Tributaries Zone, Middle and Lower Zones, based primarily on abiotic features, but with characteristic biotic communities as well.

A considerable literature on river classification systems has built up, particularly in western Europe, based on physical features. This has been reviewed *in toto* by ILLIES and BOTOSANEANU (1963) and discussed with respect to the established fish-zone systems in that region. Schemes using single criteria have been summarized by KAMLER (1965) – temperature; CUMMINS (1966), THORUP (1966), ULFSTRAND (1968a) – substrate type and current, and a number of typologies based on the distribution of a limited faunal group (see ILLIES and BOTOSANEANU and THORUP). The most widely accepted classification is that of ILLIES (1961a) which was designed for universal application based on data from a series of rivers in a range of latitudes. This divides rivers into two main zones, the Rhithron (montane) and Potamon (foothill and plain) with subdivisions of each category into three subzones, epi-, meta- and hypo-, that roughly correspond to the various zonal systems proposed by a number of authors for diverse river systems. HARRISON (1965) has suggested that the rivers of South Africa can be fitted into the ILLIES classification system with some modification of the criteria defining sections and CHUTTER (1970) has further expanded the epipotamic definition to include both stable and unstable eroding sections of the rivers of the high altitude South African plateau. With further modification at several important points, the system appears applicable to the Gombak.

On a physical basis the rhithron was defined by ILLIES as the section of a stream, starting at its source, in which the annual range of monthly mean temperature does not exceed 20 °C, current velocity is high, flow turbulent, oxygen saturation high, volume usually small, bottom formed of fixed boulders, stable cobble, gravels and sand and with no appreciable silt deposition. The potamon downstream of the rhithron was described as having an annual range of mean monthly temperature in excess of 25 ° in tropical latitudes, low current velocities, unbroken flow, seasonally variable discharge, oxygen deficiency in pooled areas on occasion, partial light extinction as a result of suspended solids and silt deposition in a bottom dominated by sand. In his 1961a paper ILLIES also presented a tentative plan correlating altitude with the boundary between rhithron and potamon at various latitudes and, on the basis of work in the Andean headwaters of the Amazon, suggested that at the equator $\pm$ 10 °, this lies above 2000 m (see also ILLIES 1964). Rigid application of these criteria would place all of the Gombak, except perhaps the uppermost tributaries, in the potamon; nowhere on the river did temperatures consistently fall below 20 °C and the highest elevation is less than 1500 m. However, HARRISON (1965) felt that for the rhithron, the range of temperature was less important than the actual maximum and that silt and turbidity were major factors limiting the rhithronic fauna in the South African context. If for the original rhithron temperature requirement is substituted, 'mean temperature less than 25 °C and diurnal fluctuation less than 2 °', the sections of the Gombak that otherwise have the physiographic characteristics (gradient, discharge and substrate) of that zone could be included in the rhithron, i.e. the Upper Zone in which the rain forest canopy moderates insolation. The Middle and Lower Zones were certainly in the potamon on the basis of temperature, turbidity, gradient and substrate criteria, although, as seen, oxygen levels were never particularly low as turbulence was considerable in all sections.

MOON (1939) advanced the idea that periphyton was relatively important in headwater areas, and that as the river aged, detritus became dominant. ULFSTRAND (1968a) partially agreed for the Lapland streams he worked on, but commented on the proportional importance of detritus even in the young streams. Here such a criterion has no meaning as algal development depends solely on light availability and forest-origin detritus is abundant even up in the highest headwater tributaries.

Faunistically, the rhithron was defined as containing cold stenothermic, rheobiontic and polyoxybiontic species often with morphological adaptations to torrential life, i.e. generally encompassing the cool-adapted montane families of ROSS (1956). The potamo-benthos, on the other hand, includes eurythermal or warm stenothermal forms, tolerant of a range of current and oxygen conditions and containing typically lenitic forms. Both zones in addition provide suitable biotopes for an extensive conglomeration of tolerant taxa with wide ecological distribution.

Most taxa of the Gombak fauna fall into this last group of eurytopic forms, but a number were apparently limited to the Upper Zone because of temperature or a requirement for silt-free conditions. Nothing is known of the thermal preferences of the Southeast Asian fauna, so no real discussion on this criterion can be made except as a generalization within a family. Many of the families listed both by ILLIES (1961a, 1964) and HARRISON (1965) as having predilection for the rhithron were found exclusively in the Upper Zone of the Gombak, e.g. Leptophlebiidae *(Isca, Choroterpes, Dipterophlebiodes)*, Heptageniidae *(Epeorus,* one *Thalerosphyrus* sp.), Nemouridae, Leuctridae, Blepharoceridae, Simuliidae, Psychodidae *(Maruina)*, some Tipulidae, some Elmidae, *Eulichas*, Helodidae, Helicopsychidae, Stenopsychidae, Rhyacophilidae, Philopotamidae and other caddis, Corydalidae *(Hermes)*, but a large group of species with apparently wide tolerance for silt and temperature were also a significant element. Such groups as Baetidae, Neoephemeridae, Chironomidae (Orthocladiinae), Perlidae, some Trichoptera (notably the various Hydropsychidae and Psychomyiidae), such genera as *Thraulus, Compsoneuriella, Caenis* and also the Psephenidae and Dryopidae, Naucoridae, *Ptilomera* and many Odonata, although not typical rhithron animals, were important in this zone, but extended right down through the Middle and some into the Lower Zone as well. In these potamon reaches, other families, notably Potamanthidae, Chironomidae (especially Chironomini), Haliplidae and the non-insect groups became prominent. Many other species were probably limited to the upper river for reasons of habitat availability *(Prosopistoma, Peltoperlodes)*, strict rheobiontic requirements (various Gerridae) or silt-free conditions *(Isonychia, Hexatoma)*.

The major problem in trying to homologize the situation in the Gombak with that in other river systems is the paramount status given to the temperature tolerance ranges of the fauna in most studies. The ranges commonly ascribed to various taxa or families obviously do not apply to their Gombak counterparts. Absolute temperature is less likely to be the limiting factor governing the distribution in tropical rivers than the amplitude of the range. As seen, the upper river is thermally regulated by the rain forest, with only small diurnal fluctuations from a high minimum. Warm stenothermy may therefore be important to those representatives of ostensibly cold stenothermal groups found here. Other habitat requirements of current, food, substrate size, or, as discussed by CHUTTER (1970), freedom from sedimentation in the microhabitat, are probably more important in delimiting the ranges of the individual species in the eurytopic families.

Overall, if any comparative categorization of the river is possible, the Upper Tributaries can be described as rhithron on the strength of physical parameters and the presence of some specialist torrent Diptera and Plecoptera. The main river of the Upper Zone, however, is intermediate between hyporhithron and epipotamon. Physically, except for temperature, it is rhithron and faunistically, it has many characteristically rhithronic species, but there is no definite faunal break between it and the epipotamonic, soft-bottomed erosive Middle (Station III) and upper Lower Zones (Station IV). Anthropogenic changes below Station V have obscured the faunal relationships there, but originally most of this section was probably epipotamon to the point where all gradient is lost and the combined Klang and Gombak rivers become deep and slow-flowing about 15 km below their confluence.

In summarizing their concept of zonation, ILLIES and BOTOSANEANU (1963) concluded that changes in zoonose are most likely to occur at points where physical parameters are suddenly modified, generally at points where watercourses of similar status meet. Such nodal points occur where stream order, as defined, changes. The 3° and lower order tributaries, which are a physically homogeneous group differing only in length, would constitute a zone. The 4° river resulting from the confluence of a number of these rhithron streams ends where Tributary D enters and forms a 5° river. This more or less corresponds to the boundary between Upper and Middle Zones as described and, coincidentally, the present edge of the forest. At this node there are both a significant change in substrate type from predominantly boulder to one with many small pools between riffles and a number of faunal changes, so the end of the hyporhithron may occur here. The epipotamon status of the river probably remains unchanged through the next two increments of order (6° and 7°). Al-

377

though the confluencing Tributaries H and K are of equal order to the main river, they carry much smaller modal discharges and only slightly modify environmental and faunal conditions.

If this can be accepted, the rhithron in this equatorial region is altitudinally lowered by some 1700–1900 m from that experienced in Amazonia! True rhithron zones undoubtedly occur in the higher rivers of the central range and on the isolated peaks of the peninsula, but until these locations are investigated and their faunas compared, the real status of the presently defined Upper Zone will remain speculative. The presence of the forest may extend montane river conditions down into the foothills provided that substrate and gradient conditions remain unchanged, with the fauna adapting to the stabilized longitudinal modifications to temperature even though these may be at higher levels than their preferred range, i.e. a species may extend its distribution downstream into reaches with marginally tolerable temperatures if other habitat conditions remain suitable and presumably, if there is no representative of the euryoecious cosmopolitan families competing for the same niche. A somewhat analogous phenomenon was described by Ross (1963) for temperate deciduous forests that superimpose ecological conditions of temperature, erosion load and allochthonous food supply on the streams running through them. Corroboration for this altitudinal lowering of the rhithron comes from recent work by Hynes (1971) on a West Indian stream. There, based on faunistic and physical criteria, he concluded that the rhithron existed to within 30 m above sea level. Whether this is a general result of insularity or peninsularity compressing stream zonation cannot be decided until further evidence is available.

From a practical viewpoint, any zonation system is only of limited value as exceptional distributions and faunas are the rule. At a general level, zones may be defined by gross physical characteristics, but at the invertebrate level, the substrate in its broadest sense determines the ecotypes present. Wherever suitable substrate conditions for a taxon occur in a river, irrespective of zone, that species most likely will be found there. Each river is therefore a specialized individual case and the type of mondial classification system advanced by Illies can really only be considered a first order definition, particularly with regard to faunistic criteria which, as pointed out by Berg et al. (1948) and Ulfstrand (1968a), have a limited regional applicability.

VII. POLLUTION

In its total effect, pollution tends to carry an ecosystem away from its natural complexity and stability to a less stable state in which diversity is decreased and the self-regulating buffering capacity against change diminished. However, any definition that implies that the slightest alteration in the biodynamic cycle by an introduced agent is pollution, irrespective of aesthetic, social or economic impact is impractical, however desirable from the strictly biological viewpoint. The preservation of all but a few isolated ecosystems in pristine condition is inexpedient so a compromise, both in the definition of pollutional damage and in the level at which the real needs of society for equitable use of the habitat are fulfilled, is necessary. An acceptable, but open-ended, definition of water pollution that would seem to cover all situations is that of DOUDOROFF and WARREN (1957): 'impairment of the suitability of water for any beneficial human use, actual or potential, by way of any foreign material added thereto'. This, as realized by them, legally pinpoints the resultant injury, not the causative agent, as pollution. Unless a water body loses some of its value as a natural resource by the addition of some factor or unless its utility is impaired, altering its diversity or changing its biological structure may not be pollution. However, the loss of biological adaptability in the form of species eliminated from a rich species assemblage by an effluent (PATRICK 1964) may in the long-term have deleterious effects on the potential uses of a water body, particularly if the needs of society change. The problem of definition is especially relevant in the Malaysian context where much of the alteration effected by 'pollutants' has little direct impact on the presently constituted utility of the watercourse.

The literature on pollution in tropical freshwaters, particularly lotic systems, is scarce and work in Malayan rivers confined to the NORRIS and CHARLTON (1962) report on the Gombak already mentioned and a recent chemical survey by TAN and PROWSE (in press) that defined the effects of factory and residential wastes and incursive marine waters on the water quality of the Malacca River. The only other references are those in the semi-popular press (JOHNSON 1961, 1968b, PROWSE 1968, WYCHERLEY 1968, ANONYMOUS 1970) that refer to the general aspects of tin-mining and domestic pollution, and to the more serious immediate problem in Malaysia, that of river erosion, already cited in the section on sediment load.

Probable effects of additives to the Gombak have been mentioned with respect to physical, chemical and biotic aspects at relevant places in earlier sections. However, the real problem of isolating effects resulting from effluent inflows and those indicative simply of the maturing of the watercourse was not satisfactorily resolved as in the river most changes in environmental quality were cumulative, the result of natural and anthropogenic modifications over extended reaches rather than at point sources.

Chemically, changes in the river were not great and although most ionic concentrations increased downstream from Station IV, the contributions from natural geologic sources were probably more important for some elements than pollution, although nutrient salt and electrolyte increases undoubtedly originated with human activities. This agricultural and domestic enrichment amounted to little more than mild fertilization by normally accepted standards. It must be pointed out, however, that determinations made on the transitory water mass are less indicative of the true state of eutrophication of a river system than analysis of the more or less static sediments which harbour the important microsaprobes (see LENHARD and DU PLOOY 1965). In the present context of high temperatures and turbulence, much of the nutrient entering the river would have been rapidly assimilated by the pelic community. The algal flora was not extensive, as seen, but production rates were high at Station V, reflecting the availability of nutrients. Toxic pollutants were not apparently common in the intensively surveyed sections as little mortality of fish or invertebrates was seen. Heavy metal ions may have contributed to the elimination of some faunal elements in the reaches below Station V. Ammonia never reached lethal concentrations because oxidative conditions prevailed and turbulence prevented the formation of areas of reduced conditions. Insecticide-herbicide pollution, particularly sodium arsenite from *Hevea* plantations, was not obvious. GILDERHUS (1966) found that continuous exposure to low sublethal concentrations stunted growth of young fish but JONES (1964) reported that many fish are not sensitive to even moderate arsenite concentrations and, as pointed out earlier, the ion is not mobile except under low redox potentials in the soil. Dilution of other chemical agents, used only sporadically as far as could be determined, minimized their effects. At Location B, fish and prawns were absent after the intermittent application of insecticide (?Malathion) to the watercress beds, but no mortality was seen, so an avoidance reaction was indicated.

Industrial effluents were not present in sufficient concentration to seriously affect the river. Rubber processing was carried out in two large factories in the lower watershed, one on Tributary K, the other just below Station V, and in several smaller establishments whose output was insignificant. The Tributary K factory wastes containing appreciable amounts of both solid latex trimmings and ammonia coagulant were discharged into the stream after sedimentation in a small lagoon that also served as a cooling-water reservoir. However, the effects of this effluent were masked by the presence, on the tributary, of a large labour settlement from which an extensive domestic waste load originated. Filamentous bacteria and reduced conditions were present except at high discharge, but the malodorous, grey-coloured water was rapidly mixed and neutralized in the main river and no immediate deleterious effects were found from the suspended load of rubber wastes. The other large factory discharged its solid and liquid wastes directly into the river and it was noticeable that below this site at Station V, little bathing or washing activity was carried on in the river by the riparian population. Whether this was a direct result of fouling of the river or of easy availability of piped (but metered) water was problematical. Sundry minor inflows from metal and engineering works and paint-shop effluents via small side drains accounted for the sporadic records made of metal salts in the river, but relative dilution was always large. Timber processing was concentrated in several areas in the lower section, and most sawmills disposed of their by-products by burning; however, a few were sited directly on the river and dumped their wood wastes into the water (see Plate 14). Only local build-up of waterlogged, slowly-decomposing, bottom-smothering sawdust occurred as floods periodically scoured the area, but such materials constituted a significant increase to the organic load, especially as, being only slowly decomposable, they carried their oxygen demand (O.D.) over long distances.

With regard to the above categories of pollutants, the isolated catastrophic combination of heavy effluent, low discharge and accompanying high temperature must be considered the limiting case that may damage the environment, not the modal situation. None of the above was considered an immediate problem in the Gombak, but with future expansion of such activities, the effects of the individual pollutants in the context of local conditions and on the local biota will have to be investigated.

Organic enrichment of the river occurred from a number of sources, the chief of which was the natural input of allochthonous plant materials from the forest and riparian vegetation. The extent of this load was indicated earlier and its almost non-existent effect on the B.O.D., D.O. and C.O.D. levels of the upper stations described. The observed increases in the O.D. and the modest decreases in available oxygen in the lower river may have several explanations. The volume of suspended 'natural' organics, including garden refuse from smallholdings, increased with passage downriver and was further elevated by materials washing out of the paddy field areas. Concomitantly with this inflow, the comminuted materials of forest origin had been in the river sufficiently long to acquire or develop a saprobic flora and fauna and, with the increased nutrient levels of the water, were probably being broken down at a greatly accelerated rate. This was exemplified by the long-term B.O.D. results (Fig. 38). The sources of the increased nutrients were, to a minor degree, direct fertilization and manuring in the rice-fields but more important, the mineralization of domestic sewage. As discussed earlier, much of this took place in side ditches or, in sections of the unsewered city, in earth closets on the banks, from which seepage was extensive. Direct defecation into the river by the rural population was the rule and contributed a minor organic load. More critically, it constituted a public health hazard as, apart from the serious bacterial and viral dangers, ascarid worms were regularly seen in the river and no doubt the viability of eggs of these and other parasites is high at the prevailing pH and temperatures. The nutrients and microbial populations furnished by this effluent (and the unknown phosphate concentrations added from detergent products) provided conditions under which decomposition of the natural organics and sawmill wastes was initiated or certainly enhanced. The separation of the contributions of each to the total load and O.D. was impossible, but it may well be that the conspicuous domestic sewage was less important on a volume basis. In spite of the organic load, oxygen concentrations in the river were never seriously depleted, always exceeding 50% saturation, so that except in local backwaters problems of odour from this source did not occur. Overall the organic load was not serious. It constituted an additional source of allochthonous food that enriched the sediments and water column, both directly as a food source and indirectly by releasing nutrients that promoted other decomposition reactions, without seriously depleting oxygen concentrations. There was no development of filamentous bacteria, perhaps because of physical conditions, but from very preliminary investigations, an extensive saprobic community of protozoa and ciliates developed in the mud.

Inorganic sediments are without question the major pollutant in most Malayan rivers, originating as a result of a wide spectrum of activities classified as 'development' (Plate 15). Historically, open cast pump and dredging tin extraction has denuded large tracts of lowland river alluvium, leaving unstable and almost sterile areas of tailings that are eroded of large loads of fine sediments with every storm. At peak floods restraining bunds frequently rupture or are eroded, releasing catastrophic loads into the watercourses (see DOUGLAS 1970 for review). More recently the chronic problem of erosion as a result of poor land use has drawn increasing attention and the periodic sediment load of many rivers has been recognized as originating in poorly-managed forest or land stripped for primary crops and urban development. In the Upper and Middle Zones, sediments from agricultural and logging activities and from damage done to the watershed by road construction were intermittently transported during and following rain. The return waters from paddy field irrigation carried a continuous suspended load into the river below Station IV and this was supplemented by large concentrations of tin-mining suspensoids below Station V. The effects of such erosional loads were minimal in the upper reaches because at the prevailing velocities little permanent deposition occurred and damage was limited to abrasion. In the river below Station III where competence was less, deposition of eroded materials added to the instability of the benthic biotopes by altering flow patterns and filling in the interstices of the riffle habitats. In many places leaf-drifts and marginal vegetation were buried, sometimes producing local areas of anaerobic conditions. Reference has been made to the diminuation of algal production as a result of increased light attenuation. Similar obvious effects of silt and sand on the lotic eco-system have been described by HYNES (1960), CORDONE and KELLEY (1961), HERBERT et al. (1961), NISBET (1961), EDWARDS (1969), GAMMON (1969) and many others. Much more serious to the river regime is the build-up of sediment banks, particularly in slow-water areas. This loss of channel capacity was the major economic consequence of poor land practice as it and the resultant blockage of channels by debris at poorly designed bridges and culverts were a primary cause of the dis-astrous 1971 floods in the lower Gombak valley and of floods in other locations (references in erosion load section).

From a biological viewpoint, the insidious effects of transported in-organics were more important than these more easily observed pheno-mena. Direct effects on fish were probably limited to a reduction in visual feeding facility as a result of turbidity. Physiologically, most fish larger than fry are little affected by turbidities less than several thousand ppm unless the particles are angular and irritative (see WALLEN 1951, HYNES 1960, JONES 1964, KATZ et al. 1970). However, deposited silt affects spawning gravels by blocking interstitial flow and oxygen exchange and may lead to high egg or larval mortality. No assessment of such effects was possible because the breeding-sites and reproduction habits of the principal fish species are as yet unknown.

The loss of interstitial pore space, burial of detritic and algal foods, silting of attachment sites, blocking of filter-nets and physical abrasion were the main changes effected at the microhabitat level by inorganic pollutants, i.e. a loss of invertebrate niche diversity, particularly in the erosional areas of the river (see ALLEN 1960, HARRISON et al. 1963, CHUTTER 1969b). Direct measurement of such changes was not possible, but, as seen, the Md and pore size were markedly reduced in the lower river. These alterations were not, of course, solely the result of elevated denudation rates, but were a function of topography, gradient and con-tinual natural erosion as well.

The effects of these changes on the flora and fauna, whether attributable to pollution or not, have been noted in detail in previous sections. Overall changes in the algal community were too irregular to permit definition of an enrichment flora as although species composition changed, the adverse silt and light conditions prevented the development of any substantial community or successional sequence of the type described for English rivers by BUTCHER (1947). Fish did not constitute a satisfactory yardstick of water quality because of their mobility and independence from localized factors. If toxic substances or low oxygen concentrations occur, the distribution of tolerant or direct-breathing fish may become important. Neither of these factors applied, at least not at Station V, as a very mixed population was found there. In the lower river at the D.I.D. site, only species that are able to breathe at the surface or have accessory breathing mechanisms were found, indicating that at that location oxygen concentration may occasionally be limiting.

The invertebrate fauna, on the other hand, showed definite alteration of structure. As has been reported by GAUFIN and TARZWELL (1956), HYNES (1960, 1965), BRINKHURST (1965a), CHOUTEAU (1968) and others, the invertebrate community provides a sensitive index of changing conditions that is more realistic than chemical or microbiological data which only indicate instantaneous or short-term conditions. The presence, but more important, the relative abundance of invertebrate taxa with moderately long life cycles reflects the longer-term, including extreme, conditions at a location that governed the development of the community. However, until the habitat requirements, tolerance ranges and inter-specific relationships between the various taxa are known from unaltered locations, the assessment of the severity of various pollutants and their interactive effects causing changes in the fauna cannot be interpreted.

The major differences in the invertebrate faunas between Stations IV and V, the proliferation of the large naidid, tubificid and leech populations, the very large increment in the Chironomidae, particularly the red, blood-gilled *Chironomus* sp. larvae, the enrichment and increase in the mollusc community and the presence of vorticellids was attributed directly to the increased food supply available in the bottom sediments (cf. WARREN et al. 1964) and, for the snails, in the marginal vegetation. Moreover, much of the rest of the community remained unaltered as long as its habitat type was still present, demonstrating that the pollutional effects were positive for those particular taxa that proliferated to high standing stocks and were not deleterious to the rest of the community. Numerically, some elements declined, e.g. Orthocladiinae, *Neoperla*, Coleoptera (Elmidae), Ephemeroptera (except *Baetis* spp. and *Potamanthodes* which became very abundant with the plenitude of food) and most Trichoptera, but whether this was indicative of adverse effects from the organic effluent or an extension of the successive elimination of taxa by changing substrate conditions could not be determined. KREIS and JOHNSON (1958) reported that Ephemeroptera, Plecoptera and Trichoptera were less common below the return of irrigation waters. However, the volume of these was not very great and salinity and silt increases were minor. Much of the silt was precipitated in the ditches in the beds of *Nitella* and only swept into the river when flow rates were increased so that damage from this source was not considered important. At the lowest site on the main river, a classical organic pollution fauna of Tubificidae *(Branchiura)*, Naididae *(Branchiodrilus, Allonais)*, gastropods, eristalid worms and some low-oxygen tolerant Odonata was found and here the rest of the taxa had been eliminated by a combination of substrate obliteration and reduced oxygen tension.

An extensive European and American literature has developed listing 'indicator species' which ostensibly define water quality wherever they occur, but as pointed out by BRINKHURST (1965b, et sqq.), many species, particularly oligochaetes, are found under a wide spectrum of gross biotic conditions, provided that at the microhabitat level, tolerable conditions for that species exist; for the large oligochaetes this means fine silt and organic matter. The 'indicator' status of these worms, the red chironomids and some Crustacea results more from their ability to tolerate and exploit conditions of smothering silt and low oxygen under which other species decline, than from a predilection for polluted conditions. Where large populations of these are found, particularly in less than optimal riffle substrates, pollution certainly may be indicated, but as here, where under 'natural' conditions their preferred habitat (BRINKHURST 1966) is found, their presence may be incidental to any pollution, i.e. in most cases, the absence of an expected species is more indicative of actual conditions (if the ecological preference ranges of that species are known), than is the presence of an association of widely tolerant species.

Comparison of effects of pollution in the Gombak with those in rivers of other geographical areas is difficult. Not only is the biota different, but the physical characters of the watercourses are dissimilar. The South African rivers described by HARRISON (1958b), OLIFF (1960 et sqq.), CHUTTER (1963, 1970) and HARRISON et al. (1963) have similar substrates and current conditions, but not the rich natural organic load from the forest, while the temperate rivers of BUTCHER (1946), HUET (1949) (Europe), HIRSCH (1958 – New Zealand), JOLLY and CHAPMAN (1966 – Australia) and of GAUFIN and TARZWELL (1956), GAUFIN and GAUFIN (1966) and many other recent North American authors have more stable flow regimes, with single or limited flood periods and generally lower temperatures. However, the gross pollutional effects are apparently the same, with the elimination of all but the ubiquitously found association of species that can tolerate low oxygen conditions. The less obvious changes in other Orders and families where some species are tolerant and others are eliminated by even mild pollution cannot be compared until more data on specific taxa are available from tropical rivers, but *Neoperla*, various elmid beetles, *Compsoneuriella* and some cased Leptoceridae, which are members of families often considered clean water forms (HARRISON 1958b), were found in significant numbers on the riffle at Station V. The severe conditions that are reported from rivers elsewhere, in which long stretches became polysaprobic, could not occur in the Gombak as, parallel to the situation in South African rivers, current velocities were sufficient in all but a few pools to provide physical aeration and to remove extensive accumulations of organics. Anaerobic conditions were never found in the surface waters of the river and probably occurred in the muds only at depth.

Summary: Within the terms of the DOUDOROFF and WARREN definition, the serious effects of pollution in the river were riparian flooding resulting from loss of channel capacity as a function of heavy inorganic load, minor nuisance caused by malodorous rubber factory effluents and an unrecognized public health hazard from the presence of raw sewage which did no more than mildly enrich the water. Biotically, however, an acceleration of the loss of ecosystem diversity with longitudinal succession took place comparable to, but not as severe as that reported for Vietnamese streams following military defoliation (WOODWELL 1970) that led in the extreme to the establishment of a typical narrow-based pollution fauna. Because natural changes generated by watershed characteristics acted concurrently with the anthropic agents in the lower reaches and tended to have similar effects on the substrate and the biota, the relative contribution of the pollutants remained problematical, particularly since both were applied over an extended area and their effects were both additive and cumulative.

Remedial action immediately required is a curb on destructive agricultural practices, rehabilitation of mined-out areas and, with the envisaged rebuilding of Route II, concerted planning to reduce the silt load entering the upper river. Careful routing of this enlarged highway across slopes of minimal gradient, immediate stabilization of stripped slopes and in-filled sections and meticulous drainage during cut and fill phases of the construction will be necessary. If these measures are not carried out, the combination of increased run-off from the deforested areas and decreased channel capacity in the vicinity of Kuala Lumpur may generate serious problems.

The other major pollutant, domestic sewage, can only be controlled by the provision of an adequate sewage collection and treatment system and education of the riparian population. Control of industrial effluents, insignificant in the Gombak at present but more severe in the lower Klang valley, must come through enforced legislation based on realistic standards derived for the local situation. These must take into consideration the high mean temperatures, variable reaeration potential and, because of the inorganic loads that prevent extensive floral development, low recovery capacity of the local rivers.

SUMMARY

The physical, chemical and biological features of the Sungai Gombak, West Malaysia, were investigated from September 1969 to July 1970 with a view to comprehensive definition of a Malayan small-river biome in its natural state and description of some effects of anthropic activities on this.

1. The physiography of the catchment, particularly as it affects the drainage system, was described in detail. Based on topographical, lithological and vegetational features and gradient changes and substrate, the river was arbitarily divided into sections for the purposes of the study. Upper Undisturbed and Lower Disturbed Catchment areas were designated for investigation of the hydrologic cycle. To facilitate physio-chemical and biotic studies, Upper, Middle and Lower Zones were delimited and five main sampling stations representative of the major sections of the river were established for routine analysis: Stations I and II in the upper forested area, Station III in the Middle Zone and Stations IV and V in the lower reaches, one above, the other below domestic, rice-field and industrial effluent inflows. Additional locations on the principal tributaries were used as subsidiary stations.

2. Light and temperature conditions were governed primarily by the presence or absence of the evergreen rain forest. Stations I and II were well-shaded, receiving only 4.5 and 7.4% of the available light while the lower three stations were almost fully exposed. Water temperature variations in the river were small, with only slight (1–2 °C) seasonal modification. The ranges of mean daily temperatures were narrow, 23.0 ± 1.7, 22.1 ± 2.0, 26.0 ± 2.0, 26.8 ± 2.3 °C for Stations I, II, IV and V respectively, but showed some longitudinal increase passing downstream, particularly in maxima which periodically exceeded 33 °C at the unshaded stations. Coefficients of thermal astatism, indicative of the stability of the environment, were less than 1.5 for all stations over the study period.

3. Precipitation was recorded by a network of nine gauges and over the 60-week intensive study period, the weighted rainfall for the whole water-shed amounted to 340 cm concentrated in the two inter-monsoon periods October–December and April–June. Considerable rain occurred in all months and rainless periods longer than 10 days were rare. Intensity was greater in the lower foothill region, but frequency was higher in the Upper Catchment. For the 1968/69 water year, the Lower Catchment

had slightly higher than proportional precipitation, 65.1% over 63.6% of the total area.

4. Discharge in the Gombak was highly variable, characterized by rapidly rising and falling flood peaks with each intense storm. Mean discharge for the study period was 5.54 cumec, with extremes of MDD at 16.60 and 2.06 cumec and maximum instantaneous discharge of 31.00 cumec. The upper forested watershed contributed only 28.6% of total annual run-off from an area of 36.4% of the catchment. This discharge amounted to about 40% of total precipitation, indicating some retention capacity and regulation of run-off. The Lower Disturbed Catchment had run-off considerably in excess of its areal proportion and E + T losses of only 46%. These effects were the result of replacement of the natural forest vegetation with monolayer crops and urban development in the lower valley that reduced its capacity to stabilize discharge. A preliminary rationalization of the hydrologic cycle for the catchment was made.

5. The bottom substrates were characterized using the Md and 1st and 3rd quartiles of the *phi* size distribution and related to the lithologies, gradients and current velocities that maintain the bottom configuration of the various secions. The upper stations were characterized by large boulders with interspersed gravels and sands and by coarse sand bottoms in the pool areas. Progressive size decrease in riffle substrate occurred downstream and depositional areas at the lower two stations accumulated some silt and mud. Interstitial pore space varied little between stations, with 21.5–27.0% void volume in riffle areas and 9.0–14.5% of depositional area sediments as free space. Substrate stability for all but the largest particles was ephemeral, with disruption of the surface layers of most depositional reaches and of some riffle areas in the forest river with every spate.

6. Dissolved and suspended erosional loads were measured and denudation processes under jungle and in deforested areas discussed. Chemical weathering contributed a significant part of the total mineral load, much of it silica, during modal flow periods, but significant natural erosion and land slumping occurred in the upper reaches, supplying a moderate suspended load. Poor logging practices, indiscriminate and abusive land clearance for cultivation and fluvial tin mining, the last two particularly in the lower catchment, added large concentrations of inorganic sediments to the river under the intense rainfall conditions. Mean total loads at normal flow ranged from 45–80 ppm from Station I–V, of which 37–50 ppm was total dissolved load with 14–24 ppm organic. However, flood loads of up to 500 ppm, almost all suspended inorganics, were common. Extremes of 3000+ ppm at Station II, 10700 at Tributary E and 5000+ at the D.I.D. gauging site were recorded. Erosion in the Gombak, although not as severe as in some other Southeast Asian rivers, is considerably elevated over that in forested, control water-

sheds. Diminished channel capacity in the lower river and concomitant flooding are already a problem.

7. The chemical environment was defined and the relative solute contributions of precipitation (20%), tributaries draining watersheds of different, limited area, lithologies, and effluent inflows were assessed. The upper, unmodified river was characterized by low conductivity; circumneutral pH; low alkalinity and buffering capacity, some of which was provided by the relatively abundant silica salts; low calcium, magnesium and nutrient salt levels; dissolved oxygen near saturation. Progressive enrichment occurred downstream for all ions except silicon, with elevated nutrient salt levels and the addition of minor concentrations of noxious elements below the urban-industrial areas. Comparisons were drawn between this impoverished environment and other tropical lotic waters.

8. Assessment of the organic environment was made by measuring the input of allochthonous materials, the dissolved and suspended organic loads and the amounts of organic material in the bottom sediments. The natural organic load of forest origin was resistant to decomposition and exerted little B.O.D. on the upper river, but with higher nutrient levels and admixture of domestic effluents in the lower catchment, conditions for break-down improved and much of the material began to degrade. Oxygen deficits from high B.O.D.s were prevented by rapid reaeration as a result of turbulence and the high surface to volume ratio of the river. The abundance of sedimentary organics was dependent on antecedent discharge conditions and the stability of the inorganic substrates and was therefore highly variable. The large leaf-twig particles of the forest were progressively comminuted downstream and supplemented by agricultural detritus and, in the Lower Zone, by sawdust, rubber scraps, paper and other refuse materials. Preliminary caloric values of the suspended and bottom organic fractions were determined.

9. Algal subcommunities were discussed in four broad categories of epilithic, epipelic, epiphytic and planktonic-culture for each river section. The affinities, spatial distribution of the important taxa and differences between stations were described in relation to physico-chemical conditions. Primary production rates were estimated by providing placebo colonization substrates of glass, polyethylene and asbestos-cement and measuring the accumulated biomass and Chl. a (net production). On the most natural asbestos-cement substrate, the mean rates of Chl. a accumulation were 0.05, 0.16, 1.13, 0.53, 0.16 mg/sq.m/day, with efficiencies of energy conversion of only 0.03, 0.05, 0.03, 0.01, 0.01%, for Stations I to V, respectively. Shading by the forest canopy and deficient nutrient supply kept periphytic production low at the upper stations and light attenuation and abrasion by silt loads below Station IV accounted for the poor development there. Only at Stations III and IV were conditions amenable to modest algal growth. Accumulation rates of organic matter (AFDW) were dependent primarily on antecedent flow

conditions, with spates often completely denuding the bottom of both algae and deposited organics. Floral community succession and stability were discussed with respect to changing pigment ratios and the taxonomic diversity at each station.

10. The invertebrate fauna of the river and tributaries was described and a preliminary list of taxa with notes on ecological preferences for substrate and food presented. The faunal characteristics of each river zone were annotated and the factors effecting specific segregation in both adult and immature stages and longitudinal discontinuity of occurence were discussed for the important groups. The Upper Tributaries and main river in the forest have similar species assemblages, with some notable differences, while the lowland river has a number of taxa, particularly in the non-insects, not found higher in the river, in addition to a block of ecotypic species found ubiquitously in gravel riffles and/or sandy pools. A large proportion of the invertebrate fauna is dependent on allochthonous leaves and detritus as a food source. Only a limited number of dependent algivores were found; specialist carnivorous taxa constituted a significant part of the fauna, particularly in the forest river.

Species diversity decreased downstream with the loss of the variety of niches afforded by the riffle and bank-root biotopes which were replaced by more homogeneous substrates modified by siltation and a marginal unstable plant biotope that was subject to periodic scour.

Life cycle data, where available from sizing of immature specimens and adult flight records, indicated continuous or semi-continuous asynchronous breeding patterns for most of the invertebrates.

Quantitative data were obtained for erosional and depositional habitats at each station for 15 four-week periods from October 1968. Numerically, the riffle areas supported a larger standing stock than the pools except at Station II where the mean numbers/sq.m were similar for the two areas. Fluctuations in benthic numbers as a result of spates were marked. In depositional areas, mean densities were between 2300 and 3900 for Stations I–IV and about 30000 for the enriched sediments at Station V where Oligochaeta and Chironomidae proliferated. The riffle fauna increased progressively in abundance downstream with mean populations/sq.m of 2600, 2900, 7200, 8900, 40350 at Stations I–V. Similar relationships were found in terms of biomass with 0.2–1.4 g dry wt/sq.m for the unenriched river and 4.1 g at Station V. Physical and ethological factors governing abundance were discussed, but the state of the substrate, both areally and as a food source, was considered the prime factor governing density. The seasonal fluctuations of populations of the major taxa were described with respect to substrate conditions and food availability. Assessment of the total lotic community was not possible because of difficulties inherent in trying to quantify the fauna of the boulder and bank-root biotopes.

12. A deep sediment sampler was described and the results obtained

on the vertical distribution of the benthic fauna in an area of uniform substrate were analysed. These data demonstrated that stream animals occur in significant numbers deep in the bottom sediments and that, at most, only half the benthos lives in the upper 10 cm usually 'quantitatively' sampled in stream surveys. The significance of a deep-living population in regulating density, recolonizing denuded areas and optimizing utilization of available food resources was discussed.

13. Secondary production could not be assessed quantitatively, but the factors affecting the productive capacity of the river were reviewed and some speculation on probable rates was made.

14. Invertebrate drift was measured at the two upstream stations as an adjunctive sampling method to determine the relationships between benthic density, river conditions and drift in a non-seasonal ecosystem and as a measure of one source of fish food. Mean drift rates (number of individuals/24h past a transect) were 15.5×10^3 and 222.8×10^3 for Stations I and II respectively, giving mean drift densities of 1.8 and 1.6 individuals/cu.m. Biomass relationships were obscured by contributions from one or two large terrestrials, but mean total drift amounted to 7 and 112 g/day at the two stations. The terrestrial component was an important constituent at all times, making up about one-fifth of the total drift by numbers, but almost 50% by weight at Station II. The endogenous aquatic drift was variable in composition, but included almost all the benthic taxa. The factors governing the amount and distance of drift and the diel drift periodicity found in many groups were examined. A nocturnal increase in drift rate was commonly found, with an overall day/night ratio for the aquatic drift component of less than 0.5. This was ascribed to an alteration of activity pattern at night that increased the propensity to drift. There was little periodicity to the exogenous drift. The percentage of the bottom fauna present in the drift at any instant was low (0.013, 0.018% at Stations I and II respectively), ruling out any volitive movement into the water column by the fauna. No apparent correlation existed between benthic density and drift density so that no basis was available for estimating a production rate by using drift as a measure of excess growth over carrying capacity of the bottom.

Vectored flight movements of the imagines of the Ephemeroptera, Trichoptera and Plecoptera failed to support the 'colonization cycle' concept. Downstream flight dominated the small day and night catches irrespective of wind direction and no deviations in behaviour were recorded for any particular sexual group. Ovigerous females apparently moved downstream with the rest of the population. The rationale for a colonization cycle in the tropical river context was discussed in the light of these findings with a general conclusion that where dispersal pressure in both adult and immature stages is minimal because of low population density, such a cycle to ensure utilization of all potential aquatic niches may be unnecessary.

15. Twenty-eight species of fish were identified from the Gombak River and data on the distributions, relative abundance and tentative preferred diets were listed. Channel size and suitable substrate determined the resident populations, but many of the cyprinid species particularly were nomadic. A large part of the ichthyofauna was indiscriminantly eury-phagous, deriving much of its food from allochthonous sources, (fruit, flowers, leaves, terrestrial invertebrates), but also taking endogenous benthos as available. The specialized feeders (except the detritivore *Osteochilus* at Station V) constituted only a minor part of the community. Successional changes were mainly additional, although replacement of several species that had ranges limited to the upper forest river occurred and others became less abundant as conditions changed. Diversity, as determined by MARGALEF's index, increased linearly with stream order, indicating the greater number and complexity of spatial and feeding niches for fish in the larger channels.

The fishery on the river was surveyed and total yield in terms of numbers and biomass and catch per unit estimated. An annual crop of 30 metric tons from a water surface area of about one-third of a sq. km was reportedly taken, representing a yield of 87 g/sq.m. Catch per unit effort was high, 35, 11, 9 fish/h for the Upper, Middle and Lower Zones respectively, but all fish caught were retained irrespective of size and the biomass/unit effort was uniform in all zones at about 250 g/h. The ability of the fish population to sustain this fishing pressure was debated, but with continuous recruitment and potentially high production rates with the surfeit of most foods, the high yield was probably accountable.

The other vertebrates that had interaction with the aquatic ecosystem were listed with brief annotations.

16. The preliminary physiographic divisions of the watercourse were discussed with respect to established classification schemes. Successional relationships within the biota were examined and tentative zonation of the river, using the system of ILLIES (1961a) was made. The fauna divided into two groups, the rhithronic but warm stenothermal, montane specialists and the more eurytopic forms for which biotic factors probably limited distribution. The Upper Tributaries were considered to be Rhithron and perhaps the whole forest zone river falls in this classification if the temperature criterion were modified. The Middle and Lower Zones were definitely Potamon on both physico-chemical and faunistic grounds. The possible role of the forest in maintaining montane con-ditions for the fauna below the normal altitudinal boundaries as long as substrate conditions remain unchanged was conjectured.

17. The effects of human activities on the river were difficult to differ-entiate from the changes resulting from natural modification of the watercourse. Chemically, pollution added only mild fertilization con-centrations of nutrients and apparently had little influence on biotic conditions. Toxic industrial wastes were present only rarely and dilution

was adequate. Domestic sewage constituted a public health hazard and added to the organic load, but oxygen saturation levels, even at the high temperatures, were never seriously depleted. Inorganic sediment loads from logging, land clearance and tin mining were the chief pollutant causing loss of channel competence and various direct and indirect effects on the aquatic community. Periphytic growth was reduced by shading and physical smothering at the lower stations and diversity of the invertebrate fauna decreased as a result of modification of the interestitial niches of the riffle and sand biotopes. A restricted, but rich worm-midge-snail pollution fauna proliferated under extreme conditions below Station V and many fish species were eliminated.

18. In this preliminary survey of a Malayan river, of necessity carried out over only a restricted period, much of the discussion has had to be conjectural and the inference drawn speculative. Until the faunistic base is better defined and more extensive work done on other streams, there is no basis for comparison except that used here, with work on other equatorial rivers where possible, but more often non-tropical references that are undoubtedly not strictly analogous. The only foundation for interpretation of ecological relationships is extensive experience with the local ecosystem. It is hoped that this initial definition of the biome may provide a reference for future work.

Disposition of Collections: At the time of writing, much of the material remained unidentified in the hands of the taxonomists listed previously. Final disposition of type materials of new species will be at their discretion. Sample specimens of most of the insect taxa and fish are deposited with the Zoology Division, School of Biological Sciences, University of Malaya, and names will be appended if and when they become available.

ACKNOWLEDGEMENTS

A study of this type involves the cooperation of a number of people and the generous assistance of the following is particularly appreciated. My wife, Judy, helped in all aspects of the work, and without her constant encouragement much would not have been possible. The various taxonomists already mentioned in the appropriate places in the text worked collections from the river. The Drainage and Irrigation Department, Malaysia (Mr. TAN HOE TIM) and BINNIE and Partners (M) (Messrs C. J. A. BINNIE and G. J. PILLAI) made available their precipitation and discharge records for the Gombak. The Rubber Research Institute of Malaya (Dr. P. R. WYCHERLEY) provided insolation data for the Kuala Lumpur region for 1968/69. The Forest Research Institute (Mr. P. F. BURGESS) allowed the use of maximum-minimum air temperature data applicable to the Lower Catchment. Dr. G. A. PROWSE made available the library facilities of the Tropical Fish Culture Research Institute, Malacca. The O.C.P.D. Selangor issued a curfew pass to allow the field work to be continued during the period of civil unrest in 1969. The Department of Geology, University of Malaya, generously permitted use of its Atomic Absorption Spectrophotometer. Dr. R. P. C. MORGAN made available sediment analysis apparatus in the Department of Geography, University of Malaya, and Dr. I. DOUGLAS, visiting lecturer to that department, provided stimulating discussion on the problems of tropical erosion and permitted use of some of his unpublished data. The riverine inhabitants of the Ulu Gombak condoned intrusions into their kampongs and tampering with their river, and cooperated in the fishing survey.

Various members of the School of Biological Sciences, University of Malaya, assisted materially. The members of the research supervisory committee, Drs. J. I. FURTADO, J. A. BULLOCK and A. J. BERRY freely offered advice. INCHE BAH TERA and INCHE MOHAMED NASIR ALANG acted as interpreters during the survey of the river fishery. INCHE MAHMUD BIN SIDER assisted with the electrofishing. Mr. N. P. SHUNMUGAM typed the original manuscript and Mr. YEE FOOK YOON prepared the final drafts of some of the figures. The provision of physical facilities by the School is gratefully acknowledged.

The final draft was typed by my mother Mrs. E. J. BISHOP whose forbearance must be recognized as must the help and advice of the staff of the publishing house: Dr. W. Junk b.v., Den Haag.

The work was supported by a scholarship from the Kementarian

Pelajaran Malaysia, under the Commonwealth Scholarship and Fellowship Plan – Malaysian Awards 1968, which generously granted a nine-month extension to the original two-year tenure period to allow completion of the work. In addition, a part-time demonstratorship in the School of Biological Sciences was held for the 1968/69, 1969/70, 1970/71 academic sessions.

Financial assistance with the publication of the illustrations was provided by an extraordinary grant from the Inland Waters Directorate of Environment Canada: this support is most gratefully acknowledged.

REFERENCES

ABELL, D. L. 1961. The role of drainage analysis in biological work on streams. *Verh. Internat. Verein. Limnol.* 14: 533–537.

ALABASTER, J. S. and HARTLEY, W. G. 1962. The efficiency of a direct current electric fishing method in trout streams. *J. Anim. Ecol.* 31: 385–388.

ALBRECHT, M.-L. 1953. Die Plane und andere Flämingbäche (Ein Beitrag zur Kenntnis der Fliessgewässer der Endmoränenzüge der Norddeutschen Tiefebene). *Z. Fisch. (N.F.)* 1: 390–473.

ALBRECHT, M.-L. 1959. Die quantitative Untersuchung der Bodenfauna fliessender Gewässer. *Z. Fisch, (N.F.)* 8: 481–550.

ALBRECHT, M.-L. 1968. Die Wirkung des Lichtes auf die Quantitative Verteilung der Fauna im Fliessgewässer. *Limnologica* 6: 71–82.

ALEXANDER, J. B. 1959. Pre-Tertiary stratigraphic succession in Malaya. *Nature, Lond.* 183: 230–232.

ALEXANDER, J. B. 1962. A short outline of the geology of Malaya with special reference to Mesozoic orogeny. *Monogr. Amer. geophys. Un.* 6: 81–86.

ALEXANDER, J. B. 1968. The geology and mineral resources of the neighbourhood of Bentong, Pahang, and adjoining portions of Selangor and Negri Sembilan. *Malaya Geol. Survey Mem. (N.S.)* 8: 1–250.

ALFRED, E. R. 1964. Notes on a collection of fresh-water fishes from Penang. *Bull. Nat. Mus. Singapore* 32: 143–154.

ALFRED, E. R. 1966. The fresh-water fishes of Singapore. *Zool. Verh.* 78: 1–68.

ALFRED, E. R. 1969. Conserving Malayan fresh-water fishes. *Malay. Nat. J.* 22: 69–74.

ALLANSON, B. R. 1961. The physical, chemical and biological conditions in the Jukskei-Crocodile river system. *Hydrobiologia* 18: 1–76.

ALLANSON, B. R. and KERRICK, J. E. 1961. A statistical method for estimating the number of animals found in field samples drawn from polluted rivers. *Verh. Internat. Verein. Limnol.* 14: 491–494.

ALLEN, K. R. 1949. Some aspects of the production and cropping of fresh waters. *Trans. R. Soc. New Zealand* 77: 222–228.

ALLEN, K. R. 1951. The Horokiwi Stream – A study of trout population. *N.Z. Mar. Dept. Fish. Bull.* 10: 1–231.

ALLEN, K. R. 1959. The distribution of stream bottom faunas. *Proc. N. Z. Ecol. Soc.* 6: 5–8.

ALLEN, K. R. 1960. Effect of land development on stream bottom faunas. *Proc. N. Z. Ecol. Soc.* 7: 20–21.

ALLEN, K. R. 1969. Distinctive aspects of the ecology of stream fishes: a review. *J. Fish. Res. Bd. Canada* 26: 1429–1438.

ALLEN, S. E., CARLISLE, A., WHITE, E. J. and EVANS, C. C. 1968. The plant nutrient content of rain water. *J. Ecol.* 56: 497–504.

AMBÜHL, H. 1959. Die Bedeutung der Strömung als ökologischer Faktor. *Schweiz. Z. Hydrol.* 21: 133–264.

AMBÜHL, H. 1961. Die Strömung als physiologischer und ökologische Faktor. *Verh. Internat. Verein. Limnol.* 14: 390–395.

(A)merican (P)uplic (H)ealth (A)ssociation 1965. Standard Methods for the Examination of Water and Wastewater including Bottom Sediments and Sludges. A.P.H.A., Inc., New York, 12th edition, 769 pp.

ANDERSON, N. H. 1966. Depressant effect of moonlight on activity of aquatic insects. *Nature, Lond.* 209: 319–320.

ANDERSON, N. H. 1967. Biology and downstream drift of some Oregon Trichoptera. *Can. Ent.* 99: 507–521.

ANDERSON, N. H. and LEHMKUHL, D. M. 1968. Catastrophic drift of insects in a woodland stream. *Ecology* 49: 198–206.

ANDERSON, V. G. 1945. Some effects of the atmospheric evaporation and transpiration on the composition of natural waters in Australia. *Proc. R. Aust. Chem. Inst.* 12: 41–68.

*ANGELIER, E. 1953. Recherches écologiques et biogéographiques sur la faune des sables submergés. *Arch. Zool. Exptl. Gen.* 90: 37–162.

ANGELIER, E. 1962. Remarques sur la répartition de la faune dans le milieu interstitiel hyporhéique. *Zool. Anzeig.* 166: 351–356.

ANONYMOUS. 1953. Methods of Chemical Analyses as applied to Sewage and Sewage Effluents. H.M.S.O. 48 pp.

ANONYMOUS. 1970. Polluted Malaysia. *Malay. Nat. J.* 23: 45–46.

ARMITAGE, K. B. 1961. Distribution of riffle insects of the Firehole River, Wyoming. *Hydrobiologia* 17: 152–174.

ARTHUR, J. W. and HORNING, W. B. II. 1969. The use of artificial substrates in pollution surveys. *Amer. Midl. Nat.* 82: 83–89.

ASCHOFF, J. 1966. Circadian activity pattern with two peaks. *Ecology* 47: 657–662.

BAALSRUD, K. and HENRIKSEN, A. 1964. Measurement of suspended matter in stream water. *J. Amer. Wat. Wks. Ass.* 56: 1194–1200.

BACKHAUS, D. 1969. Ökologische Untersuchungen an den Aufwuchsalgen der obersten Donau und ihrer Quellflüsse V. Biomassenbestimmung und Driftmessungen. *Arch. Hydrobiol. Suppl.* XXXVI: 1–26.

BACKIEL, T. and LE CREN, E. D. 1967. Some density relationships for fish population parameters. Pp. 261–293, in: GERKING, S. D. (ed.). The Biological Basis of Freshwater Fish Production, Blackwell Sci. Publ., Oxford, 495 pp.

BADCOCK, R. M. 1954. Comparative studies in the populations of streams. *Ann. Rept. Inst. Freshw. Res. Drottningholm* 35: 38–50.

BAILEY, R. G. 1966. Observations on the nature and importance of organic drift in a Devon river. *Hydrobiologia* 27: 353–367.

BAKER, R. A. 1969. Trace organic contaminant concentration by freezing. III. Ice washing. *Wat. Res.* 3: 717–730.

BANKS, N. 1931. Some neuropteroid insects from the Malay Peninsula. *J. Fed. Malay States Mus.* 16: 377–379.

BANKS, N. 1938. Further neuropteroid insects from Malaya. *J. Fed. Malay States Mus.* 18: 221–223.

BARNARD, J. L. 1970. Benthic ecology of Bahia de San Quintin, Baja California. *Smithsonian Contrib. Zoology* No. 44: 1–60.

BEAUCHAMP, R. S. A. 1953. Sulphates in African inland waters. *Nature, Lond.* 171: 769–771.

BEERS, G. D. and NEUHOLD, J. M. 1968. Measurement of stream periphyton on paraffin-coated substrates. *Limnol. Oceanogr.* 13: 559–562.

*BERG, A. 1961. Rôle écologique des eaux de la Cuvette congolaise sur la croissance de la jacinthe d'eau *(Eichhornia crassipes* (Mart.) Solms.) *Acad. roy. Sci. d'Outre-Mer, N.S.* 12: 1–120.

BERG, C. O. 1953. Sciomyzid larvae (Diptera) that feed on snails. *J. Parasit.* 39: 630–636.

BERG, C. O. 1964. Snail-killing sciomyzid flies: biology of the aquatic species. *Verh. Internat. Verein. Limnol.* 15: 926–932.

---

* not seen in the original.

BERG, K. 1943. Physiographical studies on the River Susaa. *Folia Limnol. Scand.* 1: 1–174.

BERG, K., BOISSON-BENNIKE, S. A., JONASSON, P., KEIDING, J. and NIELSON, A. 1948. Biological studies on the River Susaa. *Folia Limnol. Scand.* 4: 1–318.

BERG, K. and OCKELMANN, K. W. 1959. The respiration of freshwater snails. *J. exp. Biol.* 36: 690–708.

BERRY, A. J. 1963. An introduction to the non-marine molluscs of Malaya. *Malay. Nat. J.* 17: 1–17.

BERRY, M. J. 1956. Erosion control on Bukit Bakar, Kelantan. *Malay. For.* 19: 3–11.

BERTHÉLEMY, C. 1967. Sur l'écologie comparée des Plécoptères, des *Hydraena* et des Elminthidae des Pyrénées. *Verh. Internat. Verein. Limnol.* 16: 1727–1730.

BERTHÉLEMY, C. 1968. Contribution à la connaissance des Leuctridae. *Ann. Limnol.* 4: 175–198.

BESCH, W. and HOFMANN, W. 1968. Le macrobenthos sur des substrats de polyéthylène dans les eaux courantes. 2. La Steinach, une rivière de la zone à truite. *Ann. Limnol.* 4: 235–263.

BESCH, W., HOFMANN, W. and ELLENBERGER, W. 1967. Das Makrobenthos auf Polyäthylensubstraten in Fliessgewässern. 1. Die Kinzig ein Fluss der unteren Salmoniden und oberen Barbenzone. *Ann. Limnol.* 3: 331–367.

BINNIE and PARTNERS, 1970. Kuala Lumpur Hydrological Survey. Volume III. Streamflow and Rainfall Data 1969. Govt. of Selangor, Kuala Lumpur, 74 pp.

BINNS, W. O. 1969. Forests, rainfall and run-off. *Yb. Ass. River Auth.* 1969: 19–33.

BISHOP, J. E. 1968. Movements of the invertebrate fauna in a stream ecosystem. Unpublished M. Sc. Thesis, Univ. Waterloo, Waterloo, Ont., 187 pp.

BISHOP, J. E. 1969. Light control of aquatic insect activity and drift. *Ecology* 50: 371–380.

BISHOP, J. E. and HYNES, H. B. N., 1969a. Upstream movements of the benthic invertebrates in the Speed River, Ontario. *J. Fish. Res. Bd. Canada.* 26: 279–298.

BISHOP, J. E. and HYNES, H. B. N. 1969b. Downstream drift of the invertebrate fauna in a stream ecosystem. *Arch. Hydrobiol.* 66: 59–90.

BISWAS, K. 1929. Papers on Malayan aquatic biology. IV. Freshwater algae (with addendum). *J. Fed. Malay States Mus.* 14: 404–435, 479–481.

BLACHE, J. 1951. Aperçu sur le plancton des eaux douces du Cambodge. *Cybium* 6: 62–94.

BLAIR, K. G. 1927. Aquatic Lampyrid larva from S. Celebes. *Trans. Ent. Soc. Lond.* 1927: 43–45.

BLUM, J. L. 1956a. The ecology of river algae. *Bot. Rev.* 22: 291–341.

BLUM, J. L. 1956b. The application of the climax concept to algal communities of streams. *Ecology* 37: 603–604.

BLUM, J. L. 1957. An ecological study of the algae of the Saline River, Michigan. *Hydrobiologia* 9: 361–408.

BLUM, J. L. 1960. Algal populations in flowing water. *Spec. Publs. Pymatuning Lab. Fld. Biol.* 2: 11–21.

BOCCARDY, J. A. and COOPER, E. L. 1963. The use of rotenone and electrofishing in surveying small streams. *Trans. Amer. Fish. Soc.* 92: 307–310.

BOCHKOV, A. P. 1970. Forest influence on river flows. *Nature and Resources* 6: 10–11.

BOTOSANEANU, L. 1960. Sur quelques regularités observées dans le domaine de l'écologie des insectes aquatiques. *Arch. Hydrobiol.* 56: 370–377.

BOURNAUD, M. 1963. Le courant, facteur écologique et éthologique de la vie aquatique. *Hydrobiologia* 21: 125–165.

BOVBJERG, R. V. 1959. Density and dispersal in laboratory crayfish populations. *Ecology* 40: 504–506.

BOVBJERG, R. V. 1964. Dispersal of aquatic animals relative to density. *Verh. Internat. Verein. Limnol.* 15: 879–884.

BOYCOTT, A. E. 1936. The habitats of fresh-water mollusca in Britain. *J. Anim. Ecol.* 5: 116.

BRANDT, R. A. M. 1968. Description of new non-marine molluscs from Asia. *Arch. Molluskenkunde* 98: 213–289.

Bray, J. R. and Gorham, E. 1964. Litter production in forests of the world. *Adv. ecol. Res.* 2: 101–157.

Brinkhurst, R. O. 1964. Observations on the biology of the Tubificidae (Oligochaeta). *Verh. Internat. Verein. Limnol.* 15: 858–863.

Brinkhurst, R. O. 1965a. Observations on the recovery of a British river from gross organic pollution. *Hydrobiologia* 25: 9–51.

Brinkhurst, R. O. 1965b. The biology of the Tubificidae with special reference to pollution.
Pp. 56–65, in: Tarzwell, C. M. (ed.). Biological Problems in Water Pollution. Third Seminar 1962. U.S. Pub. Hlth. Serv. Publ. No. 999–WP–25, 424 pp.

Brinkhurst, R. O. 1966. The Tubificidae (Oligochaeta) of polluted water. *Verh. Internat. Verein. Limnol.* 16: 854–859.

Brinkhurst, R. O. 1967. Sampling the benthos. Cyclostyled MS. Great Lakes Inst. Rep. PR 32. Univ. Toronto, Toronto, 6 pp.

Brinkhurst, R. O. 1969. Observations on the food of the aquatic Oligochaeta (abstract). *Verh. Internat. Verein. Limnol.* 17: 829–830.

Brinkhurst, R. O. and Chua, K. E. 1969. Preliminary investigation of the exploitation of some potential nutritional resources by three sympatric tubificid oligochaetes. *J. Fish. Res. Bd. Canada* 26: 2659–2668.

Brinkhurst, R. O. and Kennedy, C. R. 1965. Studies on the biology of the Tubificidae (Annelida, Oligochaeta) in a polluted stream. *J. Anim. Ecol.* 34: 429–443.

Brock, T. D. 1967. Relationship between standing crop and primary productivity along a hot spring thermal gradient. *Ecology* 48: 566–571.

Brock, T. D. and Brock, M. L. 1967. The measurement of chlorophyll, primary productivity, photophosphorylation and macromoleculues in benthic algal mats. *Limnol. Oceanogr.* 12: 600–605.

Brook, A. J. 1955. The aquatic fauna as an ecological factor in studies of the occurrence of freshwater algae. *Rev. Algol.* 1: 141–145.

Brown, D. S. 1960. The ingestion and digestion of algae by *Chloeon dipterum* L. (Ephemeroptera). *Hydrobiologia* 16: 81–96.

Brown, D. S. 1961. The food of the larvae of *Chloeon dipterum* L. and *Baetis rhodani* Pictet (Insecta-Ephemeroptera). *J. Anim. Ecol.* 30: 55–76.

Brown, E. S. 1970. Nocturnal insect flight direction in relation to the wind. *Proc. R. ent. Soc. Lond.* (A) 45: 39–43.

Brundin, L. 1951. The relation of O_2-microstratification at the mud surface to the ecology of the profundal bottom fauna. *Ann. Rept. Inst. Freshw. Res. Drottningholm* 32: 32–42.

Bullard, W. E., Jr. 1965. Role of watershed management in the maintenance of suitable environments for aquatic life.
Pp. 265–269, in: Tarzwell, C. M. (ed.). Biological Problems in Water Pollution. Third Seminar 1962. U.S. Pub. Hlth. Serv. Publ. No. 999–WP–25, 424 pp.

Bullock, J. A. 1969. Population estimation in the torrent frog, *Amolops larutensis* (Anura: Ranidae). *J. Zool.* 159: 167–180.

Bullock, J. A. and Furtado, J. I. 1968. Population assessment and movement in *Ptilomera dromas* Breddin (Hemiptera-Gerridae). *Arch. Hydrobiol.* 64: 121–130.

Burgess. P. F. 1969. Ecological factors in hill and mountain forests of the States of Malaya. *Malay. Nat. J.* 22: 119–128.

Burtt Davy, J. 1938. The classification of tropical woody vegetation types. Imperial Forestry Institute, Oxford. Inst. Paper No. 13.

Buscemi, P. A. 1966. The importance of sedimentary organics in the distribution of benthic organisms. *Spec. Publs. Pymatuning Lab. Ecol.* 4: 79–86.

Butcher, R. W. 1932. Studies on the ecology of rivers. II. The microflora of rivers with special reference to the algae on the river bed. *Ann. Bot.* 46: 813–861.

Butcher, R. W. 1946. The biological detection of pollution. *J. Inst. Sew. Purif.* 2: 92–97.

Butcher, R. W. 1947. Studies on the ecology of rivers. VII. The algae of organically enriched waters. *J. Ecol.* 35: 186–191.

Butcher, R. W., Pentelow, F. T. K. and Woodley, A. 1930. Variations in composition of river water. *Int. Revue ges. Hydrobiol.* 24: 47–80.

Carlsson, G. 1967. Environmental factors influencing blackfly populations. *Bull. Wld. Hlth. Org.* 37: 139–150.

Castenholz, R. W. 1960. Seasonal changes in the attached algae of freshwater and saline lakes in Lower Grand Coulee, Washington. *Limnol. Oceanogr.* 5: 1–28.

Castenholz, R. W. 1961. An evaluation of a submerged glass method of estimating production of attached algae. *Verh. Internat. Verein. Limnol.* 14: 155–159.

Central Electricity Board. 1956. River flow in the Cameron Highlands. Typescript. Hydro-Electrical Tech. Memor. No. 3. C.E.B. Fed. Malay. Kuala Lumpur, 28 pp.

Chapman, D. W. 1966. Food and space as regulations of salmonid populations in streams. *Amer. Naturalist* 100: 345–357.

Chapman, D. W. 1967. Production in fish populations.
Pp. 3–29, in: Gerking, D. S. (ed.). The Biological Basis of Freshwater Fish Production, Blackwell Sci. Publ., Oxford. 495 pp.

Chapman, D. W. and Demory, R. 1963. Seasonal changes in the food ingested by aquatic insect larvae and nymphs in two Oregon streams. *Ecology* 44: 140–146.

Charlton, F. G. 1964. Standard catchments in the estimation of flood flows. *J. trop. Geog.* 18: 43–53.

Chaston, I. 1968. Endogenous activity as a factor in invertebrate drift. *Arch. Hydrobiol.* 64: 324–334.

Chaston, I. 1969a. A comparison of the activity patterns of the aquatic larvae *Protonemura meyeri* (Plecoptera) and *Chaoborus punctipennis* (Diptera). *Amer. Midl. Nat.* 82: 302–307.

Chaston, I. 1969b. The light threshold controlling the periodicity of invertebrate drift. *J. Anim. Ecol.* 38: 171–180.

Chebotarev, N. P. 1962. Theory of stream runoff. Israel Program for Scientific Translations. Jerusalem 1966. 464 pp.

Cheng, L. and Fernando, C. H. 1969. A taxonomic study of the Malayan Gerridae (Hemiptera: Heteroptera) with notes on their biology and distribution. *Oriental Insects* 3: 97–160.

Chia, L. S. 1967. Meteorological observations 1962–1966. Typescript. Dept. of Geography, University of Malaya, Kuala Lumpur.

Chouteau, F. 1968. Influence de la pollution industrielle et domestique sur les populations animales de la rivière Isère au cours de sa traversée de la région grenobloise. *Trav. Lab. Hydrobiol. Piscic. Univ. Grenoble* 59/60: 39–63.

Chutter, F. M. 1963. Hydrobiological studies on the Vaal River in the Vereeniging Area. Part 1. Introduction, water chemistry and studies on the fauna of habitats other than muddy bottom sediments. *Hydrobiologia* 21: 1–65.

Chutter, F. M. 1969a. The distribution of some stream invertebrates in relation to current speed. *Int. Revue ges. Hydrobiol.* 54: 413–422.

Chutter, F. M. 1969b. The effects of silt and sand on the invertebrate fauna of streams and rivers. *Hydrobiologia* 34: 57–76.

Chutter, F. M. 1970. Hydrobiological studies in the catchment of Vaal Dam, South Africa. Part 1. River zonation and the benthic fauna. *Int. Revue ges. Hydrobiol.* 55: 445–494.

Chutter, F. M. 1972. A reappraisal of Needham and Usinger's data on the variability of a stream fauna when sampled with a Surber sampler. *Limnol. Oceanogr.* 17: 139–141.

Chutter, F. M. and Noble, R. G. 1966. The reliability of a method of sampling stream invertebrates. *Arch. Hydrobiol.* 62: 95–103.

*Cianficconi, F. and Riatta, M. 1957. Impiego di pietre artificiali per l'analisi quantitativa delle colonizzazioni faunistiche dei fondi potamici. *Boll. Pesca Piscic. Idrobiol.* N.S. 12: 229–335.

CLAUS, G. 1961. Monthly ecological studies on the flora of the Danube at Vienna in 1957–58. *Verh. Internat. Verein. Limnol.* 14: 459–465.

COCHE, A. G. 1967. Fish culture in rice fields. A world-wide synthesis. *Hydrobiologia* 30: 1–44.

COCHRAN, W. G. and COX, G. M. 1959. Experimental Designs. Charles E. Tuttle Co., Tokyo., 611 pp.

COLEMAN, M. J. and HYNES, H. B. N. 1970. The vertical distribution of the invertebrate fauna in the bed of a stream. *Limnol. Oceanogr.* 15: 31–40.

COLLET, M. H. 1942. Crest-stage meter for measuring static heads. *Civil Eng.* 12: 396.

COOKE, W. B. 1956. Colonization of artificial bare areas by microorganisms. *Bot. Rev.* 22: 613–638.

COOKE, W. B. 1961. Pollution effects on the fungus population of a stream. *Ecology* 42: 1–18.

COOK, S. F. and MOORE, R. L. 1969. The effects of a rotenone treatment on the insect fauna of a California stream. *Trans. Amer. Fish. Soc.* 98: 539–544.

COOPER, W. E. 1965. Dynamics and production of a natural population of a fresh-water amphipod *Hyallela asteca*. *Ecol. Monogr.* 35: 377–394.

CORBEL, J. 1964. L'érosion terrestre, étude quantitative. *Ann. Géogr.* 73: 385–412.

*CORBET, P. S. 1956. Estimation of growth rate of aquatic insects in tropical lakes. 2nd Symp. Afr. Hydrobiol. Inland Fish. Brazzaville. C.S.A. Publ. No 25: 115 (referenced by PETR 1970).

CORBET, P. S. 1964. Temporal patterns of emergence in aquatic insects. *Can. Ent.* 96: 264–279.

CORBET, D. M. 1945. Stream-gauging procedure. *U.S. Geol. Survey Water-supply Paper* 888: 43–51.

CORDONE, A. J. and KELLEY, D. W. 1961. The influence of inorganic sediments on the aquatic life of streams. *Calif. Fish Game* 47: 189–228.

COSTA, H. H. and BALASUBRAMANIAM, S. 1965. The food of the tadpoles of *Rhacophorus cruciger cruciger* (Blyth). *Ceylon J. Sci. (Bio. Sci.)* 5: 105–109.

COSTA, H. H. and FERNANDO, E. C. M. 1967. The food and feeding relationships of the common meso and macrofauna in the Maha Oya, a small mountainous stream at Peradeniya, Ceylon. *Ceylon J. Sci. (Bio. Sci.)* 7: 74–90.

CRIDLAND, C. C. 1958. Ecological factors affecting the numbers of snails in a permanent stream. *J. trop. Med. Hyg.* 61: 16–21.

CRISP, D. T. and LeCREN, E. D. 1970. The temperature of three different small streams in northwest England. *Hydrobiologia* 35: 305–323.

CUMMINS, K. W. 1962. An evaluation of some techniques for the collection and analysis of benthic samples with special emphasis on lotic waters. *Amer. Midl. Nat.* 67: 477–504.

CUMMINS, K. W. 1964. Factors limiting the microdistribution of larvae of the caddisflies *Pycnopsyche lepida* (Hagen) and *Pycnopsyche guttifer* (Walker) in a Michigan stream (Trichoptera: Limnephilidae). *Ecol. Monogr.* 34: 271–295.

CUMMINS, K. W. 1966. A review of stream ecology with special emphasis on organism-substrate relationships. *Spec. Publs. Pymatuning Lab. Ecol.* 4: 2–51.

CUMMINS, K. W. and LAUFF, G. H. 1969. The influence of substrate particle size on the microdistribution of stream macrobenthos. *Hydrobiologia* 34: 145–181.

CUSHING, C. E. 1967. Periphyton productivity and radionuclide accumulation in the Columbia River, Washington, U.S.A. *Hydrobiologia* 29: 125–139.

DAGET, J. 1957. Données récentes sur la biologie des poissons dans le delta central du Niger. *Hydrobiologia* 9: 321–347.

DALE, W. L. 1956. Wind and drift currents in the South China Sea. *Malay. J. trop. Geog.* 8: 1–31.

DALE, W. L. 1959. The rainfall of Malaya. Part I. *J. trop. Geog.* 13: 23–37.

DALE, W. L. 1960. The rainfall of Malaya. Part II. *J. trop. Geog.* 14: 11–28.

DALE, W. L. 1963. Surface temperatures in Malaya. *J. trop. Geog.* 17: 57–71.

DARNELL, R. M. 1961. Trophic spectrum of an estuarine community based on studies of

Lake Pontchartrain, Louisiana. *Ecology* 42: 553–568.

DARNELL, R. M. 1964. Organic detritus in relation to secondary production in aquatic communities. *Verh. Internat. Verein. Limnol.* 15: 462–470.

DE BACH, P. 1966. The competitive displacement and coexistence principles. *Ann. Rev. Ent.* 11: 183–212.

DÉCAMPS, H. 1967. Écologie des Trichoptères de la vallée d'aure (Hautes Pyrénées). *Ann. Limnol.* 3: 399–577.

*DELFS, J. 1956. Können wir die Wasserlieferung aus dem Wald dursch forstliche Massnahmen beeinflussen. *Forst. u. Holzwirt.* No. 11 (referenced by RUTTER 1958).

DICKMAN, M. 1968. The effect of grazing by tadpoles on the structure of a periphyton community. *Ecology* 49: 1188–1190.

DIMOND, J. B. 1967. Evidence that drift of stream benthos is density related. *Ecology* 48: 855–857.

DITTMAR, H. 1955. Ein Sauerlandbach. *Arch. Hydrobiol.* 50: 305–552.

DODSON, S. I. 1970. Complementary feeding niches sustained by size-selective predation. *Limnol. Oceanogr.* 15: 131–137.

DOEGLAS, D. J. 1968. Grain-size indices, classification and environment. *Sedimentology* 10: 83–100.

DORIER, A. and VAILLANT, F. 1954. Observations et expériences relatives à la resistance au courant de divers invertébrés aquatiques. *Trav. Lab. Hydrobiol. Grenoble* 45/46: 9–31.

DORIER, A. and VAILLANT, F. 1955. Sur le facteur vitesse du courant. *Verh. Internat. Verein. Limnol.* 12: 593–597.

DOUDOROFF, P. and WARREN, C. E. 1957. Biological indices of water pollution with special reference to fish populations.
In: Biological Problems in Water Pollution, pp. 144–163. U.S. Public Health Service, ROBERT A. TAFT Sanitary Engineering Center, Cincinnati.

DOUGLAS, B. 1958. The ecology of the attached diatoms and other algae in a small stony stream. *J. Ecol.* 46: 295–322.

DOUGLAS, I. 1967. Man, vegetation and the sediment yield of rivers. *Nature, Lond.* 215: 925–928.

DOUGLAS, I. 1968a. Erosion in the Sungei Gombak Catchment, Selangor, Malaysia. *J. trop. Geogr.* 26: 1–16.

DOUGLAS, I. 1968b. The effects of precipitation chemistry and catchment area lithology on the quality of river water in selected catchments in Eastern Australia. *Earth Science Journal* 2: 126–144.

DOUGLAS, I. 1969. The efficiency of humid tropical denudation systems. *Trans. Inst. Br. Geogr.* 46: 1–16.

DOUGLAS, I. 1970. Measurements of river erosion in West Malaysia. *Malay. Nat. J.* 23: 78–83.

DUFFER, W. R. and DORRIS, T. C. 1966. Primary productivity in a southern Great Plains stream. *Limnol. Oceanogr.* 11: 143–151.

DUMONT, H. J. 1969. A quantitative method for the study of periphyton. *Limnol. Oceanogr.* 14: 303–307.

EATON, J. W. and MOSS, B. 1966. The estimation of numbers and pigment content in epipelic algal populations. *Limnol. Oceanogr.* 11: 584–595.

EDINGTON, J. M. 1965. The effect of water flow on populations of net-spinning Trichoptera. *Mitt. Internat. Verein. Limnol.* 13: 40–48.

EDINGTON, J. M. 1966. Some observations on stream temperature. *Oikos* 15: 265–273.

EDINGTON, J. M. 1968. Habitat preferences in net-spinning caddis larvae with special reference to the influence of water velocity. *J. Anim. Ecol.* 37: 675–692.

EDMONDSON, W. T. 1956. The relation of photosynthesis by pheoplankton to light in lakes. *Ecology* 37: 161–174.

EDWARDS, D. 1969. Some effects of siltation upon aquatic macrophytic vegetation in rivers. *Hydrobiologia* 34: 29–37.

EDWARDS, F. W. 1928. Diptera Nematocera from the Federated Malay States Museums. *J. Fed. Malay States Mus.* 14: 1–139.

EDWARDS, F. W. 1934. Deutsche limnologische Sunda-Expedition. The Simuliidae (Diptera) of Java and Sumatra. *Arch. Hydrobiol. Suppl.* XIII: 92–138.

EDWARDS, R. W. 1962. The effects of plants and animals on the conditions in fresh water streams with particular references to their oxygen balance. *Int. J. Air Wat. Poll.* 6: 505–520.

EDWARDS, R. W. and OWENS, M. 1962. The effects of plants on river conditions IV. The oxygen balance of a chalk stream. *J. Ecol.* 50: 207–220.

EDWARDS, R. W., OWENS, M. and GIBBS, J. W. 1961. Estimates of surface aeration in two streams. *J. Inst. Water Engrs.* 15: 395–405.

EDWARDS, R. W. and ROLLEY, H. L. J. 1965. Oxygen consumption of river muds. *J. Ecol.* 53: 1–19.

EFFORD, I. E. 1960. A method of studying the vertical distribution of the bottom fauna in shallow waters. *Hydrobiologia* 16: 288–292.

EGGLISHAW, H. J. 1964. The distributional relationship between the bottom fauna and plant detritus in streams. *J. Anim. Ecol.* 33: 463–477.

EGGLISHAW, H. J. 1965. Estimating and accounting for the quantity of bottom fauna in highland streams. *Rep. Challenger Soc.* 3 No. XVII.

EGGLISHAW, H. J. 1968. The quantitative relationship between bottom fauna and plant detritus in streams of different calcium concentrations. *J. Appl. Ecol.* 5: 731–740.

EGGLISHAW, H. J. 1969. The distribution of benthic invertebrates on substrata in fast-flowing streams. *J. Anim. Ecol.* 38: 19–33.

*EIGENMANN, C. H. 1912. The freshwater fishes of British Guiana, including a study of the ecological groupings of species and the relation of the fauna of the plateau to that of the lowlands.
Mem. Carneg. Mus. 5 (no. 67): 1–578 (referenced by LOWE-MCCONNELL 1969).

EINSELE, W. 1960. Die Strömungsgeschwindigkeit als beherrschender Faktor bei der limnologischen Gestaltung der Gewässer. *Oesterr. Fish. Suppl.* 1, 2: 40 p.p.

ELLIOTT, J. M. 1965a. Daily fluctuations of drift invertebrates in a Dartmoor stream. *Nature, Lond.* 205: 1127–1129.

ELLIOTT, J. M. 1965b. Invertebrate drift in a mountain stream in Norway. *Norsk entomol. Tiddskr.* 13: 97–99.

ELLIOTT, J. M. 1967a. Invertebrate drift in a Dartmoor stream. *Arch. Hydrobiol.* 63: 202–237.

ELLIOTT, J. M. 1967b. The life histories and drifting of the Plecoptera and Ephemeroptera in a Dartmoor stream. *J. Anim. Ecol.* 36: 343–362.

ELLIOTT, J. M. 1967c. The food of trout *Salmo trutta* in a Dartmoor stream. *J. appl. Ecol.* 4: 59–72.

ELLIOTT, J. M. 1968a. The life histories and drifting of Trichoptera in a Dartmoor stream. *J. Anim. Ecol.* 37: 615–625.

ELLIOTT, J. M. 1968b. The daily activity patterns of mayfly nymphs (Ephemeroptera). *J. Zool.* 155: 201–221.

ELLIOTT, J. M. 1969a. Life history and biology of *Sericostoma personatum* Spence (Trichoptera). *Oikes* 20: 110–118.

ELLIOTT, J. M. 1969b. Diel periodicity in invertebrate drift and the effect of different sampling periods. *Oikos* 20: 524–528.

ELLIOTT, J. M. 1970a. The diel activity patterns of caddis larvae (Trichoptera). *J. Zool.* 160: 279–290.

ELLIOTT, J. M. 1970b. Diel changes in invertebrate drift and the food of trout *Salmo trutta* L. *J. Fish Biol.* 2: 161–165.

ELLIOTT, J. M. 1971. Upstream movements of benthic invertebrates in a Lake District Stream. *J. Anim. Ecol.* 40: 235–252.

ELLIOTT, J. M. and MINSHALL, G. W. 1968. The invertebrate drift in the River Duddon, English Lake District. *Oikos* 19: 39–52.

ELLIS, M. M. 1936. Erosion silt as a factor in aquatic environments. *Ecology* 17: 29–42.

ENGELMANN, M. D. 1966. Energetics, terrestrial field studies and animal productivity. *Adv. Ecol. Res.* 3: 73–115.

ERIKSEN, C. H. 1963. A method of obtaining interstitial water from shallow aquatic substrates and determining the oxygen concentration. *Ecology* 44: 191–193.

ERIKSEN, C. H. 1964. The influence of respiration and substrate upon the distribution of burrowing mayfly naiads. *Verh. Internat. Verein. Limnol.* 15: 903–911.

ERIKSEN, C. H. 1966. Benthic invertebrates and some substrate-current-oxygen inter-relationships. *Spec. Publs. Pymatuning Lab. Ecol.* 4: 98–115.

ERIKSEN, C. H. 1968. Ecological significance of respiration and substrate for burrowing Ephemeroptera. *Can. J. Zool.* 42: 527–548.

ERIKSSON, E. 1956. Air-borne salts and the chemical composition of river waters. *Tellus* 7: 242–250.

EVANS, F. C. 1956. Ecosystem as the basic unit in ecology. *Science* 123: 1127–1128.

FAGER, E. W. 1957. Determination and analysis of recurrent groups. *Ecology* 38: 586–595.

FELDMETH, C. R. 1970. The respiratory energetics of two species of stream caddis fly larvae in relation to water flow. *Comp. Biochem. Physiol.* 32: 193–202.

FERNANDO, C. H. and CHENG, L. 1963. A guide to Malayan water bugs (Hemiptera-Heteroptera) with keys to the genera. Cyclostyled MS, Dept of Zoology, Univ. Singapore, 33 pp.

FERNANDO, C. H. and GATHA, S. 1963. Guide to families of Malayan aquatic Coleoptera. Cyclostyled MS, Dept of Zoology, Univ. Singapore, 29 pp.

FITTKAU, E. J. 1964. Remarks on limnology of central Amazon rain-forest streams. *Verh. Internat. Verein. Limnol.* 15: 1092–1096.

FITTKAU, E. J. 1967. On the ecology of Amazonian rain-forest streams. *Atas do Simpósio sôbre a Biota Amazônica.* Vol. 3 (Limnologia): 97–108.

FITTKAU, E. J. 1969. The fauna of South America. Pp. 624–658, in: FITTKAU, E. J. et al. (eds.). Biogeography and Ecology in South America. W. Junk Publishers, The Hague, 946 pp.

FITTKAU, E. J. 1970. Limnological conditions in the headwater region of the Xingu River, Brasil. *Trop. Ecol.* 11: 20–25.

FITTKAU, E. J., ILLIES, J., KLINGE, H., SCHWABE, G. H., SIOLI, H. (eds.) 1968/69. Biogeography and Ecology in South America. W. Junk Publishers, The Hague, 946 pp.

FJERDINGSTAD, E. 1950. The microflora of the River Mølleae with special reference to the relation of the benthal algae to pollution. *Folia Limnol. Scand.* 5: 1–124.

FJERDINGSTAD, E. 1964. Pollution of streams estimated by benthal phytomicro-organisms. I. A saprobic system based on communities of organisms and ecological factors. *Int. Revue ges. Hydrobiol.* 49: 63–131.

FLEMER, D. A. 1970. Primary productivity of the north branch of the Raritan River, New Jersey. *Hydrobiologia* 35: 273–296.

FORD, J. B. 1962. The vertical distribution of larval Chironomidae (Dipt.) in the mud of a stream. *Hydrobiolgia* 19: 262–272.

FOX, H. M. and WINGFIELD, C. A. 1938. A portable apparatus for the determination of oxygen dissolved in a small volume of water. *J. Exp. Biol.* 15: 437–443.

FRITSCH, F. E. 1929. The encrusting algal communities of certain fast flowing streams. *New Phytol.* 28: 165–196.

FREDEEN, F. J. H. 1964. Bacteria as food for blackfly larvae (Diptera-Simuliidae) in laboratory cultures and in natural streams. *Can. J. Zool.* 42: 527–548.

FURTADO, J. I. 1966. Studies on Malayan Odonata. Unpublished Ph.D. Thesis, Univ. Malaya, Kuala Lumpur, 280 pp.

FURTADO, J. I. 1969. Ecology of Malaysian odonates: Biotope and association of species. *Verh. Internat. Verein. Limnol.* 17: 863–887.

GAMBELL, A. W. and FISHER, D. W. 1964. Occurrence of sulfate and nitrate in rainfall. *J. Geophys. Res.* 69: 4203–4210.

*GAMESON, A. L. and WHEATLAND, A. B. 1958. Ultimate oxygen demand and course of oxidation of sewage effluents. *J. Inst. Inst. Sew. Purif.* 12: 106–117.

GAMMON, J. R. 1969. The effect of inorganic sediment on macroinvertebrate and fish populations of a central Indiana stream (abstract). *Proc. Indiana Acad. Sci.* 78: 203.

GARCIA, M. N. 1970. Los pigmentos como indicadores ecologicos en las aguas corrientes del centro de Espana. *Agua* 58: 34–71.

GAUFIN, A. R. 1965. Environmental requirements of Plecoptera.
Pp. 105–110, in: TARZWELL, C. M. (ed.). Biological Problems in Water Pollution. Third Seminar 1962. U.S. Pub. Hlth. Serv. Publ. No. 999–WP–25, 424 pp.

GAUFIN, R. F. and GAUFIN, A. R. 1966. Ecological aspects of organic pollution in streams. WHO/EBL/66–54, 14 pp.

GAUFIN, A. R., HARRIS, E. K. and WALTER, H. J. 1956. A statistical evaluation of stream bottom sampling data obtained from three standard samplers. *Ecology* 37: 643–648.

GAUFIN, A. R. and TARZWELL, C. M. 1956. Aquatic macro-invertebrate communities as indicators of organic pollution in Lyttle Creek. *Sewage and Industrial Wastes* 28: 906–924.

GEIJSKES, D. C. 1942. Observations on temperature in a tropical river. *Ecology* 23: 106–110.

GELDIAY, R. 1956. Studies on local populations of the freshwater limpet *Ancylus fluviatilis* Müller. *J. Anim. Ecol.* 25: 389–402.

GESSNER, F. 1960. Limnologische Untersuchungen am Zusammenfluss des Rio Negro und des Amazonas (Solimoes). *Int. Revue ges. Hydrobiol.* 45: 55–79.

GILDERHUS, P. A. 1966. Some effects of sublethal concentrations of sodium arsenite on bluegills and the aquatic environment. *Trans. Amer. Fish. Soc.* 95: 289–296.

GILLESPIE, D. M. 1969. Population studies of four species of molluscs in the Madison River, Yellowstone National Park. *Limnol. Oceanogr.* 14: 101–114.

GLASS, L. W. and BOVBJERG, R. V. 1969. Density and dispersion in laboratory populations of caddisfly larvae *(Cheumatopsyche,* Hydropsychidae*)*. *Ecology* 50: 1082–1084.

GOBBETT, D. J. 1964. The lower Palaeozoic rocks of Kuala Lumpur, Malaysia. *Fed. Mus. J.* 9: 67–79.

GOLTERMAN, H. L. 1967. Influence of the mud on the chemistry of water in relation to productivity.
Pp. 297–313, in: GOLTERMAN, H. L. and CLYMO, R. S. (eds.). Chemical Environment in the Aquatic Habitat. Proc. of an IBP Symp. 10–16 Oct. 1966. Royal Netherlands Acad. Sci., Amsterdam, 322 pp.

GOLTERMAN, H. L. and CLYMO, R. S. 1969. Methods for Chemical Analysis of Freshwaters. IBP Handbook No. 8. Blackwell Scientific Publications, Oxford, 172 pp.

GORE, A. J. P. 1968. The supply of six elements by rain to an upland peat area. *J. Ecol.* 56: 483–495.

GORHAM, E. 1958. The influence and importance of daily weather conditions in the supply of chloride, sulphate and other ions to freshwaters from atmospheric conditions. *Phil. Trans. R. Soc. Lond. (B)* 241: 147–178.

GORHAM, E. 1961. Factors influencing supply of major ions to inland waters, with special reference to the atmosphere. *Bull. geol. Soc. Amer.* 72: 795–840.

GOSE, K. 1960. On the influence of pollution by the Ashio Copper-Mine upon the stream organisms. *Jap. J. Limnol.* 21: 1–8.

GRANT, J. S. 1957. Forests and streamflow. *Malay. For.* 20: 122–126.

GRANT, P. R. and MACKAY, R. J. 1969. Ecological segregation of systematically related stream insects. *Can. J. Zool.* 47: 691–694.

GRAY, J. R. A. and EDINGTON, J. M. 1969. Effect of woodland clearance on stream temperature. *J. Fish. Res. Bd. Canada* 26: 399–403.

GREAVES, J. E. 1934. The arsenic content of soils. *Soil Science* 38: 355–362.

405

GREEN, J. 1970. Freshwater ecology in the Mato Grosso, Central Brazil. I. The conductivity of some natural waters. *J. nat. Hist.* 4: 289–299.

*GRIFFITH, M. E. 1945. The environment, life history and structure of the water boatman, *Ramphocorixa acuminata* (Uhler). *Univ. Kansas Sci. Bull.* 30: 241–365.

GROENEWALD, A. A. v. J. 1964. Observations on the food habits of *Clarias gariepinus* Burchell, the South African freshwater Barbel (Pisces: Clariidae) in Transvaal. *Hydrobiologia* 23: 287–291.

GRUBB, P. J., LLOYD, J. R., PENNINGTON, T. D. and WHITMORE, T. C. 1963. A comparison of montane and lowland rain forest in Ecuador. I. The forest structure, physiognomy and floristics. *J. Ecol.* 51: 567–601.

GRZENDA, A. R. and BREHMER, M. L. 1960. A quantitative method for the collection and measurement of stream periphyton. *Limnol. Oceanogr.* 5: 190–194.

GULLAND, J. A. 1967. The effects of fishing on the production and catches of fish. Pp. 295–313, in: GERKING, S. D. (ed.). The Biological Basis of Freshwater Fish Production, Blackwell Sci. Publ., Oxford, 495 pp.

GUMTOW, R. B. 1955. An investigation of the periphyton in a riffle of the West Gallatin River, Montana. *Trans. Amer. microscop. Soc.* 74: 278–292.

GUY, H. P. 1969. Laboratory theory and methods for sediment analysis. Techniques of Water-Resources Investigations of the U.S. Geological Survey. Book 5, Chapter C1: 1–58.

HAGEMAN, R. H. and FLESKER, D. 1960. Nitrate reductase activity in corn seedlings as affected by light and nitrate content of nutrient media. *Plant. Physiol.* 35: 700–708.

HAMILTON, A. L. 1969. On estimating annual production. *Limnol. Oceanogr.* 14: 771–782.

HAMILTON, J. D. 1961. The effect of sand-pit washings on a stream fauna. *Verh. Internat. Verein. Limnol.* 14: 435–439.

HARDING, W. A. and MOORE, J. P. 1927. Hirudinea. The Fauna of British India. Taylor and Francis, London, 302 pp.

HARGRAVE, B. T. 1969. Epibenthic algal production and community respiration in the sediments of Marion Lake. *J. Fish. Res. Bd. Can.* 26: 2003–2026.

HARINASUTA, C. and KRUATRACHUE, M. 1962. The first recognized endemic area of bilharziasis in Thailand. *Ann. trop. Med. Parasit.* 56: 314–322.

HARKER, J. E. 1953. An investigation of the distribution of the mayfly fauna of a Lancashire stream. *J. Anim. Ecol.* 22: 1–13.

HARPER, P. and MAGNIN, E. 1969. Cycles vitaux de quelques Plécoptères des Laurentides (insectes). *Can. J. Zool.* 47: 483–494.

HARREL, R. C., DAVIS, B. J. and DORRIS, T. C. 1967. Stream order and species diversity of fishes in an intermittent Oklahoma stream. *Amer. Midl. Nat.* 78: 428–436.

HARREL, R. C. and DORRIS, T. C. 1968. Stream order, morphometry, physicochemical conditions and community structure of benthic macroinvertebrates in an intermittent stream system. *Amer. Midl. Nat.* 80: 220–251.

HARRISON, A. D. 1958a. Hydrobiological studies on the Great Berg River, Western Cape Province. Part 2. Quantitative studies on sandy bottoms and further information on the fauna arranged systematically. *Trans. R. Soc. S. Africa* 35: 227–276.

HARRISON, A. D. 1958b. Hydrobiological studies on the Great Berg River, Western Cape Province. Part 4. The effects of organic pollution on the fauna of parts of the Great Berg R. system and of the Krom stream, Stellenbosch. *Trans. R. Soc. S. Africa* 35: 299–329.

HARRISON, A. D. 1965. River zonation in Southern Africa.
Arch. Hydrobiol. 61: 380–386.

HARRISON, A. D. 1966. Recolonization of a Rhodesian stream after drought. *Arch. Hydrobiol.* 62: 405–421.

HARRISON, A. D. 1968. The effects of calcium bicarbonate concentration on the oxygen consumption of the freshwater snail *Biomphalaria pfeifferi* (Pulmonata: Planorbidae). *Arch. Hydrobiol.* 65: 63–73.

HARRISON, A. D. and AGNEW, J. D. 1962. The distribution of invertebrates endemic to acid streams in the Western and Southern Cape Province. *Ann. Cape Prov. Mus.* II: 273–291.

HARRISON, A. D. and ELSWORTH, J. F. 1958. Hydrobiological studies on the Great Berg R., Western Cape Province. Part I. General description, chemical studies and main features of the flora and fauna. *Trans. R. Soc. S. Africa* 35: 125–226.

HARRISON, A. D. and FARINA, T. D. W. 1965. A naturally turbid water with deleterious effects on the egg capsules of planorbid snails. *Ann. trop. Med. Parasit.* 59: 327–330.

HARRISON, A. D., KELLER, R. and LOMBARD, W. A. 1963. Hydrobiological studies On the Vaal River in the Vereeniging Area. Part II. The chemistry, bacteriology and invertebrates of the bottom mud. *Hydrobiologia* 21: 66–89.

HARRISON, A. D. and SHIFF, C. J. 1966. Factors influencing the distribution of some species of aquatic snails. *S. Afr. J. Sci.* 62: 253–258.

HARRISON, A. D., WILLIAMS, N. V. and GREIG, G. 1970. Studies on the effects of calcium bicarbonate concentrations on the biology of *Biomphalaria pfeifferi* (Krauss) (Gastropoda: Pulmonata). *Hydrobiologia* 36: 317–327.

HARRISON, J. L. and LIM, B. L. 1957. Monitor lizards of Malaya. *Malay. Nat. J.* 12: 1–10.

HARROD, J. J. 1965. Effect of current speed on the cephalic forms of the larva of *Simulium ornatum* var *nitidifrons* Edwards (Diptera: Simuliidae). *Hydrobiologia* 26: 8–12.

HAYNE, D. W. and BALL, R. C. 1956. Benthic productivity as influenced by fish predation. *Limnol. Oceanogr.* 1: 162–175.

HENDRICKS, A., PARSONS, W. M., FRANCISCO, D., DICKSON, K., HENLEY, D. and SILVEY, J. K. G. 1969. Bottom fauna studies of the lower Sabine River. *Texas J. Sci.* 21: 175–187.

HENSON, E. B. 1965. A cage sampler for collecting aquatic fauna. *Turtox News* 43: 298–299.

HENSON, E. B. and VIBBER, J. H. 1969. Precipitation into Lake Champlain, U.S.A.: a source of dissolved minerals. *Verh. Internat. Verein. Limnol.* 17: 148–153.

HERBERT, D. W. M., ALABASTER, J. S., DART, M. C. and LLOYD, R. 1961. The effect of china-clay wastes on trout streams. *Int. J. Air Wat. Pollut.* 5: 56–74.

HERRE, A. W. C. T. 1940. Additions to the fish fauna of Malaya and notes on rare or little known Malayan and Bornean fishes. *Bull. Raffles Mus. Singapore* 16: 27–61.

HICKLING, C. F. 1962. Fish Culture. Faber and Faber, London, 295 pp.

HILSENHOFF, W. L. 1969. An artificial substrate device for sampling benthic stream invertebrates. *Limnol. Oceanogr.* 14: 465–471.

HIRANO, M. 1967. Freshwater algae collected by the moint Thai-Japanese biological expedition to Southeast Asia 1961/62. *Nature and Life in S.E. Asia* V: 1–71.

HIRSCH, A. 1958. Biological evaluation of organic pollution of New Zealand streams. *N. Z. J. Sci.* 1: 500–553.

HOHN, M. H. 1966. Artificial substrate for benthic diatoms-collection, analysis, and interpretation. *Spec. Publs. Pymatuning Lab. Ecol.* 4: 87–97.

HOHN, M. H. and HELLERMAN, J. 1963. The taxonomy and structure of diatom populations from three eastern North American rivers using three sampling methods. *Trans. Amer. microscop. Soc.* 82: 250–329.

HOLDEN, M. J. and GREEN, J. 1960. The hydrology and plankton of the river Sokoto. *J. Anim. Ecol.* 29: 65–84.

HO, R. 1960. The evolution of the Indo-Malayan region. Pp. 9–20, in: PURCHON, R. D. (ed.). Proc. Cent. Bicent. Congr. Biol. Singapore, Dec. 2–9, 1958. Univ. Malaya Press, Singapore, 302 pp.

HOPKINS, T. L. 1964. A survey of some marine bottom samplers. Pp. 213–256, in: SEARS, M. (ed.). Progress in Oceanography Vol. 2, Pergamon Press, New York, 271 pp.

HOOVER, M. D. 1944. Effect of removal of forest vegetation upon water yields. *Trans. Amer. Geophys. Union* 6: 868–975.

HORTON, P. A. 1961. The bionomics of brown trout in a Dartmoor stream. *J. Anim. Ecol.* 30: 311–338.

407

HORTON, R. E. 1945. Erosional development of streams and their drainage basins. *Bull. Geol. Soc. Amer.* 56: 275–370.

HUBENDICK, B. 1958. Factors conditioning the habitat of freshwater snails. *Bull. Wld. Hlth. Org.* 18: 1072–1080.

HUET, M. 1949. La pollution des eaux, l'analyse biologique des eaux polluées. Extrait du Bulletin du Centre Belge d'étude et de documentation des eaux. Nos 5 et 6.

*HUET, M. 1956. Aperçu de la pisciculture en Indonesie. *Bull. Agric. Congo Belge* XLVII (ref. by HICKLING 1962).

HUET, M. 1958. Correlations entre l'analyse biologique et l'analyse physico-chemique des eaux courantes rheophiles polluées par matières organiques. *Verh. Internat. Verein. Limnol.* 13: 584–589.

HUET, M. 1959. Profiles and biology of western European streams as related to fish management. *Trans. Amer. Fish Soc.* 88: 155–163.

HUET, M. 1964. The evaluation of the fish productivity in fresh waters. *Verh. Internat. Verein. Limnol.* 15: 524–528.

HUGHES, D. A. 1966a. Mountain streams of the Barberton Area, Eastern Transvaal. Part I. A survey of the fauna. Part II. The effect of vegetational shading and direct illumination on the distribution of stream fauna. *Hydrobiologia* 27: 401–459.

HUGHES, D. A. 1966b. On the dorsal light response in a mayfly nymph. *Anim. Behav.* 14: 13–16.

HUGHES, D. A. 1966c. The role of responses to light in the selection and maintenance of microhabitat by the nymphs of two species of mayfly. *Anim. Behav.* 14: 17–33.

HUGHES, D. A. 1970. Some factors affecting drift and upstream movements of *Gammarus pulex. Ecology* 51: 301–305.

HULOT, A. 1951. Observations sur la biologie de *Citharinus congicus* Blgr. et de *Citharinus gibbosus* Blgr. dans la region forestière Congolaise aux environs de Yangambi. *Verh. Internat. Verein. Limnol.* 11: 201–209.

HULTIN, L., SVENSSON, B. and ULFSTRAND, S. 1969. Upstream movements of insects in a South Swedish small stream. *Oikos* 20: 553–557.

HUMPHREY, G. F. 1963. Phytoplankton pigments in the Pacific Ocean. Proc., Conf. of Primary Productivity Measurement, Marine and Freshwater, Hawaii, Aug. 1961. U.S. Atomic Energy Comm. TID–7633: 121–141.

HUTCHINSON, G. E. 1957. A Treatise on Limnology. Volume I. Geography, Physics and Chemistry. John Wiley and Sons, Inc., New York, 1015 pp.

HYNES, H. B. N. 1941. The taxonomy and ecology of the namphs of British Plecoptera with notes on the adults and eggs. *Trans. R. ent. Soc. Lond.* 91: 459–557.

HYNES, H. B. N. 1952. The Neoperlinae of the Ethiopian Region (Plecoptera, Perlidae). *Trans. R. ent. Soc. Lond.* 103: 85–108.

HYNES, H. B. N. 1958. The effect of drought on the fauna of a small mountain stream in Wales. *Verh. Internat. Verein. Limnol.* 13: 826–833.

HYNES, H. B. N. 1960. The Biology of Polluted Waters. Liverpool Univ. Press, Liverpool, 202 pp.

HYNES, H. B. N. 1961. The invertebrate fauna of a Welsh mountain stream. *Arch. Hydrobiol.* 57: 344–388.

HYNES, H. B. N. 1963. Imported organic matter and secondary productivity in streams. Proc. of the XVI International Congr. Zool. Wash. Vol. 4: 324–329.

HYNES, H. B. N. 1965. The significance of macroinvertebrates in the study of mild river pollution.
Pp. 235–240, in: TARZWELL, C. M. (ed.). Biological Problems in Water Pollution. Third Seminar 1962. U.S. Pub. Hlth. Serv. Publ. No. 999–WP–25, 424 pp.

HYNES, H. B. N. 1968a. The Scientific Results of the Hungarian Soil Zoological Expedition to the Brazzaville Congo 36. The Plecoptera species *Neoperla spio* (Newman). *Opusc. Zool. Budapest* VIII: 353–356.

HYNES, H. B. N. 1968b. Further studies on the invertebrate fauna of a Welsh mountain stream. *Arch. Hydrobiol.* 65: 360–379.

408

HYNES, H. B. N. 1969. The ecology of flowing waters in relation to management Part I. *J. Wat. Pollution Control Federation* 42: 418–424.

HYNES, H. B. N. 1970. The ecology of stream insects. *Ann. Rev. Ent.* 15: 25–42.

HYNES, H. B. N. 1971. Zonation of the invertebrate fauna in a West Indian stream. *Hydrobiologia* 38: 1–8.

HYNES, H. B. N. and COLEMAN, M. J. 1968. A simple method of assessing the annual production of stream benthos. *Limnol. Oceanogr.* 13: 569–573.

HYNES, H. B. N. and KAUSHIK, N. K. 1969. The relationship between dissolved nutrient salts and protein production in submerged autumnal leaves. *Verh. Internat. Verein. Limnol.* 17: 95–103.

HYNES, H. B. N. and WILLIAMS, T. R. 1962. The effect of DDT on the fauna of a Central African stream. *Ann. trop. Med. Parasit.* 56: 78–91.

IDE, F. P. 1935. The effect of temperature on the distribution of the mayfly fauna of a stream. *Univ. Toronto Studies, Biol. Series* 39: 9–76. (Publ. Ont. Fish. Res. Lab. No. 50).

ILLIES, J. 1952. Die Mölle. Faunistisch-ökologische Untersuchungen an einem Forellenbach im Lipper Bergland. *Arch. Hydrobiol.* 46: 424–612.

ILLIES, J. 1955. Der biologische Aspekt der limnologischen Fliesswässer-typisierung. *Arch. Hydrobiol.* Suppl. XXII: 337–346.

ILLIES, J. 1961a. Versuch einer allgemeinen biozönotischen Gliederung der Fliessgewässer. *Int. Revue ges. Hydrobiol.* 46: 205–213.

ILLIES, J. 1961b. Gebirgsbäche in Europa und in Südamerika ein limnologischer Vergleich. *Verh. Internat. Verein. Limnol.* 14: 517–523.

ILLIES, J. 1964. The invertebrate fauna of the Huallaga, a Peruvian tributary of the Amazon River from the sources down to Tingo Maria. *Verh. Internat. Verein. Limnol.* 15: 1077–1083.

ILLIES, J. and BOTOSANEANU, L. 1963. Problems et méthodes de la classification et de la zonation écologique des eaux courantes, considerées surtout du point de vue faunistique. *Mitt. Internat. Verein. Limnol.* 12: 1–57.

IMEVBORE, A. M. A. 1970. The chemistry of the River Niger in the Kainji Reservoir Area. *Arch. Hydrobiol.* 67: 412–431.

INGER, R. F. and CHIN, P. K. 1962. The Fresh-water Fishes of North Borneo. *Fieldiana (Zool.)* 45: 1–268.

*IVANOVA, S. S. 1958. (Nutrition of some mayfly larvae). *Proc. Mikoyan Moscow. tech. Inst. Fishing Industry* 9: 102–120.

IVLEV, V. S. 1966. The biological productivity of waters. (Trans. W. E. RICKER: Fish. Res. Bd. Canada Trans. Ser. No. 394). *J. Fish. Res. Bd. Canada* 23: 1727–1759.

*JACCARD, P. 1902. Lois de distribution florale dans la zone alpine. *Bull. Soc. Vaudoise Sci. Natur.* 38: 69–130.

JENKINS, T. M. JR. 1969. Night feeding of brown and rainbow trout in an experimental stream channel. *J. Fish. Res. Bd. Canada* 26: 3275–3278.

JENKINS, T. M. JR., FELDMETH, C. R. and ELLIOTT, G. V. 1970. Feeding of rainbow trout *(Salmo gairdneri)* in relation to abundance of drifting invertebrates in a mountain stream. *J. Fish. Res. Bd. Can.* 27: 2356–2361.

JOHANNSEN, O. A. 1934. Aquatic Diptera. Part I. Nemocera, exclusive of Chironomidae and Ceratopogonidae. *Cornell Univ. Agr. Exp. Sta. Mem.* 164: 1–71.

JOHANNSEN, O. A. 1935. Aquatic Diptera. Part II. Orthorrapha-Brachycera and Cyclorrhapha. *Cornell Univ. Agr. Exp. Sta. Mem.* 177: 1–62.

JOHANNSEN, O. A. 1937a. Aquatic Diptera. Part III. Chironomidae: Subfamilies Tanypodinae, Diamesinae, and Orthocladiinae. *Cornell Univ. Agr. Exp. Sta. Mem.* 205: 1–84.

JOHANNSEN, O. A. 1937b. Aquatic Diptera. Part IV. Chironomidae: Subfamily Chironominae. *Cornell Univ. Agr. Exp. Sta. Mem.* 210: 1–56.

JOHNSON, A. 1970. Blue-green algae in Malaysian rice-fields. *J. Sing. natn. Acad. Sci.* 1: 30–36.

409

JOHNSON, D. S. 1960. Some aspects of the distribution of freshwater organisms in the Indo-Pacific area, and their relevance to the validity of the concept of an Oriental region in zoogeography. Pp. 170–181, in: PURCHON, R. D. (ed.). Proc. Cent. Bicent. Congr. Biol. Singapore Dec. 2–9, 1958. Univ. Malaya Press, Singapore, 302 pp.

JOHNSON, D. S. 1961. Freshwater life in Malaya and its conservation. Pp. 232–239, in: WYATT-SMITH, J. and WYCHERLY, P. (eds.). Malay. Nat. J. Special Issue: Nature Conservation in Western Malaysia, 1961, 261 pp.

JOHNSON, D. S. 1964. A question of nomenclature. Malay. Nat. J. 15: 160–162.

JOHNSON, D. S. 1967a. On the chemistry of freshwaters in southern Malaya and Singapore. Arch. Hydrobiol. 63: 477–496.

JOHNSON, D. S. 1967b. Distributional patterns of Malayan freshwater fish. Ecology 48: 722–730.

JOHNSON, D. S. 1968a. Malayan blackwaters. Proc. Symp. Recent Adv. Trop. Ecol. 1968: 303–310.

JOHNSON, D. S. 1968b. Water pollution in Malaysia and Singapore: some comments. Malay. Nat. J. 21: 221–222.

JOHNSON, N. M., LIKENS, G. E., BORMANN, F. H., FISHER, D. W. and PIERCE, R. S. 1969. A working model for the variation in streamwater chemistry at the Hubbard Brook Experimental Forest, New Hampshire. Water Resource Res. 5: 1353–1363.

JOHNSTONE, D. and CROSS, W. P. 1949. Elements of Applied Hydrology. The Ronald Press Co., New York, 276 pp.

JOLLY, V. H. and Chapman, M. A. 1966. A preliminary biological study of the effects of pollution on Farmer's Creek and Cox's River, New South Wales. Hydrobiologia 27: 160–192.

JÓNASSON, P. M. 1955. The efficiency of sieving techniques for sampling freshwater bottom fauna. Oikos 6: 183–207.

JÓNASSON, P. M. 1958. The mesh factor in sieving techniques. Verh. Internat. Verein. Limnol. 13: 860–866.

JONES, J. H. and HATCH, M. B. 1937. The significance of inorganic spray residue accumulation in orchard soils. Soil Science 44: 37–61.

JONES, J. R. E. 1950. A further ecological study of the river Rheidol: the food of the common insects of the main river. J. Anim. Ecol. 19: 159–174.

JONES, J. R. E. 1951. An ecological study of the River Towy. J. Anim. Ecol. 20: 68–86

JONES, J. R. E. 1958. A further study of the zinc polluted River Ystwyth. J. Anim. Ecol. 27: 1–14.

JONES, J. R. E. 1964. Fish and River Pollution. Butterworth Inc., Washington, D.C., 203 pp.

JONES, M. J. and BROMFIELD, A. R. 1970. Nitrogen in the rainfall at Samaru, Nigeria. Nature, Lond. 227: 86.

JUDAY, C. and BIRGE, E. A. 1932. Dissolved oxygen and oxygen consumed in the lake waters of northeastern Wisconsin. Trans. Wisc. Acad. Arts, Sci. and Letters 27: 415–486.

KALFF, J. 1969. A diel periodicity in the optimum light intensity for maximum photosynthesis in natural phytoplankton populations. J. Fish. Res. Bd. Canada 26: 463–468.

KAMLER, E. 1965. Thermal conditions in mountain waters and their influence on the distribution of Plecoptera and Ephemeroptera larvae. Ekol. pol. Ser. A. 14: 377–414.

KAMLER, E. 1967. Distribution of Plecoptera and Ephemeroptera in relation to altitude above mean sea level and current speed in mountain waters. Polskie Arch. Hydrobiol. 14: 29–42.

KAMLER, E. and RIEDEL, W. 1960. A method of quantitative study of the bottom fauna of Tatra streams. Polskie Arch. Hydrobiol. 8: 95–105.

KANAPATHY, K. 1968. A survey of the quality of some padi irrigation water in Malaya and its interpretation. Malay. Agric. J. 46: 286–297.

KARUNAKARAN, L. 1969. A new genus of the subfamily Tanypodinae (Diptera, Nematocera: Chironomidae) from Singapore. Proc. R. ent. Soc. Lond. (B) 38: 75–79.

KATZ, M., SJOLSETH, D. E., ANDERSON, D. R. and TYNER, L. R. 1970. Effects of

410

pollution on fish life, general aspects of water pollution as related to fish. *J. Water Pollut. Contr. Fed.* 42: 983–1002.

KAWAI, T. 1966. Survey of the aquatic insect fauna of the Tokachi River system. *Sci. Rept. Hokkaido Salmon Hatchery* 20: 65–81.

KEMMERER, A. J. and NEUHOLD, J. M. 1969. A method for gross primary productivity measurements. *Limnol. Oceanogr.* 14: 607–610.

KEMP, P. H. 1967. Hydrobiological studies on the Tugela River system. Part VI. Acidic drainage from mines in the Natal coalfields. *Hydrobiologia* 29: 393–425.

KENNEDY, C. R. 1966. The life history of *Limnodrilus hoffmeisteri* Clap. (Oligochaeta, Tubificidae) and its adaptive significance. *Oikos* 17: 158–168.

KENWORTHY, J. B. 1969. Water balance in the tropical rain forest: a preliminary study in the Ulu Gombak Forest Reserve. *Malay. Nat. J.* 22: 129–135.

KEUP, L. E. 1968. Phosphorus in flowing waters. *Wat. Res.* 2: 373–386.

KING, D. L. and BALL, R. C. 1967. Comparative energetics of a polluted stream. *Limnol. Oceanogr.* 12: 27–33.

KJENSMO, J. 1965. Some notes on the use of filters in $KMnO_4$ analyses. *Hydrobiologia* 26: 574–578.

KLAPÀLEK, F. 1909. Verläufiger Bericht über exotische Plecopteren. *Wiener ent. Ztg.* 28: 215–232.

KLEEREKOPER, H. 1955. Limnological observations in Northeastern Rio Grande do Sul, Brazil. *Arch. Hydrobiol.* 55: 553–567.

KLEIN, L. 1957. Aspects of River Pollution. Butterworths Publ., London, 621 pp.

KLINGE, H. 1968. Litter production in an area of Amazonian terra firme forest. Part I. Litter-fall, organic carbon and total nitrogen contents of litter. Part II. Mineral nutrient content of the litter. *Amazoniana* 1: 287–302, 303–310.

KLINGE, H. and OHLE, W. 1964. Chemical properties of rivers in the Amazonian area in relation to soil conditions. *Verh. Internat. Verein. Limnol.* 15: 1067–1076.

KNÖPPEL, H.-A. 1970. Food of central Amazonian fishes. Contribution to the nutrient-ecology of Amazonian rain-forest streams. *Amazoniana* 2: 257–352.

KOBAYASHI, J. 1959. Chemical investigations on river waters of South-eastern Asiatic countries. (Report I). The quality of waters of Thailand. *Ber. Ohara Instituts landwirtsch. Biologie* XI (2): 167–233.

KOBAYASHI, J. 1960. A chemical study of the average quality and characteristics of river waters of Japan. *Ber. Ohara Instituts landwirtsch. Biologie* XI (3): 313–358.

KOBAYASI, H. 1961. Chlorophyll content in sessile algal community of Japanese mountain river. *Bot. Mag. Tokyo* 74: 228–235.

KONTKANEN, P. 1957. On the delimitation of communities in research on animal biocoenotics. *Cold Spr. Harb. Symp. quant. Biol.* 22: 373–378.

KORMONDY, E. J. 1969. Comparative ecology of sandspit ponds. *Amer. Midl. Nat.* 82: 28–61.

KREIS, R. D. and JOHNSON, W. C. 1968. The response of macrobenthos to irrigation return water. *J. Water Pollut. Contr. Fed.* 40: 1614–1621.

KUBÍČEK, F. 1968. Die organische Drift des Svratka-Flusses im Abschnitt zwischen zwei Talsperren (Vír, Kníničky). *Hydrobiologia* 31: 402–416.

KUEHNE, R. A. 1962. A classification of streams, illustrated by fish distribution in an eastern Kentucky creek. *Ecology* 43: 608–614.

KUEHNE, R. A. 1966. Depauperate fish faunas in sinking creeks near Mammoth Cave, Kentucky. *Copeia* 1966: 306–311.

KULLBERG, R. G. 1968. Algal diversity in several thermal spring effluents. *Ecology* 49: 751–755.

KURASAWA, H. 1959. Studies on the biological production of fire pools in Tokyo. XII. The seasonal changes in the amount of algae attached on the wall of pools. *Misc. Rep. Res. Inst. Nat. Resources* 51: 15–21.

KURECK, A. 1969. Tagesrhythmen lappländischer Simuliiden (Diptera). *Oecologia* (Berl.) 2: 385–410.

411

LARIMORE, R. W. 1961. Fish populations and electrofishing success in a warm-water stream. *J. Wildl. Mgmt.* 25: 1–12.

LA RIVERS, I. 1956. Aquatic Orthoptera.
P. 154, in: USINGER, R. L. (ed.). Aquatic Insects of California. Univ. California Press, Berkeley and Los Angeles, 508 pp.

LEAMY, M. L. 1966. Soil classification in Malaya. Proc. 2nd Mal. Soil Conf., 79, Dept of Agric., Kuala Lumpur.

LEECH, H. B. and CHANDLER, H. P. 1956. Aquatic Coleoptera.
Pp. 293–371, in: USINGER, R. L. (ed.). Aquatic Insects of California. Univ. California Press, Berkeley and Los Angeles, 508 pp.

LEHMANN, U. 1967. Drift und Populationsdynamik von *Gammarus pulex fossarum* Koch. *Z. Morph. Ökol. Tiere* 60: 227–274.

LEHMANN, U. 1970. Stromaufwärts gerichteter Flug von *Philopotamus montanus* (Trichoptera). *Oecologia* 4: 163–175.

LENHARD, G. 1965. A note on the calculation of ultimate oxygen demand. *Hydrobiologia* 26: 1–7.

LENHARD, G. and DU PLOOY, A. 1965. Studies on bottom deposits of the Vaal river system. *Hydrobiologia* 26: 271–291.

LEONARD, J. W. 1939. Comments on the adequacy of accepted stream bottom sampling technique. Trans. 4th N.A. Wildl. Conf. 288–295.

LEOPOLD, L. B. 1962. Rivers. *Amer. Sci.* 50: 511–537.

LEOPOLD, L. B., WOLMAN, M. G. and MILLER, J. P. 1964. Fluvial Processes in Geomorphology. Freeman, San Francisco, 522 pp.

LE-VAN-DANG. 1970. Les travaux sur la biologie des eaux continentales au Vietnam-situation presenté et progres réalisés.
Pp. 59–64, in: Proc. IBP/PF Regional Meeting of Inland Water Biologists in S.E. Asia. UNESCO, Djakarta, 172 pp.

LEVANIDOV, V. YA. 1949. Biology of aquatic organisms: Significance of allochthonous material as a food resource in a water body as exemplified by the nutrition of the water louse *(Asellus aquaticus L.) Trud. vsesoyuz. gidrobiol. Obschch.* 1: 100–117.

LIBOSVÁRSKY, J. 1966. Successive removals with electrical fishing gear- a suitable method for making population estimates in small streams. *Verh. Internat. Verein. Limnol.* 16: 1212–1216.

LIBOSVÁRSKY, J. 1967. Estimates of fish population by means of electrofishing. *Vertebr. Zpr.* 2: 3–10.

LIEBERMAN, J. A. and HOOVER, M. D. 1948. The effect of uncontrolled logging on stream turbidity. *Water and Sewage Works* 95: 255–258.

LIEFTINCK, M. A. 1932. A new species of *Prosopistoma* from the Malay Archipelago (Ephemeropt.). *Tijdschr. Ent.* (Suppl.) 75: 44–55.

LIKENS, G. E., BORMANN, F. H., JOHNSON, N. M., FISHER, D. W. and PIERCE, R. S. 1970. Effects of forest cutting and herbicide treatment on nutrient budgets in the Hubbard Brook watershed-ecosystem. *Ecol. Monogr.* 40: 23–47.

LIKENS, G. E., BORMANN, F. H., JOHNSON, N. M. and PIERCE, R. S. 1967. The calcium, magnesium, potassium, and sodium budgets for a small forested ecosystem. *Ecology* 48: 772–785.

LILLEHAMMER, A. 1966. Bottom fauna investigations in a Norwegian river. The influence of ecological factors. *Nytt Mag. Zool.* 13: 10–29.

LINDUSKA, J. P. 1942. Bottom type as a factor influencing the local distribution of mayfly nymphs. *Can. Ent.* 74: 26–30.

LINSLEY, R. K., KOHLER, M. A. and PAULHUS, J. L. H. 1958. Hydrology for Engineers. McGraw-Hill Book Co. Inc., New York, 340 pp.

LOCKWOOD, J. G. 1967. Probable maximum 24-hour precipitation over Malaya. *Meteorological Mag.* 96: 11–19.

LONGHURST, A. R. 1959. The sampling problem in benthic ecology. *Proc. N.Z. Ecol. Soc.* 6: 8–12.

412

LORENZEN, C. J. 1967. Determination of chlorophyll and pheo-pigments: spectrophotometric equations. *Limnol. Oceanogr.* 12: 343–346.

LOWE-McCONNELL, R. H. 1967. Fish productivity and production in tropical freshwaters. *J. Anim. Ecol.* 36: 16P–18P.

LOWE-McCONNELL, R. H. 1969. Speciation in tropical freshwater fishes. *Biol. J. Linn. Soc.* 1: 51–75.

LUND, J. W. G. and TALLING, J. F. 1957. Botanical limnological methods with special reference to the algae. *Bot. Rev.* 23: 489–583.

MACAN, T. T. 1957. The life histories and migrations of the Ephemeroptera in a stony stream. *Trans. Soc. Brit. Ent.* 12: 129–156.

MACAN, T. T. 1958a. The temperature of a small stony stream. *Hydrobiologia* 12: 89–106.

MACAN, T. T. 1958b. Methods of sampling the bottom fauna in stony streams. *Mitt. Internat. Verein. Limnol.* 8: 1–21.

MACAN, T. T. 1961a. A review of running water studies. *Verh. Internat. Verein. Limnol.* 14: 587–602.

MACAN, T. T. 1961b. Factors that limit the range of fresh-water animals. *Biol. Rev.* 36: 151–198.

MACAN, T. T. 1962a. Ephemeroptera in Britain. XI *Verh. Internat. Kongress für Ent.*, Wien, 1960. 3: 258–262.

MACAN, T. T. 1962b. Biotic factors in running water. *Schweiz. Z. Hydrol.* 24: 386–407.

MACAN, T. T. 1962c. Ecology of aquatic insects. *Ann. Rev. Ent.* 7: 261–288.

MACAN, T. T. 1963. Freshwater Ecology. John Wiley & Sons, New York, 338 pp.

MACARTHUR, R. H. 1969. Patterns of communities in the tropics. *Biol. J. Linn. Soc.* 1: 19–30.

MACIOLEK, J. A. 1962. Limnological organic analyses by quantitative dichromate oxidation. Part IV. Methods of dichromate oxidative analysis. *U.S.D.I. Fish Wildl. Serv. Res. Rept.* 60: 17–26.

MACKAY, R. J. 1969. Aquatic insect communities of a small stream on Mont St. Hilaire, Quebec. *J. Fish. Res. Bd. Canada* 26: 1157–1183.

MACKAY, R. J. and KALFF, J. 1969. Seasonal variation in standing crop and species diversity of insect communities in a small Quebec stream. *Ecology* 50: 101–109.

MACKERETH, F. J. H. 1963. Some methods of water analysis for limnologists. *Freshw. Biol. Assn. Sci. Publ.* No. 21 (1963), 70 pp.

MACKERETH, J. C. 1960. Notes on the Trichoptera of a stony stream. *Proc. R. ent. Soc. Lond.* (A) 35: 17–23.

MADSEN, B. L. 1966. Om rytmisk aktivitet hos døgnfluenymfer. *Flora og Fauna* 72: 148–154.

MADSEN, B. L. 1968a. The distribution of nymphs of *Brachyptera risi* Mort. and *Nemoura flexuosa* Aub. (Plecoptera) in relation to oxygen. *Oikos* 19: 304–310.

MADSEN, B. L. 1968b. A comparative ecological investigation of two related mayfly nymphs. *Hydrobiologia* 31: 337–349.

MAITLAND, P. S. 1965. Notes on the biology of *Ancylus fluviatilis* in the River Endrick, Scotland. *Proc. Malac. Soc. Lond.* 36: 339–347.

MAITLAND, P. S. 1966. The fauna of the River Endrick – Studies on Loch Lomond. Glasgow, 194 pp.

MAITLAND, P. S. 1969. A simple corer for sampling sand and finer sediments in shallow water. *Limnol. Oceanogr.* 14: 151–156.

MAITLAND, P. S. and PENNEY, M. M. 1967. The ecology of the Simuliidae in a Scottish river. *J. Anim. Ecol.* 36: 179–206.

MALAISSE, F. 1969. Les faciès d'un cours d'eau tropical: la Luanza (Haut-Katanga, Rép. dém. Congo). *Verh. Internat. Verein. Limnol.* 17: 936–940.

MANN, K. H. 1955. Some factors influencing the distribution of freshwater leeches in Britain. *Verh. Internat. Verein. Limnol.* 12: 582–587.

MANN, K. H. 1956. The study of the oxygen consumption of five species of leeches. *J. exp. Biol.* 33: 615–626.

MANN, K. H. 1961. The oxygen requirements of leeches considered in relation to their habitats. *Verh. Internat. Verein. Limnol.* 14: 1009–1013.

MANN, K. H. 1964. The pattern of energy flow in the fish and invertebrate fauna of the River Thames. *Verh. Internat. Verein. Limnol.* 15: 485–495.

MANN, K. H. 1965. Energy transformations by a population of fish in the River Thames. *J. Anim. Ecol.* 34: 253–275.

MANN, K. H. 1966. Secondary production with particular reference to freshwater habitats. *J. Anim. Ecol.* 35: 11P–12P.

MANN, K. H. 1967. The cropping of the food supply. Pp. 243–257, in: GERKING, S. D. (ed.). The Biological Basis of Freshwater Fish Production. Blackwell Sci. Publ., Oxford, 495 pp.

MARGALEF, R. 1949. A new limnological method for the investigation of thin-layered epilithic communities. *Hydrobiologia* 1: 215–216.

MARGALEF, R. 1960. Ideas for a synthetic approach to ecology of running water. *Int. Revue ges. Hydrobiol.* 45: 133–153.

MARGALEF, R. 1961a. Communication of structure in planktonic populations. *Limnol. Oceanogr.* 6: 124–128.

MARGALEF, R. 1961b. Corrélations entre certains charactères synthétiques des populations de phytoplancton. *Hydrobiologia* 18: 155–164.

MARGALEF, R. 1968. Perspectives in Ecological Theory. Univ. Chicago Press, Chicago, 111 pp.

MARLIER, G. 1951. Recherches hydrobiologiques dans les rivières du Congo Oriental. I. Composition des eaux. *Hydrobiologia* 3: 217–227.

MARLIER, G. 1954. Recherches hydrobiologiques dans les rivières du Congo Oriental. II. Étude écologique. *Hydrobiologia* 6: 225–264.

MASON, W. T. JR. 1968. An introduction to the identification of chironomid larvae. Federal Water Pollution Control Administration, Cincinnati, 89 pp.

MASON, W. T., ANDERSON, J. B. and MORRISON, G. E. 1967. Limestone-filled, artificial substrate sampler-float unit for collecting macro-invertebrates in large streams. *Progr. Fish Cult.* 29: 74.

MATHEWS, C. P. 1967. The energy budget and nitrogen turnover of a population of *Gammarus pulex* in a small woodland stream. *J. Anim. Ecol.* 36: 62P.

MATHEWS, C. P. 1970. Estimates of production with reference to general surveys. *Oikos* 21: 129–133.

MATHEWS, C. P. 1971. Contribution of young fish to total production of fish in the River Thames near Reading. *J. Fish. Biol.* 3: 157–180.

MATHEWS, C. P. and KOWALCZEWSKI, A. 1969. The disappearance of leaf litter and its contribution to production in the River Thames. *J. Ecol.* 57: 543–552.

MAYR, E. 1963. Animal Species and Evolution. Harvard Univ. Press, Cambridge, 797 pp.

McCONNELL, W. J. 1968. Liminological effects of organic extracts of litter in a southwestern impoundment. *Limnol. Oceanogr.* 13: 343–349.

McCONNELL, W. J. and SIGLER, W. F. 1959. Chlorophyll and productivity in a mountain river. *Limnol. Oceanogr.* 4: 335–351.

McFARLAND, B. H. and WEBER, C. I. 1970. Seasonal changes in the periphyton of a small calcareous stream. Unpublished paper presented at 18th Ann. Meeting Midwest Benth. Soc. St. Mary's College, Winona, Minnesota, April 1–3, 1970.

McINTIRE, C. D. 1966. Some effects of current velocity on periphyton communities in laboratory streams. *Hydrobiologia* 27: 559–570.

McINTIRE, C. D., GARRISON, R. L., PHINNEY, H. K. and WARREN, C. E. 1964. Primary production in laboratory streams. *Limnol. Oceanogr.* 9: 92–102.

McLACHLAN, A. J. 1969. Substrate preferences and invasion behaviour exhibited by larvae of *Nilodorum brevibucca* Freeman (Chironomidae) under experimental conditions. *Hydrobiologia* 33: 237–249.

414

McLay, C. L. 1968. A study of drift in the Kakanui River, New Zealand. *Aust. J. mar. Freshwat. Res.* 19: 139–149.

McLay, C. 1970. A theory concerning the distance travelled by animals entering the drift of a stream. *J. Fish. Res. Bd. Canada* 27: 359–370.

Mecom, J. O. and Cummins, K. W. 1964. A preliminary study of the trophic relationships of the larvae of *Brachycentrus americanus* (Banks) (Trichoptera: Brachycentridae). *Trans. Amer. microscop. Soc.* 80: 233–243.

Medway, Lord. 1965/66. The Ulu Gombak Field Studies Centre, University of Malaya. *The Malayan Scientist* 2: 1–16.

Meyer-Waarden, P. F., Halsband, E. and Halsband, I. 1960. Bibliographie über die Electrofischerei und ihre Grundlagen. *Arch. Fischereiwiss.* 9: 1–104.

Minckley, W. L. 1963. The ecology of a spring stream Doe Run, Meade County, Kentucky. *Wildl. Monogr.* 11: 1–124.

Minshall, G. W. 1967. Role of allochthonous detritus in the trophic structure of a woodland springbrook community. *Ecology* 48: 139–149.

Minshall, G. W. 1968. Community dynamics of the benthic fauna in a woodland springbrook. *Hydrobiologia* 32: 305–339.

Minshall, G. W. and Kuehne, R. A. 1969. An ecological study of invertebrates of the Duddon, an English mountain stream. *Arch. Hydrobiol.* 66: 169–191.

Minshall, G. W. and Winger, P. V. 1968. The effect of reduction in stream flow on invertebrate drift. *Ecology* 49: 580–582.

Mitchell, B. A. 1957. A note on land erosion in the Cameron Highlands. *Malay. For.* 20: 30–32.

Mizuno, T. and Mori, S. 1970. Preliminary hydrobiological survey of some Southeast Asian inland waters. *Biol. J. Linn. Soc.* 2: 77–117.

Moon, H. P. 1935. Methods and apparatus suitable for an investigation of the littoral region of oligotrophic lakes. *Int. Revue ges. Hydrobiol.* 32: 319–333.

Moon, H. P. 1939. Aspects of the ecology of aquatic insects. *Trans. Brit. Ent. Soc.* 6: 39–49.

Moore, J. P. 1935. Leeches from Borneo and the Malay Peninsula. *Bull. Raffles Mus.* 10: 67–79.

Moretti, G. and Gianotti, F. S. 1962. Der Einfluss der Strömung auf die Verteilung der Trichopteren *Agapetus* gr. *fuscipes* Curt. und *Silo* gr. *nigricornis* Pict. *Schweiz. Z. Hydrol.* 24: 467–484.

Morgan, N. C. and Egglishaw, H. J. 1965. A survey of the bottom fauna of streams in the Scottish Highlands. Part I. Composition of the fauna. *Hydrobiologia* 25: 181–211.

Morgans, J. F. C. 1956. Notes on the analysis of shallow-water soft substrata. *J. Anim. Ecol.* 25: 367–387.

Morisawa, M. 1968. Streams: their dynamics and morphology. McGraw-Hill, New York, 175 pp.

Morisita, M. 1959. Measuring of interspecific association and similarity between communities. *Memoirs Fac. Sci., Kyushu Univ.* (E) 2: 215–235.

Moss, B. 1967a. A spectrophotometric method for the estimation of percentage degradation of chlorophylls to pheo-pigments in extracts of algae. *Limnol. Oceanogr.* 12: 335–340.

Moss, B. 1967b. A note on the estimation of chlorophyll *a* in freshwater algal communities. *Limnol. Oceanogr.* 12: 340–342.

Moss, B. 1968. The chlorophyll *a* content of some benthic algal communities. *Arch. Hydrobiol.* 65: 51–62.

Moss, B. 1969. Algal growth in African waters. *Limnol. Oceanogr.* 14: 591–601.

Mottley, C. McC., Rayner, H. J. and Rainwater, J. H. 1939. The determination of the food grade of streams. *Trans. Amer. Fish. Soc.* 68: 336–343.

Mountford, M. D., 1962. An index of similarity and its application to classificatory problems. Pp. 43–50, in: Murphy, P. W. (ed.). Progress in Soil Zoology. Butterworths, London, 398 pp.

Müller, K. 1954a. Investigations on the organic drift in North Swedish streams. *Ann. Rept. Inst. Freshw. Res. Drottningholm* 35: 133–148.

Müller, K. 1954b. Die Drift in fliessenden Gewässern. *Arch. Hydrobiol.* 49: 539–545.

Müller, K. 1963a. Diurnal rhythm in 'organic drift' of Gammarus pulex *Nature, Lond.* 198: 806–807.

Müller, K. 1963b. Tag-Nachtrhythmus von Baetidenlarven in der 'Organischen Drift'. *Naturwissenschaften* 50: 161.

Müller, K. 1963c. Temperature und Tagesperiodik der 'Organischen Drift' von Gammarus pulex. *Naturwissenschaften* 50: 410–411.

Müller, K. 1966. Die Tagesperiodik von Fliesswasserorganismen. *Z. Morph. Ökol. Tiere* 56: 93–142.

Mundie, J. H. 1971. Sampling benthos and substrate materials, down to 50 microns in size, in shallow streams. *J. Fish. Res. Bd. Canada* 28: 849–860.

Munro, J. L. 1967. The food of a community of East African freshwater fishes. *J. Zool.* 151: 389–415.

Naidu, K. V. 1965. Some fresh-water Oligochaeta of Singapore. *Bull. Nat. Mus. Sing.* 33: 13–21.

Neal, E. C., Patten, B. C. and Depoe, C. E. 1967. Periphyton growth on artificial substrate in a radioactively contaminated lake. *Ecology* 48: 918–924.

Needham, P. R. and Usinger, R. L. 1956. Variability in the macrofauna of a single riffle in Prosser Creek, California, as indicated by the Surber sampler. *Hilgardia* 24: 383–409.

Neel, J. K. 1951. Interrelations of certain physical and chemical features in a head-water limestone stream. *Ecology* 32: 368–391.

Neel, J. K. 1953. Certain limnological features of a polluted irrigation stream. *Trans. Amer. microscop. Soc.* 72: 119–135.

Neess, J. and Dugdale, R. C. 1959. Computation of production for populations of aquatic midge larvae. *Ecology* 40: 425–430.

Negus, C. L. 1966. A quantitative study of growth and production of unionid mussels in the River Thames at Reading. *J. Anim. Ecol.* 35: 513–532.

Neill, R. M. 1938. The food and feeding of the brown trout *(Salmo trutta* L.*)* in relation to the organic environment. *Trans. roy. Soc. Edinb.* 59: 481–520.

Nelson, D. J. and Scott, D. C. 1962. Role of detritus in the productivity of a rock-outcrop community in a Piedmont stream. *Limnol. Oceanogr.* 7: 396–413.

Newcombe, C. L. 1949. Attachment materials in relation to water productivity. *Trans. Amer. microscop. Soc.* 68: 355–361.

Newcombe, C. L. 1950. A quantitative study of attachment materials in Sodon Lake, Michigan. *Ecology* 31: 204–215.

Ng, S. K. 1969. Soil resources in Malaya.
Pp. 141–151, in: Stone, B. C. (ed.). Natural Resources in Malaysia and Singapore. Proceedings of the second Symposium on Scientific and Technological Research in Malaysia and Singapore, Univ. Malaya, 1–4 Feb. 1967, 256 pp.

Nielson, A. 1950. The torrential invertebrate fauna. *Oikos* 2: 177–196.

Nikolski, G. V. 1933. On the influence of the rate of flow on the fish fauna of the rivers of Central Asia. *J. Anim. Ecol.* 2: 266–281.

Nikolski, G. V. 1937. On the distribution of fishes according to the nature of their food in the rivers flowing from the mountains of middle Asia. *Proc. Int. Assn. Limnol.* 8: 169–176. (Abstract only seen).

Nilsson, N. A. 1967. Interactive segregation between fish species.
Pp. 295–313, in: Gerking, S. D. (ed.). The Biological Basis of Freshwater Fish Production. Blackwell Sci. Publ. Oxford, 495 pp.

Nisbet, M. 1961. Un example de pollution de rivière par vidange d'une retenue hydroélectrique. *Verh. Internat. Verein. Limnol.* 14: 678–680.

Norris, R. C. and Charlton, J. I. 1962. A chemical and biological survey of the Sungei Gombak. Government Printer, Kuala Lumpur, 57 pp.

416

NYE, P. H. 1961. Organic matter and nutrient cycles under moist tropical forest. *Pl. Soil* 13: 333–346.

ODUM, H. T. 1956. Primary production in flowing waters. *Limnol. Oceanogr.* 1: 102–117.

ODUM, H. T. 1957a. Trophic structure and productivity of Silver Springs. *Ecol. Monogr.* 27: 55–112.

ODUM, H. T. 1957b. Primary production measurements in eleven Florida springs and a marine turtle grass community. *Limnol. Oceanogr.* 2: 85–97.

ODUM, H. T., McCONNELL, W. and ABBOTT, W. 1958. The Chlorophyll 'A' of communities. *Pub. Inst. Mar. Sci. Texas* 5: 65–97.

OH, K. Y. 1965. Hydrology in Malaya – rainfall and river discharge. *Malay. Agric. J.* 45: 182–190.

OHLE, W. 1934. Chemische und physikalische Untersuchungen norddeutscher Seen. *Arch. Hydrobiol.* 26: 386–464, 584–658.

OHLMACHER, F. J. and GLEASON, G. R. 1964. Frozen core sampler. Unpublished paper presented at 12th Ann. Meeting Midwest Benth. Soc., Put-In-Bay, Ohio, April 2–3, 1964.

OLIFF, W. D. 1960. Hydrobiological studies on the Tugela River system. I. The main Tugela River. *Hydrobiologia* 14: 281–385.

ORMEROD, J. G., GRYNNE, B. and ORMEROD, K. S. 1966. Chemical and physical factors involved in the heterotrophic growth response to organic pollution. *Verh. Internat. Verein. Limnol.* 16: 906–910.

OWEN, G. 1951. A provisional classification of Malayan soils. *J. Soil Sci.* 2: 20–42.

OW, Y. H. C. 1965. Rivers in Malaya – their deterioration and remedial measures. *Malay. Agric. J.* 45: 17–20.

PARSONS, J. D. 1968. The effects of acid strip-mine effluents on the ecology of a stream. *Arch. Hydrobiol.* 65: 25–50.

PASTERNAK, K. 1968. Characteristics of the substratum of the River Dunajec catchment basin. *Acta Hydrobiol.*, Krakow, 10: 299–317.

PATON, J. R. 1964. The origin of the limestone hills of Malaya. *J. Trop. Geog.* 18: 134–147.

PANTON, W. P. 1964. The 1962 soil map of Malaya. *J. trop. Geog.* 18: 118–124.

PATRICK, R. 1949. A proposed biological measure of stream conditions based on a survey of the Conestoga Basin, Lancaster County, Penna. *Proc. Acad. Nat. Sci. Phila.* 101: 277–341.

PATRICK, R. 1963. The structure of diatom communities under varying ecological conditions. *Ann. New York Acad. Sci.* 108: 353–358.

PATRICK, R. 1964. A discussion of the results of the Catherwood Expedition to the Peruvian headwaters of the Amazon. *Verh. Internat. Verein. Limnol.* 15: 1084–1090.

PATRICK, R., HOHN, M. H. and WALLACE, J. H. 1954. A new method for determining the pattern of the diatom flora. *Not. Nat. Acad. Nat. Sci. Phila.* 259: 1–12.

PATTEN, B. C. 1962. Species diversity in net phytoplankton of Raritan Bay. *J. Mar. Res.* 20: 57–75.

PEARSON, W. D. and FRANKLIN, D. R. 1968. Some factors affecting drift rates of *Baetis* and Simuliidae in a large river. *Ecology* 49: 75–81.

PENMAN, H. L. 1963. Vegetation and Hydrology. Technical Communication No. 53. Commonwealth Bureau of Soils, Harpenden, England.

PENNAK, R. W. and VAN GERPEN, E. D. 1947. Bottom fauna production and physical nature of the substrate in northern Colorado trout streams. *Ecology* 28: 42–48.

PERCIVAL, E. and WHITEHEAD, H. 1929. A quantitative study of the fauna of some types of stream bed. *J. Ecol.* 17: 282–314.

PETERS, W. L. 1967. New species of *Prosopistoma* from the Oriental Region (Prosopistomatoidea: Ephemeroptera). *Tijdschr. Ent.* 110: 207–222.

PETERS, W. L. 1972. The nymph of *Dipterophleboides* sp. (Leptophlebiidae: Ephemeroptera) *Ent. News* 83: 53–56.

417

PETERS, W. L. and EDMUNDS, G. F. JR. 1970. Revision of the generic classification of the eastern hemisphere Leptophlebiidae. *Pacific Insects* 12: 157–240.

PETR, T. 1970. The bottom fauna of the rapids of the Black Volta River in Ghana. *Hydrobiologia* 35: 399–418.

PHAUP, J. D. 1968. The biology of *Sphaerotilus* species. *Wat. Res.* 2: 597–614.

*PHELPS, E. B. 1944. Stream sanitation. New York. (Refefenced by HYNES 1960).

PHILIPSON, G. N. 1954. The effect of water flow and oxygen concentration on six species of caddis fly (Trichoptera). *Proc. Zool. Soc. Lond.* 124: 547–564.

PHILIPSON, G. N. 1969. Some factors affecting the net-spinning of the caddis fly *Hydropsyche instabilis* Curtis (Trichoptera, Hydropsychidae). *Hydrobiologia* 34: 369–377.

PIECZYNSKA, E. 1968. Dependence of the primary production of periphyton upon the substrate area suitable for colonization. *Bull. Akad. pol. Sci.*, Cl. II 16: 165–169.

PIMENTAL, D. and WHITE, P. C. JR. 1959. Physiochemical environment of *Australorbis glabratus*, the snail intermediate host of *Schistosoma mansoni* in Puerto Rico. *Ecology* 40: 533–541.

POLLARD, R. A. 1955. Measuring seepage through salmon spawning gravel. *J. Fish. Res. Bd. Canada.* 12: 706–741.

POORE, M. D. 1961. River control and conservation in Malaya. Pp. 48–51, in: WYATT-SMITH, J. and WYCHERLEY, P. (eds.). *Malay. Nat. J.* Special Issue: Nature Conservation in Western Malaysia, 1961, 261 pp.

PRESTON, F. W. 1948. The commonness and rarity of species. *Ecology* 29: 254–283.

PROWSE, G. A. 1957. An introduction to the desmids of Malaya. *Malay. Nat. J.* 11: 42–58.

PROWSE, G. A. 1958. The Eugleninae of Malaya. *Gard. Bull. Sing.* 16: 136–204.

PROWSE, G. A. 1962a. Diatoms of Malayan freshwaters. *Gard. Bull. Sing.* 19: 1–104.

PROWSE, G. A. 1962b. Further Malayan freshwater Flagellata. *Gard. Bull. Sing.* 19: 105–145.

PROWSE, G. A. 1968. Pollution in Malayan waters. *Malay. Nat. J.* 21: 149–158.

PROWSE, G. A. and RATNASABAPATHY, M. 1970. A species list of freshwater algae from the Taiping Lakes, Perak, Malaysia. *Gard. Bull. Sing.* 25: 179–187.

RADFORD, D. S. and HARTLAND-ROWE, R. 1971. Subsurface and surface sampling of benthic invertebrates in two streams. *Limnol. Oceanogr.* 16: 114–117.

RAMANANKASINA, E. 1969. Première contribution à l'étude faunistique de la rivière Andriandrano. *Verh. Internat. Verein. Limnol.* 17: 941–948.

RAPP, A. 1960. Recent development of mountain slopes in Kärkevagge and surroundings, Northern Scandinavia. *Geogr. Ann. Stockh.* 42: 65–200.

RATNASABAPATHY, M. 1971. Algae from Gunong Jerai (Kedah Peak), Malaysia. *Gard. Bull. Sing.* (in press).

REISH, D. J. 1959. A discussion of the importance of the screen size in washing quantitative marine bottom samples. *Ecology* 40: 307–309.

RICHARDS, F. A. with THOMPSON, T. G. 1952. The estimation and characterization of plankton populations by pigment analysis. II. A spectrophotometric method for the estimation of plankton pigments. *J. Mar. Res.* 11: 156–172.

RICHARDSON, J. A. 1947. An outline of the geomorphological evolution of British Malaya. *Geol. Mag.* 84: 129–144.

RICHARDSON, L. R. 1971. A new australian Dineta/Barbronia-like leach, and related matters (Hirudinoidea: ? Erpobdellidae) *Proc. Limn. Soc. N.S.W. Pt.* 95: 221–231.

RICHARDS, P. W. 1957. The Tropical Rain Forest. Cambridge Univ. Press, London, 450 pp.

RICKER, W. E. 1934. An ecological classification of certain Ontario streams. *Univ. Toronto Studies, Biol. Series* 37: 1–114. (Publ. Ont. Fish. Res. Lab. No. 49).

RICKER, W. E. 1946. Production and utilization of fish populations. *Ecol. Monogr.* 16: 373–391.

RICKER, W. E. 1952. Systematic studies in Plecoptera. *Indiana Univ. Publ., Sci. Ser.* 18: 1–200.

RICKER, W. E. (ed.). 1968. Methods for Assessment of Fish Production in Fresh Waters. IBP Handbook No. 3, Blackwell Sci. Publ., Oxford, 313 pp.

ROBACK, S. S., CAIRNS, J. JR. and KAESLER, R. L. 1969. Cluster analysis of occurrence and distribution of insect species in a portion of the Potomac River. *Hydrobiologia* 34: 484–502.

ROE, F. W. 1953. The geology and mineral resources of the neighbourhood of Kuala Selangor and Rasa, Selangor, Federation of Malaya, with an account of the geology of Batu Arang coal-field. *Malaya Geol. Survey Mem.* N.S. 7: 1–163.

ROOS, T. 1957. Studies on upstream migration in adult stream-dwelling insects. I. *Ann. Rept. Inst. Freshw. Res. Drottningholm* 38: 167–193.

ROSS, H. H. 1956. Evolution and Classification of the Mountain Caddisflies. Univ. Illinois Press, Urbana.

ROSS, H. H. 1963. Stream communities and terrestrial biomes. *Arch. Hydrobiol* 59: 235–242.

ROUND, F. E. 1964. The ecology of benthic algae. Pp. 138–184, in: JACKSON, D. F. (ed.). Algae and Man. Plenum Press, New York, 434 pp.

ROUND, F. E. 1965. The Biology of the Algae. Edward Arnold, London, 269 pp.

RUPPRECHT, R. 1969. Zur Artspezificität der Trommelsignale der Plecopteren (Insecta). *Oikos* 20: 26–33.

RUTTER, A. J. 1958. The effects of afforestation on rainfall and run-off. *J. Instn. Pub. Hlth. Engrs.* 119–138.

RUTTER, A. J. 1967. Evaporation in forests. *Endeavour* 97: 39–44.

RUTTNER, F. 1931. Hydrographische und hydrochemische Beobachtungen auf Java, Sumatra und Bali. *Arch. Hydrobiol.* Suppl. VIII.

RUTTNER, F. 1963. Fundamentals of Limnology. Trans. FREY, D. G. and FRY, F. E. J. Univ. Toronto Press, Toronto, 295 pp.

RZÓSKA, J. 1961. Some aspects of the hydrobiology of the River Nile. *Verh. Internat. Verein. Limnol.* 14: 505–507.

RZÓSKA, J. 1967. Freshwater productivity in the tropics: notes on secondary level studies. *J. Anim. Ecol.* 36: 15 P.

SANDS, W. N. 1934. The coloured scums of padi fields. *Malay. Agric. J.* 22: 484.

SATTLER, W. 1962. Über einem Fall von hygropetrischen Lebensweise einen Philopotamide *(Chimarrha,* Tricoptera*)* aus dem brasilianischen Amazonasgebiet. *Arch. Hydrobiol.* 58: 125–135.

SATTLER, W. 1963. Über den Körperbau, die Ökologie und Ethologie der Larve und Puppe von *Macronema* (Hydropsychidae) ein als Larve sich von 'Micro-Drift' ernährends Trichopter aus dem Amazonasgebiet. *Arch. Hydrobiol.* 59: 26–60.

SATTLER, W. 1968. Weitere Mitteilungen über die Ökethologie einer neotropischen *Macronema* – Larvae. *Amazoniana* 1: 211–219.

SATTLER, W. and KRACHT, A. 1963. Drift-fang einer Trichopterenlarve unter Ausnutzung der Differenz von Gesamtdruck und statischen Druck des fliessenden Wassers. *Naturwissenschaften* 50: 362.

SAXENA, S. C. and CHANDY, M. 1966. Adhesive apparatus in certain Indian hill-stream fishes. *J. Zool.* 148: 315–340.

*SCHMASSMANN, H. J. 1951. Untersuchungen über den Stoffhaushalt fliessender Gewässer. *Schweiz. Z. Hydrol.* 13 (cited in RUTTNER 1963).

SCHMIDT, H.-W. 1969. Tages- und jahresperiodische Driftaktivität der Wassermilben Hydrachnellae, Acari). *Oecologia* (Berl.) 3: 240–248.

SCHMITZ, W. 1954. Grundlagen der Untersuchung der Temperaturverhältnisse in den Fliessgewässern. *Ber. Limnol. Flusst. Freudenthal* 6: 29–50.

SCHNELLER, M. V. 1955. Oxygen depletion in Salt Creek, Indiana. *Invest. Indiana Lakes and Streams.* 4: 163–175.

SCHUHMACHER, H. 1969. Kompensation der Abdrift von Köcherfliegen-Larven (Insecta, Trichoptera). *Naturwissenschaften* 56: 378.

SCHWARTZ, H. I. 1969. Hydrologic aspects of limnology in South Africa. *Hydrobiologia* 34: 14–28.

SCHWARZ, P. 1970. Autökologische Untersuchungen zum Lebenszyklus von Setipalpia-Arten (Plectoptera). Teil I und Teil II. *Arch. Hydrobiol.* 67: 103–140, 141–172.

SCHWOERBEL, J. 1961. Über die Lebensbedingungen und die Besiedlung des hyporheischen Lebensraumes. *Arch. Hydrobiol.* Suppl. XXV: 182–214.

SCHWOERBEL, J. 1964. Die Bedeutung des Hyporheals für die benthische Lebensgemeinschaft der Fliessgewässer. *Verh. Internat. Verein. Limnol.* 15: 215–226.

SCHWOERBEL, J. 1966. Methoden der Hydrobiologie. Franckh'sche Verlagshandlung, Stuttgart, 207 pp.

SCHWOERBEL, J. 1967. Das hyporheische Interstitial als Grenzbiotop zwischen oberirdischem und subterranem Ökosystem und seine Bedeutung für die Primär-Evolution von Kleinsthöhlenbewohnern. *Arch. Hydrobiol.* Suppl. XXXIII: 1–62.

SCOTT, D. 1958. Ecological studies on the Trichoptera of the River Dean, Cheshire. *Arch. Hydrobiol.* 54: 340–392.

SCOTT, D. 1960. Cover on river bottoms. *Nature, Lond.* 188: 76–77.

SCOTT, D. 1966. The substrate cover-fraction concept. *Spec. Publs. Pymatuning Lab. Ecol.* 4: 75–86.

SCOTT, D. and RUSHFORTH, J. M. 1959. Cover on river bottoms. *Nature, Lond.* 183: 836–837.

SEKI, H., STEPHENS, K. V. and PARSONS, T. R. 1969. The contribution of allochthonous bacteria and organic materials from a small river with a semi-enclosed sea. *Arch. Hydrobiol.* 66: 37–47.

SHAPIRO, J. 1961. Freezing-out, a safe technique for concentration of dilute solutions. *Science* 133: 2063.

SHAPIRO, J. 1966. The relation of humic color to iron in natural waters. *Verh. Internat. Verein. Limnol.* 16: 477–484.

SHARMA, R. E. and FERNANDO, C. H. 1961. Leeches and their ways. *Malay. Nat. J.* 15: 152–159.

SHAWINIGAN ENGINEERING COMPANY LTD. 1967. Feasibility Study, Upper Perak Hydro-Electric Development, Volume III. Hydrology V. Montreal.

SHELDON, A. L. 1968. Species diversity and longitudinal succession in stream fishes. *Ecology* 49: 193–198.

SHELDON, A. L. 1969. Size relationwhips of *Acroneuria californica* (Perlidae, Plecoptera) and its prey. *Hydrobiologia* 34: 85–94.

SHELFORD, V. E. 1911. Ecological succession. I. Stream fishes and the method of physiographic analysis. *Biol. Bull.* 21: 9–34.

SHOUP, C. S. 1943. Distribution of fresh water gastropods in relation to total alkalinity of streams. *Nautilus* 56: 130–134.

SILVA, P. C. 1962. Classification of Algae.
Pp. 827–837, in: LEWIN, R. A. (ed.). Physiology and Biochemistry of Algae. Academic Press, New York, 929 pp.

SIOLI, H. 1951. Zum Alterungsrozess von Flüssen, und Flusstypen im Amazonasgebiet. *Arch. Hydrobiol.* 45: 267–283.

SIOLI, H. 1963. Beitrage zur regionalen Limnologie des brasilianischen Amazonasgebietes. *Arch. Hydrobiol.* 59: 311–350.

SIOLI, H. 1964. General features of the limnology of Amazonia. *Verh. Internat. Verein. Limnol.* 15: 1053–1058.

SIOLI, H. 1965. Bemerkungen zur Typologie amazonischer Flüsse. *Amazoniana* 1: 74–83.

SIOLI, H. 1967a. Studies in Amazonian waters. *Atas do Simpósio sôbre a Biota Amazônica* Vol. 3 (Limnologia): 9–50.

SIOLI, H. 1967b. The Cururu region in Brazilian Amazonia: A transition zone between Hylaea and Cerrado. *J. Indian Bot. Soc.* XLVI: 452–462.

SIOLI, H. 1968a. Hydrochemistry and geology in the Brazilian Amazon region. *Amazoniana* 1: 267–277.

SIOLI, H. 1968b. Principle biotopes of primary production in the waters of Amazonia. Pp. 591–600, in: MISRA, R. and GOPAL, B. (eds.). Proc. Symp. Recent Adv. Trop. Ecol. (1968). Banaras Hindu University, Varanasi-5, India.

SIOLI H. 1969a. Ökologie im brasilianischen Amazonasgebiet. *Naturwissenschaften* 56: 248–255.

SIOLI, H. 1969b. Entwicklung und Aussichten der Landwirtschaft im brasilianischen Amazonasgebiet. *Die Erde* 100: 307–326.

SIOLI, H. and KLINGE, H. 1962. Sólos, tipos de vegetacão e águas na Amazônia. *Boletim do Museu Paraense Emilio Goeldi* 1: 27–41.

SIOLI, H. and KLINGE, H. 1966. Anthropogene vegetation im brasilianischen Amazonasgebiet. Bericht über das Internationale Symposium in Stolzenau/Weser 1961: 357–367.

SIOLI, H. SCHWABE, G. H. and KLINGE, H. 1969. Limnological outlooks on landscape-ecology in Latin America. *Tropical Ecology* 10: 72–82.

SLACK, K. V. 1955. A study of the factors affecting stream productivity by the comparative method. *Invest. Indiana Lakes and Streams* 4: 3–47.

SLACK, K. V. and FELTZ, H. R. 1968. Tree leaf control on low flow water quality in a small Virginia stream. *Environ. Sci. Technol.* 2: 126–131.

SLÁDEČEK, V. and SLÁDEČKOVÁ, A. 1963a. Limnological study of the reservoir Sedlice near Zeliv. XXIII. Periphyton production. *Sci. Papers Inst. Chem. Tech. Prague. Tech. of Water* 7: 77–133.

SLÁDEČEK, V. and SLÁDEČKOVÁ, A. 1963b. Relationship between wet weight and dry weight of the periphyton. *Limnol. Oceanogr.* 8: 309–311.

SLÁDEČEK, V. and SLÁDEČKOVÁ, A. 1964. Determination of the periphyton production by means of the glass slide method. *Hydrobiologia* 23: 125–158.

SLÁDEČKOVÁ, A. 1962. Limnological investigation methods for the periphyton ('Aufwuchs') community. *Bot. Rev.* 28: 286–350.

SLATER, J. V. 1954. The quantitative evaluation of dissolved organic matter in natural water. *Trans. Amer. microscop. Soc.* 73: 416–423.

SMART, J. and CLIFFORD, E. A. 1969. Simuliidae (Diptera) of Sabah (British North Borneo). *Zool. J. Linn. Soc.* 48: 9–47.

SMITH, H. M. 1945. The Freshwater Fishes of Siam, or Thailand. *Bull. U.S. Nat. Mus.* 188: 1–622.

SOONG, M. H. H. 1968. Aspects of reproduction in freshwater halfbeaks. *Malay. Nat. J.* 21 (Suppl.): 33–34. Abstract from Conf. Zoo. Res. in Malaysia and Singapore, 1967.

SØRENSON, T. 1948. A method of establishing groups of equal amplitude in plant sociology based on similarity of species content and its application to analyses of the vegetation of Danish commons. *Kanske vid. Selsk. Biol. Skr.* 5: 1–34.

SOWA, R. 1965. Ecological characteristics of the bottom fauna of the Wielka Puszcza stream. *Acta Hydrobiol.* 7., Suppl. 1: 61–92.

SPEER, W. S. 1963. Report to the Government of Malaysia on soil and water conservation. FAO Publ. 1788.

SPRULES, W. M. 1947. An ecological investigation of stream insects in Algonquin Park, Ontario. *Univ. Toronto Studies, Biol. Ser.* 56: 1–81 (Publ. Ont. Fish. Res. Lab. No. 69).

STEPHENSON, T. 1931. Oligochaeta from the Malay Peninsula. *J. Fed. Malay Sta. Mus.* 16: 261–285.

STEVENS, J. C. 1942. Device for measuring static heads. *Civil Eng.* 12: 103–104.

STEWART, W. D. P. 1968. Nitrogen input into aquatic ecosystems. Pp. 53–72, in: JACKSON, D. F. (ed.). Algae, Man, and the Environment. Syracuse Univ. Press, Syracuse, 554 pp.

STOCKER, Z. S. J. and WILLIAMS, D. D. 1972. A freezing core method for describing the vertical distribution of sediments in a streambed. *Limnol. Oceanogr.* 17: 136–138.

STOCKNER, J. G. 1968a. Algal growth and primary productivity in a thermal stream. *J. Fish. Res. Bd. Canada* 25: 2037–2058.

STOCKNER, J. G. 1968b. The ecology of a diatom community in a thermal stream. *Br. phycol. Bull.* 3: 501–514.

STRAHLER, A. N. 1954. Quantitative geomorphology of erosional landscapes. Compt. Rendu. 19th Internat. Geol. Congr., Sec. 13: 341–354.

STRAHLER, A. N. 1957. Quantitative analysis of watershed geomorphology. *Trans. Amer. Geophys. Union* 38: 913–920.

STRAŠKRABA, M. 1965. The effect of fish on the number of invertebrates in ponds and streams. *Mitt. Internat. Verein. Limnol.* 13: 106–127.

STRAŠKRABOVÁ-PROKEŠOVA, V. 1966. Oxidation of organic substances in the water of the reservoirs Slapy and Klicava.
Pp. 85–111, in: HRBÁČEK, J. (ed.). Hydrobiological Studies, Academic Publ. House, Prague (Part 1). (Abstract only seen).

STRICKLAND, J. D. H. 1958. Solar radiation penetrating the ocean. A review of requirements, data and methods of measurement with particular reference to photosynthetic productivity. *J. Fish. Res. Bd. Canada* 15: 453–493.

STRICKLAND, J. D. H. 1960. Measuring the production of marine phytoplankton. *Bull. Fish. Res. Bd. Canada* 122: 1–172.

STRICKLAND, J. D. H. 1963. Significance of the values obtained by primary production measurements.
Proc., Conf. of Primary Productivity Measurement, Marine and Freshwater, Hawaii, Aug. 1961. U.S. Atomic Energy Comm. TID–7633: 172–183.

STRICKLAND, J. D. H. and PARSONS, T. R. 1960. A manual of seawater analysis. *Bull. Fish. Res. Bd. Canada* 125: 1–185.

STRICKLAND, J. D. H. and PARSONS, T. R. 1968. A practical handbook of seawater analysis. *Bull. Fish. Res. Bd. Canada* 167: 1–311.

STUMM, W. and MORGAN, J. J. 1970. Aquatic Chemistry. J. Wiley and Sons, Inc., New York, 583 pp.

Subcommittee on Sedimentation of the Inter Agency Committee on Water Resources. 1957. A study of methods used in measurement and analysis of sediment loads in streams. Rept. No. 12. Some fundamentals of particle size analysis. U.S. Govt. Print. Office. 55 pp.

SYMINGTON, C. F. 1943. Foresters' manual of dipterocarps. *Mal. For. Rec.* 16: 1–244.

*SZCZEPANSKI, A. 1965. Deciduous leaves as a source of organic matter in lakes. *Bull. Acad. pol. Sci. Cl. II Sér. biol.* 12: 215–217.

SZEICZ, G. 1966. Field measurements of energy in the 0.4–0.7 micron range.
Pp. 41–51, in: BAINBRIDGE, R., EVANS, G. C. and TACKHAM, O. (eds.). Brit. Ecol. Soc. Symp. No. 6, Cambridge, March 1958, 452 pp.

TALLING, J. F. 1958. The longitudinal succession of the water characteristics in the White Nile. *Hydrobiologia* 11: 73–89.

TALLING, J. F. and DRIVER, D. 1963. Some problems in the estimation of chlorophyll-*a* in phytoplankton.
Proc. Conf. of Primary Productivity Measurement, Marine and Freshwater, Hawaii, Aug. 1961. U.S. Atomic Energy Comm. TID–7633: 142–146.

TANAKA, H. 1967. On the change of composition of aquatic insects resulting from difference in mesh size of stream bottom-samplers. *Bull. Freshw. Fish. Res. Lab., Tokyo* 17: 1–6.

TANAKA, H. 1970. On the nocturnal feeding activity of rainbow trout *(Salmo gairdnerii)* in streams. *Bull. Freshw. Fish. Res. Lab., Tokyo* 20: 73–82.

TAN, C. K. 1968. Maturity and spawning of *Trichogaster trichopterus* (Pallas). *Malay. Nat. J.* 21 (Suppl.): 34. Abstract from Conf. Zoo. Res. in Malaysia and Singapore, 1967.

TAN, H. T. 1969. Standard catchments for the derivation of rainfall runoff relation required for flood estimation of small catchments in Malaya.
Pp. 185–193, in: STONE, B. C. (ed.). Natural Resources in Malaysia and Singapore. Proceedings of the second Symposium on Scientific and Technological Research in Malaysia and Singapore, Univ. Malaya, 1–4 Feb. 1967, 256 pp.

TAN, Y. T. and PROWSE, G. A. (in press). A chemical survey of the Malacca River. *Malay. Agric. J.*

TARSHIS, I. B. and NEIL, W. 1970. Mass movement of blackfly larvae on silken threads (Diptera, Simuliidae). *Ann. Ent. Soc. Amer.* 63: 607–610.

TEBO, L. B. 1955. Effects of siltation, resulting from improper logging, on the bottom fauna of a small trout stream in the Southern Appalachians. *Progr. Fish Cult.* 17: 64–70.

TEMPLETON, W. L., DEAN, J. M., WATSON, D. G. and RANCITELLI, L. A., 1969. Freshwater ecological studies in Panama and Colombia. *Bio Science* 19: 804–808.

*THIESSEN, A. H. 1911. Precipitation for large areas. *Monthly Weather Rev.* 39: 1082–1084.

THOMAS, E. 1969a. Orientierung der Imagines von *Capnia atra* Morton (Plecoptera) II. *Oecologia* (Berl.) 2: 376–384.

THOMAS, E. 1969b. Die Drift von *Asellus coxalis septentrionalis* Herbst (Isopoda). *Oikos* 20: 231–247.

THOMAS, J. D. 1966. On the biology of the catfish *Clarias senegalensis* in a man-made lake in the Ghanaian savanna with particular reference to its feeding habits. *J. Zool.* 148: 476–514.

THOMAS, W. A. 1970. Weight and calcium losses from decomposing tree leaves on land and in water. *J. Appl. Ecol.* 7: 237–241.

THOMSEN, L. C. 1937. Aquatic Diptera. Part V. Ceratopogonidae. *Cornell Univ. Agr. Exp. Sta. Mem.* 210: 57–80.

THORUP, J. 1966. Substrate type and its value as a basis for the delimitation of bottom fauna communities in running waters. *Spec. Publs. Pymatuning Lab. Ecol.* 4: 59–74.

*TJÖNNELAND, A. 1961. Light trap catches of *Neoperla spio* (Newman) (Insecta, Plecoptera) at Jinja, Uganda. *Contrib. Fac. Sci. Univ. Coll. Addis Ababa Ser. C. Zool.* 1: 1–16.

*TRICART, J. 1965. Le modelé des regions chaudes, forêts et savanes. Paris (referenced by DOUGLAS 1969).

TRUESDALE, G. A., DOWNING, A. L. and LOWDEN, G. F. 1955. The solubility of oxygen in pure water and sea-water. *J. appl. Chem.* 5: 53–62.

TSUDA, M. 1961. Important role of net-spinning caddis-fly larvae in Japanese running waters. *Verh. Internat. Verein. Limnol.* 376–377.

TURNBULL-KEMP, P. ST. J. 1960. Quantitative estimations of populations of the river crab *Potamon (Potamonautes) perlatus* (M. Edw.) in Rhodesia trout streams. *Nature, Lond.* 185: 481.

TUSA, I. 1969. On the feeding biology of the brown trout *(Salmo trutta* m. *fario* L.*)* in the course of day and night. *Zool. Listy.* 18: 275–284. (Abstract only seen).

ULFSTRAND, S. 1967. Microdistribution of benthic species (Ephemeroptera, Plecoptera, Trichoptera, Diptera: Simuliidae) in Lapland streams. *Oikos* 18: 293–310.

ULFSTRAND, S. 1968a. Benthic animal communities in Lapland streams. *Oikos* Suppl. 10: 1–120.

ULFSTRAND, S. 1968b. Life cycles of benthic insects in Lapland streams (Ephemeroptera, Plecoptera, Trichoptera, Diptera: Simuliidae). *Oikos* 19: 167–190.

ULFSTRAND, S. 1969. Ephemeroptera and Plecoptera from River Vindelälven in Swedish Lapland. *Entomol. Ts. Årg.* 90: 145–165.

ULMER, G. 1940. Eintagsfliegen (Ephemeropteren) von den Sunda-Inseln. *Arch. Hydrobiol.* Suppl. XVI: 581–692.

ULMER, G. 1951. Köcherfliegen (Trichopteren) von den Sunda-Inseln. Teil I. *Arch. Hydrobiol.* Suppl. XIX: 1–528.

ULMER, G. 1955. Köcherfliegen (Trichopteren) von den Sunda-Inseln. Teil II. *Arch. Hydrobiol.* Suppl. XXI: 408–608.

ULMER, G. 1957. Köcherfliegen (Trichopteren) von den Sunda-Inseln. Teil III. *Arch. Hydrobiol.* Suppl. XXIII: 109–470.

*(U)nited (S)tates (D)epartment of (A)griculture 1960. Soil classification, a comprehensive system. Seventh approximation. Soil Cons. Serv. U.S. Dept of Agric., Washington, D.C.

USINGER, R. L. 1956. Aquatic Hemiptera.
Pp. 182–228, in: USINGER, R. L. (ed.). Aquatic Insects of California. Univ. California Press, Berkeley and Los Angeles, 508 pp.

VAAS, K. F. and SACHLAN, M. 1956. Cultivation of common carp in running water in West Java. Proc. 6th Session Indo-Pacific Fish Council, Tokyo, 1955. F.A.O. Bangkok.

VAIDYA, B. S. 1967. Study of some environmental factors affecting the occurrence of Charophytes in Western India. *Hydrobiologia* 29: 256–262.

VAN BEMMELEN, R. W. 1949. The Geology of Indonesia. Vol 1A. Govt. Printing Office, The Hague. 732 pp.

*VAN DIJK, J. W. and EHRENCRON, V. K. R. 1949. The different rate of erosion with two adjacent basins in Java.
Contributions Gen. Agric. Res. Sta. (Bogor, Indonesia) No. 84: 1–10.

VAN SOMEREN, V. D. 1945. Studies on the biology of rainbow trout *(Salmo irideus)* in East Africa. I. Their food in the Thiba, Keringa and Sagana Rivers, Kenya Colony. *J. E. Africa Agric. Nat. Hist. Soc.* 18: 148–157.

VAN SOMEREN, V. D. 1946. The habitats and tolerance ranges of *Lymnaea (Radix) caillaudi*, the intermediate snail host of liver fluke in East Africa. *J. Anim. Ecol.* 15: 170–197.

VAN SOMEREN, V. D. 1952. The Biology of Trout in Kenya Colony. Govt. Printer, Nairobi, 114 pp.

VAUX, W. G. 1968. Intragravel flow and interchange of water in a streambed. *Fishery Bull. U.S. Fish Wildl. Serv.* 66: 479–489.

VENKATESWARLU, T. and JAYANTI, T. V. 1968. Hydrobiological studies of the River Sabarmati to evaluate water quality. *Hydrobiologia* 31: 442–448.

VENKATESWARLU, V. 1969. An ecological study of the algae of the River Moosi, Hyderabad (India) with special reference to water pollution. I. Physico-chemical complexes. II. Factors influencing the distribution of algae. *Hydrobiologia* 33: 117–143, 352–363.

VERDUIN, J. 1969. Critique of research methods involving plastic bags in aquatic environments. *Trans. Amer. Fish. Soc.* 98: 335–336.

VERRIER, M.-L. 1953. Le rhéotropisme des larves d'éphémères. *Bull. Biol. Fr. Belg.* 87: 1–34.

VINCENT, R. E. and MILLER, W. H. 1969. Altitudinal distribution of brown trout and other fishes in a headwater tributary of the South Platte River, Colorado. *Ecology* 50: 464–466.

VOLLENWEIDER, R. A. (ed.). 1969. A Manual on Methods for Measuring Primary Production in Aquatic Environments. IBP Handbook No. 12., Blackwell Sci. Publ., Oxford, 213 pp.

WALLACE, J. B., WOODALL, W. R. and SHERBERGER, F. F. 1970. Breakdown of leaves by feeding of *Peltoperla maria* nymphs (Plecoptera, Peltoperlidae). *Ann. Ent. Soc. Amer.* 63: 562–567.

*WALLEN, I. E. 1951. The direct effect of turbidity on fishes. *Bull. Okla. Agric. Mech. Coll.* 48: 1–27.

WALSHE, B. M. 1948. The oxygen requirements and thermal resistance of chironomid larvae from flowing and still waters. *J. exp. Biol.* 25: 35–44.

WANG, W.-C. and BRABEC, D. J. 1969. Nature of turbidity in the Illinois River. *J. Am. Wat. Wks. Ass.* 61: 460–464.

WANG, W.-C. and EVANS, R. L. 1969. Variation of silica and diatoms in a stream. *Limnol. Oceanogr.* 14: 941–944.

WANNER, H. 1970. Soil respiration, litter fall and productivity of tropical rain forest. *J. Ecol.* 58: 543–547.

WARNICK, S. L. and BELL, H. L. 1969. The acute toxicity of some heavy metals to different species of aquatic insects. *J. Wat. Pollut. Control. Fed.* 41: 280–284.

WARREN, C. E., WALES, J. H., DAVIS, G. E. and DOUDOROFF, P. 1964. Trout production in an experimental stream enriched with sucrose. *J. Wildl. Mgmt.* 28: 617–660.

WATERS, T. F. 1961a. Notes on the chlorophyll method of estimating the photosynthetic capacity of stream periphyton. *Limnol. Oceanogr.* 6: 486–488.

WATERS, T. F. 1961b. Standing crop and drift of stream bottom organisms. *Ecology* 42: 532–537.

WATERS, T. F. 1962a. Diurnal periodicity in the drift of stream invertebrates. *Ecology* 43: 316–320.

WATERS, T. F. 1962b. A method to estimate the production rate of a stream bottom invertebrate. *Trans. Amer. Fish. Soc.* 91: 243–250.

WATERS, T. F. 1964. Recolonisation of denuded stream bottom areas by drift. *Trans. Amer. Fish. Soc.* 93: 311–315.

WATERS, T. F. 1965. Interpretation of invertebrate drift in streams. *Ecology* 46: 327–334.

WATERS, T. F. 1966. Production rate, population density and drift of a stream invertebrate. *Ecology* 47: 595–604.

WATERS, T. F. 1968. Diurnal periodicity in the drift of a day-active stream invertebrate. *Ecology* 49: 152–153.

WATERS, T. F. 1969a. The turnover ratio in production ecology of freshwater invertebrates. *Amer. Nat.* 103: 173–185.

WATERS, T. F. 1969a. Invertebrate drift-ecology and significance to stream fishes. Pp. 121–134, in: NORTHCOTE, T. G. (ed.). Symp. on Salmon and Trout in Streams, H. R. MacMillan Lectures in Fisheries, Univ. British Columbia, Vancouver, 1969.

WATTS, I. E. M. 1955. Equatorial Weather with Particular Reference to Southeast Asia. Univ. London Press, London.

WEBB, J. E. 1969. Biologically significant properties of submerged marine sands. *Proc. R. Soc. Lond.* (B) 174: 355–402.

WEBER, N. A. 1959. Isothermal conditions in tropical soil. *Ecology* 40: 153–154.

WEIBEL, S. R., ANDERSON, R. J. and WOODWARD, R. L. 1964. Urban land runoff as a factor in stream pollution. *J. Wat. Polln. Contr. Fed.* 36: 914–924.

WELCOMME, R. L. 1969. The biology and ecology of the fishes of a small tropical stream. *J. Zool.* 158: 485 529.

WENE, G. and WICKLIFF, E. L. 1940. Modification of a stream bottom and its effect on the insect fauna. *Can. Ent.* 72: 131–135.

WENINGER, G. 1968. Vergleichende Drift-Untersuchungen an niederösterreichischen Fliessgewässern (Flysch-, Gneis-, Kalkformation). *Schweiz. Z. Hydrol.* 30: 138–185.

WENTWORTH, C. K. 1922. A scale of grade and class terms for clastic sediments. *J. Geol.* 30: 377–392.

WETZEL, R. G. 1963. Primary productivity of periphyton. *Nature, Lond.* 197: 1026–1027.

WETZEL, R. G. 1964. A comparative study of the primary productivity of higher aquatic plants, periphyton and phytoplankton in a large, shallow lake. *Int. Revue ges. Hydrobiol.* 49: 1–61.

WHITESIDE, B. G. and McNATT, R. M. 1972. Fish species diversity in relation to stream order and physicochemical conditions in the Plum Creek drainage basin. *Amer. Midl. Nat.* 88: 90–101.

WHITFORD, L. A. 1960. The current effect and growth of freshwater algae. *Trans. Amer. microscop. Soc.* 79: 302–309.

WHITFORD, L. A. and SCHUMACHER, G. J. 1963. Communities of algae in North Carolina streams and their seasonal relations. *Hydrobiologia* 22: 133–196.

WHITFORD, L. A. and SCHUMACHER, G. J. 1964. Effect of a current on respiration and mineral uptake in *Spirogyra* and *Oedogonium. Ecology* 45: 168–170.

WHITMORE, T. C. and BURNHAM, C. P. 1969. The altitudinal sequence of forests and soils on granite near Kuala Lumpur. *Malay. Nat. J.* 22: 99–118.

WIGGINS, G. B. 1966. The critical problem of systematics in stream ecology. *Spec. Publs. Pymatuning Lab. Ecol.* 4: 52–58.

WIGGINS, G. B. 1969. Contributions to the biology of the Asian caddisfly family Limnocentropodidae (Trichoptera). *Life Sciences Contributions of the Royal Ontario Museum* 74: 1–29.

WILHM, J. L. and LONG, J. 1969. Succession in algal mat communities at three nutrient levels. *Ecology* 50: 645–652.

WILLIAMS, C. B. 1953. The relative abundance of different species in a wild animal population. *J. Anim. Ecol.* 22: 14–31.

WILLIAMS, P. M. 1968. Organic and inorganic constituents of the Amazon River. *Nature, Lond.* 218: 937–938.

WILLIAMS, T. R. 1962. The diet of freshwater crabs associated with *Simulium neavei* in East Africa. II. The diet of *Potamon berardi* from Mount Elgon, Uganda. *Ann. trop. Med. Parasit.* 56: 362–367.

WILLIAMS, T. R. 1965. The diet of freshwater crabs associated with *Simulium neavei* in East Africa. III. The diet of *Potamonautes niloticus* and of an unidentified crab species from Mount Elgon, Uganda. *Ann. trop. Med. Parasit.* 59: 47–50.

WILLIAMS, T. R. 1966. Species succession in the river fauna of Mount Elgon, Uganda. *J. Anim. Ecol.* 35: 21P–22P.

WILLIAMS, T. R., CONNOLLY, R., HYNES, H. B. N. and KERSHAW, W. E. 1961. Size of particles ingested by *Simulium* larvae. *Nature, Lond.* 189: 78.

WILLIAMS, T. R., HYNES, H. B. N. and KERSHAW, W. E. 1961. Maintenance and diet of African freshwater crabs associated with *Similium neavei*. *Ann. Soc. Belge Méd. Trop.* 4: 291–292.

WILLIAMS, W. D. 1964. Some chemical features of Tasmanian inland waters. *Austral. J. mar. freshw. Res.* 15: 107–122.

WINNER, R. W. 1969. Seasonal changes in biotic diversity and in Margalef's pigment ratio in a small pond. *Verh. Internat. Verein. Limnol.* 17: 503–510.

WISNIEWSKI, T. F. 1958. Algae and their effects on dissolved oxygen and biochemical oxygen demand. Pp. 157–180, in: Oxygen Relationships in Streams. R. A. Taft San. Eng. Center, Ohio, Tech. Rept. W 58–2.

WOODWELL, G. M. 1970. Effects of pollution on the structure and physiology of eco-systems. *Science* 168: 429–433.

WRIGHT, S. C. and MILLS, I. K. 1967. Productivity studies on the Madison River, Yellowstone National Park. *Limnol. Oceanogr.* 12: 568–577.

WU, Y. F. 1931. A contribution to the biology of *Simulium* (Diptera). *Pap. Mich. Acad. Sci.* 13: 543–599.

WYATT-SMITH, J. 1964. A preliminary vegetation map of Malaya with descriptions of the vegetation types. *J. trop. Geog.* 18: 200–213.

WYCHERLEY, P. R. 1967. Rainfall probability tables for Malaysia. R.R.I.M. Planting Manual 12, Rubber Research Institute of Malaya, Kuala Lumpur, 85 pp.

WYCHERLEY, P. R. 1968. Pollution in Malayan waters. *Malay. Nat. J.* 22: 43.

WYCHERLEY, P. R. 1969. Forests and productivity. *Malay. Nat. J.* 22: 187–197.

YAALON, D. H. 1964. Chemical changes in rain-fed marsh waters during the dry season. *Limnol. Oceanogr.* 9: 218–223.

YOUNT, J. L. 1956. Factors that control species numbers in Silver Springs, Florida. *Limnol. Oceanogr.* 1: 286–295.

ZAFAR, A. R. 1959. Taxonomy of lakes. *Hydrobiologia* 13: 287–299.

ZAHAR, A. R. 1951. The ecology and distribution of black flies (Simuliidae) in South-east Scotland. *J. Anim. Ecol.* 20: 33–62.

PLATES

Plate 1. The upper Gombak valley in the steep-sloped, jungle-clad foothills of the main range.

Plate 2. The sinuous Lower Zone river below Station IV with extensive irrigated rice fields and peripheral urban development along highway.

Plate 3. Lowland Dipterocarp Forest above Station II.

Plate 4. Forest tributary at Station I at modal flow (0.1 cumec) – note alternating riffle-pool morphology and narrow gravel flood beach at left.

Plate 5. Riffle and flowing pool at Station I with drift net in position. Dark lutaceous sands and gravels typical of this stream are visible at left. Note clarity of water and virtually complete canopy.

Plate 6. Gombak River at Station II at low discharge (1.4 cumec) over typical riffle area.

Plate 7. Gombak River at Station II during spate (discharge 4–6 cumec with high erosional load).

Plate 8. The long stabilized pool area below Station II where vertical distribution sampling was carried out. Note quartz sands, even flow and dense canopy.

Plate 9. Gombak River just below the forest zone 2 km above Station III. Gradient and mean substrate size have decreased, boulders are rare and the river is almost completely exposed to insolation.

Plate 10. Station III submerged riffle at modal flow. Riparian cultivation and low banks with emergent and trailing grasses are common.

Plate 11. A long depositing reach below Station III at minimum discharge showing partially exposed sand and gravel bars. Water is only slightly turbid.

Plate 12. Short riffle and flowing pool at Station IV during the irrigation season when discharge is much reduced. The low, frequently flooded sandy banks are partially stabilized by grasses but are subject to periodic displacement and redeposition by floods.

433

Plate 13. The Gombak River at Station V during a low discharge period showing the dense marginal vegetation and high turbidity of the water.

Plate 14. Small sawmill near Station V dumping unburned waste products directly into the river which already carries considerable silt and garbage load.

Plate 15. Alluvial tin mining, a major source of fine suspended sediments.

APPENDIX

Appendix A. Water quality at the five main stations for the period October 1968 to November 1969. (Ions, BOD, COD and DO in mg/l.)

STATION I

	pH	μmho₂₅	Ca	Mg	Na	K	Fe	Mn	Cu	Zn	eumec
22–X–68	—	44.76	4.69	—	2.40	1.03	0.22	—	—	—	0.113
21–XI–68	7.00	37.20	2.49	—	2.20	1.40	0.18	—	—	—	0.123
19–XII–68	7.25	41.50	2.56	—	2.90	1.30	0.18	—	—	—	0.148
17–I–69	7.35	37.20	3.05	—	2.90	1.20	0.10	—	—	—	0.131
13–II–69	6.65	40.80	0.80	1.02	2.70	1.70	0.13	nd	nd	nd	0.090
13–III–69	6.78	48.00	0.74	1.15	3.20	1.37	0.13	nd	nd	nd	0.065
10–IV–69	7.10	43.20	0.44	1.04	2.40	1.06	0.06	nd	nd	nd	0.059
8–V–69	6.98	39.60	0.68	1.08	1.50	1.25	0.27	nd	nd	0.15	0.108
5–VI–69	6.87	37.20	0.37	0.93	2.60	1.30	0.10	nd	nd	0.11	0.195
3–VII–69	6.76	36.00	0.59	0.95	2.50	1.25	0.10	nd	nd	nd	0.114
31–VII–69	6.95	38.40	0.52	0.93	2.08	0.98	0.02	nd	nd	nd	0.105
28–VIII–69	6.80	37.20	0.25	0.91	1.60	1.50	0.09	nd	nd	nd	0.142
25–IX–69	7.00	32.40	0.20	0.39	1.05	1.25	0.06	nd	nd	nd	0.171
23–X–69	6.92	31.92	0.30	0.70	0.79	1.30	0.10	nd	nd	nd	0.251
20–XI–69	6.85	33.18	0.55	0.54	0.65	1.20	0.06	nd	nd	0.09	0.310

	HCO₃alk	Cl	PO₄–P	NO₃–N	NH₃–N	SO₄	COD	BOD	DO	% sat.	temp.°C
22–X–68	20.13	0.52	0.024	0.11	—	4.0	1.41	—	7.55	94.3	24.0
21–XI–68	16.20	0.60	0.036	0.10	—	1.0	0.53	0.16	7.77	92.3	23.0
19–XII–68	20.68	0.32	0.025	0.07	—	3.0	0.50	0.25	7.65	93.9	22.7
17–I–69	18.18	0.80	0.005	0.10	—	0.5	0.63	0.10	7.86	95.8	22.5
13–II–69	18.79	0.47	0.005	0.09	—	1.0	0.78	0.21	7.42	92.3	23.8
13–III–69	19.70	0.59	0.010	0.07	—	1.0	1.66	0.26	7.30	91.9	24.2
10–IV–69	20.07	0.38	0.010	0.09	0.10	1.0	0.41	0.24	7.34	91.5	23.8
8–V–69	16.59	0.46	0.025	0.28	0.08	0.0	3.74	0.40	7.04	89.6	25.0
5–VI–69	17.26	0.42	0.035	0.11	0.08	0.5	0.83	0.16	7.63	94.5	23.4
3–VII–69	17.75	0.09	0.030	0.10	0.08	0.0	1.06	0.34	7.75	96.2	23.6
31–VII–69	17.75	0.38	0.020	0.07	0.07	1.0	0.68	0.08	7.55	90.4	23.4
28–VIII–69	16.71	0.49	0.020	0.10	0.03	1.0	1.35	0.12	7.70	96.0	23.8
25–IX–69	15.49	0.41	0.015	0.17	0.07	0.0	0.31	0.18	7.88	97.0	22.8
23–X–69	14.46	0.45	0.020	0.10	0.03	0.0	0.25	0.27	7.78	95.2	22.8
20–XI–69	13.54	0.20	0.005	0.10		0.0	1.35	0.14	7.60	94.6	23.7

N.B.: A.A. Spectrophotometer used for anion determinations beginning 13–II–69. Note change in calcium values.

438

STATION II

Date	pH	μmho$_{25}$	Ca	Mg	Na	K	Fe	Mn	Cu	Zn	cumec
22–X–68	—	32.88	2.54	—	2.00	1.08	0.36	—	—	—	2.60
21–XI–68	6.90	27.60	1.73	—	2.05	1.40	0.28	—	—	—	2.36
19–XII–68	7.33	28.80	1.80	—	2.40	1.38	0.16	—	—	—	1.86
17–I–69	7.39	28.32	2.38	0.64	2.50	1.45	0.04	—	—	—	1.77
13–II–69	6.70	32.16	0.60	0.74	2.40	1.90	0.13	nd	nd	nd	1.13
13–III–69	6.93	33.60	0.61	0.59	2.50	1.52	0.09	nd	nd	nd	0.92
10–IV–69	7.13	30.00	0.35	0.68	2.00	1.26	0.06	nd	nd	0.15	0.84
8–V–69	7.10	31.20	0.54	0.56	1.25	1.50	0.19	nd	nd	0.22	1.79
5–VI–69	6.90	27.60	0.31	0.65	2.15	1.50	0.13	nd	nd	nd	1.69
3–VII–69	6.90	31.20	0.51	0.65	2.12	1.50	0.10	0.007	nd	nd	1.31
31–VII–69	6.90	30.00	0.48	0.61	1.94	1.10	0.04	nd	nd	nd	1.46
28–VIII–69	6.70	30.00	0.22	0.31	1.50	1.60	0.09	nd	nd	nd	1.41
25–IX–69	7.15	32.40	0.20	0.44	1.01	1.33	0.08	trace	nd	nd	1.36
23–X–69	7.05	26.40	0.30	0.44	0.72	1.60	0.20	nd	nd	nd	2.79
20–XI–69	6.90	28.80	0.45	0.39	0.55	1.40	0.08	nd	nd	0.91?	2.71

Date	HCO$_3$alk	Cl	PO$_4$–P	NO$_3$–N	NH$_3$–N	SO$_4$	COD	BOD	DO	% sat.	temp. °C
22–X–68	16.78	0.52	0.024	0.14	—	2.0	1.22	—	7.50	91.9	23.0
21–XI–68	12.14	0.65	0.020	0.10	—	0.5	1.12	0.11	8.00	94.2	22.6
19–XII–68	15.43	0.39	0.015	0.10	—	2.5	1.11	0.06	7.72	92.9	22.3
17–I–69	14.95	0.64	0.005	0.14	—	0.5	1.10	0.05	8.00	96.1	22.2
13–II–69	14.21	0.50	0.005	0.09	—	1.0	0.72	0.49	7.44	93.6	24.5
13–III–69	15.01	0.48	0.005	0.08	—	0.0	0.78	0.14	7.44	92.2	24.0
10–IV–69	14.34	0.23	0.000	0.13	—	0.5	0.99	0.20	7.67	94.6	23.4
8–V–69	13.54	0.39	0.010	0.19	0.08	0.0	3.10	0.40	7.34	92.0	24.4
5–VI–69	11.83	0.31	0.020	0.19	0.08	0.0	1.02	0.24	7.80	95.7	23.2
3–VII–69	13.66	0.13	0.000	0.12	0.08	0.0	1.12	0.15	7.74	93.0	23.0
31–VII–69	14.34	0.26	0.020	0.09	0.06	0.0	1.10	0.06	7.63	90.9	23.1
28–VIII–69	13.18	0.36	0.005	0.10	0.08	0.0	1.13	0.20	7.84	96.6	23.3
25–IX–69	13.54	0.46	0.005	0.16	0.03	0.0	0.64	0.18	7.92	96.5	22.6
23–X–69	11.47	0.41	0.015	0.20	0.07	0.2	1.08	0.14	7.86	95.2	22.4
20–XI–69	12.26	0.23	0.005	0.14	0.03	0.0	1.35	0.28	7.84	96.0	23.0

STATION III

	pH	μmho25	Ca	Mg	Na	K	Fe	Mn	Cu	Zn	cumec	temp.°C
22-X-68	—	34.71	2.34	—	2.40	1.80	0.47	—	—	—	1.95	25.0
21-XI-68	6.85	27.60	1.61	—	2.40	2.10	0.49	—	—	—	2.84	24.7
19-XII-68	7.25	28.80	1.60	—	2.80	1.99	0.32	—	—	—	2.45	24.0
17-I-69	7.20	27.96	2.20	0.48	2.80	2.10	0.35	nd	—	—	2.19	25.0
13-II-69	6.70	30.96	0.50	0.35	3.10	2.90	0.15	nd	nd	nd	1.01	27.7
13-III-69	7.00	33.60	0.40	0.30	3.00	2.09	0.20	nd	nd	nd	1.23	28.5
10-IV-69	7.03	30.00	0.22	0.35	2.20	1.54	0.11	nd	nd	nd	0.60	27.0
8-V-69	7.10	30.00	0.34	0.35	1.50	2.00	0.39	trace	nd	0.22	1.67	26.9
5-VI-69	6.90	28.80	0.24	0.39	2.52	1.90	0.20	nd	nd	nd	1.62	25.3
3-VII-69	6.88	31.20	0.42	0.39	2.44	2.00	0.20	nd	trace	nd	1.22	24.4
31-VII-69	6.95	30.00	0.33	0.36	2.17	1.25	0.15	0.015	nd	trace	0.91	26.1
28-VIII-69	6.72	30.00	0.19	0.21	1.70	2.20	0.25	nd	nd	nd	1.67	26.1
25-IX-69	7.25	33.60	0.30	0.30	1.15	1.92	0.15	nd	nd	nd	1.70	25.7
23-X-69	7.64	26.40	0.25	0.22	0.80	2.40	0.30	0.004	nd	nd	3.94	23.8
20-XI-69	6.90	29.40	0.45	0.23	0.75	2.05	0.20	nd	nd	0.23	3.03	26.0

	HCO3alk	Cl	PO4-P	NO3-N	NH3-N	SO4	COD	BOD	DO	% sat.
22-X-68	16.78	0.52	0.031	0.14	—	2.0	1.99	—	7.55	93.9
21-XI-68	12.26	0.77	0.025	0.10	—	1.0	1.64	0.18	7.68	94.6
19-XII-68	16.47	0.41	0.020	0.10	—	3.0	1.41	0.36	7.92	96.8
17-I-69	13.60	0.82	0.010	0.17	—	0.0	1.81	0.32	7.76	96.7
13-II-69	12.99	0.37	0.005	0.10	—	0.0	1.44	0.66	7.65	99.6
13-III-69	14.34	0.58	0.035	0.13	—	0.0	1.44	0.81	7.55	99.6
10-IV-69	14.88	0.37	0.005	0.12	—	0.0	1.51	0.50	7.67	98.2
8-V-69	13.91	0.41	0.025	0.15	0.06	0.0	5.46	1.17	7.25	93.1
5-VI-69	12.69	0.20	0.020	0.17	0.08	0.0	1.55	0.37	7.74	96.6
3-VII-69	14.40	0.29	0.025	0.15	0.08	0.0	2.07	0.32	7.73	95.2
31-VII-69	14.34	0.24	0.010	0.10	0.06	0.0	1.53	0.37	7.69	96.5
28-VIII-69	13.66	0.31	0.020	0.14	0.07	0.0	1.59	0.40	7.64	96.3
25-IX-69	14.88	0.51	0.015	0.19	0.03	0.0	0.70	0.26	7.76	97.6
23-X-69	11.41	0.29	0.030	0.26	0.07	0.0	1.88	0.23	7.64	93.1
20-XI-69	12.75	0.33	0.005	0.17	0.03	0.0	1.75	0.46	7.46	94.4

STATION IV

	pH	µmho25	Ca	Mg	Na	K	Fe	Mn	Cu	Zn	cumec	temp.°C
22–X–68	—	36.54	2.34	—	2.40	1.83	0.61	—	—	—	4.27	26.0
21–XI–68	6.60	27.60	1.56	—	2.55	2.10	0.76	—	—	—	4.12	26.0
19–XII–68	7.10	28.80	1.45	—	3.20	2.03	0.50	—	—	—	3.67	25.0
17–I–69	6.85	26.12	2.10	0.05	2.90	2.20	0.46	—	—	—	4.21	25.5
13–II–69	6.30	30.96	0.20	0.33	2.90	2.80	0.30	nd	nd	nd	2.08	29.0
13–III–69	6.70	33.60	0.40	0.33	3.20	2.17	0.40	nd	nd	nd	1.89	29.9
10–IV–69	6.60	30.00	0.21	0.22	2.40	1.61	0.37	nd	nd	0.22	2.02	28.2
8–V–69	6.60	30.00	0.34	0.33	2.00	2.00	0.65	0.002	nd	nd	1.15	29.8
5–VI–69	6.54	30.00	0.30	0.32	2.74	2.10	0.43	nd	trace	nd	3.01	26.6
3–VII–69	6.57	31.20	0.40	0.35	2.58	2.00	0.40	trace	nd	0.10	2.35	24.4
31–VII–69	6.68	30.00	0.35	0.33	2.46	1.50	0.35	0.015	nd	nd	1.64	27.0
28–VIII–69	6.55	28.80	0.16	0.30	1.80	2.40	0.37	trace	nd	nd	3.16	27.2
25–IX–69	6.80	30.00	0.10	0.14	1.23	1.83	0.19	nd	nd	trace	2.15	27.0
23–X–69	6.70	25.20	0.15	0.20	0.87	2.70	0.50	0.007	nd	trace	5.73	24.4
20–XI–69	6.71	28.20	0.35	0.18	0.85	2.15	0.28	nd	nd	1.10?	5.11	27.0

	HCO₃alk	Cl	PO₄-P	NO₃-N	NH₃-N	SO₄	COD	BOD	DO	% sat.
22–X–68	16.63	1.29	0.036	0.16	—	2.0	2.50	—	7.25	91.5
21–XI–68	11.96	0.75	0.036	0.16	—	1.5	2.01	0.42	7.37	92.4
19–XII–68	15.31	0.38	0.030	0.15	—	3.5	1.71	0.61	7.52	92.5
17–I–69	12.81	0.90	0.005	0.18	—	0.0	2.50	0.59	7.36	92.1
13–II–69	12.38	0.48	0.020	0.10	—	0.0	1.44	0.86	7.02	92.2
13–III–69	14.21	0.87	0.030	0.13	—	0.0	1.71	0.95	7.04	94.9
10–IV–69	13.66	0.42	0.030	0.14	—	0.0	1.74	0.91	7.10	92.8
8–V–69	14.34	0.32	0.020	0.13	0.05	0.0	2.08	0.86	6.86	91.9
5–VI–69	12.44	0.27	0.030	0.19	0.07	0.0	1.72	0.75	7.16	90.6
3–VII–69	12.99	0.19	0.040	0.16	0.07	0.0	2.56	0.64	7.25	88.8
31–VII–69	14.52	0.35	0.020	0.08	0.05	0.0	1.52	0.54	6.98	87.8
28–VIII–69	12.81	0.46	0.020	0.11	0.07	0.0	2.01	0.72	7.28	93.6
25–IX–69	12.81	0.53	0.010	0.18	0.03	0.0	0.99	0.62	7.36	94.2
23–X–69	10.43	0.43	0.045	0.21	0.07	0.0	2.23	0.41	7.15	86.7
20–XI–69	12.14	0.43	0.020	0.20	0.03	0.0	1.89	0.40	7.10	90.9

Appendix A. continued

STATION V

Date	pH	μmho_{25}	Ca	Mg	Na	K	Fe	Mn	Cu	Zn	temp. °C	cumec
22-X-68	—	46.58	3.91	—	2.80	2.75	0.94	—	—	—	26.5	4.36
21-XI-68	6.60	37.20	3.33	—	2.90	2.50	1.04	—	—	—	25.5	3.44
19-XII-69	6.95	40.80	2.47	—	3.35	2.73	0.70	—	—	—	25.6	3.29
17-I-69	6.78	37.20	3.77	0.45	3.20	2.50	0.72	nd	nd	nd	26.2	5.05
13-II-69	6.23	46.56	0.80	0.58	3.40	4.00	0.45	nd	nd	nd	28.8	1.84
13-III-69	6.63	44.40	0.69	0.55	3.70	2.70	0.50	trace	nd	0.31	30.5	1.87
10-IV-69	6.55	42.96	0.66	0.78	3.00	1.75	0.68	0.004	nd	0.67	29.0	1.58
8-V-69	6.45	50.40	1.00	0.50	2.50	2.50	1.45	trace	nd	trace	29.8	1.98
5-VI-69	6.47	39.60	0.54	0.62	3.00	2.30	0.80	trace	trace	trace	29.1	4.05
3-VII-69	6.51	44.40	0.84	0.68	2.92	2.25	0.70	0.022	nd	0.20	24.8	3.57
31-VII-69	6.60	50.40	0.73	0.44	2.67	2.50	0.69	trace	nd	nd	29.0	1.92
28-VIII-69	6.52	37.20	0.25	0.26	2.00	2.75	0.58	0.086	nd	nd	28.5	3.82
25-IX-69	6.70	37.20	0.30	0.30	1.46	2.10	0.31	0.015	nd	nd	27.8	2.75
23-X-69	6.64	35.76	0.50	0.30	1.00	3.00	0.85	nd	nd	trace	25.0	7.29
20-XI-69	6.70	35.40	0.75	0.30	0.90	2.55	0.50	nd	nd	0.82?	27.2	6.20

Date	HCO_3alk	Cl	PO_4-P	NO_3-N	NH_3-N	SO_4	COD	BOD	DO	% sat.
22-X-68	22.88	1.29	0.930	0.19	—	0.0	2.76	—	6.15	78.1
21-XI-68	15.56	1.05	0.265	0.22	—	0.0	3.48	2.37	6.53	81.1
19-XII-68	20.80	0.55	0.600	0.16	—	1.0	3.46	3.98	6.14	76.8
17-I-69	17.39	1.76	0.230	0.17	—	0.0	3.29	2.65	6.67	84.2
13-II-69	17.75	0.75	0.575	0.12	—	0.0	2.16	>5.76	5.76	75.0
13-III-69	19.28	1.15	0.650	0.17	—	0.0	3.04	4.68	5.80	78.6
10-IV-69	18.91	0.86	0.150	0.19	—	0.0	1.97	1.51	6.04	79.8
8-V-69	22.27	0.92	0.480	0.22	0.16	0.0	3.00	3.98	4.78	64.0
5-VI-69	16.35	0.76	0.330	0.28	0.11	0.0	2.31	2.16	6.12	80.6
3-VII-69	18.61	0.65	0.560	0.19	0.09	0.0	3.60	2.45	6.08	74.9
31-VII-69	22.57	1.03	1.040	0.16	0.04	0.5	2.96	6.58	5.10	67.1
28-VIII-69	15.49	0.47	0.400	0.20	0.11	0.0	1.95	2.00	6.30	82.6
25-IX-69	16.41	0.81	0.235	0.20	0.04	0.0	1.53	1.72	6.32	81.8
23-X-69	15.62	0.49	0.300	0.23	0.06	0.0	3.33	2.76	6.24	77.2
20-XI-69	15.13	0.63	0.245	0.33	0.03	0.5	3.31	2.34	6.50	83.7

Appendix A. continued

C.O.D. ($KMnO_4$ oxidation) results from supplementary determinations made in conjunction with erosional load analyses

	STATION I	STATION II	STATION III	STATION IV	STATION V
8–XI–68	0.94	1.69	2.07	2.61	3.00
5–XII–68	0.50	0.81	1.74	1.99	2.61
2–I–69	0.73	0.93	1.60	1.60	1.40
30–I–69	0.47	1.21	2.15	2.00	2.24
27–II–69	0.54	1.32	1.74	1.57	2.47
27–III–69	0.64	1.63	2.21	2.50	3.68
24–IV–69	0.81	1.18	1.56	1.83	2.96
23–V–69	1.35	2.54	3.86	4.29	5.00
19–VI–69	2.34	2.24	2.04	1.88	2.65
17–VII–69	0.91	1.46	1.93	1.96	3.20
14–VIII–69	0.28	0.65	1.28	1.49	1.75
11–IX–69	0.18	0.53	0.94	1.12	2.90
9–X–69	0.05	0.58	0.99	1.41	2.09
6–XI–69	0.31	0.73	1.26	1.35	1.82

443

CYANOPHYTA

Chroococcales
  Chroococcaceae
    Aphanothece stagnina (Spreng.) A. Braun
    Merismopedia ?glauca (Ehr.) Nageli
  Entophysalidaceae
    Entophysalis sp.
Chamaesiphonales
  Chamaesiphonaceae
    Chamaesiphon sp.
  Dermocarpaceae
    Stichosiphon sansabaricus Drout et Daily
Nostocales
  Oscillatoriaceae
    Lyngbya ?allorgei Fremy
    Lyngbya sp.
    Plectonema sp.
    Microcoleucus sp.
    Oscillatoria 4 spp.
    Schizothrix sp.
  Nostocaceae
    Anabaena sp.
    Nostoc sp.
  Scytonemataceae
    Scytonema sp.
    Tolypothrix tenuis (Kütz.) J. Schmidt ex.
    Tolypothrix sp.
  Rivulariaceae
    Calothrix sp.
Stigonematales
  Stigonemataceae
    Stigonema ocellatum (Dillw.) Thuret.

RHODOPHYTA

Bangiales
  Erythrotrichiaceae
    Compsopogon ?coeruleus (Balbis) Mont.
Nemalionales
  Achrochaetiaceae
    Audouinella sp.
  Batrachospermaceae
    Batrachospermum 2 spp.
Ceramiales
  Delesseriaceae
    Caloglossa leprieurii Mont. *forma typica* Post

CRYPTOPHYTA

Cryptomonadales
  Cryptomonadaceae

Chilomonas paramecium Ehr.
Chilomonas sp.
Cryptomonas 2 spp.

BACILLARIOPHYTA

Eupodiscales
  Coscinodiscaceae
    Melosira roseana Raben.
    Melosira ruettneri Hust.
    Melosira italica (Ehr.) Kütz.
Biddulphiales
  Biddulphiaceae
    Hydrosera whampoensis Schwarz
Fragilariales
  Fragilariceae
    Fragilaria vaucheriae (Kütz.) Boye Petersen
    Fragilaria sp.
    Synedra ulna (Ehr.) var *amphirhynchus* Grunow
    Synedra ?ulna (Nitzsch) Ehr.
    Synedra sp.
  Eunotiaceae
    Eunotia ?monodon Ehr.
    Eunotia pectinalis (Kütz.) Raben.
    Eunotia robusta Ralfs
    Eunotia 2 spp.
Achnanthales
  Achnanthaceae
    Achnanthes brevipes C. A. Agardh
    Achnanthes crenulata Grunow
    Achnanthes inflata A. Boyer
    Achnanthes lanceolata (Bréb.) Grunow
    Achnanthes sp.
    Cocconeis ?thumensis Mayer
    Cocconeis sp.
Naviculales
  Cymbellaceae
    Cymbella javanica Hust.
    Cymbella sumatrensis Hust.
    Cymbella lanceolata (Ehr.) van Heurck
    Cymbella tumida (Bréb.) van Heurck
    Cymbella ventricosa Kütz.
    Cymbella sp.
    Gomphonema gracile Ehr.
    Gomphonema longiceps Ehr. var *subclavata* Grunow
    Gomphonema parvulum (Kütz.) van Heurck
    Gomphonema subventricosum Hust.
    Gomphonema sp.
  Naviculaceae
    Frustulia javanica Hust.
    Frustulia rhomboides (Ehr.) de Toni
    Frustulia saxonica Raben.
    Diploneis ?ovalis (Hilse) P. T. Cleve

      Pleurosigma sp.
      Stauroneis sp.
      Neidium sp.
      Navicula amphibola P. T. Cleve
      Navicula cancellata Donkin
      Navicula elegans Ehr.
      Navicula feurborni Hust.
      Navicula confervacea Kütz.
      Navicula ?pupula Kütz.
      Navicula sp.
      Pinnularia biceps Gregory var *minor* (Boye Petersen) P. T. Cleve
      Pinnularia viridis (Nitzsch) Ehr.
      Pinnularia gibba W. Smith
      Pinnularia legumen Ehr.
      Pinnularia microstauron (Ehr.) P. T. Cleve
      Pinnularia ?trigonocephala P. T. Cleve

Bacillariales
  Nitzschiaceae
      Hantzschia sp.
      Nitzschia palea (Kütz.) W. Smith
      Nitzschia sigma (Kütz.) W. Smith
      Nitzschia sp.

Surirellales
  Surirellaceae
      Surirella tenuissima Hust.
      Surirella angusticostata Hust.
      Surirella capronii Bréb.
      Surirella biserata Bréb.
      Surirella muelleri Hust.
      Surirella robusta Ehr.
      Surirella robusta Ehr. var *splendida* (Ehr.) van Heurck
      Surirella tenera Gregory

Epithemiales
  Epithemiaceae
      Epithemia gibberula Kütz.
      Epithemia gibberula var *producta* Grunow

CHRYSOPHYTA

Chrysomonadales
  Ochromonadaceae
      Ochromonas sp.
  Mallomonadaceae
      Mallomonas 2 spp.
Heterotrichales
  Tribonemataceae
      Bumilleria sp.

PYRROPHYTA

Peridiniales
  Gymnodiniaceae
      Gymnodinium sp.

EUGLENOPHYTA

Euglenales
  Euglenaceae
    Euglena agilis Carter
    Euglena elongata Schewiakoff
    Euglena fusca (Klebs) Lemm.
    Euglena mutabilis Schmitz
    Euglena ?vivida Playfair
    Euglena sp.
    Lepocinclis ovum (Ehr.) Lemm.
    Lepocinclis salina Fritsch
    Phacus ?onyx Pochmann
    Phacus platalea Drezepolski
    Phacus stokesii Lemm.
    Phacus sp.
    Strobomonas ?australica (Playfair) Defl.
    Trachelomonas curta Da Cunha emend Defl.
    Trachelomonas dubia Swirenko emend Defl.
    Trachelomonas hispida (Perty) Stein emend Defl.
    Trachelomonas ?oblonga Lemm.
    Trachelomonas ?similis Stokes
    Trachelomonas volvocina Ehr.
    Trachelomonas volzii Lemm. var *cylindrica* Playfair
  Astasiaceae
    Astasia sp.
  Peranemataceae
    Entosiphon ovatum Skuja
    Heteronema leptosmum Skuja
    Heteronema polymorphum Defl.
    Heteronema sp.
    Notosolenus stenochismos Skuja
    Notosolenus sp.
    Peranema cuneatum Playfair
    Peranema trichophorum (Ehr.) Stein
    Peranema 3 spp.
    Petalomonas heptaptera Prowse
    Petalomonas mediocanellata Stein
    Petalomonas sp.

CHLOROPHYTA

Volvocales
  Chlamydomonadaceae
    Chlamydomonas sp.
    Sphaerellopsis sp.
  Tetrasporales
  Palmellaceae
    Palmella sp.
    Asterococcus sp.
Chlorococcales
  Hydrodictyaceae
    Pediastrum sp.
  Characiaceae
    Dictyosphaerum pulchellum Wood

Dimorphococcus sp.
Coelastraceae
   Coelastrum sp.
Selenastraceae
   Ankistrodesmus falcatus (Corda) Ralfs
   Ankistrodesmus falcatus (Corda) Ralfs var *spirilliformis* G. S. West
Scenedesmaceae
   Scenedesmus quadricauda (Turp.) Bréb.
   Scenedesmus sp.
Ulotrichales
  Ulotrichaceae
   Ulothrix subtilissima Raben.
   Ulothrix tenuissima Kütz.
   Uronema sp.
  Cylindrocapsaceae
   Cylindrocapsa ?geminella Wolle
   Cylindrocapsa ?conferta G. S. West
   Schizomeris leibleinii Kütz.
  Chaetophoraceae
   Stigeoclonium sp.
Cladophorales
  Cladophoraceae
   Cladophora sp.
   Rhizoclonium ?hieroglyphicum (C. A. Agardh) Kütz.
Oedogoniales
  Oedogoniaceae
   Oedocladium sp.
   Oedogonium sp.
Zygnematales
  Zygnemataceae
   Mougeotia/Debarya sp.
   Spirogyra 2 spp.
  Desmidiaceae
   Actinotaenium sp.
   Cosmarium obsoletum (Hantzsch) Reinsch
   Cosmarium pseudoconnatum Nordst. var *ellipsoideum* W. et G. S. West
   Cosmarium granatum Bréb.
   Cosmarium ?pseudogranatum Nordst.
   Closterium ehrenbergii Meneghini
   Closterium libellula Focke
   Closterium ?rostratum Ehr.
   Closterium acutum (Lyngbye) Bréb.
   Closterium moniliferum (Bory) Ehr.
   Closterium sp.
   Euastrum ?spinulosum Delp.
   Euastrum binale (Turp.) Ehr. var *brevius* (Bernard) Hirano
   Micrasterias crux-melitensis (Ehr.) Ralfs.
   Micrasterias foliacea Bailey
   Penium sp.
   Pleurotaenium ovatum Nordst.
   Pleurotaenium cylindricum (Turner) W. et G. S. West
   Pleurotaenium 2 spp.
   Staurastrum sp.

Appendix B. continued

CHAROPHYTA

Charales
  Characeae
    Chara sp.
    Nitella acuminata A. Braun
    Nitella sp.

Appendix C. List of the macro-invertebrate taxa recorded from all stations during the study.

HYDROIDEA

Hydra sp.

TRICLADIDA

Planariidae
  Dugesia lindbergi de Beauchamp
  Dugesia sp.

NEMATODA

Labronema sp.
Diplogasterid type
Other Nematoda

GORDIIDA

Gordiidae
  Gordius sp.
Chordodidae
  Paragordius type
  ?*Beatogordius* type

PELECYPODA

Sphaeriidae
  Pisidium (Neopisidium) javanum van Bentham Jutting
Unionidae
  Unid. lamellibranch

GASTROPODA (Prosobranchia)

Viviparidae
  Siamopaludina martensi (Frauenfeld) (= *Bellamya javanica* V.d. Büsch)
Ampullaridae
  Pila scutata (Mousson)
Thiaridae
  Brotia costula (Rafinesque)
  Thiara scabra (Müller)
  Melanoides tuberculata (Müller)
  Melanoides var A
  Melanoides var B

GASTROPODA (Pulmonata)

Lymnaeidae
  Lymnaea rubiginosa Michelin
Planorbidae
  Gyraulus convexiusculus (Hutton)
  Indoplanorbis exustus (Deshayes)

450

Ancylidae
  Ferrissia javana (Martens)

OLIGOCHAETA

Naididae
  Branchiodrilus semperi (Bourne)
  Allonais inaequalis Stephenson
  Pristina proboscoidea Beddard
  Unid. *Pristina* sp.
  Chaetogaster sp.
  Aulophorus sp.
  Stylaria sp. nr *fossularis* Leidy
  Other Naididae
Haplotaxidae
  Haplotaxis sp.
Phreodrilidae
  Phreodrilus sp.
Tubificidae
  Limnodrilus hoffmeisteri Claparède
  Limnodrilus silvani Eisen
  Bothrioneurum sp.
  Branchiura sowerbyi Beddard
Enchytraeidae
  Unidentified type
Megascolecidae
  Eukerria kukenthali (Michaelsen)
  Pheretima sp(p).
  Ocnerodrilus occidentalis Eisen
Lumbricidae
  Pontoscolex corethrurus (Müller)

HIRUDINEA

Glossiphonidae
  Glossiphonia weberi Blanchard
  Batrachobdella reticulata (Kaburaki)
  Helobdella sp. nr *nociva* Harding
  Unid. *Helobdella* sp.
  Placobdella ?inleana Oka
  Unid. *Placobdella* 2 spp.
Erpobdellidae
  Barbronia weberi (Blanchard)
  Barbronia sp. ?var of *weberi*
  Herpobdelloidea lateroculata Kaburaki
Hirudidae
  Myxobdella ?annandalei Oka

ACARINA

Torrenticolidae/Atractididae
  Types 1, 2, 3, 4, 5
  Hydrachnellid type H

Appendix C. continued

Unid. Hydracarina
Mesostigmata/Sarcoptiformes/Oribatei (many)

CLADOCERA

Daphnidae
  Moinodaphnia ?macleayii (King)

OSTRACODA

Cypridae
  Types 1, 2, 3

COPEPODA

Cyclopoida
Harpacticoida

AMPHIPODA

Talitridae
  Orchestia anomala Chevreux

DECAPODA

Palaemonidae
  Macrobrachium malayanus Roux
  Macrobrachium pilimanus (de Man)
  Macrobrachium geron Holthuis
Atyidae
  Atya spinipes Newport
Potamonidae
  Potamon johorense Roux
  Paratelphusa maculata de Man

COLLEMBOLA

Poduroidea
Isotomidae
  Unid. type

ORTHOPTERA

Epilampridae
  Epilampra sp.
  Rhicnoda sp.
  Unid. type
Tridactylidae
  Unid. type

MEGALOPTERA

Corydalidae

452

Appendix C. continued

Hermes sumatrensis Weele

EPHEMEROPTERA

Siphlonuridae
  Isonychia sp. A
Baetidae
  Baetis spp. A, B, C, D, E, F, H, J, Km, Ko, L, M, X
  Pseudocloeon spp. G. B. C.
  Baetidae *genus incertus* (?*Baetopus*)
Oligoneuriidae
  Chromarcys sp. A
Heptageniidae
  Epeorus sp. A
  Thalerosphyrus spp. A, B, C
  Compsoneuriella sp. A
Leptophlebiidae
  Thraulus bishopi Peters and Tsui sp. nov.
  Isca sp. A or spp.?
  Habrophlebiodes sp. A
  Dipterophlebiodes sp. A
  Choroterpes s.s. sp. A
Ephemerellidae
  Ephemerella (*Drunella*) spp. A, B
  Ephemerella (*Crinitella*) sp. A
  Ephemerella s.s. type
  Teloganodes sp.
Tricorythidae
  Tricorythus sp. A
  Tricorythidae *genus incertus*
Potamanthidae
  Potamanthodes sp. A
Ephemeridae
  Ephemera (*Dierephemera*) sp. A
Neoephemeridae
  Neoephemeropsis spp. A, B
Caenidae
  Caenis spp. A, B
  Caenidae *genus incertus*
Prosopistomatidae
  Prosopistoma wouterae Lieftinck

ODONATA

Protostictidae
  Drepanosticta 2 spp.
Protoneuridae
  Prodasineura sp.
  Prodasineura laidlawii (Förster)
  Prodasineura verticalis (Selys)
Platycnemididae
  Copera marginipes (Rambur)
  Coeliccia albicauda (Förster)

453

Appendix C. continued

    Calicnemia chaseni (Laidlaw)
    Indocnemis orang (Selys)
Coenagrionidae
    Pseudagrion spp.
    Pseudagrion perfuscatum Lieftinck
    ?Ceriagrion sp.
Amphipterygidae
    Devadatta a. argyoides (Selys)
Chlorocyphidae
    Rhinocypha spp.
    Rhinocypha perforata limnata Selys
    Rhinocypha fenestrella Rambur
    Libellago l. lineata (Burm.)
Epallagidae
    Euphaea o. ochracea Selys
    Dysphaea dimidiata Selys
Calopterygidae
    Neurobasis c. chinensis (Linnaeus)
    Vestalis amoena Selys
Aeschnidae
    Tetracanthagyna sp.
Gomphidae
    Acrogomphus sp.
    Burmagomphus divaricatus Lieftinck
    Gomphidia perakensis Laidlaw
    Gomphidia a. abboti Williamson
    Heliogomphus kelantanensis (Laidlaw)
    Ictinogomphus decoratus melaenops (Selys)
    Leptogomphus ?risi Laidlaw
    Macrogomphus sp.
    Megalogomphus sumatranus (Kruger)
    Merogomphus femoralis Laidlaw
    Microgomphus c. chelifer Selys
    Microgomphus sp. nov.
    Onychogomphus fruhstoferi Lieftinck
    Paragomphus capricornis (Förster)
    Phaenandrogomphus asthenes Lieftinck
    Unid. Gomphidae
Cordulegasteridae
    Chlorogomphus dyak (Laidlaw)
Corduliidae
    Idionyx ?yolanda Selys
    Macromia type
    Macromia gerstaeckeri Laidlaw
    Macromia arachnomia Lieftinck
    Macromia callista Laidlaw
Libellulidae
    ?Zygonyx ida Selys
    Zygonyx iris malayana (Laidlaw)
    Neurothemis fluctans (Fabr.)
    Neurothemis fulvia (Drury)
    Trithemis type
    Trithemis aurora (Burm.)

Appendix C. continued

    Trithemis festiva (Rambur)
    Orthetrum type
    Orthetrum t. testaceum (Burm.)
    Orthetrum glaucum (Brauer)
    Orthetrum chrysis (Selys)
    Onychothemis type
    Onychothemis culminicola Förster
    Onychothemis coccinea Lieftinck
    Onychothemis testacea Laidlaw

PLECOPTERA

Peltoperlidae
    Neopeltoperla fraterna (Banks)
    Peltoperlodes biseata Kawai
Nemouridae
    Amphinemura minuta Kawai
    Protonemura jacobsoni (Klap.)
    Nemoura sp.
Leuctridae
    Rhopalopsole spinata Kawai sp. nov.
Perlidae
    Perla kelantonica (Klap.)
    Paragnetina sp.
    Unidentified Perlidae
    Neoperla luteola (Burm.)
    Neoperla nitida Kimmins
    Neoperla sp. nov.
    Tylopyge helvus Kawai sp. nov.
    Phanoperla clarissa (Banks)
    Neoperline type J
    Kiotina sp.
    Etrocorema nigrogeniculata (Enderl.)
    Etrocorema sp. (?*ahenobarba* Klap.)
    Etrocorema trapeza Kawai sp. nov.
    Unid. neoperline nymphs

HEMIPTERA

Hydrometridae
    Hydrometra sp.
Veliidae
    Rhagovelia femorata Dover
    Tetraripis doveri Lundb.
    Microvelia sp.
Gerridae
    Amemboa horvathi Esaki
    Limnogonus fossarum (Fabr.)
    Limnometra anadyomene (Kirk.)
    Metrocoris nigrofasciatus Dist.
    Pleciobates tuberculatus Esaki
    Ptilomera lundbladi Hung. & Mat.
    Rheumatogonus intermedius Hung.

Appendix C. continued

 Ventidius sp.
 Esakia sp.
Dipsocoridae
 Cryptostemma sp.
Nepidae
 Cercometus pilipes Dallas
 Ranatra varipes Stal.
Notonectidae
 Enithares malayensis Brookes
 Enithares mandalayensis Dist.
Pleiidae
 Plea liturata (Fieb.)
Helotrephidae
 Helotrephes corporaali China
Belostomatidae
 Diplonychus (*Sphaerodema*) *rusticus* (Fabr.)
Naucoridae
 Aphelocheirus gularis Horv.
 Laccocoris nervicus Mont.
 Ctenipocoris asiaticus Mont.
Corixidae
 Micronecta sp.

LEPIDOPTERA

Pyralidae
 Cataclysta type 88
 Cataclysta type 132
 Cataclysta type 350±
 Paragyractis type
 Nymphula (*Paraponyx*) type R3
 Nymphula (*Paraponyx*) type R5
 Nymphula (*Paraponyx*) type D5
 Gill-less Nymphulini type F1
 Gill-less Nymphulini type R1

TRICHOPTERA

Philopotamidae
 Chimarra spp. a, b, c, d, e
 Chimarra sp. nr. *fulmecki*
Rhyacophilidae
 Rhyacophila spp. a, b
 Rhyacophila type
Glossosomatidae
 Glossosoma sp.
 Synagapetus sp.
Hydropsychidae
 Hydropsyche spp. S, a, b
 Hydropsychodes spp. a, b
 Unid. Hydropsychinae
 Diplectrona spp. a, P, B.
 Diplectrona sp. nr *aurivittata*

456

    Cheumatopsyche sp.
    Macronemum sp.
    Unid. Macronematinae 3 spp.
    Amphipsyche sp.
Psychomyiidae
    Psychomyia sp. a
    Psychomyiella sp.
    Psychomyiella sp. a
    Ecnomus spp. a, b, c, d, e, f
    Psychomyiinae types a, b, c, d
Polycentropodidae
    Hyalopsychella sp.
    Pseudoneureclipsis sp.
    Tinodes spp. a, b, c, d
    Nyctiophylax sp.
    Nyctiophylax sp. a
    Polyplectropus sp.
    Polycentropus sp.
    Polycentropodinae type a
Stenopsychidae
    Stenopsyche sp.
Odontoceridae
    Marilia sp.
Goeridae
    Goera sp.
Helicopsychidae
    Helicopsyche sp.
Calamoceratidae
    Ganonema sp.
Limnocentropodidae
    Limnocentropus sp.
Leptoceridae
    Tagalopsyche sp.
    Oecetis spp. a, b, c
    Setodes spp. a, b
    Adicella spp. a, b, c, d
    ?*Adicella* sp.
    Leptocella sp.
    Triaenodes sp.
    Unid. Mystacidini
    Leptocerinae types a, b, c, d, e
    Unid. Leptoceridae
Lepidostomatidae
    Lepidostomatinae spp. a, b, c, d
    Dinarthropsis sp.
    Neolepidostoma sp.
Hydroptilidae
    Plethus sp.
    Hydroptila spp. G, H, X

COLEOPTERA

Dystiscidae

    Hydrovatus sp.
    Unid. Dystiscidae 3 types
Gyrinidae
    Dinuetes type
    Orectochilus sp.
Haliplidae
    Haliplus sp.
Hydrophilidae
    Amphiops sp.
    Enochrus sp.
    Helochares sp.
    Psalitrus sp.
    Unid. Hydrophilidae 3 types
Hydraenidae
    Unid. Hydraenidae
Helodidae
    Scirtes type
Ptilodactylidae
    Ptilodactyla sp.
    Eulichas sp.
Psephenidae
    Eubrianax type
    Psephenus type
    Psephenoides type
Dryopidae
    Types 1, 2, 3
Elmidae
    20 spp.
Torrindicolidae
    Torrindicola sp.
Lampyridae
    Types 1, 2

DIPTERA

Chironomidae
(Tanypodinae)
    Procladius sp.
    Pentaneura spp. a, b
(Chironominae-Chironomini)
    Chironomus s.s. types, 1, 2, 3, 4, 5, 6
    Larva A
    Larva A types 1, 2, 3, D, B
    Cryptochironomus sp.
    Stenochironomus sp.
    Xenochironomus sp.
    Stichtochironomus sp.
    Glyptotendipes sp.
    Microtendipes sp.
    Chironomini types B, 527
    Polypedilum spp. 1, 2
    Paralauterborniella sp.
(Chironominae-Tanytarsini)

Appendix C. continued

   Tanytarsus spp. 1, 2, 3, 4, 5, 6, 7, 8, 9
   Micropsectra sp.
(Orthocladiinae)
  Larva C
  Corynoneura spp. 1, 2
  Trichocladius sp.
  Psectrocladius spp. 1, 2
  Cardiocladius sp.
  Cricotopus spp. 1, 2
  Orthocladiinae types 1, 2, 3, 4, 5, 6, 7, 8, 9, 10, 11, 12, 13
(Diamesinae)
  ?Prodiamesa type
Simuliidae
  Simulium (Gomphostilbia) flavocinctum Edw.
  S. (Gomphostilbia) sundaicum Edw. (= *rayohense* Smart & Clifford)
  S. (Gomphostilbia) pegalanense Smart & Clifford
  S. (Gomphostilbia) nr *sundaicum* cf. *varicorne* Edw.
  S. (Gomphostilbia) spp. B, 2
  Similium (Eusimulium) aureohirtum Brun. (= *tuararense* Smart & Clifford)
  S. (Eusimulium) spp. A, 3
  Simulium s.s. *kiuliense* Smart & Clifford (? = *nobile* de Meig.)
  Simulium kinabaluense Smart & Clifford
  Simulium nitidithorax Puri
  Simulium laterale Edw.
  Simulium digitatum Puri
  Simulium striatum Brun.
  Simulium spp. 1a, 3, 4
Ceratopogonidae
  Bezzia group
  Culicoides type
  Forcipomyia type
Dixidae
  Dixa sp.
Tipulidae
  Hexatoma types 1, 2
  Antocha type
  Megistocera type
  Elliptera type
  Limonia type
  Unid. Tipulidae
Blepharoceridae
  Apistomyia sp.
  Genus incertus nr *Blepharocera*
Psychodidae
  Pericoma type
  Maruina ?indica
  ?Maruina sp.
Rhagionidae
  Atherix type
Empididae
  Types 1, 2
Dolichopodidae
  Type 1

Stratiomyiidae
  Stratiomys type
Tetanoceridae
  Scyiomyzid types 1, 2
Syrphidae
  Eristalis type
  Unid. Cyclorrhapha

A = adults	E = emergents	P = pupae		L = larvae	
		STATION I		STATION II	
		Nos	% of total drift	Nos	% of total drift

EPHEMEROPTERA

Baetis spp.		10597	19.75	6764	9.46
Pseudocloeon spp.		3818	7.11	15460	21.62
Isonychia sp.		75	0.14	87	0.12
Caenis sp.		2067	3.85	1394	1.95
Neoephemeropsis sp.		33	0.06	370	0.52
Tricorythus sp.		129	0.24	256	0.36
Thalerosphyrus spp.		225	0.42	349	0.49
Compsoneuriella sp.		16	0.03	3	0.00
Epeorus sp.		32	0.06	64	0.09
Ephemerella spp.		96	0.18	163	0.23
Thraulus bishopi		144	0.27	257	0.36
Choroterpes sp.		144	0.27	162	0.23
Isca sp.		529	0.99	448	0.63
Habrophlebiodes sp.		369	0.69	225	0.31
Other Leptophlebiidae		16	0.03	—	—
Ephemera (*Dierephemera*) sp.		55	0.10	4	0.01
Prosopistoma ?uuouterae		176	0.33	176	0.25
Ephemeroptera E		65	0.12	326	0.46
Total Ephemeroptera		18586	34.64	26508	37.09

DIPTERA

Orthocladiinae/Diamesinae		946	1.76	720	1.01
Chironominae		790	1.47	1140	1.59
Tanypodinae		484	0.90	307	0.43
Ceratopogonidae		16	0.03	—	—
Simulium spp.		2723	5.07	4055	5.67
Psychodidae		16	0.03	32	0.04
Atherix type		16	0.03	1	0.00
Other Diptera L		976	1.82	1115	1.56
Diptera P		3741	6.97	2550	3.57
Total Diptera		9708	18.08	9920	13.87

PLECOPTERA

Peltoperlodes sp.		144	0.27	32	0.04
Nemouridae		1076	2.00	840	1.17
Perlidae		487	0.91	1730	2.42
Total Plecoptera		1707	3.18	2602	3.63

COLEOPTERA

Hydrophilidae L		—	—	111	0.16

	STATION I		STATION II	
	Nos	% of total drift	Nos	% of total drift
Dytiscidae L	126	0.23	80	0.11
Elmidae L and A	3208	5.98	6881	9.62
Gyrinidae L	16	0.03	58	0.08
Psephenidae L	1	0.00	16	0.02
Helodidae L	199	0.37	197	0.28
Eulichas sp.	5	0.01	—	—
Other A	117	0.22	33	0.05
Total Coleoptera	3672	6.84	7376	10.32

TRICHOPTERA

Chimarra spp.	367	0.68	432	0.60
Rhyacophila sp.	65	0.12	274	0.38
Glossosomatidae	—	—	49	0.07
Hydropsyche spp.	514	0.96	778	1.09
Diplectrona spp.	876	1.63	960	1.34
Cheumatopsyche sp.	65	0.12	18	0.02
Macronemum spp.	32	0.06	21	0.03
Other Hydropsychidae	820	1.53	384	0.54
Psychomyiidae/Polycentropodidae	16	0.03	780	1.09
Marilia sp.	1	0.00	—	—
Helicopsyche sp.	—	—	16	0.02
Ganonema sp.	48	0.09	32	0.04
Tagalopsyche sp.	101	0.19	34	0.05
Oecetis spp.	32	0.06	321	0.45
Mystacidini	32	0.06	—	—
Other Leptoceridae	467	0.87	629	0.88
Neolepidostoma spp.	99	0.19	340	0.48
Other Hydroptilidae	560	1.04	320	0.45
Plethus sp.	—	—	16	0.02
Trichoptera E	66	0.12	2	0.00
Other Trichoptera?	—	—	32	0.04
Total Trichoptera	4161	7.75	5438	7.59

ODONATA

Zygoptera	217	0.40	244	0.34
Anisoptera	261	0.49	140	0.20
Total Odonata	478	0.89	384	0.54

HEMIPTERA

Veliidae	80	0.15	—	—
Ptilomera lundbladi	183	0.34	12	0.02
Metrocoris nigrofasciatus	1	0.00	—	—
Pleciobates tuberculatus	—	—	1	0.00
Cercometus pilipes	—	—	1	0.00

	STATION I		STATION II	
	Nos	% of total drift	Nos	% of total drift
Micronecta sp.	16	0.03	—	—
Plea liturata	240	0.45	432	0.60
Naucoridae	17	0.03	136	0.19
Total Hemiptera	537	1.00	582	0.81

OTHER AQUATICS

Dugesia spp.	—	—	16	0.02
Nematoda	19	0.04	33	0.05
Gordiida	10	0.02	12	0.02
Naididae	352	0.66	266	0.37
Other Oligochaeta	20	0.04	3	0.00
Copepoda	16	0.03	—	—
Macrobrachium spp.	61	0.11	36	0.05
Collembola	834	1.55	338	0.47
Hermes sumatrana	—	—	32	0.04
Epilampridae	83	0.15	7	0.01
Cataclysta spp.	34	0.06	86	0.12
Hydracarina	1097	2.04	1056	1.48
Anura L	309	0.58	895	1.25
Fish	716	1.33	178	0.25
Total Other Aquatics	3551	6.62	2958	4.13
TOTAL AQUATIC DRIFT	42400	79.00	55768	77.98

TERRESTRIALS

Ephemeroptera A	193	0.36	484	0.68
Plecoptera A	16	0.03	19	0.03
Trichoptera A	130	0.24	278	0.39
Hymenoptera	5020	9.35	8066	11.28
Lepidoptera L	196	0.37	279	0.39
Coleoptera	1069	1.99	895	1.25
Isoptera	841	1.57	875	1.22
Blattidae	82	0.15	1	0.00
Others	3723	6.94	4849	6.78
TOTAL TERRESTRIAL DRIFT	11270	21.00	15746	22.02
TOTAL DRIFT	53670	100.00	71514	100.00

Appendix E. Sample questionnaire

SURVEY OF LOCAL FISHING ON THE GOMBAK RIVER

Location (Kampong name or mile reference).....................................

Number of people (from headman)............. Number of fisherman...........

Fishing frequency... per week... per month... per year Duration.............. h

Methods........ rod........ seine........ cast net....... spear...... harpoon

.... 'Tuba'.. other

CATCH COMPOSITION

Type	Number caught	Size		How often caught*
		range	average	
1. *Rasbora*	....	....	....	....
2. *Hampala*	....	....	....	....
3. *Puntius*	....	....	....	....
4. *Osteochilus*	....	....	....	....
5. *Acrossocheilus*	....	....	....	....
6. *Tor*	....	....	....	....
7. *Glyptothorax*	....	....	....	....
8. *Clarias*	....	....	....	....
9. *Fluta*	....	....	....	....
10. *Macrones*	....	....	....	....
11. *Mastacembelus*	....	....	....	....
12. *Channa*	....	....	....	....
13. *Silurichthys*	....	....	....	....
14. *Anabas*	....	....	....	....
15. *Trichogaster*	....	....	....	....
16. *Dermogenys*	....	....	....	....
Other	....	....	....	....
....	....	....	....	....
....	....	....	....	....

* everytime, often, rarely.

Do you fish in the paddy fields?............ What method?......................

What do you catch?................................. How many?............

Do outsiders fish in river?............................ How often?............

What method?............................ How much do they catch?.........

Remarks ....

....

....

....

Appendix F. Checklist of the fishes of the Gombak River (1968/69)

EVANTOGNATHI

Cyprinidae

Rasbora sumatrana (Bleeker)
Mystacoleucus marginatus (Valenciennes)
Hampala macrolepidota van Hasselt
Puntius binotatus (Valenciennes)
Osteochilus hasseltii (Cuvier and Valenciennes)
Tor soro (Valenciennes)
Acrossocheilus deauratus (Cuvier and Valenciennes)
Acrossocheilus sp.?

NEMATOGNATHI

Siluridae

Silurichthys hasseltii Bleeker

Clariidae

Clarias batrachus (Linnaeus)
Clarias teijsmanni Bleeker

Bagridae

Macrones wyckii (Bleeker)

Bagaridae

Glyptothorax platypogonoides (Bleeker)
Glyptothorax major (Boulenger)

OPISTHOMI

Mastacembelidae

Mastacembelus armatus (Lacépède)
Mastacembelus maculatus Cuvier and Valenciennes

LABYRINTHICI

Channidae

Channa gachua (Hamilton)
Channa melasoma (Bleeker)
Channa striata (Bloch)
Channa lucius (Cuvier)

Anabantidae

Anabas testudineus (Bloch)
Betta pugnax (Cantor)
Trichogaster trichopterus (Pallas)

SYNBRANCHI

Flutidae

Fluta alba (Zuiew)

SOLENICHTHYS

Syngnathidae

Doryichthys deokhathoides (Bleeker)

MICROCYPRINI

Poeciliidae

Poecilia reticulata Peters

SYNENTOGNATHI

Hemiramphidae

Hemiramphodon pogonognathus (Bleeker)
Dermogenys pusillus Kuhl and van Hasselt

Cyclocheilichthys apogon (Valenciennes) (Cyprinidae) and *Belontia* sp. (Anabantidae) have been reported from the watershed but were not seen in this study.

465

SUBJECT INDEX

A

Accessory respiratory apparatus in fish 354, 384
Adiabatic lapse rate 37
Aerial photographs 6
Africa(n) 1, 83, 95, 187, 223, 240, 275, 347, 361
Algae 118, 121, 142–186, 201, 212, 214, 225, 228, 230–231, 236, 237, 240, 245, 256, 262, 294, 374, 380
  accumulated ash-free dry weight 169–170, 179, 389
  biomass determination 168-171, 178
  communities of 143, 146–152, 156–158, 184, 389
  composition of the flora 142–143
  effect of discharge and velocity on 155–156, 161, 164, 171, 184, 186
  effects of pollution on 153–154, 162, 164, 171, 185, 389, 393
  effects of substrate on 142, 155–156, 171, 183, 186
  effects of water chemistry on 145, 153–5
  light as a controlling factor 142, 154, 155, 163, 177, 178, 183, 389
  limitation by grazing 155, 178, 186
  methods of sampling 143–144, 163–168
  net and gross production by 163–186, 383, 389
  phytoplankton, absence of 142
Alkalinity 85, 95, 96, 113, 117, 255, 389
Allochthonous organics 15, 98, 125, 134, 137, 138, 249, 359, 361, 378, 381, 390, 392
Alluvial plain 7, 68, 198, 201
Alluviation processes 6
Alluvium 98
Aluminium 86, 102, 111, 125
Amazonia 1, 84, 95, 102, 105, 109, 112, 115, 125, 129, 249, 290, 347, 351, 361, 378
Amazon-Orinoco headwaters 347
Amazon R. and tributaries 80, 83, 88, 98, 142, 230, 266, 375
Ammonia 85, 88, 118, 132, 380
Anion excess, discussion of 113, 114

Anthropogenic effects 1, 377, 386, 388, 392
Arakawa R. 182
Arenaceous deposits 6, 70
Argillaceous deposits 6
Arsenic 86, 111, 117, 380
Artificial substrates for algae 164, 165, 166–181, 389
Asbestos-cement substrates 166–168, 172–173, 176–177, 183–184, 186, 389
Asia 348
Asian rivers, other 105, 137
Atomic absorption spectrophotometry 86, 98–99, 109, 394
'Aufwuchs' community 165, 184, 352
Australia(n) 92, 103, 115, 385

B

Bamboo groynes 8, 26
Bandung 122
Bank erosion 25, 26
Barium 86, 111
Batu Caves 5, 13, 71, 115
Batu R. 7, 19, 29
Bed-load transport 32, 70
Benthoplankton 316
Benthos 3, 254, 306, 333, 341, 392
  vertical distribution of 299–309
  composition of 188–298
Benthos box-sampler 30, 189
Bicarbonate 95, 254, 255
Bifurcation ratio of drainage network 11
Bilharziasis 201
Binnie and Partners (Malaysia) 5, 47, 394
  B.P. gauging weir 47, 50, 52, 54, 78, 220, 352
Biochemical oxygen demand, B.O.D. 85, 118, 126, 128–132, 381, 389
Biotope classification 195, 250, 259, 266, 278–290
Biotopic associations 259–260
Black (Upper) Volta R. 270, 286, 289, 290
Blongkong, Sg. 52, 54
B.O.D./C.O.D. ratio 132
Braiding of channel 8
Brazil 240, 347

Bristol, University of 188

INDEX OF ORGANISMS

Citations are for the body of the text, appendices omitted. Where logical, common and anglicized names have been indexed by their scientific form i.e. 'riffle-beetles', 'elmids' are referenced by Elmidae.

476